Edited by
Oliver Brand
Isabelle Dufour
Stephen M. Heinrich
Fabien Josse

Resonant MEMS

Related Titles

Briand, D., Yeatman, E., Roundy, S. (eds.)

Micro Energy Harvesting

2015
Print ISBN: 978-3-527-31902-2; also available
in electronic formats

Bechtold, T., Schrag, G., Feng, L. (eds.)

**System-level Modeling of
MEMS**

2013
Print ISBN: 978-3-527-31903-9; also available
in electronic formats

Korvink, J.G., Smith, P.J., Shin, D. (eds.)

**Inkjet-based
Micromanufacturing**

2012
Print ISBN: 978-3-527-31904-6; also available
in electronic formats

Erturk, A.A.

**Piezoelectric Energy
Harvesting**

2011
Print ISBN: 978-0-470-68254-8; also available
in electronic formats

Bolic, M.M. (ed.)

**RFID Systems - Research
Trends and Challenges**

2010
Print ISBN: 978-0-470-74602-8; also available
in electronic formats

Ramm, P., Lu, J.J., Taklo, M.M. (eds.)

Handbook of Wafer Bonding

2012
Print ISBN: 978-3-527-32646-4; also available
in electronic formats

Garrou, P., Bower, C., Ramm, P. (eds.)

Handbook of 3D Integration
**Volumes 1 and 2: Technology and
Applications of 3D Integrated Circuits**

2012
Print ISBN: 978-3-527-33265-6; also available
in electronic formats

Garrou, P., Koyanagi, M., Ramm, P. (eds.)

Handbook of 3D Integration
Volume 3: 3D Process Technology

2014
Print ISBN: 978-3-527-33466-7; also available
in electronic formats

Edited by Oliver Brand, Isabelle Dufour, Stephen M. Heinrich, and Fabien Josse

Resonant MEMS

Fundamentals, Implementation and Application

WILEY-VCH

Verlag GmbH & Co. KGaA

The Editors

Prof. Oliver Brand
School Electrical/Comp.Eng.
Georgia Inst. of Technology
777 Atlantic Drive
Atlanta, GA
United States

Prof. Isabelle Dufour
Université de Bordeaux
Laboratoire IMS
Bâtiment CBP
16 av. Pey Berland
33607 Pessac cedex
France

Prof. Stephen M. Heinrich
Marquette University
Civil, Construction and Environmental
Engineering
Haggerty Hall 265
Milwaukee, WI
United States

Prof. Fabien Josse
Marquette University
Electrical & Computer Eng.
Haggerty Hall 294
Milwaukee, WI
United States

■ All books published by **Wiley-VCH** are carefully produced. Nevertheless, authors, editors, and publisher do not warrant the information contained in these books, including this book, to be free of errors. Readers are advised to keep in mind that statements, data, illustrations, procedural details or other items may inadvertently be inaccurate.

Library of Congress Card No.: applied for

British Library Cataloguing-in-Publication Data
A catalogue record for this book is available from the British Library.

Bibliographic information published by the Deutsche Nationalbibliothek
The Deutsche Nationalbibliothek lists this publication in the Deutsche Nationalbibliografie; detailed bibliographic data are available on the Internet at <http://dnb.d-nb.de>.

© 2015 Wiley-VCH Verlag & Co. KGaA, Boschstr. 12, 69469 Weinheim, Germany

Print ISBN: 978-3-527-33545-9
ePDF ISBN: 978-3-527-67636-1
ePub ISBN: 978-3-527-67635-4
Mobi ISBN: 978-3-527-67634-7
oBook ISBN: 978-3-527-67633-0

Cover Design Schulz
Typesetting Laserwords Private Limited, Chennai, India
Printing and Binding Markono Print Media Pte Ltd., Singapore

Printed on acid-free paper

Contents

Series Editor Preface

You hold in your hands the eleventh volume of our book series *Advanced Micro & Nanosystems*, dedicated to the field of *Resonant MEMS*. We have been very fortunate to enlist Prof. Oliver Brand, Prof. Isabelle Dufour, Prof. Stephen Heinrich, and Prof. Fabien Josse as Volume Editors. All four have extensive expertise in different aspects of *Resonant MEMS* and, as a team, actually have collaborated in recent years, resulting in a number of joint research publications. In a similar way, this book project turned out to be a true team project, from establishing the desired table of contents; to selecting an international team of experts as chapter authors; to assembling, editing, and fine-tuning the contents.

You might ask, why a book on *Resonant MEMS*? Clearly, resonant devices fabricated using MEMS (MicroElectroMechanical Systems) technologies are not new; in fact, one of the early MEMS devices is the *Resonant Gate Transistor*, published by Harvey C. Nathanson and co-workers in the *IEEE Transactions on Electron Devices* in 1967. Over the years, a resonant sensor version of just about every sensor imaginable has been investigated. In general, *Resonant MEMS* (and in particular resonant sensors) promise very high sensitivities, but often come at the expense of a more complicated device design and fabrication. In recent years, modern numerical modeling tools, in particular finite element modeling (FEM) software, and a number of fundamental theoretical studies have helped design better *Resonant MEMS*, and, as a result, first commercial devices based on *Resonant MEMS* have been developed. The best example might be the success of MEMS-based resonant gyroscopes in consumer electronic devices, such as smart phones and gaming consoles. As the field matures, we found a book that summarizes all aspects of *Resonant MEMS*, ranging from the *Fundamentals* to *Implementation* and *Application*, to be very timely. You have the result in your hands, and we hope that you enjoy reading this book as much as we do.

This book would not have been possible without a significant time commitment by the volume editors as well as the chapter authors. We want to thank them most heartily for their effort! Our thanks also go to the Wiley staff for their strong support of this project. The final printed result once again speaks for itself!

Atlanta, Pittsburgh,   *Oliver Brand, Gary K. Fedder,*
Zurich, Freiburg,   *Christofer Hierold, Jan G. Korvink,*
Kyoto, January 2015   *Osamu Tabata, Series Editors*

Preface

As the editing team for Vol. 11 of the *Advanced Micro & Nanosystems* series, entitled *Resonant MEMS: Fundamentals, Implementation and Application*, we hope that you benefit from this significant collaborative effort among the experts who have kindly contributed to this project. The book's *raison d'être* is to elucidate the various aspects of MEMS resonators, to identify the state of the art in this rapidly changing field, and to serve as a valuable reference tool to the readership, including serving as a springboard for future advances in this discipline.

Given the breadth of the resonant MEMS field, we have elected to group the various chapters of this volume into three parts as indicated by the book's subtitle. Part I, *Fundamentals*, comprises five chapters, each of which focuses on the theoretical description of the underlying physical phenomena that are relevant to virtually all resonant MEMS devices. This part includes detailed treatments on the fundamental theory of mechanical resonance; the effects of viscous fluids (a surrounding gas or liquid) on vibrating microcantilevers; a broad-based examination of various sources of damping (energy dissipation mechanisms); resonant response caused by parametric excitation, i.e., variations in resonator properties as opposed to direct (e.g., force) excitation; and an overview of the fundamentals of the finite element method with specific applications to MEMS resonators. Having laid the fundamental groundwork in Part I, the eight chapters of Part II, *Implementation*, examine how the fundamentals are applied in a practical setting to yield specific types of resonant MEMS devices and how these devices are designed to reliably perform a specific function. In particular, this group of chapters includes detailed discussions of resonant MEMS devices on the basis of the following materials and device designs: capacitive transducers, piezoelectric materials, nanoelectromechanical systems (NEMS), and organic materials (polymers). Also included in Part II are chapters treating the following practical implementation topics: electrothermal excitation methods; the use of embedded channels to overcome challenges posed by liquid-phase applications; hermetic packaging to protect the resonator and to ensure its long-term stability and reliability; and the development of compensation, tuning, and trimming techniques for the realization of high-precision resonators by accounting for variations in material properties, fabrication processes, and environmental operating conditions. Finally, in Part III, *Application*, we have included chapters that are dedicated to particular functionalities. Part III comprises four chapters on resonant MEMS for sensing applications, including the following: inertial sensing (motion detection); chemical detection in both gaseous and liquid environments; biochemical sensing for label-free, quantitative measurement of biomolecules such as proteins and nucleic acids, or even entire cells and viruses; and resonant MEMS-based rheometers for measuring the physical properties of fluids. The final chapter of Part III focuses on energy harvesting applications for converting ambient mechanical vibrations into useful electrical energy.

Finally, we would like to extend a sincere expression of gratitude to all of the chapter authors and their associated institutions, to the editorial staff at Wiley-VCH, especially Martin Preuss and Martin Graf-Utzmann, and to Sangeetha Suresh and the production staff at Laserwords. Without the tireless efforts of all of these people, this book would not have been possible. Also, all four co-editors gratefully acknowledge the financial support of *CNRS* (*France, Projet PICS*, 2012–2014) for the international collaboration required to plan and realize this volume, while three of the co-editors (Brand, Heinrich, Josse) gratefully acknowledge research funding from the National Science Foundation (U.S.) over the period 2008–present. The support provided by both of these funding agencies was instrumental in bringing this book to fruition.

Atlanta, Pessac,
Milwaukee, January 2015

Oliver Brand
Isabelle Dufour
Stephen M. Heinrich
Fabien Josse

Co-Editors

About the Volume Editors

Oliver Brand, PhD Oliver Brand received his diploma degree in Physics from Technical University Karlsruhe, Germany, in 1990 and his PhD degree from ETH Zurich, Switzerland, in 1994. From 1995 to 1997, he worked as a postdoctoral fellow at Georgia Tech. From 1997 to 2002, he was a lecturer at ETH Zurich and deputy director of the Physical Electronics Laboratory. In 2003, he joined the Electrical and Computer Engineering faculty at the Georgia Institute of Technology where he is currently a Professor. Since 2014, he serves as the Executive Director of Georgia Tech's Institute for Electronics and Nanotechnology. He has co-authored more than 190 publications in scientific journals and conference proceedings. He is a co-editor of the Wiley-VCH book series *Advanced Micro & Nanosystems*, a member of the editorial board of Sensors and Materials, and has served as General Co-Chair of the 2008 IEEE International Conference on Micro Electro Mechanical Systems (MEMS 2008). Dr. Brand is a senior member of the IEEE and a co-recipient of the 2005 IEEE Donald G. Fink Prize Paper Award. His research interests are in the areas of silicon-based microsystems, microsensors, MEMS fabrication technologies, and microsystem packaging.

Isabelle Dufour, PhD Isabelle Dufour graduated from Ecole Normale Supérieure de Cachan in 1990 and received her PhD and HDR degrees in engineering science from the University of Paris-Sud, Orsay, France, in 1993 and 2000, respectively. She was a CNRS research fellow from 1994 to 2007, first in Cachan working on the modeling of electrostatic actuators (micromotors, micropumps) and then after 2000 in Bordeaux working on microcantilever-based chemical sensors. She is currently a Professor of electrical engineering at the University of Bordeaux, and her research interests are in the areas of microcantilever-based sensors for chemical detection, rheological measurements, material characterization, and energy harvesting.

Stephen M. Heinrich, PhD Stephen M. Heinrich earned the BS degree *summa cum laude* from Penn State in 1980 and the MS and PhD degrees from the University of Illinois at Urbana-Champaign in 1982 and 1985, all in civil engineering. Hired as an Assistant Professor at Marquette University in 1985, he was promoted to his current rank of Professor in 1998. In 2000, Prof. Heinrich was awarded the *Rev. John P. Raynor Faculty Award for Teaching Excellence*, Marquette's highest teaching honor, while in 2006 he was a awarded a Fulbright Research Scholar Award to support research collaboration at the *Université de Bordeaux*. Dr. Heinrich's research has focused on structural mechanics applications in microelectronics packaging and analytical modeling of cantilever-based chemical/biosensors and, more recently, MEMS energy harvesters. The investigations performed by Dr. Heinrich and his colleagues have resulted in more than 100 refereed publications and three best paper awards from IEEE and ASME. His professional service activities include membership on the ASCE Elasticity Committee, Associate Editor positions for the *IEEE Transactions on Advanced Packaging* and the *ASME Journal of Electronic Packaging*, and technical review activities for more than 40 journals, publishers, and funding agencies.

Fabien Josse, PhD Fabien Josse received the MS and PhD degrees in Electrical Engineering from the University of Maine in 1979 and 1982, respectively. He has been with Marquette University, Milwaukee, WI, since 1982 and is currently Professor of Electrical, Computer and Biomedical Engineering. He is also an Adjunct Professor with the Department of Electrical Engineering, Laboratory for Surface Science and Technology, University of Maine. He has been a Visiting Professor with the University of Heidelberg, Germany, the Laboratoire IMS, University of Bordeaux, France, and the Physical Electronics Laboratory, ETH Zurich, Switzerland, and IMTEK, University of Freiburg, Germany. His research interests include solid state sensors, acoustic wave sensors, and MEMS devices for liquid-phase biochemical sensor applications, investigation of novel sensor platforms, and smart sensor systems. Prof. Josse is a senior member of IEEE and associate editor (2002–2009) of the *IEEE Sensors Journal*.

List of Contributors

Gabriel Abadal
Universitat Autonòma de
Barcelona (UAB)
Escola d'Enginyeria
Department d'Enginyeria
Electrònica
Campus UAB 08193
Bellaterra
Spain

Reza Abdolvand
University of Central Florida
Department of Electrical
Engineering and Computer
Sciences
4000 Central Florida Blvd.
Building 116 – Room 346
Orlando, FL 32816-2362
USA

Vaida Auzelyte
Microsystem Laboratory
Ecole Polytechnique Federal de
Lausanne (EPFL)
EPFL STI IMT IMT-LS-GE
BM 3107 (Batiment BM)
Station 17, 1015 Lausanne
Switzerland

Farrokh Ayazi
Georgia Institute of Technology
School of Electrical and
Computer Engineering
777 Atlantic Drive
Atlanta, GA 30332-0250
USA

and

Qualtré Inc
225 Cedar Hill St
Marlborough, MA 01752
USA

N. Barniol
Universitat Autonòma de
Barcelona (UAB)
Escola d'Enginyeria
Department d'Enginyeria
Electrònica
Campus UAB 08193
Bellaterra
Spain

Luke A. Beardslee
SUNY College of Nanoscale
Science and Engineering
546 Mercer Street
Albany, NY 12208
USA

Stephen P. Beeby
Department of Electronics and
Computer Science
University of Southampton
Highfield
SO17 1BJ
Southampton
UK

Oliver Brand
Georgia Institute of Technology
School of Electrical and
Computer Engineering
777 Atlantic Drive
Atlanta, GA 30332-0250
USA

Jürgen Brugger
Microsystem Laboratory
Ecole Polytechnique Federal de
Lausanne (EPFL)
EPFL STI IMT IMT-LS-GE
BM 3107 (Batiment BM)
Station 17, 1015 Lausanne
Switzerland

Thomas P. Burg
Max Planck Institute for
Biophysical Chemistry
Am Fassberg 11
37077 Göttingen
Germany

Isabelle Dufour
Université de Bordeaux
Laboratoire IMS
Bâtiment CBP
16 av. Pey Berland
33607 Pessac cedex
France

Cornelis Anthony van Eysden
KTH Royal Institute of
Technology and Stockholm
University
NORDITA
Roslagstullsbacken 23
10691 Stockholm
Sweden

Gary K. Fedder
Carnegie Mellon University
The Robotics Institute
Department of Electrical and
Computer Engineering
5000 Forbes Avenue
Pittsburgh, PA 15213-3890
USA

Shirin Ghaffari
Stanford University
Mechanical Engineering
Department
Building 530
440 Escondido Mall
Stanford, CA 94305-3030
USA

Jonathan Gonzales
Oklahoma State University
Department of Electrical and
Computer Engineering
700 North Greenwood Ave.
Tulsa, OK 74106-0702
USA

Andrew Bradley Graham
Stanford University
Department of Mechanical
Engineering
Building 530
440 Escondido Mall
Stanford, CA 94305-3030
USA

Congzhong Guo
Carnegie Mellon University
Department of Electrical and
Computer Engineering
5000 Forbes Avenue
Pittsburgh, PA 15213-3890
USA

Martin Heinisch
Johannes Kepler University Linz
Institute for Microelectronics
and Microsensors
Altenbergerstraße 69
4040 Linz
Austria

Stephen M. Heinrich
Marquette University
Department of Civil,
Construction
and Environmental Engineering
265 Haggerty Hall
P.O. Box 1881
Milwaukee, WI 53201-1881
USA

Gavin Ho
NanoFab Corporation
800 West El Camino Real
Suite 180
Mountain View, CA 94040-2586
USA

Bernhard Jakoby
Johannes Kepler University Linz
Institute for Microelectronics
and Microsensors
Altenbergerstraße 69
4040 Linz
Austria

Blake N. Johnson
Drexel University
Department of Chemical and
Biological Engineering
3141 Chestnut Street
Philadelphia, PA 19104
USA

Fabien Josse
Marquette University
Department of Electrical and
Computer Engineering
294 Haggerty Hall
P.O. Box 1881
Milwaukee, WI 53201-1881
USA

Thomas William Kenny
Stanford University
Department of Mechanical
Engineering
Building 530
440 Escondido Mall
Stanford, CA 94305-3030
USA

Matthew William Messana
Stanford University
Department of Mechanical
Engineering
Building 530
440 Escondido Mall
Stanford, CA 94305-3030
USA

Raj Mutharasan
Drexel University
Department of Chemical and
Biological Engineering
3141 Chestnut Street
Philadelphia, PA 19104
USA

Liviu Nicu
LAAS-CNRS
7 avenue du Colonel Roche
31077 Toulouse
France

Francesc Perez-Murano
Institut de Microelectrònica de
Barcelona (IMB-CNM CSIC)
Campus de la UAB
08193 Bellaterra
Spain

Gianluca Piazza
Carnegie Mellon University
Department of Electrical and
Computer Engineering
Roberts Engineering Hall
Room 333, 5000 Forbes Avenue
Pittsburgh, PA 15213
USA

Siavash Pourkamali
The University of Texas at Dallas
Department of Electrical
Engineering
800 W. Campbell Road
Richardson, TX 75080
USA

Erwin K. Reichel
Johannes Kepler University Linz
Institute for Microelectronics
and Microsensors
Altenbergerstraße 69
4040 Linz
Austria

Jeffrey F. Rhoads
Purdue University
School of Mechanical
Engineering
Birck Nanotechnology Center
Ray W. Herrick Laboratories
585 Purdue Mall
West Lafayette, IN 47907-2088
USA

John Elie Sader
The University of Melbourne
Department of Mathematics
and Statistics
3010 Victoria
Australia

Veronica Savu
Microsystem Laboratory
Ecole Polytechnique Federal de
Lausanne (EPFL)
EPFL STI IMT IMT-LS-GE
BM 3107 (Batiment BM)
Station 17, 1015 Lausanne
Switzerland

Silvan Schmid
Technical University of Denmark
DTU Nanotech
Department of Micro- and
Nanotechnology
Ørsteds Plads
Building 345Ø
Room 158
2800 Kgs. Lyngby
Denmark

Diego Emilio Serrano
Georgia Institute of Technology
School of Electrical and
Computer Engineering
777 Atlantic Drive
Atlanta, GA 30332-0250
USA

and

Qualtré Inc
225 Cedar Hill St
Marlborough, MA 01752
USA

Roozbeh Tabrizian
Georgia Institute of Technology
School of Electrical and
Computer Engineering
777 Atlantic Drive
Atlanta, GA 30332-0250
USA

Herre S. J. van der Zant
Delft University of Technology
Kavli Institute of Nanoscience
Lorentzweg 1
2628 CJ Delft
The Netherlands

Warner J. Venstra
Delft University of Technology
Kavli Institute of Nanoscience
Lorentzweg 1
2628 CJ Delft
The Netherlands

Luis Guillermo Villanueva
California Institute of
Technology
1200 E California Blvd.
MC 149-33
Pasadena, CA
USA

Part I
Fundamentals

Resonant MEMS – Fundamentals, Implementation and Application, First Edition.
Edited by Oliver Brand, Isabelle Dufour, Stephen M. Heinrich and Fabien Josse.
© 2015 Wiley-VCH Verlag GmbH & Co. KGaA. Published 2015 by Wiley-VCH Verlag GmbH & Co. KGaA.

1
Fundamental Theory of Resonant MEMS Devices

Stephen M. Heinrich and Isabelle Dufour

1.1
Introduction

Resonators based on MEMS (micro-electromechanical systems) and NEMS (nano-electromechanical systems) span a broad spectrum of important current applications, including detection of chemical [1−7] and biological substances [2−4, 6−10], measurement of rheological properties of fluids [11−14], and energy harvesting [15−17], to name only a few. While the devices that perform these diverse functions span an equally broad range in geometric layout, material properties, circuitry, fabrication techniques, packaging, and so on, they all have one aspect in common: the phenomenon of "resonance" forms the basis of their operating principle. More specifically, they usually perform their desired functions by monitoring how interactions with the environment (with various "measurands") influence the resonant behavior (e.g., the resonant frequency) of the device. Conversely, how effectively the device performs its function will depend to a large degree on the underlying resonant characteristics of the device (e.g., its quality factor, which determines the resonant peak "sharpness" on a plot of response vs driving frequency). Since all such devices rely on resonant vibrations to accomplish one or more tasks effectively, a firm understanding of this highly interdisciplinary field requires that one be familiar with the fundamental theory of mechanical vibrations. To facilitate such familiarity is therefore the primary goal of this initial chapter.

In the sections that follow, an attempt will be made to achieve several specific objectives. In Section 1.2, a glossary of the major notation and terminology of the chapter will be presented, followed by a summary of the theory of single-degree-of-freedom (SDOF) damped oscillators in Section 1.3 for the cases of free vibration and forced harmonic vibration. This summary, which also includes definitions of the resonator's quality factor Q, methods for estimating its value experimentally, and a brief discussion of how multiple dissipation sources contribute to Q, is intended to familiarize the reader with the fundamental concepts associated with mechanical resonant phenomena. The SDOF section lays the groundwork for the understanding of multiple-degree-of-freedom (MDOF) dynamic system

Resonant MEMS – Fundamentals, Implementation and Application, First Edition.
Edited by Oliver Brand, Isabelle Dufour, Stephen M. Heinrich and Fabien Josse.
© 2015 Wiley-VCH Verlag GmbH & Co. KGaA. Published 2015 by Wiley-VCH Verlag GmbH & Co. KGaA.

behavior, which is the focus of Section 1.4. The mechanical behavior of such systems is introduced by means of two simple, yet highly relevant, examples for a cantilever beam, including its free-vibration response and its response due to a sinusoidal end force. The solutions presented for the cantilever will be based on a "continuous-systems" (distributed-parameters) modeling approach and will serve as a vehicle for (i) introducing the key concepts of natural/resonant frequencies and mode shapes for MDOF systems and (ii) showing how the resonant response of such systems may often be interpreted and approximated as that associated with an SDOF system. Section 1.5 furnishes a list of potentially useful natural frequency formulas for some of the more common geometries and vibration modes used in resonant MEMS/NEMS applications, while the chapter concludes with a brief summary (Section 1.6).

1.2
Nomenclature

A summary of the primary notation and terminology used in this chapter is given below. Note that all of the "ω" frequency quantities indicated below are "circular" or "angular" frequencies in that all have units of radians per second. Any of these frequencies may be converted to their corresponding frequencies, denoted by "f," having units of cycles per second, or hertz, through the relationship $f = \omega/2\pi$.

FRF = frequency response function = a plot of a particular response quantity (e.g., displacement amplitude at a point) vs the actuation frequency when the resonator is excited by a sinusoidal force;

ω = actuation (exciting) frequency;

λ = dimensionless actuation frequency;

ω_0 = undamped natural frequency of an SDOF system (referred to as simply "natural frequency" by some authors);

ω_d = damped natural frequency of an SDOF system;

ω_{res} = resonant frequency = the exciting frequency that results in a resonant state, defined as a vibrational state corresponding to a relative maximum (resonant peak) on the *FRF* for displacement amplitude; note that some authors define the resonant frequency as being identical to the undamped natural frequency, not as the exciting frequency causing peak displacement response; for small damping levels, the difference in the two values is insignificant;

ω_n = undamped natural frequency of nth mode of an MDOF system ($n = 1, 2, \ldots$);

λ_n = dimensionless undamped natural frequency parameter of nth mode of an MDOF system ($n = 1, 2, \ldots$);

r = frequency ratio, that is, ratio of exciting frequency to undamped natural frequency;

ζ = damping ratio;

Q = quality factor;

$\varphi_n(\xi)$ = nth mode shape of a cantilever beam, where the mode shapes represent the set of possible constant-shape free vibrations, ($n = 1, 2, \ldots$);

$\psi(\xi)$ = vibrational shape of a cantilever beam when excited by a harmonic tip force;

D = dynamic magnification factor for an SDOF system = ratio of dynamic displacement amplitude to quasi-static (zero-frequency) value;

θ = lag angle by which the steady-state displacement of an SDOF system trails the applied harmonic force;

D_{tip} = dynamic magnification factor for harmonically loaded cantilever tip = ratio of dynamic displacement amplitude at beam tip to quasi-static (zero-frequency) value;

θ_{tip} = lag angle by which the steady-state displacement at the tip of a cantilever trails the applied harmonic tip force.

The quantities listed above will be examined and discussed in greater detail in the sections that follow.

1.3
Single-Degree-of-Freedom (SDOF) Systems

A large number and variety of MEMS/NEMS resonators may be accurately modeled as SDOF damped oscillators (Figure 1.1) because their vibrational response may be described in terms of a single time-dependent position coordinate. Even for those devices for which the harmonically excited vibrational response requires multiple degrees of freedom to describe, a single mode of vibration tends to dominate at or near a resonant state, thus permitting one to model the response via SDOF theory without a significant loss of accuracy. For these reasons, a review of elementary SDOF vibration theory is appropriate. (More detailed treatments may be found in any elementary vibrations textbook such as [18–21].) The discussion will begin with a review of free-vibration results, including the important concepts of natural frequency (undamped and damped), damping ratio, and quality factor, all of which may be interpreted as inherent dynamic *properties* of the damped oscillator. The review will then continue with a summary of results for the steady-state response of the damped SDOF oscillator when excited by an applied harmonic force, that is, one that varies sinusoidally in time. This will include a mathematical description of the resonant response of the SDOF system, including

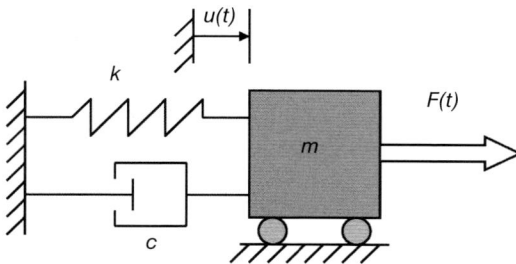

Figure 1.1 Schematic representation of a damped SDOF oscillator.

an alternative definition of the quality factor and a resonance-based experimental method (bandwidth method) for measuring Q.

In the summary that follows, several assumptions are implicit:

- energy dissipation is due to a viscous damping mechanism (damping force is proportional to velocity);
- the effective mass (m), effective damping coefficient (c), and the effective stiffness (k) of the system are constant, that is, they do not depend on time or on the frequency of oscillation;
- the system is linear, which necessitates that the physical system being modeled as an SDOF oscillator involves linear elastic and linear dissipative forces (the spring and dashpot in Figure 1.1 exhibit material linearity) and small deformations (no geometric nonlinearities).

In certain cases of practical interest, not all damping mechanisms will be of the viscous variety, nor will resonators always respond linearly or with frequency-independent properties [19, 21]. Nevertheless, an understanding of simple SDOF resonator behavior based on the above assumptions will provide an important foundation for understanding resonator behavior and a logical point of departure for grasping some of the more complex issues that arise when the aforementioned assumptions are not met. These more advanced aspects of MEMS resonator response will be treated in many of the chapters that follow.

1.3.1
Free Vibration

The assumptions of linear spring and dashpot behavior in the SDOF system of Figure 1.1 enable one to derive the following equation of motion by performing a simple force balance:

$$m\ddot{u}(t) + c\dot{u}(t) + ku(t) = F(t) \tag{1.1}$$

where m, c, and k represent, respectively, the effective mass, effective damping coefficient, and effective stiffness of the system and $u(t)$ is the displacement response of the SDOF oscillator. The "dot notation" has been used in Eq. (1.1) to represent differentiation with respect to time t. The effective externally applied force, $F(t)$, is in general related to the excitation force that is applied to the resonator by one of several actuation methods (e.g., electrostatic, electrothermal, piezoelectric). The specific manner in which the effective properties (m, c, k), the effective applied force $F(t)$, and the displacement $u(t)$ of the idealized system of Figure 1.1 are related to the physical geometry, material properties, and actuation details of a particular device may be derived by the application of first principles for a system whose vibrational shape remains constant with time. (See, e.g., [22–24] for examples involving modeling of MEMS resonators.) For the case of free vibration considered here, the system vibrates in the absence of any externally applied force, that is, $F(t) \equiv 0$, resulting in the homogeneous form of Eq. (1.1):

$$m\ddot{u}(t) + c\dot{u}(t) + ku(t) = 0 \tag{1.2}$$

The coefficients in Eq. (1.2) describe the dynamic properties of the system. From these, one may also define the following dynamic properties, known as the undamped natural frequency (ω_0) and the damping ratio (ζ):

$$\omega_0 \equiv \sqrt{\frac{k}{m}} \tag{1.3}$$

$$\zeta \equiv \frac{c}{2\sqrt{km}} \tag{1.4}$$

For the case in which the energy dissipation is sufficiently small so that the free-vibration response of the system is oscillatory, the damping ratio will be less than unity and the system is referred to as *underdamped*. This is the case for most MEMS resonators; hence, this assumption will be employed here. The free-vibration response of an underdamped SDOF system takes the form

$$u(t) = e^{-\zeta\omega_0 t}(A\cos\omega_d t + B\sin\omega_d t) \tag{1.5}$$

in which A and B are constants that depend on the initial values of $u(0)$ and $\dot{u}(0)$ that set the system into free vibration and ω_d is the *damped natural frequency*, defined as

$$\omega_d \equiv \omega_0\sqrt{1 - \zeta^2} \tag{1.6}$$

Note that the physical meaning of ω_d is that it represents the frequency of oscillation (in radians per second) of the free vibration of the damped SDOF system as indicated in Figure 1.2. Also note that the damped natural frequency is less than its undamped counterpart; however, for small-to-moderate values of the damping ratio (ζ less than, say, 0.2), the damped natural frequency is approximately the same as the undamped natural frequency:

$$\omega_d \approx \omega_0 = \sqrt{k/m} \tag{1.7}$$

In lieu of the damping ratio, an alternative way that one could characterize the energy dissipation inherent in the SDOF system is to define the *quality factor Q* in terms of the damping ratio as follows:

$$Q \equiv \frac{1}{2\zeta} = \frac{\sqrt{km}}{c} \tag{1.8}$$

From Eq. (1.5) and Figure 1.2 it is clear that smaller values of the damping ratio (larger values of Q) correspond to systems whose free-vibration response is sustained for longer durations. An approximate rule-of-thumb is that a damping ratio of 10% ($Q = 5$) corresponds to roughly a 50% decrease in amplitude over one complete cycle of free vibration. The correlation between the damping ratio and the rate of free-vibration decay serves as the basis for the logarithmic-decrement method for measuring ζ (or Q) by performing a free-vibration experiment

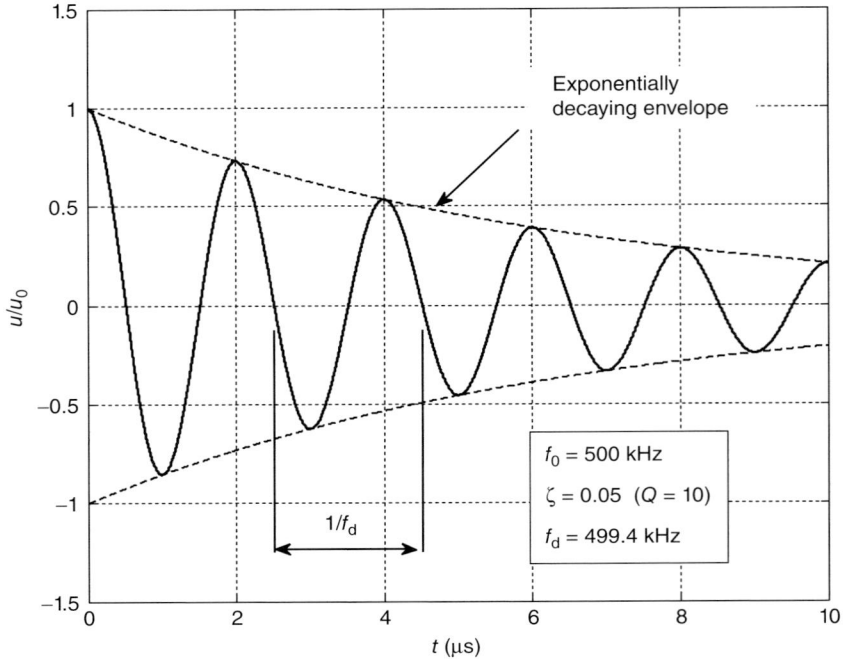

Figure 1.2 Normalized free-vibration displacement response of a damped SDOF system having undamped natural frequency $f_0 = 500\,\text{kHz}$ and 5% damping ratio ($Q = 10$). For definiteness the initial conditions have been chosen as $u(0) = u_0$, $\dot{u}(0) = 0$.

[18–21]. One may view high-Q systems as exhibiting more energy-efficient free vibrations. As indicated in the following section, this efficiency will also be displayed when high-Q (low-ζ) resonators are excited harmonically at or near a resonant state, resulting in enhanced device performance in a variety of resonant MEMS applications. To further clarify the efficiency aspect of high-Q resonators, an alternative, energy-based definition of Q, equivalent to that given by Eq. (1.8), shall be introduced within the context of harmonically forced vibrations.

1.3.2
Harmonically Forced Vibration

The equation of motion Eq. (1.1) is now considered for the special case in which oscillations are driven by an applied harmonic actuation force of amplitude F_0 and frequency ω:

$$m\ddot{u}(t) + c\dot{u}(t) + ku(t) = F_0 \sin \omega t \tag{1.9}$$

The steady-state solution of Eq. (1.9) may be written as

$$u(t) = \frac{F_0}{k} D(r, \zeta) \sin[\omega t - \theta(r, \zeta)] \tag{1.10}$$

where

$$D(r, \zeta) \equiv \frac{1}{\sqrt{(1 - r^2)^2 + (2\zeta r)^2}} \tag{1.11}$$

$$\theta(r, \zeta) \equiv \arctan\left(\frac{2\zeta r}{1 - r^2}\right) \in [0, \pi] \tag{1.12}$$

$$r \equiv \frac{\omega}{\omega_0} \tag{1.13}$$

The coefficient F_0/k in Eq. (1.10) represents the quasi-static displacement amplitude that the system would experience if the load were applied at extremely low frequencies; hence the quantity D appearing in Eq. (1.10) and defined in Eq. (1.11) is simply the ratio of the dynamic displacement amplitude (u_{max}) to the quasi-static amplitude (F_0/k) and is therefore referred to as the *dynamic magnification factor*. As may be seen from Eq. (1.10), the quantity θ represents the angle by which the displacement response lags the actuation force; it is therefore called the *lag angle* of the displacement with respect to the applied force. Both D and θ depend on r and ζ, the former being the *frequency ratio* defined by Eq. (1.13). The dependence of D and θ on the frequency and damping ratios is indicated graphically in Figure 1.3.

An examination of Figure 1.3 leads to the following observations:

- The "exact" value of resonant frequency, as defined in Section 1.2, is less than ω_0. Maximizing Eq. (1.11) with respect to r results in

$$\omega_{res} = \omega_0 \sqrt{1 - 2\zeta^2} \tag{1.14}$$

  Thus, the resonant frequency (i.e., the actuation frequency that causes maximum displacement response) is less than ω_d, which is less than ω_0. For sufficiently high Q values, these differences are of little practical importance (typically the case for most MEMS resonators); however, in resonator applications involving larger energy dissipation (e.g., biochemical detection in liquids [8–10] or rheological applications for measuring the properties of highly viscous and/or complex fluids [11–14]), the differences may be important and could necessitate that distinctions be made among the values of these various frequencies.

- The "exact" value of D_{max} (and, thus, the corresponding displacement amplitude at resonance) may be obtained by evaluating Eq. (1.11) at $r_{res} = \sqrt{1 - 2\zeta^2}$:

$$D_{max} = \frac{1}{2\zeta\sqrt{1 - \zeta^2}} \tag{1.15}$$

  Thus, for an undamped system, the displacement amplitude is theoretically infinite and occurs at $r = 1$, or when the driving frequency coincides with the undamped natural frequency. When the effects of damping are included, D_{max} occurs at $r < 1$; however, when $\zeta \leq 0.2$ ($Q \geq 2.5$), D_{max} occurs very close to $r = 1$, in which case D_{max} may be accurately estimated as

$$D_{max} \approx D|_{r=1} = \frac{1}{2\zeta} = Q \tag{1.16}$$

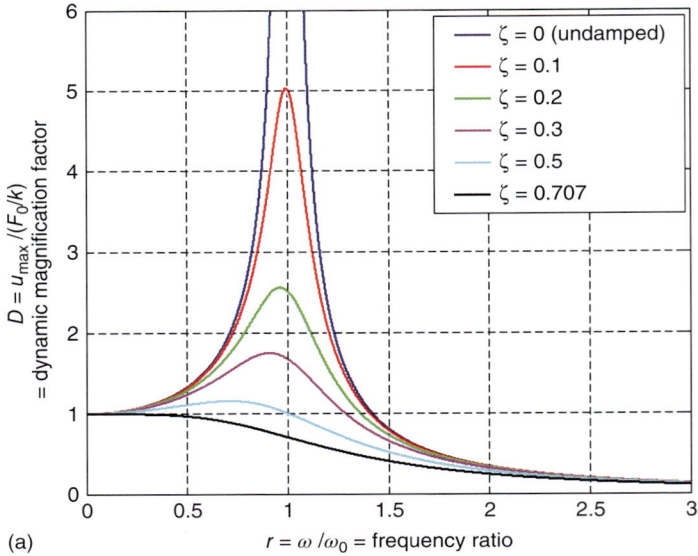

(a)

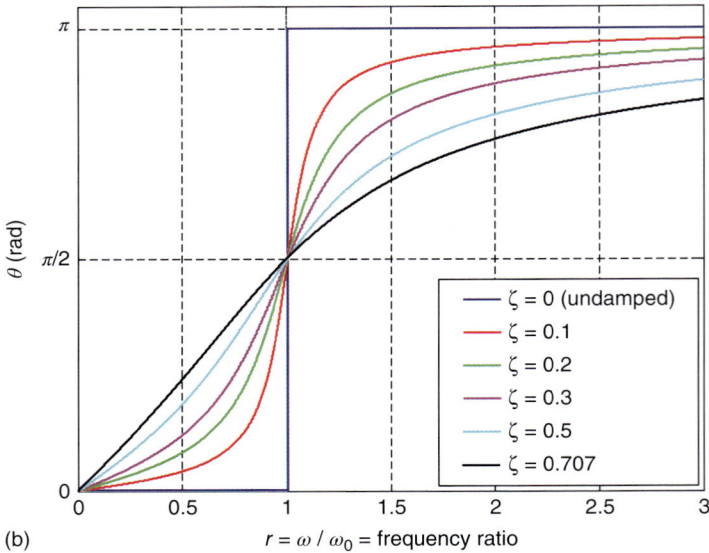

(b)

Figure 1.3 Frequency response functions characterizing the steady-state harmonic response of a damped SDOF oscillator due to a harmonic actuation force: (a) normalized displacement amplitude (dynamic magnification factor) and (b) lag angle of displacement with respect to applied force.

- Equation (1.16) indicates that in theory, for small-to-moderate damping, the value of Q may be extracted from an experimentally generated plot of D versus exciting frequency. This method is known as the "resonant amplification method" [19]. However, in practice it may be difficult to experimentally determine the quasi-static scaling factor (F_0/k), that is, the low-frequency limit for the displacement amplitude, which is needed to relate the measured displacement amplitude to D [19–20]. [See Eq. (1.10).] This limitation is usually overcome by employing the "half-power method" or "−3 dB bandwidth method" in MEMS/NEMS applications, as will be described shortly.
- D approaches 1 as $r \to 0$ (as expected) and it approaches 0 at the high-frequency limit ($r \to \infty$).
- Sharper D peaks correspond to lower ζ or higher Q. Therefore, higher Q values are desirable in, for example, sensors based on the use of MEMS resonators, for which shifts in the resonant frequency (the sensor signal) are directly related to the sensor measurand (e.g., concentration of a target substance).
- The resonant peak (relative maximum) no longer exists if $\zeta \geq \sqrt{2}/2 = 0.707$ (if $Q \leq \sqrt{2}/2 = 0.707$).
- For an undamped system, the response is completely in phase ($\theta = 0$) with the harmonic exciting force when $r < 1$ and completely out-of-phase ($\theta = \pi$) when $r > 1$.
- For $r = 1$ the lag angle is $\pi/2$, regardless of the value of the damping ratio.

A common definition employed for the quality factor of a resonator is based on a ratio of energies when the resonator is excited harmonically at its undamped natural frequency (e.g., [25]):

$$Q \equiv 2\pi \left.\frac{U_{max}}{\Delta W}\right|_{r=1} \tag{1.17}$$

where U_{max} is the maximum elastic energy (stored in the spring) and ΔW is the dissipated energy per cycle of steady-state vibration. (The numerator in Eq. (1.17) may be replaced by the value of the total mechanical energy – elastic energy plus the kinetic energy of the mass – as this sum is identical to U_{max} when $r = 1$.) The energy-based definition in Eq. (1.17) may be shown to be identical to the "property-based" definition in Eq. (1.8) by evaluating the two energies appearing in Eq. (1.17). Using Eq. (1.10),

$$U_{max} = \frac{1}{2}ku_{max}{}^2 = \frac{1}{2}k\left[\frac{F_0}{k}D(r,\zeta)\right]^2 = \frac{F_0{}^2}{2k}[D(r,\zeta)]^2 \tag{1.18}$$

Also, since the steady-state displacement is periodic in time, the total mechanical energy (of the spring and mass) will be as well. Thus, the energy dissipated by the dashpot per cycle must be the same as the work done by the applied force over one cycle. Utilizing Eqs. (1.10) and (1.12), the dissipated energy is obtained as follows:

$$
\begin{aligned}
\Delta W &= \int_{1 \text{ cycle}} F(t)\mathrm{d}u = \int_0^{2\pi/\omega} F(t)\dot{u}(t)\mathrm{d}t \\
&= \int_0^{2\pi/\omega} F_0 \sin \omega t \left[\frac{F_0}{k} D\omega \cos(\omega t - \theta)\right] \mathrm{d}t \\
&= \dots = \frac{2\pi \zeta r F_0{}^2 [D(r,\zeta)]^2}{k}
\end{aligned}
\tag{1.19}
$$

Substituting Eqs. (1.18) and (1.19) into Eq. (1.17) yields

$$
Q \equiv 2\pi \left.\frac{U_{\max}}{\Delta W}\right|_{r=1} = \left.\frac{1}{2\zeta r}\right|_{r=1} = \frac{1}{2\zeta} = \frac{\sqrt{km}}{c}
\tag{1.20}
$$

that is, the energy-based definition of Q (Eq. (1.17)) is equivalent to the definition given by Eq. (1.8).

A common experimental method used to measure Q in MEMS/NEMS resonators (and in other resonating structures) is the *bandwidth method*, also known as the −3 dB bandwidth method or the half-power method [2, 20]. This method is based on taking advantage of the Q-dependence of the shape of the *FRF* for the response amplitude in the vicinity of the resonant peak (Figure 1.3a). The *FRF* may be for the displacement amplitude or any output signal "R" that is proportional to the displacement amplitude (e.g., electrical voltage). Having experimentally determined the shape of the resonant peak, the value of Q may be estimated by the formula

$$
Q \approx \frac{\omega_{\text{res}}}{\Delta \omega} = \frac{f_{\text{res}}}{\Delta f}
\tag{1.21}
$$

in which ω_{res} is the resonant frequency (defined by the peak response value $R_{\max}$ of the *FRF*), $\Delta \omega \equiv \omega_2 - \omega_1$ is the frequency bandwidth and ω_1 and ω_2 are the frequencies corresponding to a response value of $R_{\max}/\sqrt{2}$. Analogous definitions apply to the f quantities in Eq. (1.21) for the common case in which frequency values are plotted in hertz, kilohertz, and so on. An example of a numerical calculation of Q by the bandwidth method is shown in Figure 1.4 for the case of an SDOF oscillator having a 10% damping ratio or an "exact" Q value of 5 according to the definition given by Eq. (1.8) or Eq. (1.17). In this example, the bandwidth-based Q value extracted from the shape of the *FRF* curve is 4.90, resulting in only a 2.0% difference when compared with the value furnished by Eq. (1.8) or Eq. (1.17). Most well-designed MEMS resonators will have quality factors well in excess of 5; for these cases, the percent difference will be even smaller. Thus, in most cases of practical interest in resonant MEMS applications, the value of Q based on the bandwidth method is essentially the same as those associated with the property-based (Eq. (1.8)) and energy-based (Eq. (1.17)) definitions.

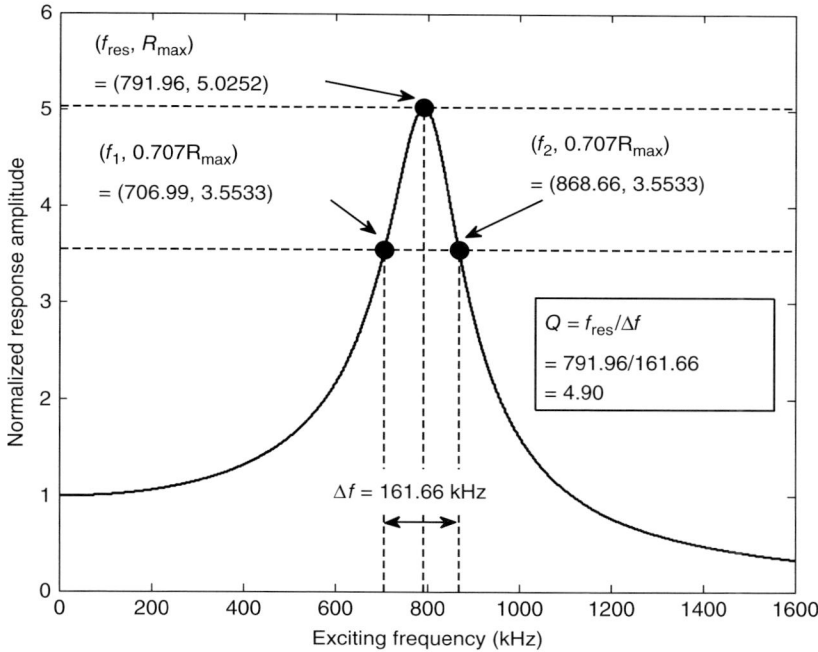

Figure 1.4 Example of Q calculation using the bandwidth method, resulting in $Q = 4.90$. Note that the "exact" Q value in this example using the definition given by Eq. (1.8) or Eq. (1.17) is 5, so that the discrepancy is only 2%. The difference will be even smaller for higher-Q systems typically used as MEMS/NEMS resonators.

1.3.3
Contributions to Quality Factor from Multiple Sources

There exist multiple energy dissipation mechanisms in MEMS/NEMS resonators, many of which are not well understood. For example, researchers have identified and attempted to model and measure the damping associated with viscous dissipation in an ambient fluid, support ("anchor") losses, thermoelastic dissipation, viscoelastic losses, and various surface-related dissipation phenomena. (See Chapter 3 for more details.) Although any or all of the various dissipation mechanisms may be acting simultaneously, in many cases one of these may dominate so that the others will have a negligible effect on the overall Q-factor that is exhibited by the resonator. When multiple loss sources have an impact on the total Q, one may use the definition of Q to derive the relationship between the total Q and the Q_i, $i = 1, 2, \ldots$, due to the individual contributions. One may easily derive this relationship, as will now be shown, by starting with the energy-based definition of Q given by Eq. (1.17). (One could also begin with the definition given by Eq. (1.8) and arrive at the same result.)

Assuming that the various sources of energy dissipation in the resonator are independent, Eq. (1.17) may be written as

$$Q \equiv 2\pi \left. \frac{U_{\max}}{\Delta W_1 + \Delta W_2 + \cdots} \right|_{r=1} \tag{1.22}$$

in which ΔW_i, $i = 1, 2, \ldots$, represents the dissipated energy per cycle due to the individual damping mechanisms. Inverting Eq. (1.22) gives

$$\frac{1}{Q} = \frac{1}{2\pi} \left. \frac{\Delta W_1 + \Delta W_2 + \cdots}{U_{\max}} \right|_{r=1} = \frac{1}{Q_1} + \frac{1}{Q_2} + \cdots \tag{1.23}$$

or

$$Q = \frac{1}{\frac{1}{Q_1} + \frac{1}{Q_2} + \cdots} \tag{1.24}$$

which is the desired result relating the total Q of the system to the individual Q_i values. Clearly, it follows from Eq. (1.24) that

$$Q \approx Q_{\min} \tag{1.25}$$

for the case in which the smallest of the Q_i, denoted by $Q_{\min}$, is much smaller than the Q_i corresponding to all other damping sources. This is a situation that may occur, for example, for in-vacuum or low-pressure gas applications in which support losses could dominate or for liquid-phase applications in which viscous dissipation in the liquid tends to be the major damping mechanism in many cases.

1.4
Continuous Systems Modeling: Microcantilever Beam Example

In Section 1.3, some important fundamentals of both free vibration and harmonically forced vibration of an idealized SDOF oscillator were summarized. However, in reality all dynamic systems have the potential to respond in a manner that requires a MDOF description in order to capture multiple "modes" of vibration that may occur, including the possible interaction of these modes during a free or forced vibration. Two approaches exist for modeling the MDOF response of such systems: a discrete-coordinate description and a continuous modeling approach.

In the discrete-coordinate approach, the system properties are often idealized in such a manner that the inertial properties and the stiffness properties are uncoupled; in other words, those portions that have mass are assumed rigid, while those parts having flexibility are assumed to be massless. As a result, the system position at any time may be expressed in terms of the displacements (and/or rotations) of a finite number of discrete locations in the system. In fact, modeling using the finite element method (FEM) falls into this category as it is based on a systematic approach for lumping mass and stiffness characteristics of the system at the "nodes" of the finite element model. (This modeling approach is discussed in more detail in Chapter 5.) The mathematical model resulting from the discrete approach is a set of ordinary differential equations (ODEs) in time, usually written in a matrix form.

The continuous modeling approach aims to maintain the distributed nature of the system's mass and stiffness characteristics and does so by representing the vibrational response in terms of continuous variables – for example, beam deflection as a function of a continuous coordinate x ranging from 0 to L (beam length) – in lieu of a discrete representation. This type of model therefore comprises an infinite number of degrees of freedom, yielding a mathematical description involving one or more partial differential equations (PDEs).

The aim of the present section is to provide a concise overview of the continuous systems modeling approach by examining the vibration of a cantilever beam. The motivation for this particular focus is threefold:

1) While obtaining solutions of the PDE(s) of the continuous modeling approach tends to be, in general, more difficult than solving the ODEs of the discrete method, in the case of many MEMS resonators, the geometries tend to be relatively simple and therefore amenable to the continuous modeling approach.

2) Micro- and nanocantilevers are quite prominent in a variety of resonant MEMS applications due to their ease of fabrication, portability, and versatility [4, 7, 10]. Hence, the example furnished by a cantilever beam will yield specific results that are highly relevant to many cantilever-based resonant devices.

3) The simple example of a cantilever serves as an ideal vehicle for presenting the general modeling approach, terminology, and concepts that are equally applicable to cases of other resonator geometries and boundary conditions (BCs) (e.g., doubly clamped "bridge" beams, membrane disks), the details of which may be found elsewhere (e.g., [26] and the references cited therein).

1.4.1
Modeling Assumptions

The continuous model for the vibration of a cantilever beam (Figure 1.5) will be based on the following assumptions:

- The mass and stiffness properties are constant (independent of time and frequency);
- The beam is prismatic (the cross section is uniform along the beam length);
- The cross section has an axis of symmetry and the beam vibration occurs along the direction of this axis;
- The system is linear: the beam is composed of a linear elastic isotropic material and the beam deformation is small (slopes of the bent beam are much less than unity);
- The kinematic assumptions of Bernoulli-Euler beam theory apply: cross sections of the beam remain planar during the deformation (no cross-sectional warping) and they remain normal to the deformed beam axis (no transverse

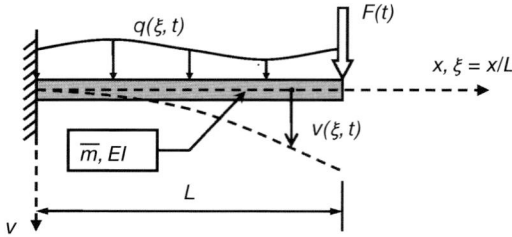

Figure 1.5 Schematic diagram of a cantilever beam modeled as a continuous system having distributed mass per unit length ($\overline{m}$) and distributed flexibility (*EI*), where E = Young's modulus and I = second moment of area (sometimes called the "moment of inertia") of the cross section.

shear strain). These assumptions tend to be applicable to beams that are relatively long and slender (length L is much larger than the largest cross-sectional dimension) [27].

- Energy dissipation (damping) is not included, although an analogous approach may be used to incorporate viscous damping (see, e.g., Chapter 2).

While the results presented here are "classical" in that they apply to cantilevers of all scales (provided that the above assumptions are met), the terminology of "mirocantilever" or "nanocantilever" will be used at times due to the fact that many of the complicating issues germane to MEMS/NEMS resonators (treated in subsequent chapters) arise due to device miniaturization.

1.4.2
Boundary Value Problem for a Vibrating Microcantilever

For the assumptions stated above, the time-dependent motion of the cantilever beam of Figure 1.5 under a general distributed transverse load, $q(\xi, t)$ (force per unit length), and a general end force, $F(t)$, is governed by a boundary value problem (BVP) in terms of the transverse deflection $v(\xi, t)$. The BVP includes the equation of motion (resulting from equilibrium of a differential slice of the beam) [18],

$$v''''(\xi, t) + \frac{\overline{m}L^4}{EI}\ddot{v}(\xi, t) = \frac{q(\xi, t)L^4}{EI} \tag{1.26}$$

and the BCs,

$$v(0, t) = v'(0, t) = v''(1, t) = 0, \quad v'''(1, t) = -\frac{L^3 F(t)}{EI} \tag{1.27}$$

In Eq. (1.26), the prime and dot notations denote differentiation with respect to the spatial (ξ) and time (t) coordinates, respectively, where $\xi \equiv x/L$ is the dimensionless spatial coordinate. The other relevant notation is defined in Figure 1.5. In the following sub-sections, solutions of this BVP will be presented for the cases of (i) a free vibration and (ii) a forced vibration due to a sinusoidal end force.

1.4.3
Free-Vibration Response of Microcantilever

The BVP reduces to its homogeneous form in the case of a free vibration:

$$v''''(\xi, t) + \frac{\overline{m}L^4}{EI} \ddot{v}(\xi, t) = 0 \tag{1.28}$$

$$v(0, t) = v'(0, t) = v''(1, t) = v'''(1, t) = 0 \tag{1.29}$$

Solutions are assumed to be in the form of "modal vibrations," that is, free vibrations of a constant shape:

$$v(\xi, t) = \varphi_n(\xi)(A \cos \omega_n t + B \sin \omega_n t), \quad n = 1, 2, \dots \tag{1.30}$$

in which ω_n is the *natural frequency of the nth mode* and $\varphi_n(\xi)$ is the corresponding *mode shape*, both of which have yet to be determined. Placing Eq. (1.30) into Eqs. (1.28) and (1.29) yields the following eigenvalue problem for determining the dimensionless natural frequencies (eigenvalues) λ_n and the associated mode shapes (eigenfunctions or eigenmodes) $\varphi_n(\xi)$:

$$\varphi_n''''(\xi) - \lambda_n{}^4 \varphi_n(\xi) = 0, \left(\lambda_n{}^4 \equiv \frac{\overline{m}L^4 \omega_n{}^2}{EI} \right) \tag{1.31}$$

$$\varphi_n(0) = \varphi_n'(0) = \varphi_n''(1) = \varphi_n'''(1) = 0 \tag{1.32}$$

The general solution of Eq. (1.31) is of the form

$$\varphi_n(\xi) = A_1 \cosh \lambda_n \xi + A_2 \cos \lambda_n \xi + A_3 \sinh \lambda_n \xi + A_4 \sin \lambda_n \xi \tag{1.33}$$

which, when placed into Eqs. (1.32), results in a linear algebraic system of the form

$$[G(\lambda_n)]\{A\} = \{0\} \tag{1.34}$$

for determining the coefficients in Eq. (1.33). Non-trivial solutions for the vector $\{A\}$ only exist if the λ_n-dependent coefficient matrix is singular, thus requiring that $\det[G(\lambda_n)] = 0$ or, after simplifying,

$$1 + \cosh \lambda_n \cos \lambda_n = 0 \tag{1.35}$$

Equation (1.35) is referred to as the *frequency equation* of the cantilever beam. Its roots, which by convention are numbered in increasing order, are given by (four significant figures)

$$\lambda_1 \equiv 1.875, \quad \lambda_2 \equiv 4.694, \quad \lambda_3 \equiv 7.855, \quad \lambda_n \approx \frac{(2n-1)\pi}{2} \quad \text{for } n > 3 \tag{1.36}$$

The corresponding (undamped) natural frequencies of the cantilever are given by the definition listed with Eq. (1.31), that is,

$$\omega_n = \lambda_n{}^2 \sqrt{\frac{EI}{\overline{m}L^4}} \quad \text{or} \quad f_n = \frac{\lambda_n{}^2}{2\pi} \sqrt{\frac{EI}{\overline{m}L^4}} \tag{1.37}$$

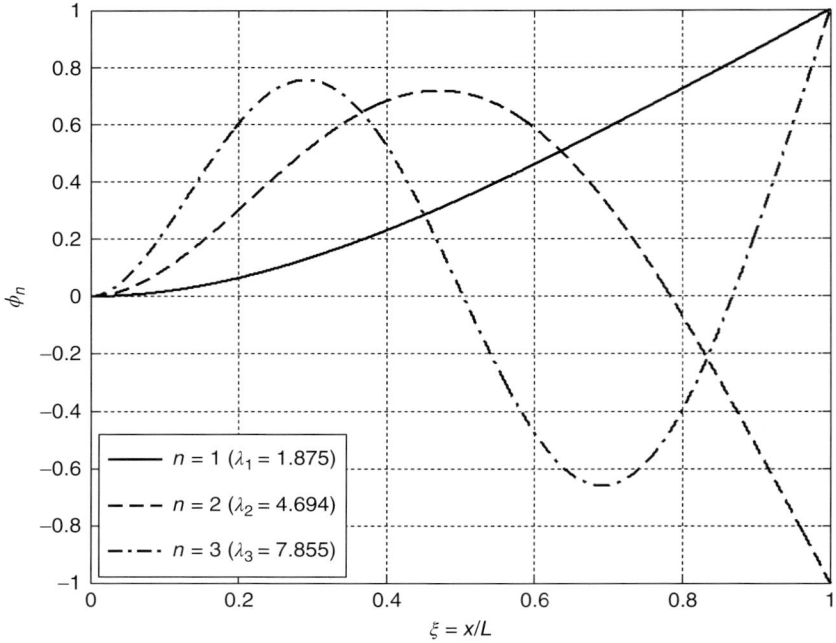

Figure 1.6 First three mode shapes of a cantilever beam. The plots have been normalized such that the tip deflection is $(-1)^{n+1}$, where n is the mode number.

The mode shapes are determined by placing each eigenvalue listed in Eq. (1.36) into Eq. (1.34), solving Eq. (1.34) for the constants A_2, A_3, A_4 in terms of A_1, and substituting the result into Eq. (1.33). This yields the individual modes shapes, $\varphi_n(\xi), n = 1, 2, \ldots$, corresponding to each of the natural frequencies:

$$\varphi_n(\xi) = A_1^{(n)} \left[\cosh \lambda_n \xi - \cos \lambda_n \xi - \left(\frac{\cosh \lambda_n + \cos \lambda_n}{\sinh \lambda_n + \sin \lambda_n} \right) (\sinh \lambda_n \xi - \sin \lambda_n \xi) \right]$$

$$(1.38)$$

The constant $A_1^{(n)}$ for each mode is arbitrary; its value may be chosen to scale the mode shape in any manner that is deemed convenient. Plots of the first three mode shapes for a cantilever beam are shown in Figure 1.6. Each mode shape shown has one or more locations at which the beam displacement is zero. Such points that experience no movement during a modal vibration are known as vibrational *nodes*. In general, the number of vibrational nodes will increase as the mode number n increases. The location of the nodes has important practical implications in resonant MEMS applications in that a designer may minimize the energy losses associated with supporting structures by placing the resonator supports at or near the vibrational nodes of the resonator. (See Chapters 3 and 5 for more details.)

1.4.4

Steady-State Response of a Harmonically Excited Microcantilever

Next, the particular case of a cantilever actuated by a harmonic end force, $F(t) = F_0 \sin \omega t$, is considered as an example of a forced vibration. The general BVP of Section 1.4.2 becomes

$$v''''(\xi, t) + \frac{\overline{m}L^4}{EI} \ddot{v}(\xi, t) = 0 \tag{1.39}$$

$$v(0, t) = v'(0, t) = v''(1, t) = 0, \quad v'''(1, t) = -\frac{F_0 L^3}{EI} \sin \omega t \tag{1.40}$$

A steady-state solution of the form

$$v(\xi, t) = \frac{F_0 L^3}{3EI} \psi(\xi) \sin \omega t \tag{1.41}$$

where $\psi(\xi)$ is the unknown vibrational shape, is assumed since the response of the undamped system is expected to be in phase (or perfectly out of phase) with the excitation force, as was the case with the undamped SDOF oscillator (Section 1.3.2). Note that the coefficient introduced in Eq. (1.41) is the tip displacement associated with a quasi-static ($\omega \to 0$) application of the tip force [27]; thus, $|\psi(\xi)|$ may be interpreted as the spatially varying normalized amplitude of beam deflection, scaled with respect to the quasi-static tip value. In particular, one may view $D_{\text{tip}} \equiv |\psi(1)|$ as being the dynamic magnification factor at the loaded end of the beam. (This is analogous to the factor D introduced in the SDOF analysis of Section 1.3.2.) Placing Eq. (1.41) into Eqs. (1.39) and (1.40) leads to the BVP governing $\psi(\xi)$:

$$\psi''''(\xi) - \lambda^4 \psi(\xi) = 0, \quad \left(\lambda^4 \equiv \frac{\overline{m}L^4 \omega^2}{EI} \right) \tag{1.42}$$

$$\psi(0) = \psi'(0) = \psi''(1) = 0, \quad \psi'''(1) = -3 \tag{1.43}$$

Unlike parameter λ_n appearing in Eq. (1.31) of the previous section, which was a system property to be determined, here parameter λ is a *specified* dimensionless driving frequency. The general solution of Eq. (1.42) is

$$\psi(\xi) = A_1 \cosh \lambda \xi + A_2 \cos \lambda \xi + A_3 \sinh \lambda \xi + A_4 \sin \lambda \xi \tag{1.44}$$

in which the constants (A_i), in order to meet the BCs (Eqs. (1.43)), must satisfy the following non-homogeneous system in which $[G(\cdot)]$ is the same matrix function that appeared in Eq. (1.34):

$$[G(\lambda)]\{A\} = \{0 \ 0 \ 0 \ -3\}^T \tag{1.45}$$

Solving Eq. (1.45), substituting the result into Eq. (1.44), and simplifying, one may arrive at the solution for the beam's vibrational shape:

$$\psi(\xi) = \frac{3\{[S(\lambda) + s(\lambda)][C(\lambda\xi) - c(\lambda\xi)] - [C(\lambda) + c(\lambda)][S(\lambda\xi) - s(\lambda\xi)]\}}{2\lambda^3[1 + C(\lambda)c(\lambda)]} \quad (1.46)$$

where the following shorthand notation has been introduced: $C(\cdot) = \cosh(\cdot), S(\cdot) = \sinh(\cdot), c(\cdot) = \cos(\cdot), s(\cdot) = \sin(\cdot)$. The magnitude of this expression evaluated at the beam tip ($\xi = 1$) results in the dynamic magnification factor D_{tip} for the tip displacement of a cantilever beam loaded by a harmonic tip force:

$$D_{tip} \equiv |\psi(1)| = \left| \frac{3\,[C\,(\lambda)\,s(\lambda) - S(\lambda)c(\lambda)]}{\lambda^3[1 + C(\lambda)c(\lambda)]} \right| \quad (1.47)$$

A plot of Eq. (1.47) versus the exciting frequency parameter is shown in Figure 1.7a, while in Figure 1.7b the lag angle θ_{tip} of the tip displacement with respect to the applied force is shown. In addition, Figure 1.8 displays the vibrational shape, given by Eq. (1.46), for different exciting frequencies. An examination of these figures within the context of the results of the free-vibration cantilever analysis (Section 1.4.3) leads to the following observations:

- The forced-vibration continuous-system model of the undamped cantilever exhibits multiple resonant frequencies (peaks in Figure 1.7a), each of which corresponds to one of the natural frequencies associated with the eigenvalues in Eq. (1.36). When damping is included, the resonant frequencies will shift to the left of the undamped natural frequencies as was seen in the SDOF analysis, but for sufficiently high Q values the undamped natural frequencies will furnish excellent estimates of the resonant frequencies.
- The dynamic magnification factor for the harmonically driven undamped cantilever approaches infinity at each of the resonant peaks (Figure 1.7a). This is analogous to the SDOF behavior of an undamped oscillator. (See $\zeta = 0$ curve in Figure 1.3a.) When damping is included, the peak magnitudes will be finite and will decrease as the amount of damping increases (similar to the SDOF behavior) and the corresponding Q values may be estimated by applying the bandwidth method to each peak.
- Figure 1.7a shows that D_{tip} approaches 1 as $\lambda \to 0$, that is, dynamic effects are negligible at low driving frequencies, as expected. This is also reflected in Figure 1.8 in which, as the driving frequency approaches 0, the vibrational shape approaches the cantilever's (cubic) deflected shape due to a static end force [27].
- For the harmonically driven undamped cantilever, the lag angle θ_{tip} of the tip displacement with respect to the applied force is either 0 or π depending on the sign of $\psi(1)$ (Figure 1.7b). Similar to the SDOF behavior of Figure 1.3b, the tip response changes from in-phase ($\theta_{tip} = 0$) to completely out-of-phase ($\theta_{tip} = \pi$) when crossing a resonant peak from left to right in Figure 1.7b. When damping is included, the transition from in-phase to out-of-phase response is more gradual, that is, the slopes of the lag angle plot are finite at the various resonances and these slopes become smaller as the level of damping is increased or as Q is decreased.

(a)

(b)

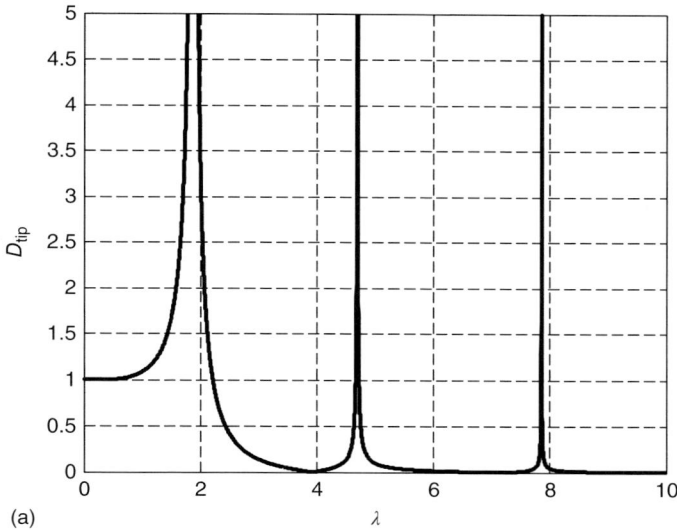

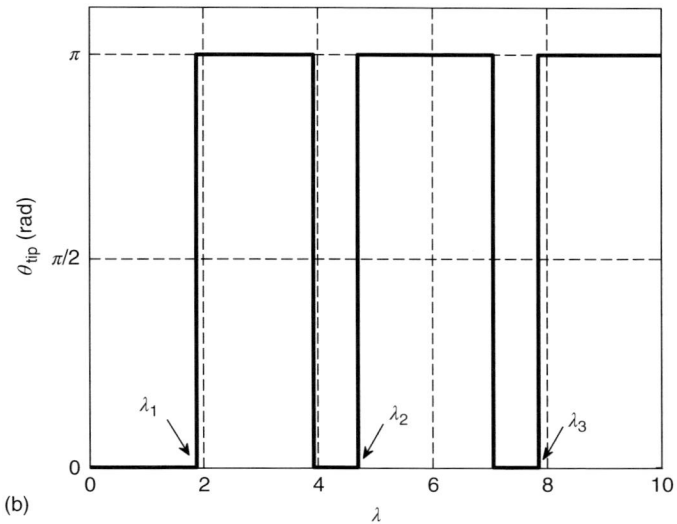

Figure 1.7 Frequency response functions characterizing the steady-state harmonic response of an undamped cantilever due to a harmonic tip force: (a) normalized tip displacement amplitude (dynamic magnification factor) and (b) lag angle of tip displacement with respect to applied force.

- Zoomed views of Figure 1.7a show that, between consecutive resonant peaks, D_{tip} attains a zero value at a particular driving frequency. This means that the beam tip is stationary when the cantilever is vibrating at these particular frequencies. This is indicative of the lag angle switching from $\theta_{tip} = \pi$ to $\theta_{tip} = 0$ as the forcing frequency is increased. (See two instances in Figure 1.7b.)

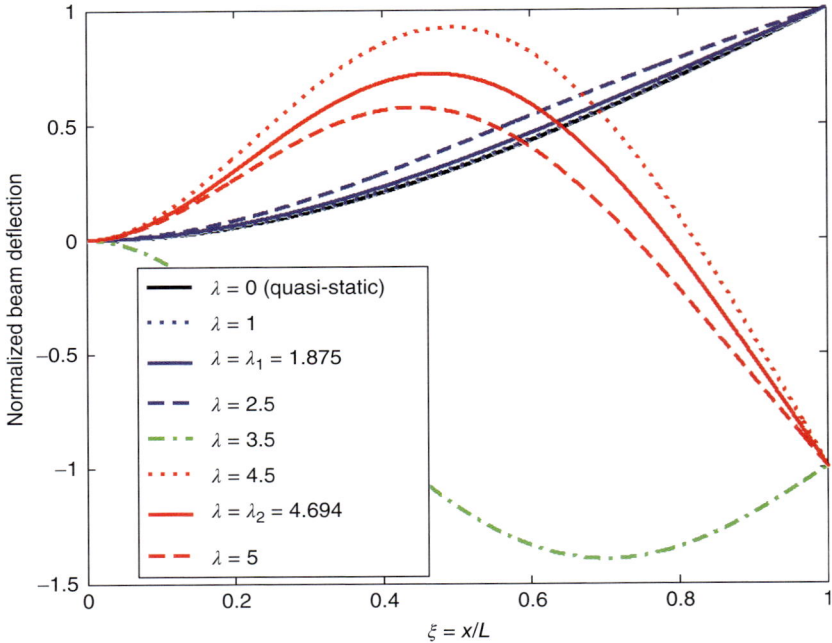

Figure 1.8 Vibrational shapes of a tip-force-actuated cantilever beam for various exciting frequencies.

- Figure 1.8 shows that, when the exciting frequency is near one of the resonant frequencies, the vibrational shape of the beam is very similar to the corresponding mode shape. This is especially true for the fundamental mode ($n = 1$). The implication is that the system behaves essentially as an SDOF system in the vicinity of a resonant peak and the vibrational shape is approximately given by the corresponding mode shape. This observation may be used to develop simple, yet accurate, SDOF models of various types of MEMS resonators (e.g., [24, 28]). However, at a driving frequency sufficiently far from the two neighboring resonant peaks, the beam shape may differ significantly from any single mode shape (e.g., see $\lambda = 3.5$ shape in Figure 1.8), indicating that the response receives contributions from multiple modes. (This may be seen more explicitly if one uses the "mode superposition method" to solve the forced-vibration problem [19].)

1.5
Formulas for Undamped Natural Frequencies

As seen in the previous sections, knowing the value of the undamped natural frequency of a resonator for a particular mode of vibration is important in several respects: (i) the relative magnitude of the driving frequency to this value determines the degree to which that mode will be excited; (ii) when damping is relatively small, the undamped natural frequency furnishes a good estimate of the resonant frequency; (iii) the undamped natural frequency yields an upper bound on the resonant frequency when damping and/or the effective mass of any surrounding fluid is significant; and (iv) in more detailed theoretical pursuits the undamped natural frequency may serve as a convenient reference frequency for normalizing the system's actual resonant frequency. All of these reasons provide the motivation in this section to catalog several formulas for determining the undamped natural frequencies for some of the more common device geometries and vibration modes that are encountered in resonant MEMS applications. All formulas listed are for circular frequency (units of radians per second) and, as noted earlier, may be converted to hertz (cycles per second) by dividing by 2π. Within each class of structure type and vibration mode considered, the range of the mode number n corresponding to the ω_n formula is $n = 1, 2, \ldots$, where $\omega_1 < \omega_2 < \ldots$. Rigid-body (zero-frequency) modes are not considered; thus, in all cases $\omega_1 > 0$.

For conciseness, the derivations of the formulas presented here are not included, but the reader is encouraged to seek out details regarding the origin of these formulas as well as the details associated with the corresponding mode shapes. Such details may be found in [26] and in the sources cited therein. All of the formulas listed are based on the assumptions that the material is linear elastic and isotropic with Young's modulus E, shear modulus G, and Poisson's ratio ν. Therefore, when applying the results to an anisotropic material, such as silicon, care should be taken in specifying the equivalent isotropic elastic constants corresponding to the appropriate direction(s). (See [29] for guidance in such cases.) Also, each device is assumed to have a uniform mass density ρ (per unit volume), all support conditions are considered "ideal" (perfectly "clamped," perfectly "free," etc.), effects of any surrounding fluid are neglected, and, unless indicated otherwise, all devices are assumed to have a uniform thickness h. Definitions of other parameters appearing in the formulas are given in Figure 1.9. Note that results for "free" (i.e., unsupported) conditions are included, despite the fact that all MEMS/NEMS resonators involve some type of supporting structure(s); the justification is that a strategic placement of supports near the resonator's vibrational nodes will result in the free BCs being approximately satisfied and, as stated earlier, minimal energy dissipation via support losses.

1.5.1
Simple Deformations (Axial, Bending, Twisting) of 1D Structural Members: Cantilevers and Doubly Clamped Members ("Bridges")

The relevant device parameters in this section are defined in Figure 1.9a.

1.5.1.1   Axial Vibrations (Along x-Axis)

$$\omega_n = \lambda_n \sqrt{\frac{E}{\rho L^2}} \qquad (1.48)$$

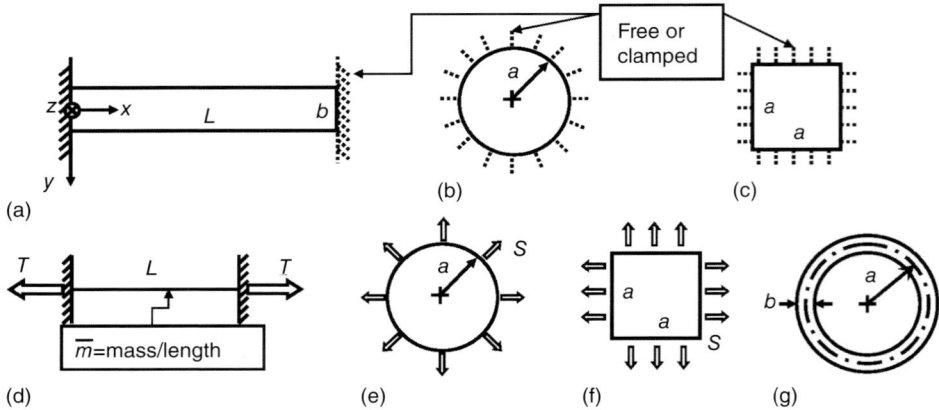

Figure 1.9 Notation for the natural frequency formulas listed in Section 1.5: (a) bars of rectangular cross section (clamped-free or clamped-clamped); (b) circular plate (free or clamped); (c) square plate (free or clamped); (d) string structure (fixed ends) under axial tension force T; (e) circular membrane (fixed periphery) under uniform tension S = force per unit length; (f) square membrane (fixed periphery) under uniform tension S = force per unit length; and (g) circular ring (free). Thickness h is considered uniform in all cases except for case (d) in which only the cross section must be uniform.

$$\lambda_n = \begin{cases} \frac{(2n-1)\pi}{2} \,, & \text{clamped-free (cantilever)} \\ \\ n\pi \,, & \text{clamped-clamped (“bridge”)} \end{cases} \tag{1.49}$$

1.5.1.2 Torsional Vibrations (Based on $h \ll b$) (Twist About x-Axis)

$$\omega_n = \lambda_n \sqrt{\frac{4Gh^2}{\rho b^2 L^2}}, \quad \lambda_n \text{ given by Eq. (1.49)} \tag{1.50}$$

1.5.1.3 Flexural (Bending) Vibrations

$$\omega_n = \lambda_n{}^2 \begin{cases} \sqrt{\frac{Eh^2}{12\rho L^4}} \,, & \text{transverse or out-of-plane } \left(\text{along } z\text{-axis}\right) \\ \\ \sqrt{\frac{Eb^2}{12\rho L^4}} \,, & \text{lateral or in-plane (along } y\text{-axis)} \end{cases} \tag{1.51}$$

in which the λ_n values are given in Table 1.1 for both the cantilever and doubly-clamped cases. If the beam's "width" dimension (b for the transverse case and h for the lateral case of Eq. (1.51)) is not small relative to length L, this wide-beam effect may be taken into account in an approximate manner by replacing E with an "effective Young's modulus" of $E_{\text{eff}} = E/(1 - v^2)$ or in a more exact manner using the formula derived in [30].

Table 1.1 Dimensionless coefficients for calculating the natural frequencies for the flexural modes of a beam using Eq. (1.51).

	Clamped-free (cantilever)	Clamped-clamped ("bridge")
λ_1	1.875	4.730
λ_2	4.694	7.853
λ_3	7.855	10.996
λ_n, $n > 3$	$(2n - 1)\pi/2$	$(2n + 1)\pi/2$

Table 1.2 Dimensionless coefficients for calculating the natural frequencies for the transverse deflection of circular and square plates under fully free and fully clamped conditions using Eq. (1.52).

	Circular plate		Square plate	
	Freely supported along periphery ($v = 0.33$)	Clamped along periphery (arbitrary v)	Freely supported along periphery ($v = 0.3$)	Clamped along periphery (arbitrary v)
$\lambda_1{}^2$	5.253	10.22	13.49	35.99
$\lambda_2{}^2$	9.084	21.26	19.79	73.41
$\lambda_3{}^2$	12.23	34.88	24.43	108.3
$\lambda_4{}^2$	20.52	39.77	35.02	131.6

1.5.2
Transverse Deflection of 2D Structures: Circular and Square Plates with Free and Clamped Supports

$$\omega_n = \lambda_n{}^2 \sqrt{\frac{Eh^2}{12(1 - v^2)\rho a^4}} \tag{1.52}$$

in which the $\lambda_n{}^2$ values are given in Table 1.2 for the four lowest modes for each case. The relevant device parameters are defined in Figures 1.9b and 1.9c for the circular and square plate cases, respectively.

1.5.3
Transverse Deflection of 1D Membrane Structures ("Strings")

Here the cross section of the string is arbitrary (i.e., it need not be rectangular with uniform thickness h) provided that the section has a plane of symmetry that coincides with the string's plane of vibration. Also, the effect of any sag in the initial string configuration is neglected. The natural frequencies for the fixed-fixed string

Table 1.3 Dimensionless coefficients for calculating the natural frequencies for the transverse vibrations of circular and square membranes supported along their periphery using Eq. (1.54).

	Circular membrane	Square membrane
λ_1	$1.357\sqrt{\pi} = 2.405$	$\sqrt{2\pi} = 4.443$
λ_2	$2.162\sqrt{\pi} = 3.832$	$\sqrt{5\pi} = 7.025$
λ_3	$2.897\sqrt{\pi} = 5.135$	$\sqrt{8\pi} = 8.886$
λ_4	$3.114\sqrt{\pi} = 5.519$	$\sqrt{10\pi} = 9.935$

(Figure 1.9d) are given by the following:

$$\omega_n = n\pi\sqrt{\frac{T}{\overline{m}L^2}} \quad , \quad (\overline{m} = \text{mass per unit length}, \ T = \text{axial tensile force}) \quad (1.53)$$

1.5.4
Transverse Deflection of 2D Membrane Structures: Circular and Square Membranes under Uniform Tension and Supported along Periphery

$$\omega_n = \lambda_n\sqrt{\frac{S}{\rho h a^2}} \tag{1.54}$$

in which S is the membrane tension (per unit length of the membrane's periphery, as indicated in Figures 1.9e and 1.9f and the λ_n values are given in Table 1.3 for the four lowest modes for both circular and square membranes.

1.5.5
In-Plane Deformation of Slender Circular Rings

The relevant device parameters for a circular ring are defined in Figure 1.9g. The results that follow are based on the assumption that the ring is slender ($b \ll a$).

1.5.5.1 Extensional Modes
The extensional modes of a slender circular ring involve only local elongation or shortening of the ring circumference without any in-plane bending of the ring occurring. The natural frequencies of these modes are given by

$$\omega_n = \sqrt{1 + (n-1)^2}\sqrt{\frac{E}{\rho a^2}} \tag{1.55}$$

1.5.5.2 In-Plane Bending Modes
The in-plane bending modes of a slender circular ring involve no extension or contraction of the ring along its circumferential direction, that is, they are uncoupled from the extensional modes. The natural frequencies of the in-plane bending modes are

$$\omega_n = \frac{n(n+1)(n+2)}{\sqrt{12[(n+1)^2+1]}} \frac{b}{a} \sqrt{\frac{E}{\rho a^2}} \tag{1.56}$$

As can be seen from the b/a factor in Eq. (1.56) in comparison with Eq. (1.55), the in-plane bending modes tend to be of much lower frequency than the extensional modes (recall that $b/a \ll 1$ for a slender ring) due to the ring's in-plane bending stiffness being much smaller than its extensional stiffness.

1.6
Summary

This chapter provided an introduction to the fundamental theory of mechanical vibration on which all MEMS/NEMS resonant devices are based. Key concepts related to resonators in general were introduced by means of specific examples, namely, the free and forced vibration of (i) the classical SDOF damped oscillator and (ii) an undamped cantilever beam. Also included was a listing of formulas for calculating the undamped natural frequencies of devices whose geometries are often utilized in MEMS/NEMS resonator applications.

The fundamental concepts introduced here necessarily neglected numerous complicating issues that are often encountered in practical applications, many of which will be treated in detail in the more specialized chapters that follow.

Acknowledgment

The authors are grateful for interesting discussions with Martin Heinisch on the chapter content and for his suggestions for improving the chapter.

References

1. Goeders, K.M., Colton, J.S., and Bottomley, L.A. (2008) Microcantilevers: sensing chemical interactions via mechanical motion. *Chem. Rev.*, **108**, 522–542.
2. Campanella, H. (2010) *Acoustic Wave and Electromechanical Resonators*, Artech House, Norwood, MA.
3. Hunt, H.K. and Armani, A.M. (2010) Label-free biological and chemical sensors. *Nanoscale*, **2**, 1544–1559.
4. Boisen, A., Dohn, S., Keller, S.S., Schmid, S., and Tenje, M. (2011) Cantilever-like micromechanical sensors. *Rep. Prog. Phys.*, **74**, 036101, 30 pp.
5. Fanget, S., Hentz, S., Puget, P., Arcamone, J., Matheron, M., Colinet, E., Andreucci, P., Duraffourg, L., Myers, E., and Roukes, M.L. (2011) Gas sensors based on gravimetric detection – A review. *Sens. Actuators B*, **160**, 804–821.
6. Eom, K., Park, H.S., Yoon, D.S., and Kwon, T. (2011) Nanomechanical resonators and their applications in biological/chemical detection: nanomechanics principles. *Phys. Rep.*, **503**, 115–163.
7. Zhu, Q. (2011) Microcantilever sensors in biological and chemical detections. *Sens. Transducers*, **125**, 1–21.

8. Braun, T., Barwich, V., Ghatkesar, M.K., Bredekamp, A.H., Gerber, C., Hegner, M., and Lang, H.P. (2005) Micromechanical mass sensors for biomolecular detection in a physiological environment. *Phys. Rev. E*, **72**, 031907, 9 pp.

9. Arlett, J.L., Myers, E.B., and Roukes, M.L. (2011) Comparative advantages of mechanical biosensors. *Nat. Nanotechnol.*, **6**, 203–215.

10. Johnson, B.N. and Mutharasan, R. (2012) Biosensing using dynamic-mode cantilever sensors: A review. *Biosens. Bioelectron.*, **32**, 1–18.

11. Belmiloud, N., Dufour, I., Colin, A., and Nicu, L. (2008) Rheological behavior probed by vibrating microcantilevers. *Appl. Phys. Lett.*, **92**, 041907, 3 pp.

12. Riesch, C., Reichel, E.K., Keplinger, F., and Jakoby, B. (2008) Characterizing vibrating cantilevers for liquid viscosity and density sensing. *J. Sens.*, **2008**, 697062, 9 pp.

13. Dufour, I., Maali, A., Amarouchene, Y., Ayela, C., Caillard, B., Darwiche, A., Guirardel, M., Kellay, H., Lemaire, E., Mathieu, F., Pellet, C., Saya, D., Youssry, M., Nicu, L., and Colin, A. (2012) The microcantilever: A versatile tool for measuring the rheological properties of complex fluids. *J. Sens.*, **2012**, 719898, 9 pp.

14. Dufour, I., Lemaire, E., Caillard, B., Debéda, H., Lucat, C., Heinrich, S.M., Josse, F., and Brand, O. (2014) Effect of hydrodynamic force on microcantilever vibrations: applications to liquid-phase chemical sensing. *Sens. Actuators B*, **192**, 664–672.

15. Anton, S.R. and Sodano, H.A. (2007) A review of power harvesting using piezoelectric materials (2003–2006). *Smart Mater. Struct.*, **16**, R1–R23.

16. Priya, S. and Inman, D.J. (eds) (2009) *Energy Harvesting Technologies*, Springer, New York.

17. Beeby, S. and White, N. (2010) *Energy Harvesting for Autonomous Systems*, Artech House, Norwood, MA.

18. Timoshenko, S. and Young, D.H. (1955) *Vibration Problems in Engineering*, 3rd edn, Van Nostrand Company, Inc., New York.

19. Clough, R.W. and Penzien, J. (1993) *Dynamics of Structures*, 2nd edn, McGraw-Hill, New York.

20. Ginsberg, J.H. (2001) *Mechanical and Structural Vibrations*, John Wiley & Sons, Inc., New York.

21. Inman, D.J. (2008) *Engineering Vibration*, Pearson Education, Inc., New Jersey.

22. Roundy, S. and Wright, P.K. (2004) A piezoelectric vibration based generator for wireless electronics. *Smart Mater. Struct.*, **13**, 1131–1142.

23. Lefeuvre, E., Badel, A., Richard, C., and Guyomar, D. (2005) Piezoelectric energy harvesting device optimization by synchronous charge extraction. *J. Intell. Mater. Syst. Struct.*, **16**, 865–876.

24. Heinrich, S.M., Maharjan, R., Dufour, I., Josse, F., Beardslee, L.A., and Brand, O. (2010) An analytical model of a thermally excited microcantilever vibrating laterally in a viscous fluid. Proceedings, IEEE Sensors 2010 Conference, Waikoloa, HI, pp. 1399-1404.

25. Yasumura, K.Y., Stowe, T.D., Chow, E.M., Pfafman, T., Kenny, T.W., Stipe, B.C., and Rugar, D. (2000) Quality factors in micron- and submicron-thick cantilevers. *J. Microelectromech. Syst.*, **9**, 117–125.

26. Blevins, R.D. (1979) *Formulas for Natural Frequency and Mode Shape*, Van Nostrand Reinhold Company Inc., New York.

27. Beer, F., Johnston, E.R., DeWolf, J., and Mazurek, D. (2011) *Mechanics of Materials*, 6th edn, McGraw-Hill, New York.

28. Dietl, J.M., Wickenheiser, A.M., and Garcia, E. (2010) A Timoshenko beam model for cantilevered piezoelectric energy harvesters. *Smart Mater. Struct.*, **19**, 055018, 12 pp.

29. Hopcroft, M., Nix, W., and Kenny, T. (2010) What is the Young's modulus of silicon? *J. Microelectromech. Syst.*, **19**, 229–238.

30. Yahiaoui, R. and Bosseboeuf, A. (2001) Improved modelling of the dynamical behaviour of cantilever microbeams. Proceedings, Micromechanics and Microsystems Europe Conference, MME 2001, Cork, Ireland, Septmber 16–18, 2001, pp. 281–284.

2
Frequency Response of Cantilever Beams Immersed in Viscous Fluids

Cornelis Anthony van Eysden and John Elie Sader

2.1
Introduction

The dynamic response of microcantilever beams is used in a broad range of applications, including ultra-sensitive mass measurements and imaging with molecular and atomic scale resolution [1–7]. Importantly, many of these applications are performed in a fluid environment (gas or liquid), which can significantly affect the dynamic response of a microcantilever. To explain this behavior, theoretical models have been developed that rigorously account for the effect of the surrounding fluid, which have been validated by detailed experimental measurements; models exist for flexural, torsional and extensional vibrational modes of cantilever beams [8–19]. These studies establish that viscosity plays an essential role in the frequency response of cantilevers of microscopic size (~100 μm in length), such as those used in the atomic force microscope (AFM) and in micro-electromechanical systems (MEMS). This contrasts to macro-scale cantilevers (~1 m in length), which are insensitive to the effects of fluid viscosity [20, 21].

The most common approach used to model the frequency response of a cantilever beam immersed in a viscous fluid is to approximate the fluid motion by a two-dimensional flow field due to local displacement of the beam [8]. Significantly, for cantilever beams of a high-aspect ratio (length/width), such as those used in the AFM, this approach has proven to be highly successful in predicting the frequency response of the fundamental mode and the next few higher order modes. Of recent interest has been the use of higher order cantilever modes [11, 22, 23], which enables operation at high-quality factors in highly viscous fluids. However, as mode order increases, the spatial wavelength of the modes decreases, ultimately leading to the violation of the two-dimensional flow field approximation and breakdown in the validity of these models. This limitation is also present in the well-known theory of Chu [20] for the resonant frequencies of a cantilever beam immersed in an inviscid fluid, which is valid for cases where fluid viscosity can be ignored. Models that account for three-dimensional effects,

Resonant MEMS – Fundamentals, Implementation and Application, First Edition.
Edited by Oliver Brand, Isabelle Dufour, Stephen M. Heinrich and Fabien Josse.
© 2015 Wiley-VCH Verlag GmbH & Co. KGaA. Published 2015 by Wiley-VCH Verlag GmbH & Co. KGaA.

and thus are valid for all modes, have been developed for both inviscid and viscous fluids [15, 24, 25].

Operation at higher mode numbers not only increases the importance of (incompressible) three-dimensional effects discussed above, but fluid compressibility can also play a major role. The reason for this feature is that the acoustic wavelength in the fluid decreases with increasing mode number and ultimately becomes comparable to and smaller than the dominant hydrodynamic length scale of the beam. Models that rigorously account for the three-dimensional effects of fluid compressibility are reported in [16].

In this chapter, we present an overview of the above-mentioned theoretical models for the fluid-structure interaction of cantilever beams immersed in fluid. The focus is on beams whose lengths greatly exceeds their widths, as is often encountered in practice. In Section 2.2.1, the two-dimensional theory for flexural modes at low mode number is presented, where fluid viscosity plays an essential role. Models for torsional, in-plane and extensional modes, also at low mode numbers, are summarized in Sections 2.2.2–2.2.4, respectively. The general, three-dimensional theory for arbitrary mode order is presented in Section 2.3, with a focus on the flexural modes; models for torsional modes are reported in [15, 16] and not detailed here. Both incompressible (Section 2.3.1) and compressible (Section 2.3.2) flows are discussed, accompanied by scaling analyses for determining the regimes where fluid compressibility is expected to be important.

2.2
Low Order Modes

2.2.1
Flexural Oscillation

In many applications, such as those found in AFM and MEMS, the fundamental mode and the next few modes are typically interrogated. In such cases, flow around the cantilever can be well approximated by one that is two-dimensional and explicit formulas for the frequency response derived. The theory pertaining to a rectangular cantilever beam immersed in a viscous fluid was first derived by Sader [8]. This model is summarized in this section, and contains the following principal assumptions:

1) Cross section of the beam is uniform along its entire length and is rectangular in geometry;
2) Length of the beam L greatly exceeds its width b, which in turn greatly exceeds it thickness h, see Figure 2.1;
3) Amplitude of oscillation is far smaller than any geometric length scale of the beam, which allows the Navier-Stokes equations to be linearized;

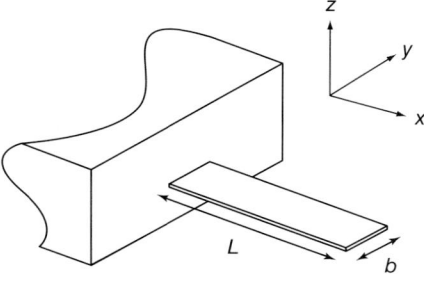

Figure 2.1 Schematic illustration showing the plan-view dimensions of a rectangular cantilever. Origin of the coordinate system is at the centroid of the beam cross section at its clamped end.

4) Internal (structural) dissipative effects are negligible in comparison with those of the fluid; and
5) Fluid is incompressible in nature and is unbounded in space.

The governing equation for the elastic deformation of the beam executing flexural oscillations is [26]

$$EI\frac{\partial^4 w(x,t)}{\partial x^4} + \mu\frac{\partial^2 w(x,t)}{\mathrm{d}t^2} = F(x,t) \tag{2.1}$$

where $w(x,t)$ is the deflection function of the beam in the z-direction, E is Young's modulus, I is the second moment of area (often referred to as the *moment of inertia*) of the beam cross section, μ is the mass per unit length of the beam, $F(x,t)$ is the external applied force per unit length in the z-direction, x is the spatial coordinate along the length of the beam, and t is time; see Figure 2.1. For a cantilever beam of uniform rectangular cross section, $I = bh^3/12$ [26] and $\mu = \rho_c bh$, where ρ_c is the cantilever density. The corresponding boundary conditions for the (clamped-free) cantilever beam are

$$\left[w(x,t) = \frac{\partial w(x,t)}{\partial x}\right]_{x=0} = \left[\frac{\partial^2 w(x,t)}{\partial x^2} = \frac{\partial^3 w(x,t)}{\partial x^3}\right]_{x=L} = 0 \tag{2.2}$$

To calculate the frequency response of the beam, we take the Fourier transform of the governing equation (2.1). Scaling the spatial variable, x, by the length L gives

$$\frac{EI}{L^4}\frac{\partial^4 \tilde{w}(x|\omega)}{\partial x^4} - \mu\omega^2 \tilde{w}(x|\omega) = \tilde{F}(x|\omega) \tag{2.3}$$

where x now denotes the scaled length coordinate, and

$$\tilde{X}(\omega) = \int_{-\infty}^{\infty} X(t)e^{i\omega t}\,\mathrm{d}t \tag{2.4}$$

is the Fourier transform of any function X of time and i is the usual imaginary unit. For simplicity, we shall henceforth omit the superfluous "~" notation in Eq. (2.4), while noting that all dependent variables refer to their Fourier space counterparts.

Solving Eq. (2.3) in the absence of any applied load, leads to the well-known result for the vacuum radial resonant frequencies

$$\omega_{\text{vac},n} = \frac{C_n^2}{L^2} \sqrt{\frac{EI}{\mu}} \qquad (2.5)$$

where $n = 1, 2, \dots$ is the mode order and C_n is the nth positive root of

$$1 + \cos C_n \cosh C_n = 0 \qquad (2.6)$$

which is well approximated by $C_n \approx (n - 1/2)\pi$ for $n \geq 2$.

We note in passing that while the present analysis focuses on cantilever beams, it is equally applicable to clamped-clamped beams provided (i) the eigenvalues C_n for clamped-clamped beams are used, and (ii) the deflection function $w(x, t)$, and corresponding homogeneous solutions $\phi_n(x)$, for a clamped-clamped beam are implemented in subsequent analyses. These are obtained by applying the appropriate boundary conditions for a clamped-clamped beam in Eq. (2.1). This yields the eigenvalues C_n satisfying $-1 + \cos C_n \cosh C_n = 0$, which are well approximately by $C_n \approx (n + 1/2)\pi$ for all mode numbers $n \geq 1$ [26]. This feature also holds for other theoretical models, including [9, 15, 25, 27–29].

Next, we turn our attention to the analysis of the load $F(x|\omega)$ applied to the cantilever. This load is decomposed into two components:

$$F(x|\omega) = F_{\text{hydro}}(x|\omega) + F_{\text{drive}}(x|\omega) \qquad (2.7)$$

corresponding to the hydrodynamic load $F_{\text{hydro}}(x|\omega)$ exerted by the fluid on the cantilever beam due to its motion, and a driving force $F_{\text{drive}}(x|\omega)$ that excites the cantilever. The hydrodynamic load is expressed generally as

$$F_{\text{hydro}}(x|\omega) = \frac{\pi}{4}\rho\omega^2 b^2 \Gamma^{\text{f}}(\omega)w(x|\omega) \qquad (2.8)$$

where ρ is the fluid density and $\Gamma^{\text{f}}(\omega)$ is the normalized hydrodynamic load, termed the *hydrodynamic function*. The superscript f refers to the flexural modes and will be used to distinguish between results for the torsional modes. It is important to emphasize that Eq. (2.8) is derived under the assumption that the length of the cantilever greatly exceeds its width, and as such, is formally consistent with the underlying assumptions of the beam equation, Eq. (2.1). The hydrodynamic function $\Gamma^{\text{f}}(\omega)$ is dimensionless and depends on the radial frequency ω through the dimensionless parameter

$$Re = \frac{\rho\omega b^2}{\eta}, \qquad (2.9)$$

where η is the fluid shear viscosity. The parameter Re is commonly termed the *Reynolds number*[1] and indicates the importance of inertial forces in the fluid relative to viscous forces.

---

1) The convention adopted for the Reynolds number conforms with [30]. The Reynolds number is often associated with the nonlinear convective inertial term in the Navier-Stokes equation. This latter convention has not been adopted here.

Cantilevers used in the AFM and MEMS applications are small enough to be significantly excited by the thermal motion of molecules in the fluid in which they are immersed; they undergo Brownian motion, in accordance with the fluctuation-dissipation theorem [31]. Equipartition of energy dictates that an energy of $k_B T/2$ is imparted to each mode of oscillation, where k_B is Boltzmann's constant and T is the absolute temperature. For the case where Brownian motion of the fluid molecules excite the cantilever, the thermal noise spectrum of the squared magnitude of the displacement function is [16, 32]

$$|w(x|\omega)|_s^2 = \frac{3\pi k_B T}{2k} \frac{\rho b}{\rho_c h} \frac{C_1^4}{\omega_{vac,1}^2} \sum_{n=1}^{\infty} \frac{\omega \Gamma_i^f(\omega)}{|C_n^4 - B_n^4(\omega)|^2} \phi_n^2(x) \tag{2.10}$$

The corresponding result for the slope is

$$\left| \frac{\partial w(x|\omega)}{\partial x} \right|_s^2 = \frac{3\pi k_B T}{2k} \frac{\rho b}{\rho_c h} \frac{C_1^4}{\omega_{vac,1}^2} \sum_{n=1}^{\infty} \frac{\omega \Gamma_i^f(\omega)}{|C_n^4 - B_n^4(\omega)|^2} \left(\frac{d\phi_n(x)}{dx} \right)^2 \tag{2.11}$$

where the subscript s refers to the spectral density, k is the normal spring constant of the cantilever beam [33], the subscript i refers to the imaginary component, and the function $B_n(\omega)$ is defined as

$$B_n(\omega) = C_1 \left(\frac{\omega}{\omega_{vac,1}} \right)^{1/2} \left(1 + \frac{\pi \rho b}{4 \rho_c h} \Gamma^f(\omega) \right)^{1/4} \tag{2.12}$$

In the limit of small dissipative effects, that is, when the quality factor greatly exceeds unity, the modes of the cantilever are approximately uncoupled and the frequency response of each mode of the beam is well approximated by that of a simple harmonic oscillator (SHO). We obtain the following expressions for the resonant frequency ω_R^f (defined as the undamped in-fluid natural frequency) and quality factor Q^f for the nth mode of vibration [8]:

$$\frac{\omega_{R,n}}{\omega_{vac,n}} = \left(1 + \frac{\pi \rho b}{4 \rho_c h} \Gamma_r^f(\omega_{R,n}) \right)^{-1/2} \tag{2.13}$$

$$Q_n = \frac{\frac{4 \rho_c h}{\pi \rho b} + \Gamma_r^f(\omega_{R,n})}{\Gamma_i^f(\omega_{R,n})} \tag{2.14}$$

where the subscript r refers to the real component. Note that for a driven cantilever, the resonant frequency corresponds to the frequency at which the driving force is $90°$ out of phase with the displacement of the cantilever; in general, this does not coincide with the frequency of maximum amplitude of each resonance peak. We emphasize that the expressions in Eqs. (2.13) and (2.14), which are formally derived in the limit of large quality factor, are also valid in the Stokes limit ($Re \to 0$), where the added apparent mass and damping coefficient of the cantilever are frequency independent.

To use the results Eqs. (2.10)−(2.14), the hydrodynamic function, $\Gamma^f(\omega)$, must be determined. This is obtained from solution of the linearized Navier Stokes equation for a rigid beam of infinite length, with identical cross section to that of the cantilever beam, undergoing transverse oscillatory motion. For a cantilever of thin rectangular cross section, this is well approximated by that of an infinitely thin blade:

$$\Gamma^f_{\text{rect}}(\omega) = \Omega(\omega)\Gamma^f_{\text{circ}}(\omega) \tag{2.15}$$

where the real and imaginary parts of the correction $\Omega(\omega)$ are given by

$$\begin{aligned}
\Omega_r(\omega) = (&0.91324 - 0.48274\tau + 0.46842\tau^2 - 0.12886\tau^3 \\
&+ 0.044055\tau^4 - 0.0035117\tau^5 + 0.00069085\tau^6) \\
&\times (1 - 0.56964\tau + 0.48690\tau^2 - 0.13444\tau^3 + 0.045155\tau^4 \\
&- 0.0035862\tau^5 + 0.00069085\tau^6)^{-1} \\
\Omega_i(\omega) = (&-0.024134 - 0.029256\tau + 0.016294\tau^2 - 0.00010961\tau^3 \\
&+ 0.000064577\tau^4 - 0.000044510\tau^5) \times (1 - 0.59702\tau \\
&+ 0.55182\tau^2 - 0.18357\tau^3 + 0.079156\tau^4 - 0.014369\tau^5 \\
&+ 0.0028361\tau^6)^{-1}, \\
\tau = \log&_{10}(Re/4) \tag{2.16}
\end{aligned}$$

and $\Gamma_{\text{circ}}(\omega)$ is the hydrodynamic function for a circular cylinder:

$$\Gamma^f_{\text{circ}}(\omega) = 1 + \frac{4iK_1(-i\sqrt{iRe/4})}{\sqrt{iRe/4}\,K_0(-i\sqrt{iRe/4})} \tag{2.17}$$

where $K_n(x)$ are modified Bessel functions of the second kind [34], and the Reynolds number is defined in Eq. (2.9)[2]. Equation (2.15) is accurate to within 0.1% over the entire range $0.1 < Re < 1000$ for both real and imaginary parts, and possesses the correct asymptotic forms $Re \to 0$ and $Re \to \infty$. In the absence of any simple exact analytical formula for the hydrodynamic function of an infinitely thin rectangular beam, Eq. (2.15) serves as a useful, practical expression.

Using the expression for the hydrodynamic function (Eq. (2.15)), and the above theoretical model, the frequency response of the cantilever can be explored. The frequency response is completely characterized by two-dimensionless parameters:

$$\overline{Re} = \frac{\rho\omega_{\text{vac},1}b^2}{\eta}, \quad \overline{T} = \frac{\rho b}{\rho_c h} \tag{2.18}$$

which enable us to study cantilevers of arbitrary dimensions and material properties, immersed in fluids of arbitrary density and viscosity. The first parameter $\overline{Re}$ is a normalized Reynolds number, which indicates the relative importance of inertial forces in the fluid to viscous forces. If this parameter greatly exceeds unity, then viscous forces exert a relatively weak effect and the flow will be predominantly

2) Note that the definition of Re in Eq. (2.9) differs from that used in [8] by a factor of 4. This is because Eq. (2.9) is the form that appears naturally in the exact solution to $\Gamma^f(\omega)$ around an infinitely flat blade; see [29, 35]. For consistency, this definition is adopted throughout this chapter.

inviscid in nature. The second parameter $\overline{T}$ is the ratio of the added-apparent mass due to inertial forces in the fluid (in the absence of viscous effects), to the total mass of the beam. This parameter indicates the relative strength of fluid inertia and is used to distinguish between immersion in gas and liquid, as will be discussed below.

Note that the values of $\overline{T}$ for beams immersed in gases and liquids typically differ by 3 orders of magnitude. This is a direct result of the difference in densities of gases relative to those of liquids. In contrast, values for $\overline{Re}$ in gases and liquids differ by only 1 order of magnitude, since the kinematic viscosities of gases are typically 1 order of magnitude greater than those of liquids. Consequently, we shall present separate results for gases and liquids that account for these differences.

In Figure 2.2 we present results for the peak frequency ω_p (i.e., that causing the peak response) and quality factor Q of the fundamental mode, as a function of both $\overline{Re}$ and $\overline{T}$. The peak frequency is numerically calculated from the frequency response, Eq. (2.13), whereas the quality factor is obtained directly from Eq. (2.14). Consequently, results presented for Q give quantitative information about the resonance peak provided $Q \gg 1$, since the analogy with the frequency response of a

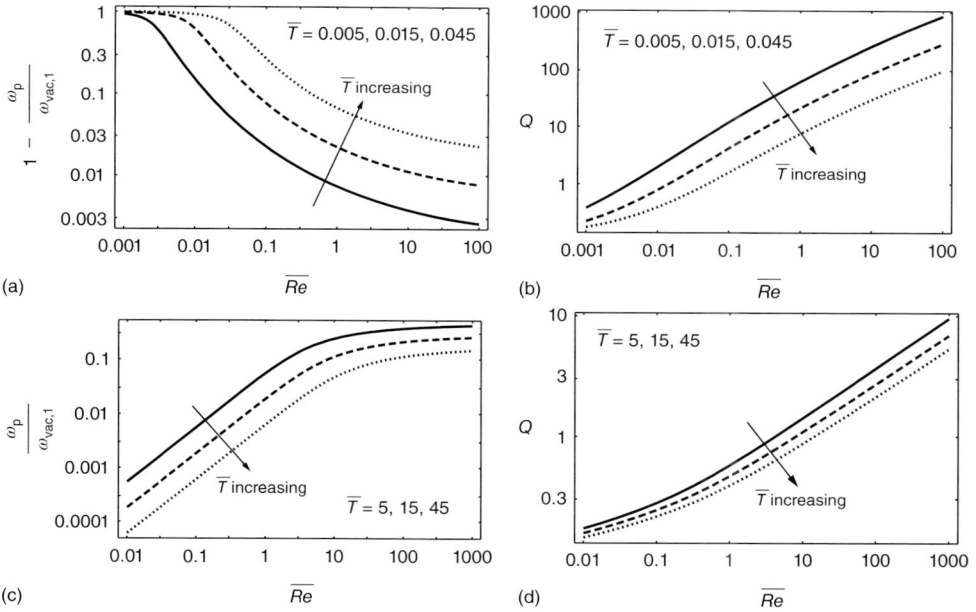

Figure 2.2 Peak frequency ω_p of fundamental resonance relative to frequency in vacuum $\omega_{vac,1}$, and quality factor $Q = Q_1$, Eq. (2.14), for the fundamental mode. (a, b) $\overline{T} = 0.045$ (short-dashed line); $\overline{T} = 0.015$ (dashed line); $\overline{T} = 0.005$ (solid line). (c, d) $\overline{T} = 45$ (short-dashed line); $\overline{T} = 15$ (dashed line); $\overline{T} = 5$ (solid line). Results in the limit $\overline{Re} \to \infty$ for $\overline{T} = 0.045, 0.015, 0.005$ are $(1 - \omega_p/\omega_{vac,1}) = 0.0172, 0.00584, 0.00196$, respectively. Results in the limit $\overline{Re} \to \infty$ for $\overline{T} = 45, 15, 5$ are $\omega_p/\omega_{vac,1} = 0.166, 0.280, 0.451$, respectively.

SHO is only valid in those cases. For $Q \leq O(1)$, however, no such analogy exists and Q only presents qualitative information about the resonance peak. In particular, for $Q \leq O(1)$ one can conclude that substantial broadening of the resonance peak is present, and that the modes are significantly coupled in the frequency domain. A reduction in Q will then result in further broadening of the peak and an increased coupling of the modes. Finally, it is interesting to note that a non-zero peak frequency is observed in all cases presented. Such behavior is not observed in a SHO model, where the peak frequency is found to be identically zero for all quality factors $Q \leq 1/\sqrt{2}$. These results demonstrate that for $Q \leq O(1)$ the frequency response of a cantilever beam is not analogous to that of a SHO.

2.2.2
Torsional Oscillation

We now turn our attention to the torsional modes of oscillation, as presented by Green and Sader [9]. This analysis follows along analogous lines to that presented above for flexural oscillation. Due to this similarity, we only summarize the key results of this model. To begin, the vacuum frequencies of a cantilever beam executing torsional oscillations are given by

$$\omega_{\mathrm{vac},n} = \frac{D_n}{L} \sqrt{\frac{GK}{\rho_c I_p}} \tag{2.19}$$

$$D_n = \frac{\pi}{2}(2n - 1) \tag{2.20}$$

where G is the shear modulus of the cantilever and $n = 1, 2, 3, \ldots$. For a thin rectangular beam $I_p = b^3 h/12$ and K is the torsional constant of the cross-section, which in the case of a thin rectangle reduces to $bh^3/3$ [33].

The total moment per unit length $M(x|\omega)$ applied to the beam is decomposed into a term corresponding to the driving moment $M_{\mathrm{drive}}(x|\omega)$ and the moment due to hydrodynamic loading of the cantilever $M_{\mathrm{hydro}}(x|\omega)$, that is,

$$M(x|\omega) = M_{\mathrm{hydro}}(x|\omega) + M_{\mathrm{drive}}(x|\omega) \tag{2.21}$$

The hydrodynamic load is then expressed generally as

$$M_{\mathrm{hydro}}(x|\omega) = -\frac{\pi}{8}\rho\omega^2 b^4 \Gamma^{\mathrm{t}}(\omega)\Phi(x|\omega) \tag{2.22}$$

where $\Phi(x|\omega)$ is the deflection angle about the longitudinal axis of the cantilever and $\Gamma^{\mathrm{t}}(\omega)$ is the hydrodynamic function. The expression for the thermal noise spectrum of the cantilever is [16, 32]

$$|\Phi(x|\omega)|_s^2 = \frac{6\pi k_B T}{k^\Phi} \frac{\rho b}{\rho_c h} \frac{D_1^2}{\omega_{\mathrm{vac},1}^2} \sum_{n=1}^{\infty} \frac{\omega \Gamma_i^{\mathrm{t}}(\omega)}{|D_n^2 - A_n^2(\omega)|^2} \gamma_n^2(x) \tag{2.23}$$

where the subscript s again refers to the spectral density, k^ϕ is the torsional spring constant [33] and

$$A_n(\omega) = \frac{\pi}{2} \frac{\omega}{\omega_{\mathrm{vac},1}} \left(1 + \frac{3\pi \rho b}{2\rho_c h}\Gamma^{\mathrm{t}}(\omega)\right)^{1/4} \tag{2.24}$$

In the limit of small dissipative effects, the conditions for which are discussed above, the resonant frequency $\omega_{R,n}$ and quality factor Q_n for the nth mode of vibration are given by

$$\frac{\omega_{R,n}}{\omega_{\text{vac},n}} = \left(1 + \frac{3\pi\rho b}{2\rho_c h}\Gamma_r^t(\omega_{R,n})\right)^{-1/2} \tag{2.25}$$

$$Q_n = \frac{\frac{2\rho_c h}{3\pi\rho b} + \Gamma_r^t(\omega_{R,n})}{\Gamma_i^f(\omega_{R,n})} \tag{2.26}$$

The hydrodynamic function $\Gamma^t(\omega)$ is obtained from the solution of the linearized Navier Stokes equation for flow around a rigid thin blade of infinite length that is executing torsional oscillation; this is analogous to the flexural oscillation case. The required result is given in [9], and is derived using the boundary integral technique of [36]. An approximate numerical formula is provided for this result, which has the correct asymptotic behavior as $Re \to 0$ and $Re \to \infty$, and is accurate to better than 0.1% over the range $10^{-4} \leq Re \leq 10^4$. For details, the reader is referred to [9].

2.2.3
In-Plane Flexural Oscillation

The model in Section 2.2.1 can be directly applied to cantilever beams of arbitrary cross section. All that is required is specification of the hydrodynamic function for the geometry under consideration. Brumley *et al.* [17] calculated the hydrodynamic function for a rectangular cylinder of arbitrary width-to-thickness ratio. It was found that the hydrodynamic function is weakly affected by the beam thickness, for width-to-thickness ratios greater than unity—this shows that approximation of a beam of small, but finite thickness, by one that is infinitely thin is a good approximation for *out-of-plane* flexural oscillation; see Section 2.2.1. The situation is very different when the beam width is smaller than its thickness, corresponding to *in-plane* flexural oscillation. In this latter case, the width-to-thickness ratio can strongly affect the hydrodynamic load—this geometric ratio must be specified to yield accurate results, in general. To facilitate practical application, tabulated data for the hydrodynamic function, spanning a wide range of width-to-thickness ratios, is presented in [17]. Cox *et al.* [19] also examined the frequency response of the in-plane flexural modes of a cantilever beam of rectangular cross section.

2.2.4
Extensional Oscillation

Theoretical models for the extensional modes of cantilever beams and free-free beams have also been derived [18, 37 – 39]. These models build on the same principle developed for flexural oscillations, and approximate the hydrodynamic load at any position along the beam by that of a rigid cylinder. Since these modes typically operate at frequencies much higher than those generated by the fundamental flexural mode, the viscous boundary layer near the surface is much smaller than the cantilever width. This, in turn, enables approximation of the hydrodynamic load by a local solution derived from Stokes' second problem for the oscillation of a half-space in a viscous fluid [40]. These models have been successfully compared to experimental measurements.

2.3
Arbitrary Mode Order

2.3.1
Incompressible Flows

In the previous section, we examined the frequency response of cantilever beams of a high aspect ratio (length/width) operating at a low mode order, where the flow field is well approximated as being two-dimensional. As mode order increases, however, the spatial wavelength of the modes decreases, which ultimately leads to violation of this two-dimensional approximation and breakdown in these models. Due to the importance of higher order modes in AFM and MEMS applications, a rigorous analysis of the three-dimensional flow field around a cantilever is vital. This is relevant to applications that utilize the higher order modes of cantilever beams [22, 41–43].

Finite-element fluid-structure models for cantilevers immersed in viscous fluids have been developed, for example, see [10, 11, 13]. Some of these models enable cantilever plates of arbitrary aspect ratio (length/width) to be investigated; however, all such models rely on sophisticated and computationally intensive numerical techniques. Since many cantilevers used in practice possess large aspect ratios, it is desirable to have theoretical models of an analytical nature that can be readily implemented, and this forms one aim of the present chapter. In addition to developing such models, we systematically investigate the effect of increasing mode number on the general characteristics of the frequency response, which is relevant to the design and operation of cantilevers in practice. Analytical solutions for the hydrodynamic functions and their use in development of analytical models for the frequency response of cantilevers in fluid, at arbitrary mode order, are reported in [15, 35]; these models are summarized here.

We consider a beam of high-aspect ratio (length/width) whose width greatly exceeds its thickness, as before. The hydrodynamic function is given by that of an infinitely long oscillating thin blade with deflection function $w(x|\omega) = Z_0 \exp(i\kappa x)$; see [15]. This yields the exact analytical solution for an incompressible viscous fluid [35]:

$$\Gamma^f(\omega, n) = 8a_1 \tag{2.27}$$

where the coefficient a_1 is obtain by solving the system of linear equations

$$\sum_{m=1}^{M}(A_{q,m}^{\kappa} + A_{q,m}^{Re})a_m = \begin{cases} 1 : & q = 1 \\ 0 : & q > 1 \end{cases} \tag{2.28}$$

for $1 \leq q \leq M$ and

$$A_{q,m}^{\kappa} = -\frac{4^{2q-1}}{\sqrt{\pi}} G_{13}^{21}\left(\frac{\kappa^2}{16}\bigg|\begin{matrix} & \frac{3}{2} \\ 0 & q+m-1 & q-m+1 \end{matrix}\right) \tag{2.29}$$

$$A_{q,m}^{Re} = -\kappa^2\frac{2^{4q-5}}{\sqrt{\pi}} G_{13}^{21}\left(\frac{\kappa^2 - iRe}{16}\bigg|\begin{matrix} & \frac{1}{2} \\ 0 & q+m-2 & q-m \end{matrix}\right)$$
$$-\frac{2^{4q-1}}{\sqrt{\pi}} G_{13}^{21}\left(\frac{\kappa^2 - iRe}{16}\bigg|\begin{matrix} & \frac{1}{2} \\ 0 & q+m-1 & q-m+1 \end{matrix}\right) \tag{2.30}$$

in terms of Meijer G-functions [44]. The integer M is to be increased until convergence in the solution is obtained; features of these solutions are discussed by van Eysden and Sader [35]. The hydrodynamic function $\Gamma^f(\omega, n)$ depends on the radial frequency ω through the Reynolds number (Eq. (2.9)), and on the mode order n through the dimensionless parameter

$$\kappa = C_n\frac{b}{L} \tag{2.31}$$

where C_n is as defined in Eq. (2.6). This parameter gives the relative aspect ratio of each spatial mode, and as such, is referred to as the *normalized mode number*. The asymptotic result for $\Gamma^f(\omega, n)$ in the limit as $\kappa \to 0$ is identical to Eq. (2.15). For the complementary limit where $\kappa \gg 1$, the hydrodynamic function has the asymptotic form,

$$\Gamma^f(\omega, n) = \frac{8}{\pi|\kappa|}\frac{\sqrt{\kappa^2 - iRe}}{\sqrt{\kappa^2 - iRe} - |\kappa|}, \kappa \to \infty \tag{2.32}$$

In the inviscid limit, the matrix elements $A_{q,m} \equiv A_{q,m}^{\kappa} + A_{q,m}^{Re}$ are given by van Eysden and Sader [16]

$$A_{q,m} = -\frac{4^{2q-1}}{\sqrt{\pi}} G_{13}^{21}\left(\frac{\kappa^2}{16}\bigg|\begin{matrix} & \frac{3}{2} \\ 0 & q+m-1 & q-m \end{matrix}\right) \tag{2.33}$$

The dependence on ω and mode number, n, in Eqs. (2.29) and (2.30) is contained in the argument of the Meijer G-function, and subsequently within the coefficients a_n. Similar formulas exist for the torsional modes of oscillation [29].

Importantly, the hydrodynamic function now has an explicit dependence on the mode number n, and we make the replacements $\Gamma^f(\omega) \to \Gamma^f(\omega, n)$ and $\Gamma^t(\omega) \to \Gamma^t(\omega, n)$ in all equations in the previous section. Consequently, Eq. (2.8) now places no restriction on the mode order n and rigorously accounts for the three-dimensional flow field around the beam (while ignoring end effects).

We first turn our attention to investigate the effect of increasing mode number on the frequency response of cantilever beams immersed in fluid. Importantly, the fundamental torsional resonant frequency greatly exceeds the fundamental flexural resonant frequency [9], as is commonly observed in measurements [45]. Since the higher order torsional modes are rarely probed in practice, we focus our discussion exclusively on the flexural modes. Nonetheless, we note that the influence of higher order mode number on the torsional frequency response will be very weak in comparison with the flexural modes, since the torsional hydrodynamic function $\Gamma^t(\omega, n)$ is much more weakly dependent on mode number, n, in comparison with the complementary dependence of the flexural hydrodynamic function $\Gamma^t(\omega, n)$ [35]. For further details regarding the torsional hydrodynamic function, the reader is referred to [35].

To completely characterize the frequency response for higher order modes, the following dimensionless parameter is required in addition to $\overline{Re}$ and $\overline{T}$ defined in Eq. (2.18),

$$\overline{\kappa} = C_1 \frac{b}{L} \tag{2.34}$$

This third parameter is a normalized aspect ratio and accounts for the induced three-dimensional hydrodynamic flow around the beam. In the singular limit of $\overline{\kappa} \to 0$, the low mode number theory of Section 2.2.1 is recovered.

In Figure 2.3, we present results for the thermal noise spectrum of the slope of the cantilever at its end point $(x = 1)$, as this is the quantity of interest in AFM measurements. Results are presented in gas (a, c) and liquid (b, d) for $\overline{\kappa} = 0$, 0.125, 0.25, 0.5, which corresponds to a geometric aspect ratio of $L/b = \infty$, 15, 7.5, 3.75, respectively. In gases, we see that the frequency response in the neighborhood of the fundamental resonance peak is very weakly dependent on the aspect ratio of the cantilever; see Figure 2.3a, c. This is a result of the weak dependence of the hydrodynamic function on $\overline{\kappa}$ for the value of normalized Reynolds number considered; see Table 2.1. However, as the mode number increases, the thermal noise spectra begin to deviate significantly from the $\overline{\kappa} = 0$ solution, as seen in Figure 2.3c. This can be understood in terms of the normalized mode number κ, which is increasing (see Eq. (2.31)), resulting in a significant change in the hydrodynamic function $\Gamma^f(\omega, n)$ [see Table 2.1 and Eq. (2.32)], and this is reflected in the frequency response.

Comparing Figure 2.3a, c and b, d, we find that immersion in liquid has a dramatic effect in comparison with that of gas. The resonance peaks broaden and shift significantly to lower frequencies. Note that the peak frequency of the fundamental mode is now approximately four times lower than the resonant frequency in gas. Nonetheless, inspecting Figure 2.3b we find that the fundamental resonance peak is very weakly affected by changing the aspect ratio of the cantilever between $L/b = 3.75$ and ∞. The same is not true of the next mode, however, where a significant effect on both the peak frequency and quality factor is observed for the largest value of $\overline{\kappa} = 0.5$ (corresponding to the smallest value of $L/b = 3.75$). Interestingly, it is found that as the normalized aspect ratio $\overline{\kappa}$ is increased (resulting in a decrease in L/b), the normalized peak frequencies shift to higher frequencies and approach the vacuum frequencies. This effect increases with increasing mode number n. Importantly, even the smallest nonzero value of $\overline{\kappa}$ considered

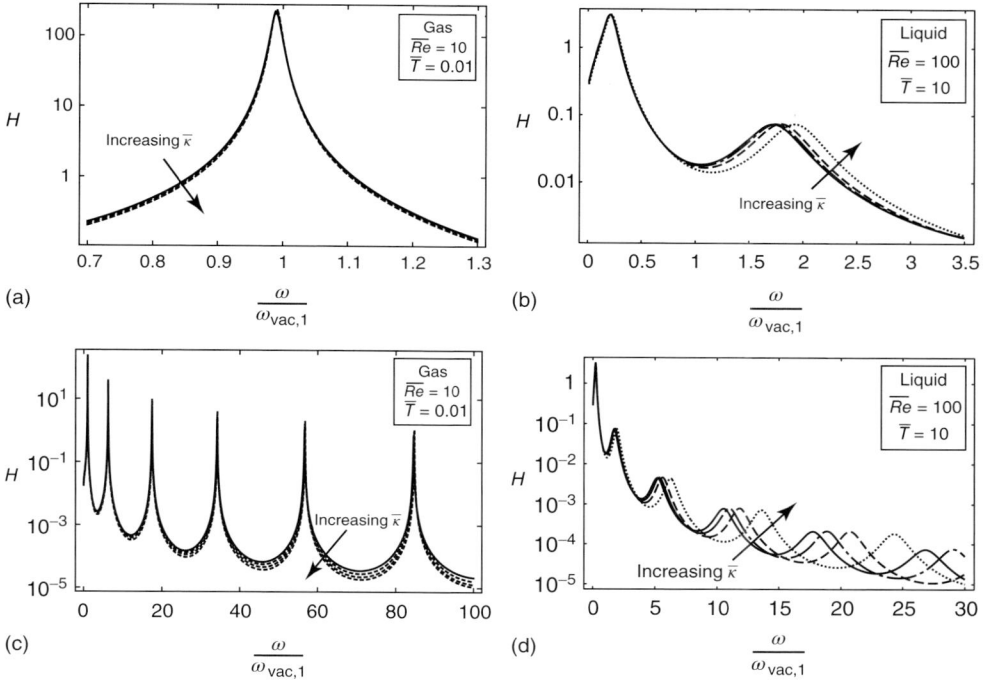

Figure 2.3 Normalized thermal noise spectrum (slope) $H = |w'(x, \omega)|_s^2 k\omega_{vac,1}/(k_B T)$, Eq. (2.11), of the flexural modes in gas. The $'$ refers to the derivative with respect to x. Normalized mode numbers $\overline{\kappa} = 0, 0.125, 0.25,$ and 0.5 corresponding to $L/b = \infty$, 15, 7.5, and 3.75, respectively.

Solid line corresponds to $\overline{\kappa} = 0$ result, and is identical to model in Section 2.2.1. (a, b) First mode (a), first and second modes (b); (c, d) First 6 modes. Results given for a gas with $Re = 10, \overline{T} = 0.01$ (a, c) and liquid with $Re = 100, \overline{T} = 10$ (b, d).

(corresponding to the largest finite value of $L/b = 15$) has a significant effect, as is strikingly evident in Figure 2.3d. From these results, it is clear that the frequency response of the higher modes of practical cantilevers can strongly depend on the normalized aspect ratio $\overline{\kappa}$. This finding for the higher modes contrasts directly to the fundamental mode that is found to be virtually independent of $\overline{\kappa}$.

The effect of varying aspect ratio on the resonant frequencies and quality factors is presented in Figures 2.4 and 2.5, respectively, where results for the first 20 modes are calculated using (i) the *arbitrary mode number model*, (ii) the *low mode number model* presented in Section 2.2.1, and (iii) the *large κ asymptotic model* that is obtained using the hydrodynamic function in Eq. (2.32). For immersion in gas, we find that the resonant frequencies are very weakly affected by the surrounding fluid, in agreement with previous studies [8–11, 13, 14, 27]; see Figure 2.4. We also observe that the arbitrary mode number model matches the low mode number model in the limit of $\overline{\kappa} = 0$ (corresponding to $L/b \to \infty$), but can diverge significantly as the aspect ratio, L/b, of the cantilever is reduced. Nonetheless, we

Table 2.1 Incompressible hydrodynamic function $\Gamma^f(\omega, n)$ for flexural modes as a function of Reynolds number Re and normalized mode number κ, accurate to six significant figures.

$\log_{10}Re^{a)}$	κ											
	0	0.125	0.25	0.5	0.75	1	2	3	5	7	10	20
−4	3919.41	59.3906	22.4062	9.13525	5.62175	4.05204	1.93036	1.2764	0.764081	0.545683	0.381972	0.190986
−3.5	1531.9	59.3861	22.4061	9.13525	5.62175	4.05204	1.93036	1.2764	0.764081	0.545683	0.381972	0.190986
−3	613.426	59.342	22.4052	9.13523	5.62174	4.05204	1.93036	1.2764	0.764081	0.545683	0.381972	0.190986
−2.5	253.109	58.9094	22.3962	9.13504	5.62172	4.05203	1.93036	1.2764	0.764081	0.545683	0.381972	0.190986
−2	108.429	55.2882	22.3078	9.13319	5.62153	4.05199	1.93036	1.2764	0.764081	0.545683	0.381972	0.190986
−1.5	48.6978	40.7883	21.5187	9.11481	5.6196	4.0516	1.93035	1.2764	0.764081	0.545683	0.381972	0.190986
−1	23.2075	22.7968	17.5378	8.9437	5.60057	4.04771	1.93027	1.27639	0.76408	0.545683	0.381972	0.190986
−0.5	11.8958	11.9511	11.0719	7.89716	5.43378	4.01051	1.92942	1.27629	0.764074	0.545682	0.381972	0.190986
0	6.64352	6.64381	6.47227	5.65652	4.64017	3.746	1.92114	1.27536	0.764012	0.545671	0.38197	0.190986
0.5	4.07692	4.0594	3.99256	3.72963	3.37543	3.00498	1.85532	1.26646	0.763397	0.545564	0.381953	0.190985
1	2.74983	2.73389	2.69368	2.5639	2.39884	2.22434	1.61821	1.20592	0.757637	0.54452	0.381786	0.190981
1.5	2.02267	2.0108	1.98331	1.9004	1.79834	1.69086	1.31175	1.04626	0.721165	0.535593	0.38018	0.190932
2	1.6063	1.59745	1.57723	1.5169	1.44276	1.36416	1.08036	0.878177	0.635443	0.496169	0.368548	0.190459
2.5	1.3623	1.35532	1.33934	1.29142	1.23203	1.16842	0.932812	0.759965	0.551349	0.435586	0.334279	0.186672
3	1.21727	1.21141	1.19792	1.15718	1.10624	1.05117	0.84292	0.686229	0.493924	0.387183	0.295972	0.172722
3.5	1.13038	1.12518	1.11316	1.07668	1.03073	0.980721	0.78879	0.641744	0.458699	0.356289	0.268907	0.15445
4	1.07814	1.07334	1.06221	1.02827	0.985314	0.938346	0.756309	0.615164	0.437743	0.337813	0.252327	0.140852
∞	1	0.995787	0.985974	0.955852	0.917424	0.875076	0.708262	0.576362	0.407863	0.311839	0.229184	0.120958

$\log_{10} Re^{b)}$												
−4	27984.8	44628.5	55176.1	71754	86311.5	100062	152411	203623	305570	407436	560225	1069521
−3.5	9816.33	14113.1	17448.2	22690.6	27294.1	31642.3	48196.5	64391.4	96629.7	128843	177159	338212
−3	3482.47	4464.16	5517.72	7175.41	8631.15	10006.2	15241.1	20362.3	30557	40743.6	56022.5	106952
−2.5	1252.66	1415.42	1745.17	2269.09	2729.42	3164.23	4819.65	6439.14	9662.97	12884.3	17715.9	33821.2
−2	458.386	457.863	552.862	717.635	863.138	1000.63	1524.11	2036.23	3055.7	4074.36	5602.25	10695.2
−1.5	171.397	160.951	177.702	227.205	273.013	316.449	481.967	643.915	966.297	1288.43	1771.59	3382.12
−1	65.8679	62.2225	61.626	72.6542	86.5364	100.144	152.418	203.625	305.57	407.436	560.225	1069.52
−0.5	26.2106	25.21	24.1432	24.7484	27.9459	31.8957	48.2199	64.3973	96.6308	128.843	177.159	338.212
0	10.8983	10.6158	10.1909	9.7009	9.91067	10.648	15.3139	20.381	30.5604	40.7448	56.0229	106.952
0.5	4.78389	4.69492	4.53952	4.24925	4.09701	4.09433	5.01844	6.49605	9.67379	12.8879	17.7171	33.8214
1	2.23883	2.20681	2.14583	2.0088	1.89659	1.82463	1.85993	2.17718	3.08849	4.08581	5.60598	10.6956
1.5	1.12164	1.10851	1.08208	1.01654	0.953355	0.901676	0.81464	0.844519	1.04394	1.32116	1.78306	3.38349
2	0.596697	0.590686	0.578118	0.545082	0.510467	0.479247	0.403803	0.383595	0.409256	0.469688	0.589749	1.07377
2.5	0.332285	0.329276	0.32283	0.305262	0.285953	0.26763	0.216732	0.194409	0.186218	0.195634	0.221631	0.349855
3	0.191043	0.189434	0.185931	0.176166	0.165118	0.154323	0.122124	0.105573	0.0938839	0.0925686	0.09682	0.126835
3.5	0.112082	0.111181	0.109199	0.103595	0.0971392	0.0907188	0.0707736	0.059728	0.0505049	0.0476557	0.0471326	0.0534759
4	0.0665172	0.0659974	0.0648471	0.0615627	0.0577366	0.0538889	0.0416384	0.0345727	0.0282418	0.025856	0.024611	0.0252877

a) Real component. $Re \to \infty$ results obtained using inviscid theory (Eq. (2.33)).

b) imaginary component.

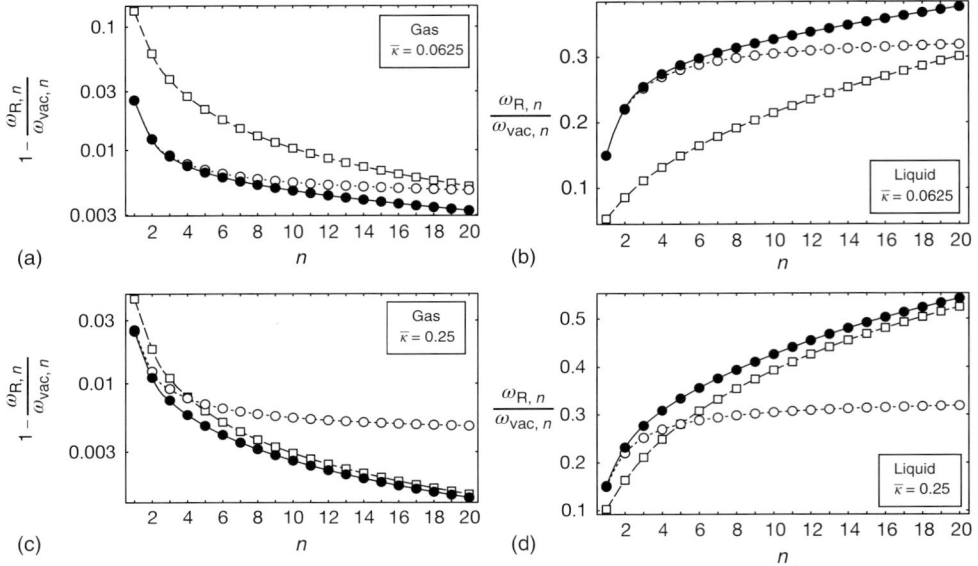

Figure 2.4 Normalized resonant frequency, Eq. (2.13), as a function of mode number n. Lines given as a guide. Low mode number model (open circle), arbitrary mode number model (solid circles), and large κ asymptotic model (open rectangles). (a, b) $\overline{\kappa} = 0.0625$ corresponding to $L/b = 30$. (c, d) $\overline{\kappa} = 0.25$ corresponding to $L/b = 7.5$. Results given for gas $Re = 1, \overline{T} = 0.01$ (a, c) and liquid $Re = 10, \overline{T} = 10$ (b, d).

find that the low mode number model is always valid for the fundamental mode for the range of normalized aspect ratios $\overline{\kappa}$ considered; this was also found to be the case up to $\overline{\kappa} = 0.5$, beyond which the underlying assumptions in all models must be drawn into question.

In Figure 2.4, the arbitrary mode number model is found to give identical results to the low mode number model for the fundamental mode then departs from this solution as the mode number n is increased, ultimately approaching the large κ asymptotic model. Similar results are also obtained for the resonant frequencies in liquid, although the deviation is much more pronounced due to the dramatic effect of liquid on the frequency response in comparison with gas, cf. Figure 2.4a, c and b, d.

Complementary results to Figure 2.4 are presented for the quality factor Q_n in Figure 2.5. For all panels it is evident that as the mode number n increases, the quality factor approaches infinity. This is a consequence of the diminishing hydrodynamic load at high mode number [see Table 2.1 and Eq. (2.32)], an effect which is also present in Figure 2.4, where the resonant frequencies in fluid approach the vacuum results. Again, we find that the low mode number model agrees well with the arbitrary mode number model for the fundamental mode always, but as the mode number n is increased these results begin to depart from the low mode number model and approach the large κ asymptotic model. It is interesting to

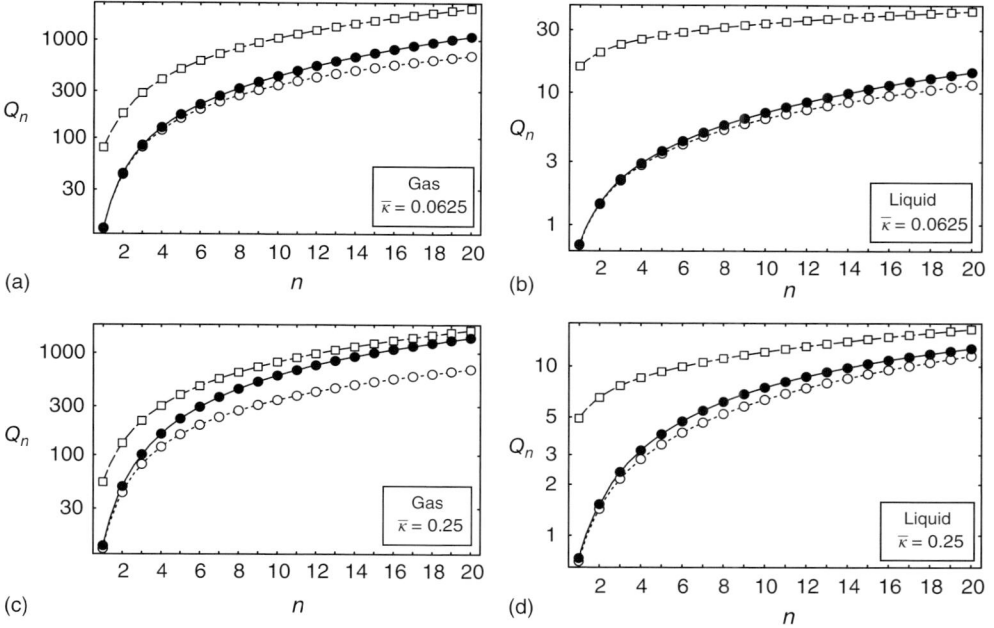

Figure 2.5 As for Figure 2.4, except results given for quality factor Q_n, Eq. (2.14), as a function of mode number n.

note, however, that the deviation from the low mode number model for immersion in gas is much more pronounced than the deviation in liquid, cf. Figure 2.5a, c and b, d. For all cases presented, it is found that the quality factor in liquid is well approximated by the low mode number model, regardless of the mode number; see [15].

Finally, we comment on the validity of the above model for practical cantilevers. In the $\kappa = 0$ limit, the low mode number model of Section 2.2.1 is valid for large L/b provided the viscous boundary layer, defined as $b/\sqrt{Re}$, does not become comparable to the length of the beam, at which point the flow field is directly affected by the three-dimensional nature of the cantilever geometry. For the higher order modes, the relevant length scale is the spatial length scale of oscillations L/C_n, and the requirement that this is much less than the viscous boundary layer thickness can be expressed as

$$Re > \kappa^2 \tag{2.35}$$

Approximating the resonant frequency of the cantilever by that in vacuum, we find that the condition (2.35) is independent of mode number and is equivalent to $\overline{Re} > \overline{\kappa}^2$. This condition is typically satisfied by practical microcantilevers, supporting the validity of the low mode number model of Section 2.2.1, for the fundamental mode. To examine the condition (2.35) carefully, we can evaluate the ratio

$$\frac{Re}{\kappa^2} = h \frac{\omega_{R,n}}{\omega_{vac,n}} \frac{\rho}{\eta} \sqrt{\frac{E}{12\rho_c}} \tag{2.36}$$

which must be greater than unity. It is interesting that this ratio is independent of the cantilever length and width and only depends on its thickness. For cantilever materials and fluids typically used in AFM and MEMS applications [2, 3, 6, 8–11, 13, 14, 27, 46, 47], Eq. (2.36) can be estimated to give $Re/\kappa^2 \sim 10^8 h\omega_{R,n}/\omega_{vac,n}$, where h is the thickness in meters. This indicates that the above ratio greatly exceeds unity for microcantilevers of practical relevance. The ratio $\omega_{R,n}/\omega_{vac,n}$ can be calculated using Eq. (2.13). Only in cases of very heavy fluid loading, or very thin cantilevers, where $\omega_{R,n}/\omega_{vac,n} \ll 1$ will the ratio in Eq. (2.36) equal unity. Thus, for most cantilevers of practical relevance, the condition $Re \gg \kappa^2$ will be satisfied, indicating the arbitrary mode number model captures the true frequency response at resonance of these cantilever beams. Complete three-dimensional computational fluid dynamics simulations of the cantilever dynamics are not essential in such situations, even for very small cantilevers. Nonetheless, by tuning the material properties of the cantilever and fluid, it may be possible to reach this limit practically; see Eq. (2.36).

2.3.2
Compressible Flows

In general, the flow around a cantilever can be considered incompressible provided the acoustic wavelength greatly exceeds the dominant length scale of the flow [29, 48]. For practical microcantilevers with high aspect ratios (length/width) and width-to-thickness ratios, operating at low mode numbers (i.e., the fundamental flexural mode and the next few modes), the flow is approximately two-dimensional and its dominant length scale is the beam width, b. In such cases, the acoustic wavelength greatly exceeds the dominant length scale, and hence, the incompressibility assumption is easily satisfied [8]. This has been well established by the experimental validation of numerous theoretical models based on incompressible flow, for example, see [13, 23, 27]. For operation at higher order modes, however, it is important to reassess how this incompressibility condition is affected.

To investigate the effects of fluid compressibility, the hydrodynamic function $\Gamma^f(\omega, n)$ must be generalized to include compressibility effects. A solution can be obtained using the model presented in the previous sections, where the incompressibility condition (assumption 5 on page 31) above is relaxed. The exact analytical solution for the flow around an infinitely long oscillating flat blade in a compressible medium is presented in [29]. The result is obtained from Eq. (2.27) under the following replacement for Eq. (2.29):

$$A_{q,m}^\kappa = -\frac{4^{2q-1}}{\sqrt{\pi}} G_{13}^{21} \left[\frac{1}{16}\left(\kappa^2 - \varsigma^2 \frac{Re}{Re - 4/3i\varsigma^2}\right) \bigg|_{0}^{\frac{3}{2}} \quad q+m-1 \quad q-m+1 \right] \tag{2.37}$$

while Eq. (2.30) is unchanged. In the inviscid limit, Eq. (2.33) generalizes to

$$A_{q,m} = -\frac{4^{2q-1}}{\sqrt{\pi}} G_{13}^{21} \left(\frac{\kappa^2 - \varsigma^2}{16} \middle| \begin{array}{ccc} & \frac{3}{2} & \\ 0 & q+m-1 & q-m \end{array} \right) \tag{2.38}$$

The hydrodynamic function $\Gamma^f(\omega, n)$ is dimensionless and depends on fluid compressibility through the dimensionless parameter

$$\varsigma = \frac{\omega b}{c} \tag{2.39}$$

where c is the sound speed of the fluid. Therefore, ς is a *normalized wave number* for the propagation of sound in the medium and is a measure of the fluid compressibility. Note that for the continuum approximation to be valid, we require $\varsigma \ll Re$ [40] and the argument of the Meijer G-function in the first term of Eq. (2.37) reduces to $(\kappa^2 - \varsigma^2)/16$ for cases of practical interest; see [29]. In the limit as $\varsigma \to 0$, the incompressible result for $\Gamma^f(\omega, n)$ presented above is obtained.

2.3.2.1 Scaling Analysis

The conditions for which fluid compressibility becomes important can be determined using a scaling analysis. For an elastic beam, the dominant length scale of the flow is the minimum of (i) the beam width b and (ii) the length scale of spatial oscillations of the beam, given by

$$\lambda_{beam} = \frac{2\pi L}{C_n} \approx \frac{4L}{2n-1} \tag{2.40}$$

Since C_n increases with increasing mode number n, the length scale of spatial oscillations reduces until it becomes smaller than the beam width. Thus, spatial oscillations of the beam will eventually set the dominant hydrodynamic length scale, as the mode number is increased.

The wavelength of acoustic oscillations generated by the beam in a fluid is c/f_n, where f_n is the resonant frequency. For the purpose of a scaling analysis, the resonant frequency can be approximated by the corresponding result in vacuum, that is, Eq. (2.5). The wavelength of sound is then

$$\lambda_{sound} = \left(\frac{C_1}{C_n}\right)^2 \frac{c}{f_{vac,1}} \approx \frac{1.425}{(2n-1)^2} \frac{c}{f_{vac,1}} \tag{2.41}$$

where $\omega_{vac,1} = 2\pi f_{vac,1}$. Therefore, the acoustic wavelength decreases in inverse proportion to the square of the mode number n, for high mode numbers. Critically, this has a stronger dependence on mode number than the length scale of spatial oscillations. Therefore, as the mode number increases, the acoustic wavelength eventually becomes comparable to the spatial wavelength of the beam, and compressibility can no longer be ignored. This so-called coincidence point corresponds to the onset of significant radiation damping of the beam, where energy is carried away in the form of sound waves [29, 48].

Equating the expressions in Eqs. (2.40) and (2.41), determines the critical mode number n_c at which coincidence is reached,

$$n_c = \frac{0.178c}{f_{vac,1}L} \tag{2.42}$$

For a cantilever 250 μm in length with a vacuum frequency of 20 kHz, immersed in air, coincidence is reached at $n_c \approx 12$. In liquids, however, the speed of sound is a factor of ~ 5 larger than in air and the critical mode number is approximately $n_c \approx 60$, which is unlikely to be probed in practice. The precise regime for which the effects of fluid compressibility become manifest is clarified below using the numerical results of the above model.

2.3.2.2 Numerical Results

To completely characterize the frequency response for higher order modes, the following dimensionless parameter is required in addition to $\overline{Re}, \overline{T}$ and $\overline{\kappa}$ defined in Eqs. (2.18) and (2.34),

$$\overline{\varsigma} = \frac{b\omega_{\text{vac},1}}{c} \tag{2.43}$$

The parameter $\overline{\varsigma}$ is the normalized wave number for the propagation of sound through the medium, and is a measure of the fluid compressibility. When the acoustic wavelength is large, this parameter approaches zero and the incompressible limit is attained.

Another dimensionless parameter of fundamental importance in compressible flow is $\overline{\varsigma}/\overline{\kappa}$, which is the ratio of the beam length to the acoustic wavelength, for the fundamental mode. Since Eq. (2.42) can be written as

$$n_c = \frac{0.596\overline{\kappa}}{\overline{\varsigma}} \tag{2.44}$$

it then follows that $\overline{\varsigma}/\overline{\kappa}$ determines the mode number at which coincidence is reached. The dimensionless parameter $\overline{\varsigma}/\overline{\kappa}$ will be used in the following results to vary the effects of compressibility.

In Figure 2.6, we present the thermal noise spectrum for the slope (angle) at the end of the cantilever for compressible flow. Results are given for practical

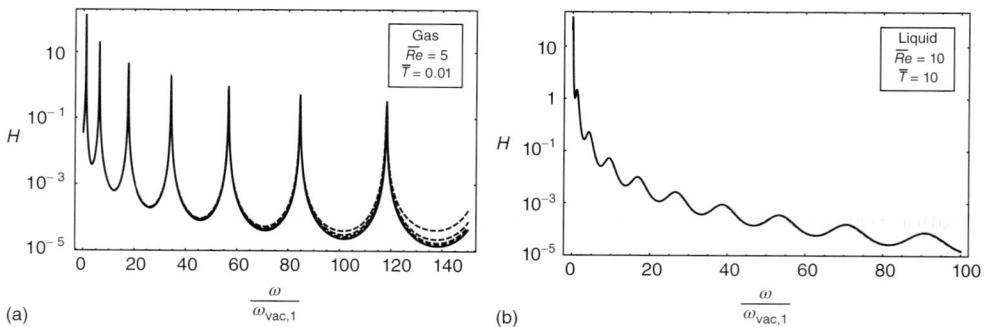

(a)

(b)

Figure 2.6 Normalized thermal noise spectrum (slope) $H = |w'(x,\omega)|_s^2 k\omega_{\text{vac},1}/(k_B T)$ for the flexural modes, showing the effects of fluid compressibility. The ' refers to the derivative with respect to x. Results given for (a) gas: $\overline{Re} = 5, \overline{T} = 0.01, \overline{\kappa} = 0.125$ and $\overline{\varsigma}/\overline{\kappa} = 0$, 0.025, 0.05, 0.075, 0.1; (b) liquid: $\overline{Re} = 10, \overline{T} = 10, \overline{\kappa} = 0.125$ and $\overline{\varsigma}/\overline{\kappa} = 0$, 0.005, 0.01, 0.015, 0.02.

values of $\overline{\kappa}, \overline{\varsigma}$, and $\overline{Re}$ in gases (Figure 2.6a) and liquids (Figure 2.6b), which clearly demonstrate the effect of increasing compressibility. Figure 2.6a shows that increasing the gas compressibility broadens the resonance peaks, which corresponds to enhanced dissipation. This effect is due to the generation of sound waves that radiate from the cantilever, which manifests itself more strongly with increasing mode number. This enhancement in dissipation is in direct contrast to the incompressible limit, where the effects of increasing mode number reduce dissipation (see Figure 2.4); this latter effect is due purely to the increasing importance of three-dimensional flow. Compressibility therefore counteracts the effects of three-dimensional flow with increasing mode number. This feature is evident in Eq. (2.37), where the variables κ and ς appear in combination.

From Figure 2.6b, we observe that compressibility has virtually no effect on the thermal noise spectrum for cantilevers operating in a liquid, as the spectra are indistinguishable from the incompressible result (all five curves coincide). This supports the above scaling argument that compressibility does not become significant in liquids until very high mode numbers are attained; for typical micro-cantilevers, this corresponds to mode numbers in excess of 20. We therefore focus the following discussion on the behavior of cantilevers in gas.

In Figure 2.7, the resonant frequencies and quality factors as a function of mode number are presented. In panels (b, d), values of $\overline{\kappa}$ and $\overline{\varsigma}$ correspond approximately to those used in the above scaling analysis. In panels (c, d) a larger aspect ratio of $L/b = 30$, corresponding to $\overline{\kappa} = 0.0625$, is presented. Three solutions are given: (i) the *viscous incompressible* solution (open circles) of Section 2.3.1, (ii) the *inviscid compressible* solution (open rectangles), Eq. (2.38), and (iii) the *complete (viscous compressible)* solution (closed circles), Eq. (2.37). For low mode numbers, compressibility effects are weak and the viscous incompressible and complete solutions coincide. However, as the mode number increases, the viscous incompressible solution begins to deviate from the complete solution. In panels (a, c), coincidence is reached at mode number 14, where the inertial component of the hydrodynamic load reaches a maximum, in agreement with predictions from the scaling analysis. However, in panels (b, d), the coincidence point now occurs at $n \sim 17$, despite Eq. (2.44) predicting the same result; the discrepancy arising from the approximate nature of the scaling analysis, which neglects the fluid loading on the beam. Above coincidence, sound waves are strongly generated in the medium, and a further increase in mode number results in rapid diminishing of this (inertial) hydrodynamic load. For large mode numbers, the complete solution coincides with inviscid compressible result, as the viscous penetration depth over which vorticity diffuses decreases with increasing mode number.

Corresponding results for the quality factor in gas are given in Figure 2.7c, d. As for the frequency response, the quality factor is dominated by incompressible viscous effects at low mode numbers, and approaches the inviscid compressible solution for large mode numbers, with the transition occurring approximately at coincidence. Importantly, above coincidence we find that the quality factor reduces significantly with increasing mode number. This finding differs greatly from the incompressible result and is due to the strong generation of acoustic

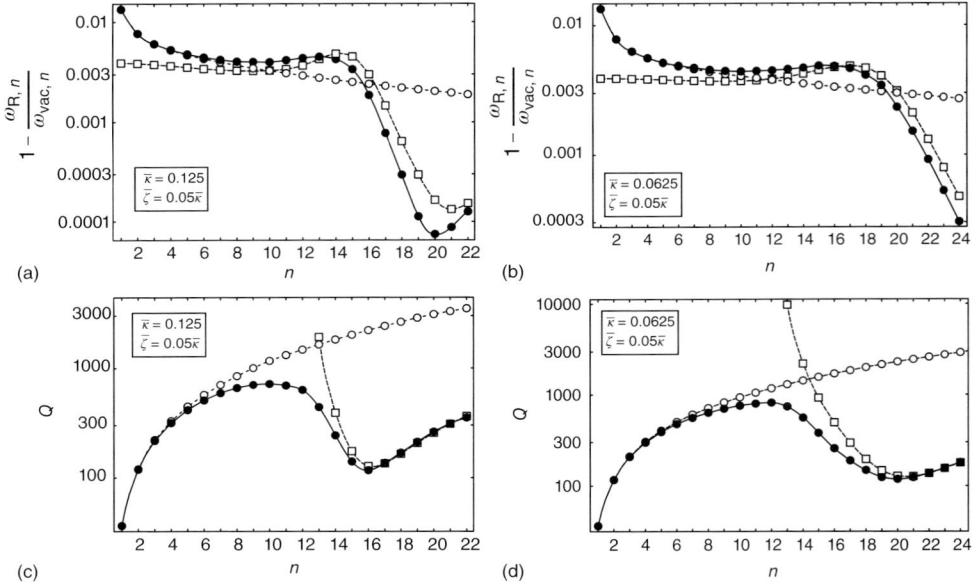

Figure 2.7 (a, b) Normalized resonant frequency $1 - \omega_{R,n}/\omega_{vac,n}$. (c, d) Quality factor Q_n as a function of mode number n. Results given for a gas with $\overline{Re} = 5, \overline{T} = 0.01, \overline{\varsigma}/\overline{\kappa} = 0.05$ and $\overline{\kappa} = 0.125$ (a, c) and $\overline{\kappa} = 0.0625$ (b, d). Lines are given as a guide only. Complete solution (solid circles), viscous incompressible solution (open circles), and inviscid compressible solution (open rectangles).

waves that greatly enhance dissipation, that is, radiation damping. This non-monotonic behavior of the quality factor with increasing mode number should be observable experimentally.

The above theoretical models for compressible flow have been derived using the hydrodynamic function for an infinitely long beam executing uniform oscillations with discrete spatial wave numbers. We conclude by commenting on the practical implications of these models to cantilever beams of finite length. In the limit where the acoustic wavelength is much greater than the length of the cantilever beam, the flow is incompressible and the validity of these models is well established [9–11, 13–15, 22–25, 27]; this is the practical case for operation in the low order modes. In the opposite situation where the acoustic wavelength is smaller than the spatial wavelength of beam oscillations, that is, above coincidence, the effects of compressibility are localized and hence unaffected by the finite length of the beam; this corresponds to the high mode number limit. Therefore, the cantilever beam model is strictly valid in the two limits of small and large acoustic wavelengths relative to the dominant hydrodynamic length scale. In the intermediate regime where the acoustic wavelength is larger than, but comparable to, the spatial wavelength of beam oscillations, the flow probes a spectrum of spatial wave numbers due to finite beam length; this has not been rigorously considered here.

Nonetheless, since the above solution holds rigorously in the bounding asymptotic limits of small and large wavelengths, it is expected to yield a good approximation in the intermediate regime.

References

1. Newell, W.E. (1968) Miniaturization of tuning forks. *Science*, **161**, 1320.
2. Binnig, G., Quate, C.F., and Gerber, C. (1986) Atomic force microscope. *Phys. Rev. Lett.*, **56**, 930.
3. Berger, R., Gerber, C., Lang, H.P., and Gimzewski, J.K. (1997) Micromechanics: a toolbox for femtoscale science: "Towards a Laboratory on a Tip". *Microelectron. Eng.*, **35**, 373.
4. Craighead, H.G. (2000) Nanoelectromechanical systems. *Science*, **290**, 1532.
5. Sader, J.E. (2002) Calibration of atomic force microscope cantilevers, in *Encyclopedia of Surface and Colloid Science*, CRC Press, pp. 846–856.
6. Lavrik, N.V., Sepaniak, M.J., and Datskos, P.G. (2004) Cantilever transducers as a platform for chemical and biological sensors. *Rev. Sci. Instrum.*, **75**, 2229.
7. Ekinci, K.L. and Roukes, M.L. (2005) Nanoelectromechanical systems. *Rev. Sci. Instrum.*, **76**, 061101.
8. Sader, J.E. (1998) Frequency response of cantilever beams immersed in viscous fluids with applications to the atomic force microscope. *J. Appl. Phys.*, **84**, 64.
9. Green, C.P. and Sader, J.E. (2002) Torsional frequency response of cantilever beams immersed in viscous fluids with applications to the atomic force microscope. *J. Appl. Phys.*, **92**, 6262.
10. Paul, M.R. and Cross, M.C. (2004) Stochastic dynamics of nanoscale mechanical oscillators immersed in a viscous fluid. *Phys. Rev. Lett.*, **92**, 235501.
11. Maali, A., Hurth, C., Boisgard, R., Jai, C., Cohen-Bouhacina, T., and Aimé, J.-P. (2005) Hydrodynamics of oscillating atomic force microscopy cantilevers in viscous fluids. *J. Appl. Phys.*, **97**, 074907.
12. Clarke, R.J., Cox, S.M., Williams, P.M., and Jensen, O.E. (2005) The drag on a microcantilever oscillating near a wall. *J. Fluid Mech.*, **545**, 397.
13. Basak, S., Raman, A., and Garimella, S.V. (2006) Hydrodynamic loading of microcantilevers vibrating in viscous fluids. *J. Appl. Phys.*, **99**, 114906.
14. Paul, M.R., Clark, M.T., and Cross, M.C. (2006) The stochastic dynamics of micron and nanoscale elastic cantilevers in fluid: fluctuations from dissipation. *Nanotechnology*, **17**, 4502.
15. van Eysden, C.A., Sader, J.E. (2007) Frequency response of cantilever beams immersed in viscous fluids with applications to the atomic force microscope: arbitrary mode order. *J. Appl. Phys.*, **101**, 044908.
16. van Eysden, C.A. and Sader, J.E. (2009) Frequency response of cantilever beams immersed in compressible fluids with applications to the atomic force microscope. *J. Appl. Phys.*, **106**, 094904.
17. Brumley, D.R., Willcox, M., and Sader, J.E. (2010) Oscillation of cylinders of rectangular cross section immersed in fluid. *Phys. Fluids*, **22**, 052001.
18. Castille, C., Dufour, I., and Lucat, C. (2010) Longitudinal vibration mode of piezoelectric thick-film cantilever-based sensors in liquid media. *Appl. Phys. Lett.*, **96**, 154102.
19. Cox, R., Josse, F., Heinrich, S.M., Brand, O., and Dufour, I. (2012) Characteristics of laterally vibrating resonant microcantilevers in viscous liquid media. *J. Appl. Phys.*, **111**, 014907.
20. Chu, W.H. (1963) Vibration of fully submerged cantilever plates in water, Tech. Rep. No. 2, DTMB, Contract NObs-86396(X). Southwest Research Institute, San Antonio, TX.
21. Lindholm, U.S., Kana, D.D., Chu, W.H., and Abramson, H.N. (1965) Elastic vibration characteristics of cantilever plates in water. *J. Ship Res.*, **9**, 11.

22. Braun, T., Barwich, V., Ghatkesar, M.K., Bredekamp, A.H., Gerber, C., Hegner, M., and Lang, H.P. (2005) Micromechanical mass sensors for biomolecular detection in a physiological environment. *Phys. Rev. E*, **72**, 031907.

23. Ghatkesar, M.K., Braun, T., Barwich, V., Ramseyer, J.-P., Gerber, C., Hegner, M., and Lang, H.P. (2008) Resonating modes of vibrating microcantilevers in liquid. *Appl. Phys. Lett.*, **92**, 043106.

24. Elmer, F.-J. and Dreier, M. (1997) Eigenfrequencies of a rectangular atomic force microscope cantilever in a medium. *J. Appl. Phys.*, **81**, 7709.

25. van Eysden, C.A. and Sader, J.E. (2006) Resonant frequencies of a rectangular cantilever beam immersed in a fluid. *J. Appl. Phys.*, **100**, 114916.

26. Landau, L.D. and Lifshitz, E.M. (1970) *Theory of Elasticity*, Pergamon, Oxford.

27. Chon, J.W.M., Mulvaney, P., and Sader, J.E. (2000) Experimental validation of theoretical models for the frequency response of atomic force microscope cantilever beams immersed in fluids. *J. Appl. Phys.*, **87**, 3978.

28. Green, C.P. and Sader, J.E. (2005) Frequency response of cantilever beams immersed in viscous fluids near a solid surface with applications to the atomic force microscope. *J. Appl. Phys.*, **98**, 114913.

29. van Eysden, C.A. and Sader, J.E. (2009) Compressible viscous flows generated by oscillating flexible cylinders. *Phys. Fluids*, **21**, 013104.

30. Batchelor, G.K. (1974) *An Introduction to Fluid Dynamics*, Cambridge University Press, Cambridge.

31. Landau, L.D. and Lifshitz, E.M. (1969) *Statistical Physics. Part 1*, Pergamon, Oxford.

32. van Eysden, C.A. and Sader, J.E. (2008) Erratum: "Frequency response of cantilever beams immersed in viscous fluids with applications to the atomic force microscope: arbitrary mode order". *J. Appl. Phys.*, **101**, 044908; (2007) *J. Appl. Phys.*, **104**, 109901.

33. Roark, R.J. (1943) *Formulas for Stress and Strain*, McGraw-Hill, New York.

34. Abramowitz, M. and Stegun, I.A. (1972) *Handbook of Mathematical Functions*, Dover, New York.

35. van Eysden, C.A. and Sader, J.E. (2006) Small amplitude oscillations of a flexible thin blade in a viscous fluid: exact analytical solution. *Phys. Fluids*, **18**, 123102.

36. Tuck, E.O. (1969) Calculation of unsteady flows due to small motions of cylinders in a viscous fluid. *J. Eng. Math.*, **3**, 29.

37. Pelton, M., Sader, J.E., Burgin, J., Liu, M., Guyot-Sionnest, P., and Gosztola, D. (2009) Damping of acoustic vibrations in gold nanoparticles. *Nat. Nanotechnol.*, **4**, 492.

38. Pelton, M., Wang, Y., Gosztola, D., and Sader, J.E. (2011) Mechanical damping of longitudinal acoustic oscillations of metal nanoparticles in solution. *J. Phys. Chem. C*, **115**, 23732.

39. Chakraborty, D., van Leeuwen, E., Pelton, M., and Sader, J.E. (2013) Vibration of nanoparticles in viscous fluids. *J. Phys. Chem. C*, **117**, 8536.

40. Landau, L.D. and Lifshitz, E.M. (1959) *Fluid Mechanics*, Pergamon, Oxford.

41. van Noort, S.J.T., Willemsen, O.H., van der Welf, K.O., de Grooth, B.G., and Greve, J. (1999) Mapping electrostatic forces using higher harmonics tapping mode atomic force microscopy in liquid. *Langmuir*, **15**, 7101.

42. Ulcinas, A. and Snitka, A. (2001) Intermittent contact AFM using the higher modes of weak cantilever. *Ultramicroscopy*, **86**, 217.

43. Stark, R.W. (2004) Optical lever detection in higher eigenmode dynamic atomic force microscopy. *Rev. Sci. Instrum.*, **75**, 5053.

44. Luke, Y.L. (1969) *The Special Functions and Their Approximations*, Academic Press, New York.

45. Green, C.P., Lioe, H., Cleveland, J.P., Proksch, R., Mulvaney, P., and Sader, J.E. (2004) Normal and torsional spring constants of atomic force microscope cantilevers. *Rev. Sci. Instrum.*, **75**, 1988.

46. Ho, C.M. and Tai, Y.C. (1998) Microelectro-mechanical-systems (MEMS) and fluid flows. *Annu. Rev. Fluid Mech.*, **30**, 579.

47. Sader, J.E., Chon, J.W.M., and Mulvaney, P. (1999) Calibration of rectangular atomic force microscope cantilevers. *Rev. Sci. Instrum.*, **70**, 3967.

48. Morse, P.M. and Ingard, K.U. (1987) *Theoretical Acoustics*, Princeton University Press, Princeton, NJ.

3
Damping in Resonant MEMS

Shirin Ghaffari and Thomas William Kenny

3.1
Introduction

Micromechanical resonators can all be modeled as second-order systems, including an "effective mass" and "effective stiffness," a resonant frequency, and damping. In the most general definition, damping represents all of the processes by which the energy associated with the vibration of the resonator (average of kinetic and potential energy over a complete cycle) decays over time. There are many mechanisms by which the energy of the resonator can be lost, and this chapter will review them in detail.

It is common to describe the energy loss mechanisms as contributing to the "quality factor" of the resonator, or the Q, which is defined according to:

$$\frac{1}{Q} = \frac{1}{2\pi} \cdot \frac{\text{Energy}_{\text{dissipated per cycle}}}{\text{Energy}_{\text{stored in the resonator}}}$$

An equivalent and more practical definition of the Q relates to the frequency response of the resonator:

$$\frac{1}{Q} = \frac{\text{Bandwidth}_{3\,\text{dB}}}{F_{\text{Resonant}}}$$

where F_{Resonant} is the resonant frequency of the resonator. In this definition, we understand that resonators with very sharp peaks in the frequency response, and therefore small 3 dB Bandwidth are resonators with high Q. For time references, a high-Q equates to a narrow frequency response, and a very well-defined frequency of oscillation, which is useful for stable references. For a gyro, high-Q equates to reduced thermomechanical noise and high sensitivity. In general, designers of resonant MEMS are interested in achieving a high Q.

The dissipation mechanisms that we will discuss in this chapter include losses to the surrounding air, losses through the anchors, losses to various and unspecified surface phenomena, losses to electronic dissipation, losses through thermoelastic dissipation, and losses through phonon scattering, which we will refer to as Akheizer dissipation. Because these decay phenomena are independent, and some

Resonant MEMS – Fundamentals, Implementation and Application, First Edition.
Edited by Oliver Brand, Isabelle Dufour, Stephen M. Heinrich and Fabien Josse.
© 2015 Wiley-VCH Verlag GmbH & Co. KGaA. Published 2015 by Wiley-VCH Verlag GmbH & Co. KGaA.

or all of them may be present in a particular resonator, we will consider them as independent contributions to the energy loss. In this case, each dissipation mechanism can individually contribute to the total Q such that,

$$\frac{1}{Q_{total}} = \Sigma \frac{1}{Q_{individual}}$$

In this manner, the total Q of a resonator is usually determined by the strongest dissipation mechanism, or the mechanism with the lowest individual Q.

3.2
Air Damping

Because of the small size of the resonating structure in an MEMS resonator, there can be a relatively large amount of surface area relative to the volume. Therefore, the effect of the forces exerted on the surfaces of the resonator cannot be neglected.

Unless the resonator is packaged at extremely low pressure, the ambient gas surrounding the resonator impedes the motion of the resonating element. Because of the movement of the resonating element relative to the surrounding air molecules, energy can be transferred from the resonator to the surrounding air, leading to a loss mechanism known as air damping. Figure 3.1 provides an illustration.

Energy transfer between the resonator and fluid molecules happens at a rate related to the relative velocity between the air and the resonator, and so the energy loss rate is approximately constant as the resonator amplitude decays. The energy transfer also occurs at a rate proportional to the frequency of the collisions between the resonator and the air molecules. As a result, air damping is generally proportional to the air pressure.

Historically, one of the first barriers to development and use of MEMS resonators was air damping because damping increases in proportion to the square of the length to thickness ratio [1]. This damping arises due to the finite viscosity and compressibility of the surrounding fluid medium. In such a situation, the viscosity generates a drag force that is proportional to the velocity of resonator motion, similar to a conventional viscous damper, while the compressibility induces an additional "spring-like" restoring force [2–8].

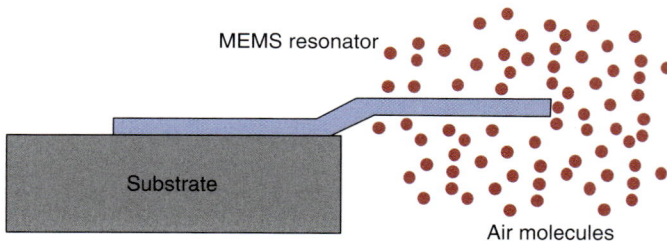

Figure 3.1 Illustration of cantilever-beam resonator vibrating amid surrounding gas.

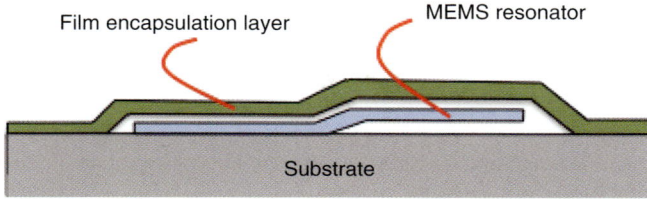

Figure labels: Film encapsulation layer, MEMS resonator, Substrate

Figure 3.2   Illustration of MEMS resonator mounted within an encapsulation. In this instance, the gaps between the moving elements of the resonator and the surroundings can be of order 1 µm.

MEMS resonators are occasionally mounted such that the moving elements are positioned inside an open chamber far from the surroundings, such as in Figure 3.1. However, in modern MEMS resonating devices, such as gyroscopes or timing elements, the resonator is usually positioned close to fixed surfaces, such as the substrate, surrounding electrodes, or encapsulation layers, such as shown in Figure 3.2.

Depending on the resonator motion being normal to or parallel to the nearby surfaces, damping may be described as "squeeze film damping" or "shear film damping" respectively [1, 4]. Many MEMS resonators utilize "comb drives" for actuation and sensing, and these three-dimensional structures combine elements of squeeze film damping and shear film damping. In general, analytical modeling of the air damping in 3-D structures with near-unity aspect ratios can be very difficult.

Depending on the dimensions, velocities, and pressures, the precise mechanism for damping can vary significantly. For many of the MEMS resonators described through this book, for example, gyroscopes and resonators for timing, they are normally operating in a regime of relatively low pressure and low damping, which is often referred to as the "free molecule" regime. In this regime, the mean free path for the molecules is generally comparable to or larger than the separation between the moving resonator and the surrounding structures. The ratio of these length scales is called the Knudsen number, and most MEMS resonators have K_n equal to 1 or much greater than 1.

A number of researchers have worked to produce detailed computational models for air dissipation in the $K_n > 1$ regime because of its importance to the operation of MEMS resonators. For instance, high K_n modifications to Christian's original free molecular model have been developed to produce an accurate quantitative comparison with experimental results [9]. Bao's approach adds the effects of molecular collisions with nearby walls and the resonator [2]. The model keeps track of elastic collisions between gas molecules and the resonator and calculates the velocity change of the molecules within the constraint of conservation of linear momentum and kinetic energy. Intermolecular collisions are ignored in this model. More recent molecular dynamic simulations that relax some of Bao's constrains such as "constant particle velocity" and "constant beam position" have resulted in better accuracy at pressures between 0.1 and 1 Torr [9]. All of these models result in qualitative scaling prediction of $Q \sim \sqrt{T/P}$ in the range of pressures that are commonly found in MEMS resonators. From their results, it is clear that precise predictions over a range of pressures is not yet achieved, probably owing to the very complex and unaccounted-for shapes of the real resonators, as well as issues like surface roughness. Nevertheless, it is true that the order-of-magnitude estimates of air damping are available for MEMS resonators in the $K_n > 1$ regime.

For real devices, the sharp dependence of the air damping on the pressure is undesirable – slight changes in pressure can occur over time, and would lead to significant changes in the dynamical properties of the resonator. As a result, resonator designers routinely work hard to select designs and package pressures such that the air damping is NOT a significant source of dissipation.

In order to determine the pressure of the air inside of a package, it is common to measure the Q of the resonator within the package, and then create an opening to vent the package. The vented resonator is then placed in a pressure test chamber and the Q is carefully measured as a function of the pressure in the test chamber. An example of data from such a measurement is shown in Figure 3.3 [10].

In the case of the device shown in Figure 3.3 (a dual-ring, dual-bar resonator), the Q measured prior to venting of the package was 210 000. In this experiment, all we can say is that the pressure inside the package, prior to venting, was less than 100 Pa, and that the Q of this resonator is probably limited by other damping mechanisms. If the Q prior to venting was lower, say around 140 000, then we would know that the pressure inside the package prior to venting was approximately 1 kPa.

At levels of pressure near 1 atm, that is, 10^5 Pa, viscous forces on the resonator become significant because the mean free path of the air molecules becomes much smaller ($\sim$100 nm) than the typical dimensions of the resonator and the

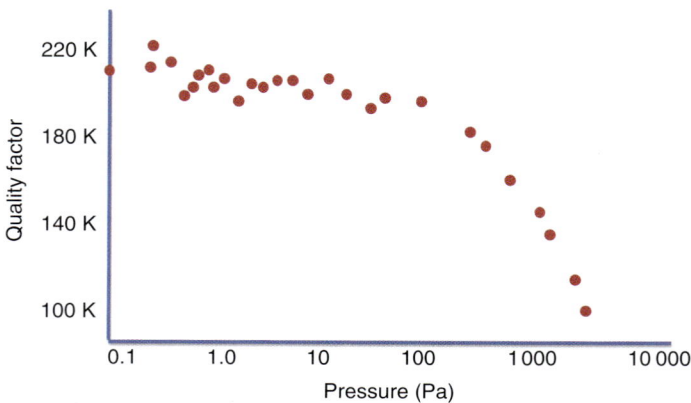

Figure 3.3 Typical measurement of Q as a function of pressure for vented resonator inside package.

cavities surrounding the resonator. In this regime, the viscosity of the medium dominates, causing energy loss purely to the surrounding medium independent of the location of the surrounding structures. In this viscous regime, the air in the vicinity of the moving surfaces of the resonator resists motion and exerts a force on the resonator. The viscous forces may be computed from the Navier-Stokes equations [3, 4]. If resonance frequency and the gap size are low enough so that the inertia forces can be neglected compared to viscous forces, damping is almost entirely viscous and depends on viscosity, which may be an assumed constant with ambient pressure. Quality factor in this case is nearly constant with pressure.

3.3
Surface Damping

In recent years, a number of researchers have developed methods for fabrication and characterization of ultra-thin MEMS structures, such as cantilever beams with thickness near $0.1\,\mu m$ [11–14]. For many of these devices, experiments showed that the dissipation scaled with the reciprocal of the thickness of the resonator.

Consider a series of similar resonators operating in high vacuum. For this series of resonators, we assume that one dimension, called the "thickness", is significantly smaller than the other two dimensions (called "length" and "width"). Further, we assume that the entire series of resonators has the same length and width, and that the only difference between them is a variation in the thickness. Without a detailed model, we can assume that the energy stored in the resonator will scale with the volume of the resonator – for this series of resonators, the volume is proportional to the thickness. If we postulate some unspecified energy dissipation that arises from some phenomena localized to the surface of the resonator, and we assert that the amount of surface area is roughly the same for this series of resonators, we can conclude that the energy loss rate should be the same for this series of resonators.

As shown in Figure 3.4, several researchers have seen exactly this trend for different series of thin resonators, and these observations have led to the dominant dissipation being assigned to a "surface loss mechanism." Some researchers have worked to assign a detailed physical mechanism to the phenomenon of surface loss. A variety of different models have been proposed to explain this scaling of loss, depending on the material that the resonator is made from and also depending on the presence or absence of metal or dielectric films on the surfaces of the resonator [10–18].

For example, Seoanez *et al.* [11] has constructed a "unified model" of surface dissipation mechanisms based on interactions between two-level energy systems at the surfaces and the energy of the entire resonator. These two-level systems can arise due to vibrational modes of adsorbed molecules, configurational rearrangements of structures at the surfaces, or electrons trapped near defects, and

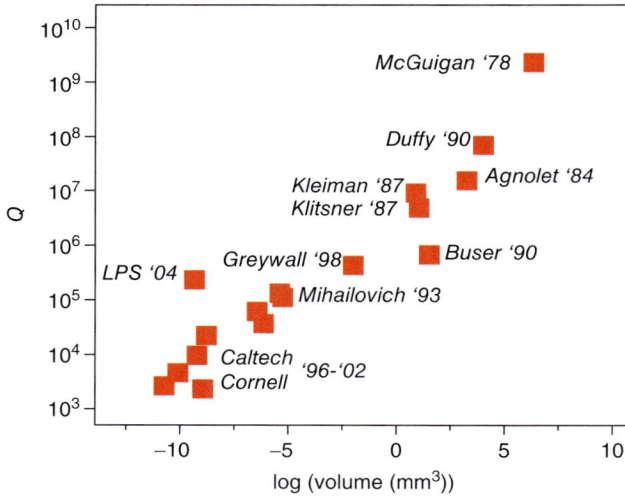

Figure 3.4 Variation of reported quality factors in monocrystalline silicon mechanical resonators with size, Figure 7 of Ref. [18].

all should give rise to specific temperature-dependent signatures in Q measurements, as well as to particular dependences between the resonant frequency of the resonator and the dissipation. One important aspect of these two-level systems as dissipation sources is that the dissipation can increase at low temperatures, making this a significant challenge for researchers exploring fundamental limits to mechanical force detection at low temperatures. These authors also paid particular attention to losses that can arise from interactions between free electrons in metal films on the surfaces of the resonator and trapped charges near those surfaces.

Esashi [12, 13] and Yasumura [15] explored the quality factors of ultrathin silicon cantilever-beam resonators, and found a proportionality to thickness, as well as significant enhancements after thermal desorption of the native oxide films on the surfaces. In both cases, there was evidence for Q increasing, decreasing, and increasing repeatedly after heating to high temperatures, exposure to air, and then heating to high temperatures again.

For many MEMS devices, such as gyroscopes or timing resonators, there is no particular advantage to operating with extremely thin structures. In fact, the sensitivity of gyroscopes will scale with the size of the moving mass, so preferred designs are usually based on relatively thick structures. For time reference resonators, frequency stability, and phase noise are both improved by using a larger device that is not so sensitive to single-monolayer phenomena on the surfaces. Of course, resonant chemical sensors are optimized to be sensitive to phenomena on surfaces, and generally are designed with high surface-to-volume ratios.

3.4
Anchor Damping

All resonators must be suspended from one or more mechanical connections to the underlying substrate or surrounding structures. These connections are called the "Anchors," and provide the mechanical connection to the rest of the system, as well as the electrical connections for biasing and sensing of the resonator. Unlike a free-free beam vibrating in space, these anchor points can also provide a pathway for energy loss from the resonator to the surroundings [19]. All loss mechanisms that refer to energy propagation from the resonator through the anchors are called "anchor damping."

Anchor damping can be considered from a macroscopic perspective or from a microscopic perspective. In the macroscopic perspective, the movement of the resonator through its cycle of oscillation can exert forces on the anchor point, and these forces can do "work" on the anchor, depending on the dimensions, leading to energy loss from the resonator. One easy way to see the macroscopic perspective is to consider a cantilever beam resonator as shown in Figure 3.5. As it vibrates up and down, there are forces applied to the anchor. Jimbo and Itoh [19] computed the energy transport into the anchor of such a cantilever, and arrived at quantitative predictions of anchor loss as a function of the length and thickness of the cantilever.

$$Q_{\text{clamping}} \approx 2.17 \, l^3 / t^3$$

One important way to reduce the energy loss through the anchors is to minimize the forces applied to the anchors during the oscillation cycle. This can be achieved in various ways, but in most cases, the key idea is to locate the anchor at a node of the vibrational mode. For instance, Nguyen has shown reduced anchor damping for a free–free resonating beam supported at the nodes of the free–free mode such as is shown in Figure 3.6 [20]. If the anchors are placed at the exact nodal points of the resonator, anchor damping is reduced greatly, but it will always be nonzero due to the finite dimension of the anchors. Experiments have shown that Q of resonators anchored at resonance node points increases with decreasing anchor size [19–23] Misalignment from the node points reduces Q [24]. This concept is very powerful, and has been widely used [21–27] by most designers seeking high-Q by reducing anchor damping.

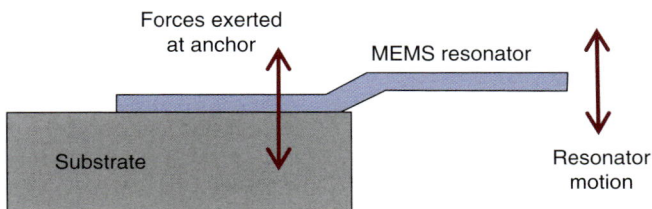

Figure 3.5 Illustration of resonator motion causing forces at anchor.

MEMS resonator

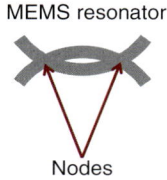

Figure 3.6 Illustration of a free-standing beam oscillating in the first bending mode. This "free–free beam" can be anchored at the nodes of the free–free mode to reduce anchor damping [21].

At a microscopic perspective, the propagation of energy from the resonators through the anchors and into the substrate is viewed as a stress wave-propagation phenomenon. Simply put, if the forces exerted at the anchors are not canceled (due to symmetry or connection at a node), these anchor forces and moments cause stress waves to propagate into the substrate, as shown in Figure 3.7. The energy carried by the stress waves is dissipated as the waves scatter and eventually die out in the substrate. The stress waves are of two kinds; surface acoustic waves (SAWs) that vibrate near the surface and shear and pressure bulk acoustic waves (BAWs) that penetrate the substrate. Both wave types undergo dissipation in the bulk and reflection at free boundaries of the substrate. Reflection can be effectively used to retrieve some of the transmitted energy [19, 21, 28].

For megahertz frequency devices, the acoustic wavelength in the silicon substrate can be larger than the substrate thickness; so some designers have focused on mounting the die in a soft mount to trap the energy in the substrate. At higher frequencies, the wavelengths can be much smaller. In these instances, some designers are seeking to reduce loss from anchor damping by exploring the introduction of "Bragg reflectors" in the substrate [19]. These "reflectors" are like the mirrors in an optical cavity, designed with a goal of trapping the acoustic energy in a region close to the resonator, and establishing standing waves at the boundaries between the resonator and the anchor, so as to prevent the transport of energy into the substrate.

Detailed models of specific anchor geometries have been constructed [21, 25] and researchers are finding that it is increasingly possible to model the details of anchor damping during resonator design.

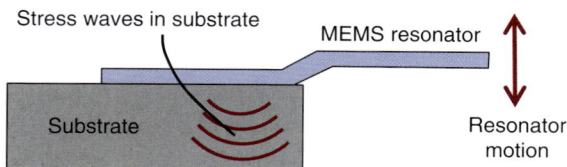

Figure 3.7 Illustration of stress waves launched into substrate as a pathway for energy loss due to anchor damping.

3.5
Electrical Damping

Whenever a structure with some trapped charges is positioned near a conductive surface, there are image charges induced on the conductive surface. If the structure is moving, such as a micromechanical resonator, the movement of the trapped charges will cause a corresponding movement of the image charges in the adjacent surface. If there is finite electrical resistivity in that adjacent surface, the movement of the charges will be resisted by ohmic losses, and these losses will reduce the Q of the resonator. A simple illustration is shown in Figure 3.7.

Electronic damping can be especially important for the particular arrangement shown in Figure 3.8, which is common for certain kinds of resonant-based imaging of surface profiles, such as in non-contact atomic force microscope (AFM). For measurement of extremely small forces, such as in magnetic resonance force microscopy, this dissipation mechanism can dominate and substantially degrade the resolution of the force sensor [29, 30]. In fact, this dissipation mechanism can be used as a method for determining the conductivity of the adjacent substrate; when combined with a scanning platform, imaging of the substrate conductivity is possible [31].

This general mechanism can be important anytime there are trapped charges in a resonating structure or in a nearby structure – the trapped charges always induce image charges, and the movement of the trapped charges will induce movement of the image charges, leading to ohmic losses. Generally, this mechanism is less important than the other mechanisms discussed in this chapter, except for cases where the amplitude of the motion is large, and in ultra-high vacuum and ultra-low temperatures, so that all other mechanisms are suppressed.

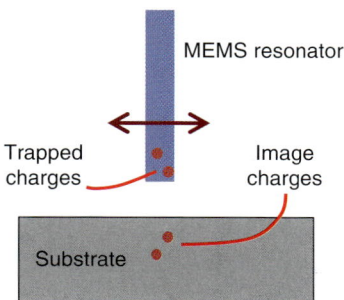

Figure 3.8 Illustration of a resonator vibrating near surface, where trapped charges in the cantilever cause ohmic dissipation in the adjacent surface.

3.6
Thermoelastic Dissipation (TED)

In 1937, Zener [32] described a dissipation mechanism in which the bending of a structure induces temperature gradients, followed by thermal transport to neutralize the gradient. This mechanism is called thermo-elastic dissipation (TED) and has been widely studied for almost 80 years. In vacuum-encapsulated MEMS resonators, this mechanism is often the dominant dissipation mechanism.

To describe TED, we will think about the crystal lattice of the resonator, and consider certain anharmonic effects. During the bending of a resonator beam, the two opposing surfaces undergo tension and compression, as shown in Figure 3.9. For the part of the resonator that experiences tension, there is an expansion of the crystal lattice, which causes a slight increase in the wavelengths of the phonon modes of the crystal. Because of anharmonicity in the crystal, these larger lattice spacings also cause a slight reduction in the energy associated with each of the phonon modes. After this adiabatic expansion of this part of the crystal, the population distribution of the phonon modes is not exactly the same as an equilibrium Bose-Einstein distribution, so various phonon-scattering processes take place to bring the system into equilibrium. At the completion of this process, the local temperature is slightly lower.

On the opposite side of the resonator beam, the opposite effects are happening – and the compression of the lattice leads to a local temperature increase. Therefore, the bending of a resonator leads to the formation of temperature gradients wherever there are strain gradients. These gradients induce thermal transport, loss, and damping in the resonator. This damping is strongest in the situations where the thermal time constant for transport across the beam is comparable to the period of the mechanical vibration.

The coupling between expansion and cooling is familiar from similar macroscopic effects, such as the expansion of an ideal gas, and can also be understood as the reverse effect of thermal expansion. In fact, the coupling between the mechanical and thermal domains can be understood by examination of the governing equations for mechanical and thermal behavior in solids.

$$\rho\frac{\partial^2 u}{\partial t^2} = E\frac{\partial^2 u}{\partial x^2} \quad c\rho\frac{\partial T}{\partial t} = \kappa\frac{\partial^2 T}{\partial x^2}$$

In these equations, u is the displacement, t is the time, x is the position, ρ is the density, E is the Young's Modulus (which should be represented by the full tensor for crystalline materials such as silicon [33]), c is the specific heat of the material, κ is the thermal conductivity of the material, and T is the temperature.

Figure 3.9 Finite element simulation of the strains in a beam clamped at both ends, and displaced in the shape of the first vibrational mode. Blue is tension and red is compression.

The first of these equations is the ordinary dynamical wave equation for solids, and the second is the Fourier heat conduction equation for solids. Most systems can be fully analyzed on the basis of these equations. However, for systems that include significant coupling between the mechanical and thermal domains, we must consider these equations with the added coupling terms [34, 35]:

$$\rho\frac{\partial^2 u}{\partial t^2} = E\frac{\partial^2 u}{\partial x^2} + \underbrace{\frac{\alpha E}{(1-2\upsilon)}\frac{\partial T}{\partial x}}_{\text{Temperature-induced force}}$$

$$c\rho\frac{\partial T}{\partial t} = \kappa\frac{\partial^2 T}{\partial x^2} - \underbrace{\frac{\alpha E T_0}{(1-2\upsilon)}\frac{\partial}{\partial x}\frac{\partial u}{\partial t}}_{\text{Strain-induced heat flow}}$$

In these equations, α is the thermal expansion coefficient and υ is poisson's ratio. In Zener's original analysis [32], the geometry of the resonator was assumed to be a simple cantilever beam. A further simplification was to ignore the coupling term in the mechanical equation, which allowed an algebraic solution to find the modes. These mechanical modes were then used as input to the heat transport equation, and dissipation was calculated, giving a simple result for the TED-limited Q for this system [32]:

$$Q_{\text{TED}} = \left(\frac{c\rho}{E\alpha^2 T_0}\right)\frac{1+(\omega\tau)^2}{\omega\tau} \quad \tau = \left(\frac{b}{\pi}\right)^2\frac{c\rho}{k}$$

In these expressions, ω is the resonant frequency of the resonator, and τ is the thermal time constant for transport across the width of the beam b. This relatively simple analytical result provides an intuitive understanding of TED in cantilevers, and has been extended to efforts at optimization of other, more complicated resonators. In particular, there has been much discussion about how certain materials may have larger or smaller values of the constants in the pre-factor to the TED-limited Q expression above – having a high specific heat and a low thermal expansion coefficient are certainly helpful for improving Q in TED-limited resonators.

The frequency-dependent term in the approximate Zener result is largest when the period of the oscillation is the same as the decay time for the thermal gradient. This coupling has also led to intuitive understanding about how to minimize TED.

However, it is important to be very careful in over-extending the intuition that arises from the Zener approximations. In fact, for more complicated resonators, there is no simple expression for Q, and the relation between the materials properties, the geometry, and the Q is extremely complex. There can even be situations where a reduction in the thermal expansion coefficient would lead to a *reduction* in the Q! Some researchers are working to develop microfabrication processes for new materials, such as fused silica, primarily because of the reduced thermal expansion coefficient in these materials.

Fortunately, with modern computers and modern finite element solvers, it is possible to produce numerical solutions to the complete fully-coupled equations

[34, 36], without making the significant approximations offered in the early work by Zener [32].

For MEMS resonators, the first observation of the TED limit was made by Roszhart in 1990 [37]. There are many instances of MEMS resonators built from bending beams, where the thermal time constant for heat transport across the structure is comparable to the mechanical resonant frequency, and this can lead to strong thermoelastic dissipation [38].

There have been several attempts to construct improved approximation methods, with the goal of providing some useful design intuition for complicated resonators. Houston *et al.* [39] worked to identify a parameter described as a "participation factor" that would be used to scale the Zener-like dissipation associated with a given flexural mode. This participation factor depends on the fraction of the total strain energy of the vibrational mode that arises in regions of high strain gradient. This approximation is relatively easy to compute, and was shown to be effective at predicting the Q values at room temperature for a specific class of double-paddle resonators. Chandorkar *et al.* [40] computed the complete spectra of thermal modes, and then identifies weighting factors representing the coupling from the vibrational mode to each of the thermal modes – by identifying the shape of the thermal mode responsible for the strongest coupling, it is possible to think about design changes that reduce this coupling (such as slots), and make steady intuitive progress toward reduced TED in MEMS resonators. This approach is applied to a series of common resonator families (beams, rings, plates, etc.), and statements about the scaling of TED with dimensions and frequency are produced [40].

The main advantage to designing resonators so that they are limited by thermoelastic dissipation is that the dynamical characteristics of the MEMS resonator can be accurately modeled, and are not strongly influenced by slight changes in the pressure inside the package. Unfortunately, the strength of TED is itself a strong function of temperature. For simple tuning-fork resonators, the TED-limited Q can scale as T^{-2} [38]. As a result, the characteristics of the device have a large and possibly undesirable temperature coefficient, which can make it very difficult or expensive to achieve high-performance oscillators over a wide range of temperature.

3.7
Akhiezer Effect (AKE)

At the beginning of our description of thermoelastic dissipation in the previous section, we considered the relationship between the deformation of the crystal lattice and the formation of localized increases or decreases in temperature. In the case of a bending beam, the stress gradient in the beam created a temperature gradient, and the thermal transport driven by this gradient caused energy loss.

There are two distinct situations in which the same general phenomenon does not lead to TED-related loss. For example, consider the case of a resonator that expands and contracts without inducing strain gradients – such as the extensional modes of a bar. In such a device, the strains are uniform, and there are no temperature gradients to drive loss. Even without transport, there is dissipation, called the Akhiezer effect (AKE) [41].

Let us again imagine a crystal lattice undergoing expansion, perhaps during the first part of the oscillation cycle for an extensional mode. As described before, this expansion leads to an increase in the wavelength for all the phonon modes of the crystal, and anharmonicity of the lattice bonds leads to a slight reduction in the frequencies and the energies of all these modes. The coupling between strain and energy levels is called the Grüneisen parameter, and it is represented as a full tensor, as the coupling is dependent on the polarization of the phonon and the propagation direction in the crystal. Because this expansion happens on a timescale that is much slower than the phonon frequencies ($\sim 10^{14}$ Hz), the initial effect is an adiabatic change – the energy of all the modes change, but the occupancy of the modes remains the same as before the expansion.

After the expansion, when the occupancy is preserved, but the energies have been slightly lowered, the distribution of population is no longer represented by an equilibrium distribution function of the Bose-Einstein type. In effect, the system is not in thermodynamic equilibrium. For the next period, driven by the second law of thermodynamics, phonon-phonon scattering proceeds in a manner that brings the system into equilibrium at a new, slightly lower temperature. During this process, entropy is increasing, and loss is occurring, even though there has been no transport.

If we now consider the other half of the resonator's cycle – the transition from extension to compression, all of the same phenomena repeat – except, now the energies of all the modes are increased, but followed as before by scattering to bring the system into a new equilibrium at a new, higher temperature.

In order for this phenomenon to take place, the timescale for thermalization must be faster than the period of the oscillation. For silicon MEMS resonators, this thermalization timescale is of the order 0.1 ns or faster; so we can expect AKE dissipation to be present for resonators with frequencies below 10 GHz.

This qualitative description explains how the cycles of extension and compression, each followed by a period of thermalization to a new temperature, leads to continuous generation of entropy and continuous loss, even without strain gradients and transport.

A quantitative analysis of AKE dissipation leads to an expression for the AKE limit in resonators [41–43]:

$$Q^* f = \frac{3\rho c^2}{2\pi \gamma_{\text{avg}}^2 cT\tau} = \frac{\rho c^4}{2\pi \gamma_{\text{avg}}^2 kT}$$

In this equation, we have the density (ρ), the speed of sound (c), the Grüneisen parameter (γ_{ave}), and the temperature (T). In this case, γ_{ave} is the average of the Grüneisen parameter over polarizations and directions. The very interesting aspect of this result is the prediction for the absolute upper limit to the Q achievable in resonators that only depends on the materials properties and the

Table 3.1 Akheiser limit of $Q–f$ product for common resonator materials and the corresponding Grüneisen parameter.

Material	$Q–f$ ($\times 10^{13}$)	γ
Si	2.3	0.51
Quartz	3.2	0.87
AlN	2.5	0.91
Diamond	3.7	0.94
Sapphire	11.3	1.1
SiC	64	0.3

temperature. All of the geometric dependencies in this model cancel out during the computation of the loss terms, and the final result is exceptionally simple and useful. The meaning is simple – for a resonator in a perfect vacuum, designed with perfectly lossless anchors, and without any electronic frictional losses, and without any strain gradients that can trigger TED, there is still an upper limit to the Q, which is determined only by the selection of the material and the temperature.

The AKE limits to MEMS resonators have been calculated by various researchers for the materials most likely to be used for MEMS resonators [43–47]; some of these results are presented in Table 3.1. We see that silicon is a relatively good choice for high-Q resonators, but that there is some improvement available by building devices in silicon carbide, sapphire or diamond. Of course,

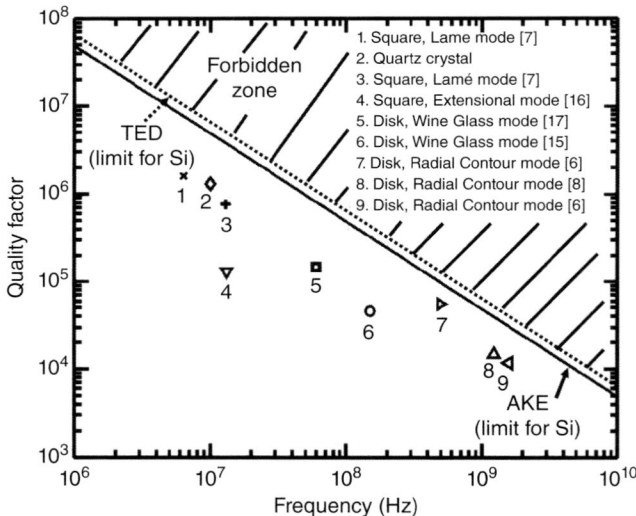

Figure 3.10 Quality factor as a function of frequency for recently reported high frequency MEMS SCS bulk resonators, compared to the upper limit resulting from materials properties. (From Ref. [44].)

there are other important considerations – the cost of the material and the availability of high-volume, low-cost manufacturing methods.

Because the Q^*f product is an upper limit to the performance of resonators, it is possible to make a plot of Q as a function of f for the highest-Q resonators. The AKE upper limit can be represented as a diagonal line on this plot. One early such plot is shown in Figure 3.10 [44].

As shown in Figure 3.10, many recently fabricated MEMS resonators are approaching this limit; as a result, there is less and less opportunity for further improvements in the Q of many high-Q MEMS resonators.

For very high-frequency resonators ($f > 10\,\text{GHz}$), there is insufficient time during a period of vibration for scattering to alter the distribution of phonons in the crystal, so thermalization is avoided and entropy is not generated. In these very high-frequency cases, the prediction is that AKE should be suppressed. For these resonators, the $Q-f$ product may exceed the normal Akheizer limit to a new limit known as Landau-Rumer regime [42, 47]. MEMS resonators with $f > 10\,\text{GHz}$ are just now being developed, and Qs above the AKE limit have not yet been observed.

References

1. Newell, W. (1968) Miniaturization of tuning forks. *Science*, **161**, 1320–1326.
2. Bao, M. and Yang, H. (2007) Squeeze film air damping in MEMS. *Sens. Actuators, A*, **136**, 3.
3. Wu, G.Q., Xu, D.H., Xiong, B., and Wang, Y.L. (2013) Effect of air damping on quality factor of bulk mode microresonators. *Microelectron. Eng.*, **103**, 86.
4. Cho, Y.-H., Pisano, A.P., and Howe, R.T. (1994) Viscous damping model for laterally oscillating microstructures. *J. Microelectromech. Syst.*, **3**, 81.
5. Griffin, W.S., Richardson, H.H., and Yamanami, S. (1966) A study of fluid squeeze-film damping. *J. Fluids Eng.*, **88** (2), 451–456.
6. White, F.M. (1974) *Viscous Fluid Flow*, McGraw-Hill Book Company.
7. Langlois, W.E. (1962) Isothermal squeeze films. *Q. Appl. Math.*, **20** (2), 131–150.
8. Zhang, C. et al. (2004) Characterization of the squeeze film damping effect on the quality factor of a microbeam resonator. *J. Micromech. Microeng.*, **14**, 1302.
9. Hutcherson, S. and Ye, W. (2004) On the squeeze-film damping of micro-resonators in the free-molecule regime. *J. Micromech. Microeng.*, **14**, 1726.
10. Kim, B. (2008) Stability and performance of wafer-scale thin-film encapsulated MEMS resonators, Stanford University.
11. Seoanez, C., Guinea, F., and Castro Neto, A.H. (2008) Surface dissipation in nanoelectromechanical systems: Unified description with the standard tunneling model and effects of metallic electrodes. *Phys. Rev. B*, **77** (12), 125107.
12. Yang, J., Ono, T., and Esashi, M. (2000) Surface effects and high quality factors in ultrathin single-crystal silicon cantilevers. *Appl. Phys. Lett.*, **77** (23), 3860.
13. Yang, J., Ono, T., and Esashi, M. (2002) Energy dissipation in submicrometer thick single-crystal silicon cantilevers. *J. Microelectromech. Syst.*, **11**, 775–783.
14. Ru, C.Q. (2009) Size effect of dissipative surface stress on quality factor of microbeams. *Appl. Phys. Lett.*, **94**, 051905.
15. Yasumura, K.Y., Stowe, T.D., Chow, E.M., Pfafman, T., Kenny, T.W., Stipe, B.C., and Rugar, D. (2000) Quality factors in micron- and submicron-thick cantilevers. *J. Microelectromech. Syst.*, **9** (1), 117–125.

16. Zener, C. (1948) *Elasticity and Anelasticity of Metals*, University of Chicago Press, Chicago, IL.

17. Wang, D.F. *et al.* (2004) Thermal treatments and gas adsorption influences on nanomechanics of ultra-thin silicon resonators for ultimate sensing. *Nanotechnology*, **15**, 1851.

18. Ekinci, K.L. and Roukes, M.L. (2005) Nanoelectromechanical systems. *Rev. Sci. Instrum.*, **76**, 061101.

19. Park, Y.-H. and Park, K.C. (2004) High-fidelity modeling of MEMS resonators. Part I. Anchor loss mechanisms through substrate. *J. Microelectromech. Syst.*, **13** (2), 238, 247.

20. Wang, K., Wong, A.-C., and Nguyen, C.T.-C. (2000) VHF free-free beam high-Q micromechanical resonators. *J. Microelectromech. Syst.*, **9**, 347.

21. Jimbo, Y. and Itao, K. (1968) Energy loss of a cantilever vibrator. *J. Horological Inst. Jpn.*, **47**, 1–15, (in Japanese).

22. Harrington, B.P. and Abdolvand, R. (2011) In-plane acoustic reflectors for reducing effective anchor loss in lateral–extensional MEMS resonators. *J. Micromech. Microeng.*, **21**, 085021.

23. Lin, Y.-W., Lee, S.-S., Li, Y., and Xie, R. (2004) Zeying and C.T-C Nguyen, series-resonant VHF micromechanical resonator reference oscillators. *IEEE J. Solid-State Circuits*, **39** (12), 2477–2491.

24. Pandey, M., Reichenbach, R.B., Zehnder, A.T., Lal, A., and Craighead, H.G. (2009) Reducing anchor loss in MEMS resonators using mesa isolation. *J. Microelectromech. Syst.*, **18** (4), 836–844.

25. Y-H Park and K.C. Park, High-fidelity modeling of MEMS resonators. Part II. Coupled beam-substrate dynamics and validation, *J. Microelectromech. Syst.*, **13** (2) 248––257 (2004).

26. Tas, V., Olcum, S., Aksoy, M., and Atalar, A. (2010) Reducing anchor loss in micromechanical extensional mode resonators. *IEEE Trans. Ultrason. Ferroelectr. Freq. Control*, **57** (2), 448–454.

27. Wang, J., Zeying, R., and Nguyen, C.T.-C. (2004) 1.156-GHz self-aligned vibrating micromechanical disk resonator. *IEEE Trans. Ultrason. Ferroelectr. Freq. Control*, **51** (12), 1607–1628.

28. Magnusson, P.C., Weisshaar, A., Tripathi, V.K., and Alexander, G.C. (2002) *Transmission Lines and Wave Propagation*, CRC Press.

29. Rugar, D., Budakin, R., Mamin, H.J., and Chui, B.W. (2004) Single spin detection by magnetic resonance force microscopy. *Nature*, **430**, 329.

30. Stipe, B.C., Mamin, H.J., Stowe, T.D., Kenny, T.W., and Rugar, D. (2001) Non-contact friction and force fluctuations between closely-spaced bodies. *Phys. Rev. Lett.*, **87**, 96801.

31. Stowe, T.D., Kenny, T.W., Thompson, D.J., and Rugar, D. (2001) Silicon dopant imaging by dissipation force microscopy. *Appl. Phys. Lett.*, **75**, 2785.

32. Zener, C. (1937) Internal friction in solids: I. Theory of internal friction in reeds. *Phys. Rev.*, **52**, 230–235.

33. Hopcroft, M.A., Nix, W.D., and Kenny, T.W. (2010) What is the Young's modulus of silicon? *IEEE J. Microelectromech. Syst.*, **19**, 229.

34. Duwel, A., Candler, R.N., Kenny, T.W., and Varghese, M. (2006) Engineering MEMS resonators with low thermoelastic damping. *J. Microelectromech. Syst.*, **15** (6), 1437–1445.

35. Lifshitz, R. and Roukes, M. (2000) Thermoelastic dissipation in micro- and nano-mechanical resonators. *Phys. Rev. B*, **61**, 5600.

36. Candler, R.N., Duwel, A., Varghese, M., Chandorkar, S.A., Hopcroft, M.A., Park, W.-T., Kim, B., Yama, G., Partridge, A., Lutz, M., and Kenny, T.W. (2006) Impact of geometry on thermoelastic dissipation in micromechanical resonant beams. *J. Microelectromech. Syst.*, **15** (4), 927–934.

37. Roszhart, T.V. (1990) The effect of thermoelastic internal friction on the Q of micromachined silicon resonators. Proceedings of the 1990 Solid-State Sensor and Actuator Workshop, Hilton Head Island, SC, p. 13.

38. Kim, B., Hopcroft, M.A., Candler, R.N., Jha, C.M., Agarwal, M., Melamud, R., Chandorkar, S.A., Yama, G., and Kenny, T.W. (2008) Temperature dependence of quality factor in MEMS resonators. *J. Microelectromech. Syst.*, **17**, 755.

39. Houston, B.H., Photiadis, D.M., Marcus, H.M., Bucaro, J.A., Liu, X., and Vignola, J.F. (2002) Thermoelastic loss in microscale resonators. *Appl. Phys. Lett.*, **80**, 1300.

40. Chandorkar, S.A., Candler, R.N., Duwel, A., Melamud, R., Agarwal, M., Goodson, K.E., and Kenny, T.W. (2009) Multimode thermoelastic dissipation. *J. Appl. Phys.*, **105**, 043505.

41. Akhieser, A. (1939) On the absorption of sound in solids. *J. Phys. (Akademiia Nauk-Leningrad)*, **1**, 277–287.

42. Landau, L.D. and Rumer, G. (1937) On the absorption of sound in solids. *Phys. Z. Sowjetunion*, **11**, 18.

43. Ghaffari, S., Chandorkar, S.A., Wang, S., Ng, E.J., Ahn, C.H., Hong, V., Yang, Y., and Kenny, T.W. (2013) Quantum limit of quality factor in silicon micro and nano mechanical resonators. *Sci. Rep.*, **3**, 3244. doi: 10.1038/srep03244

44. Chandorkar, S.A., Agarwal, M., Melamud, R., Candler, R.N., Goodson, K.E., and Kenny, T.W. (2008) Limits of quality factor in bulk-mode micromechanical resonators. Proceedings of the 2008 IEEE MEMS Conference, Tucson, AZ, p. 74.

45. Tabrizian, R., Rais-Zadeh, M., and Ayazi, F. (2009) Effect of phonon interactions on limiting the fQ product of micromechanical resonators. *IEEE Transducers*, 2131–2134.

46. Hwang, E. and Bhave, S.A. (2011) Experimental verification of internal friction at GHZ frequencies in doped single-crystal silicon. IEEE 24th International Conference on Micro Electro Mechanical Systems (MEMS), January 2011, pp. 424–427, 23–27, doi: 10.1109/MEMSYS.2011.5734452

47. Duwel, A.E., Lozow, J., Fisher, C.J., Phillips, T., Olsson, R.H., and Weinberg, M. (2011) Thermal energy loss mechanisms in micro- to nano-scale devices. *Proc. SPIE*, **8031**.

4
Parametrically Excited Micro- and Nanosystems

Jeffrey F. Rhoads, Congzhong Guo, and Gary K. Fedder

4.1
Introduction

Though an electrical or mechanical system is traditionally excited into resonance through the application of a time-periodic force whose frequency is commensurate with its natural frequency (i.e., through the application of so-called *direct excitation*), it is also possible to elicit large response amplitudes in an engineering system by periodically modulating its effective impedance, or providing so-called *parametric excitation*. This fact is manifested in the example of a base-excited pendulum. While engineering intuition alone allows one to visualize the likelihood for large angular motions in this system under resonant, horizontal excitation of the pivot, less intuitive is the fact that vertical base excitations (which lead to a time-varying effective stiffness), under certain amplitude/frequency conditions, can also render large angular deflections. Of course, this fact is less intuitive only until one considers that as a child they may have used analogous changes in their effective length to achieve large amplitude motions on playground swings!

In scientific and engineering contexts, parametrically excited systems have received considerable research attention, beginning with the early, nineteenth century research endeavors of Faraday [1], Melde [2], Mathieu [3], and Rayleigh [4, 5], and continuing with the twentieth century works of the nonlinear dynamics community (see, for example, [6−9]). Collectively, these works not only introduced the phenomena, but also brought mathematical formalism to its study. This formalism eventually led to the so-called Mathieu Equation, which is commonly used to describe the salient dynamics of parametrically excited systems:

$$\ddot{x} + 2\zeta\omega_0\dot{x} + (\omega_0^2 + \beta\cos\omega t)\,x = 0 \tag{4.1}$$

where ω_0 is the natural frequency of the resonator, ζ captures the effects of linear dissipation, and β and ω represent the amplitude and frequency of the parametric excitation, respectively.

Of particular importance in this chapter is the fact that parametric excitations can lead to large-amplitude responses, or so-called *parametric resonances*, when

Resonant MEMS – Fundamentals, Implementation and Application, First Edition.
Edited by Oliver Brand, Isabelle Dufour, Stephen M. Heinrich and Fabien Josse.
© 2015 Wiley-VCH Verlag GmbH & Co. KGaA. Published 2015 by Wiley-VCH Verlag GmbH & Co. KGaA.

the excitation frequency satisfies the condition $\omega = 2\omega_0/n$, where n is an integer greater than or equal to unity [8]. In most electrical and mechanical systems, only the *principal parametric resonance* associated with $n = 1$ can be observed in experimentation. In the presence of very weak damping, additional resonances can be observed as well, provided proper selection of ω and β [10].

Research investigations of parametric excitation in the context of resonant micro- and nanosystems trace their roots to the 1990's efforts of Rugar and Grütter [11] and Turner *et al.* [10]. The former investigated the utility of combined direct and parametric excitations in a microcantilever actuated by both electrostatic and piezoelectric means, while the latter leveraged a unique electrostatic comb drive to explore parametric effects in a torsional microelectromechanical systems (MEMS) resonator. Though distinct in research focus, these two studies paved a pathway for those that followed by not only highlighting the possibility of parametrically excited micro/nanosystems, but, more importantly, demonstrating that the dynamics associated with such systems were exceedingly rich and could be advantageously exploited in practical application, such as chemical and biological sensing, inertial sensing, signal processing, and scanning probe microscopy.

The present chapter seeks to explore this dynamical richness *and* practical utility by providing an overview of parametrically excited micro/nanosystems. To this end, the remainder of this chapter is organized as follows: Section 4.2 discusses the common sources of parametric excitation in resonant micro/nanosystems, and introduces a series of representative devices. The chapter then continues in Sections 4.3 and 4.4 with a discussion of device modeling and analysis. Sections 4.5 and 4.6 describe the linear and nonlinear dynamics associated with a representative parametrically excited system, placing emphasis on the salient features of the response that can be leveraged in practical application. The work then proceeds in Sections 4.7 and 4.8 with brief discussions of advanced dynamical phenomena and combined, direct and parametric excitations. Section 4.9 discusses emergent applications for parametrically excited micro/nanosystems, and the chapter ultimately concludes in Section 4.10 with a discussion of future prospects in this technical area.

4.2
Sources of Parametric Excitation in MEMS and NEMS

Due to the inherent multi-physical nature of micro/nanoresonators, time-varying impedances can be realized in these systems through a variety of distinct mechanisms. The most common among these are tailored electrostatic, piezoelectric, electromagnetic, and thermal actuations. Interestingly, parametric excitations can also arise inadvertently in many MEMS/NEMS (nanoelectromechanical systems). While these excitations can be exploited for practical gain, they can also stymie an inexperienced engineer if care is not taken in the course of design.

4.2.1

Parametric Excitation via Electrostatic Transduction

Consider the force associated with a variable-gap parallel-plate electrostatic actuator:

$$F_{pp} = \frac{\epsilon A V^2}{2(g-x)^2} \tag{4.2}$$

where ϵ represents the permittivity of the gap medium (typically air), A is the effective area of the actuator electrode, $V(t)$ is the applied actuation voltage, g is the nominal gap width, and x is the displacement of the micro/nanoresonator. Expanding this force in a Taylor Series about the neutral state $x = 0$ renders:

$$F_{pp} = \frac{\epsilon A V^2}{2g^2}\left[1 + 2\frac{x}{g} + 3\left(\frac{x}{g}\right)^2 + 4\left(\frac{x}{g}\right)^3\right] + \mathcal{O}(x^4) \tag{4.3}$$

As evident from this expression, if the applied voltage changes with time, that is, $V = V(t)$, the force renders both a direct excitation from the first term and a parametric excitation from the other terms, which depend on the displacement, x. Accordingly, though it is often unintentional, *all* micro/nanoresonators with variable-gap electrostatic actuators are parametrically excited! This fact has been demonstrated experimentally with both microbeam [12–16] and nanowire [17–19] geometries.

Though parametric excitations arise naturally in variable-gap, electrostatically-actuated systems, the combined (direct and parametric) nature of the excitation *can* be sub-optimal in practical contexts. To alleviate this concern, device designers have developed novel electrode configurations that are capable of yielding purely-parametric excitations. These designs include the non-interdigitated comb drives detailed in [20–22], the curved comb finger geometries presented in [23, 24] and the fringe-field actuators presented in [25], some of which are highlighted in Figure 4.1.

Non-interdigitated comb drives are planar actuators that leverage fringing electrostatic fields to create purely-parametric forces of the approximate form:

$$F_{cd} = (r_1 x + r_3 x^3)V^2(t) \tag{4.4}$$

where x represents the planar displacement of the resonator, $V(t)$ specifies the applied voltage, and r_1 and r_3 are geometry-dependent coefficients defined as electrostatic spring constants, which can be positive and/or negative depending on the exact configuration of the drive. The flexibility attendant to the selection of r_1 and r_3 was exploited extensively in the works of Turner *et al.* in the development of a series of highly tunable planar microresonators with applications in mass sensing, inertial sensing, and signal processing [28–30], as well as in the works of Urey *et al.* emphasizing the development of microscanners [31, 32].

One constraint associated with the aforementioned non-interdigitated comb drives is that their stroke is inherently limited to a displacement equal to $\pm 1/2$ of the comb drive's finger pitch, due to the nature of the fringing fields used for

(a)

(b)

(c)

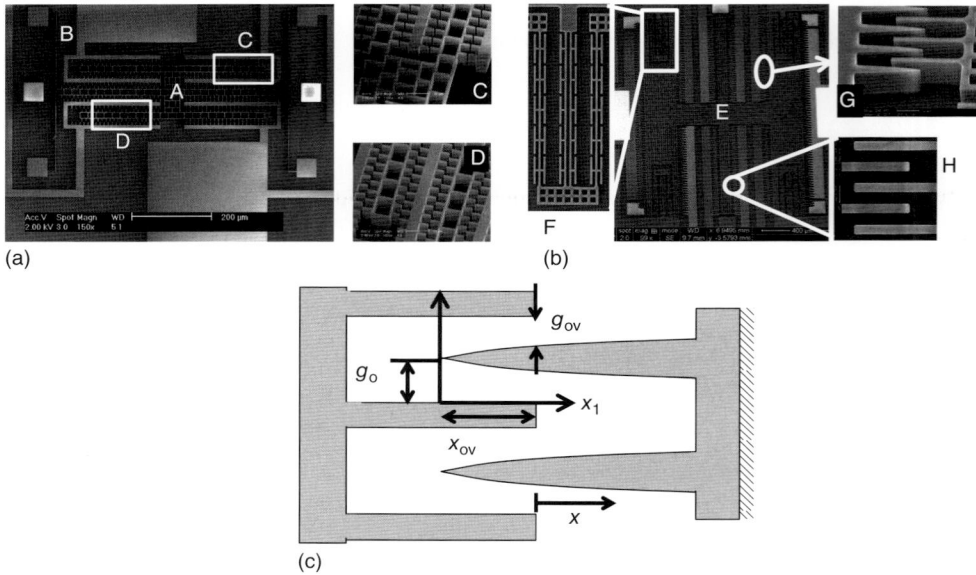

Figure 4.1 (a) A parametrically excited microresonator actuated through the use of a non-interdigitated, electrostatic comb drive and designed in a silicon-on-insulator (SOI) MEMS process. The shuttle mass (A) is suspended by crab-leg flexures (B), and driven/tuned by aligned and misaligned non-interdigitated comb drives (C and D). (Courtesy of K. Turner, University of California at Santa Barbara.) (b) A SOI-MEMS parametrically excited microresonator actuated through the use of a quadratic-shaped-finger comb drive. The shuttle mass (E) is suspended by folded flexures (F), and driven by the shaped-finger comb drive (G). The displacement is measured via the straight-finger comb drive (H). (c) The shaped-finger comb drive design parameters [23, 26, 27].

actuation. To circumvent this constraint, Guo and Fedder [23, 27] built upon the prior work of Jensen *et al.* [33] and Hirano *et al.* [34] and synthesized the shape profile of interdigitated comb fingers to obtain parametric excitation. The resulting tailored "shaped-finger" comb geometry, highlighted in Figure 4.1c, utilized a varying electrode gap to provide the displacement-dependent electrostatic spring constant required to excite parametric resonance and facilitate large displacement motion. The net result was a planar actuator that created a purely parametric force of the approximate form:

$$F_{cd}(x) = \frac{2N\epsilon h}{g_o}\left(\frac{x}{x_{o1}}\right)V^2(t) = (r_1 x)V^2(t) \tag{4.5}$$

where N and $N + 1$ are the numbers of interdigitated fingers on two sides of the comb drive, respectively, h is the thickness of the comb drive, x_{o1} is defined according to

$$x_{o1} = \frac{x_{ov}}{g_o/g_{ov} - 1} \tag{4.6}$$

and all other parameters are defined as in Figure 4.1c.

4.2.2
Other Sources of Parametric Excitation

Though the research literature is ripe with examples of electrostatically actuated, parametrically excited micro/nanosystems, many works have adopted alternative approaches capable of yielding time-varying impedances at the microscale. Interestingly, the vast majority of these have associated force models of the form presented in Eq. (4.4). The alternate sources of parametric excitation in resonant micro/nanosystems include:

• **Piezoelectric Elements:** Piezoelectric elements, which provide periodic excitations proportional to the potential difference across the element, can be used to provide parametric excitation through a number of distinct paths. Mahboob, Thomas, and their respective coworkers, for example, used piezoelectric elements to periodically modulate the stiffness of microbeams [35 – 37] (An electrostatic analog of this has also been presented in [38]). In contrast, Kaajakari and Lal, utilized a piezoelectric element to provide base excitation, essentially creating a microscale analog of the base-excited pendulum referenced in the introduction of this work [39].
• **Lorentz Forces:** Lorentz forces result from the interaction between a current-carrying loop and an external magnetic field. Should such a loop be embedded on the surface of a microresonator, or be formed from the resonating body of a nanostructure, it becomes possible to generate direct, parametric, or combined excitations, depending on the exact orientation of the current and magnetic field. This has been demonstrated in the works of Requa and Rhoads [40 – 42].
• **Thermal Interactions:** Periodic variations in thermal characteristics can also be utilized to realize a time-varying impedance. Zalalutdinov and collaborators demonstrated the effectiveness of this within the context of a laser-excited silicon microresonator [43], while Krylov and collaborators investigated the utility of parametric excitations arising from Joule heating [44].
• **Feedback Control:** Parametric excitations can also be realized artificially in systems that do not naturally exhibit the phenomena through the use of feedback control. This was demonstrated within the context of scanning probe microscopy by Moreno-Moreno, Prakash, and their respective collaborators [45, 46], parametric oscillators by Villanueva and collaborators [47], and resonant mass sensors by Prakash and collaborators [48].

4.3
Modeling the Underlying Dynamics – Variants of the Mathieu Equation

As with conventional micro/nanoresonators, the dynamic response of parametrically excited microsystems can be characterized using a variety of modeling and analysis techniques. Early studies in this area generally leveraged experimental observations and prior experience to develop phenomenological, lumped-mass models that were amenable to systematic parameter identification [30, 42, 49 – 52].

More recently, these analyses have been complemented with first-principles, distributed-parameter models, which are not only more amenable to predictive design, but when used in conjunction with model reduction techniques (e.g., modal expansions) can be used to distill physically consistent lumped-mass analogs.

Regardless of the adopted modeling approach, the lumped-mass equation of motion of a representative microresonator undergoing symmetric and periodic deflection can be written as:

$$m\ddot{x} + c\dot{x} + k_1 x + k_3 x^3 = f(x, t) \qquad (4.7)$$

where m represents the system's effective mass, c captures the effects of various linear dissipation mechanisms (e.g., viscous fluid damping, anchor loss, or thermoelastic dissipation), k_1 is the system's linear spring constant, k_3 defines the strength of the cubic nonlinearity in the system's stiffness characteristics that become significant at large displacements, and $f(x, t)$ accounts for an applied external force. The exact form of $f(x, t)$ is strongly dependent of the physical mechanism utilized for transduction, but is generally approximated by one of the following forms, depending on the salient dynamics and practical application, of interest:

Case 1: $f(x, t) = -d_1 \cos \omega t$

Case 2: $f(x, t) = -b_1 x \cos \omega t$

Case 3: $f(x, t) = -b_1 x \cos \omega t - b_3 x^3 \cos \omega t$

Case 4: $f(x, t) = -b_1 x \cos 2\omega t - d_1 \cos (\omega t + \phi)$ $\qquad (4.8)$

The first of these cases corresponds to the classical example of a directly excited resonator—a topic studied in appreciable depth in prior chapters. The second and third cases are canonical examples of force models used to explore purely parametric excitations in systems undergoing comparatively small or large deformations, respectively. Likewise, the last case is commonly used to explore combined parametric and direct excitations, which with proper phasing can be utilized for *parametric amplification or attenuation* and *noise squeezing*. Here, an emphasis is placed on Case 2. Additional information related to Cases 3 and 4 is provided in Sections 4.7 and 4.8, respectively.

Evoking the force model given in Case 2, the lumped-mass model of interest assumes the form

$$m\ddot{x} + c\dot{x} + (k_1 + b_1 \cos \omega t)x + k_3 x^3 = 0 \qquad (4.9)$$

For the sake of analysis, it proves convenient to mass normalize this model, which yields:

$$\ddot{x} + \frac{c}{m}\dot{x} + (\omega_0^2 + \hat{b}_1 \cos \omega t)x + \hat{k}_3 x^3 = 0 \qquad (4.10)$$

where, as highlighted in Table 4.1, ω_0 represents the system's purely mechanical natural frequency, $\hat{b}_1$ is the mass-normalized parametric excitation amplitude, and

Table 4.1 The dimensional and nondimensional parameters associated with Eqs. (4.10) and (4.12).

$\hat{b}_1 = \frac{b_1}{m}$	$\hat{k}_3 = \frac{k_3}{m}$	$\omega_0 = \sqrt{\frac{k_1}{m}}$
$(\bullet)' = \frac{d(\bullet)}{d\tau}$	$\varepsilon\zeta = \frac{c}{2\omega_0 m}$	$\Omega = \frac{\omega}{\omega_0}$
$\varepsilon\lambda = \frac{\hat{b}_1}{\omega_0^2} = \frac{b_1}{k_1}$		$\varepsilon\chi = \frac{\hat{k}_3 x_0^2}{\omega_0^2} = \frac{k_3 x_0^2}{k_1}$

$\hat{k}_3$ is the mass-normalized cubic stiffness. To ensure broad, cross-scale applicability, it proves convenient to nondimensionalize both space and time according to

$$z = \frac{x}{x_0}, \qquad \tau = \omega_0 t \qquad (4.11)$$

where x_0 is a characteristic length associated with the system (e.g., the thickness of the resonator), and further assume that the salient dynamics of the system are a *small* perturbation from those of an undamped harmonic oscillator. This yields a final lumped-mass representation for the system given by:

$$z'' + 2\varepsilon\zeta z' + (1 + \varepsilon\lambda\cos\Omega\tau)z + \varepsilon\chi z^3 = 0 \qquad (4.12)$$

The parameters associated with this nondimensional equation of motion are included in Table 4.1. Note that ε is a *small* parameter introduced to facilitate analysis.

The equation of motion described above represents a nonlinear variant of the Mathieu Equation, which, as highlighted in the introduction, has received considerable research attention over the past two centuries. These studies have revealed parametrically excited resonators to be deceptively complex from a dynamical systems perspective. Some of this complexity is explored in the subsequent section.

4.4
Perturbation Analysis

Given that the analysis of the equation of motion presented in Eq. (4.12) is largely intractable without the use of special functions, it proves convenient to analyze its dynamic behaviors through the use of perturbation techniques [8]. The Method of Averaging is adopted here [8, 53].

To this end, it is prudent to first introduce a constrained coordinate transformation of the form:

$$z(\tau) = a(\tau)\cos\left[\frac{\Omega\tau}{2} + \psi(\tau)\right], \qquad z'(\tau) = -a(\tau)\frac{\Omega}{2}\sin\left[\frac{\Omega\tau}{2} + \psi(\tau)\right] \quad (4.13)$$

where the amplitude $a(\tau)$ and the phase $\psi(\tau)$ are assumed to be slow functions of time when compared to $\cos(\Omega\tau/2)$ and are referred to as *slow-flow variables*. Furthermore, since dynamics near the principal parametric resonance ($\Omega \approx 2$) offer the most utility in practical application, a frequency detuning parameter σ is introduced:

$$\sigma = \frac{\Omega - 2}{\varepsilon} \tag{4.14}$$

Substituting each of these expressions into Eq. (4.12), solving for the slow-varying variables, and averaging the result over one period of the response $(2\pi/\Omega)$ renders the system's slow-flow equations:

$$a' = \frac{1}{4}\varepsilon a \left(-4\zeta + \lambda \sin 2\psi\right) + \mathcal{O}(\varepsilon^2) \tag{4.15}$$

$$\psi' = \frac{1}{8}\varepsilon \left(3\chi a^2 - 4\sigma + 2\lambda \cos 2\psi\right) + \mathcal{O}(\varepsilon^2) \tag{4.16}$$

With these equations in hand, the system's dynamic behavior can be estimated by either solving the coupled first-order differential equations that govern amplitude and phase (which is suitable for transient analysis), or by setting $(a', \psi') = (0,0)$ and solving the resulting algebraic equations (which is suitable for steady-state analysis). The present work focuses on the latter, given that the steady-state behaviors are more commonly utilized in application.

4.5
Linear, Steady-State Behaviors

Analysis of Eqs. (4.15) and (4.16) reveals an interesting fact: In the absence of nonlinearity (i.e., $\chi = 0$), the resonator has only a trivial steady-state solution: $a_1 = 0$. The stability of this solution can be determined by linearizing the system's averaged equations about the steady-state and examining the eigenvalues associated with the Jacobian of the linearized system. Due to the indeterminate nature of the phase associated with the trivial solution, it is prudent to do this following a conversion to Cartesian coordinates. Analysis of the eigenvalues associated with the trivial solution reveals that the stability of the system transitions at the critical detuning values of

$$\sigma_1 = \frac{1}{2}\sqrt{\lambda^2 - 16\zeta^2}, \qquad \sigma_2 = -\frac{1}{2}\sqrt{\lambda^2 - 16\zeta^2} \tag{4.17}$$

respectively. Specifically, when $\sigma < \sigma_2$ and $\sigma > \sigma_1$ the trivial solution is stable, and when $\sigma_2 < \sigma < \sigma_1$ the solution is unstable.

The aforementioned stability characteristic renders the "wedge of instability" in the $\lambda - \sigma$ parameter space, which is commonly used to characterize the dynamics of the Mathieu Equation and related systems. As highlighted by Figure 4.2, the parametrically excited system remains quiescent when driven with excitations that lie outside of the wedge. However, when driven by excitation conditions that lie inside of the wedge, the system response grows without bound. That is, the system enters *parametric resonance*. In directly excited systems, this growth would be bounded by linear damping mechanisms. However, in parametrically excited systems these mechanisms simply shape the instability wedge and create a threshold

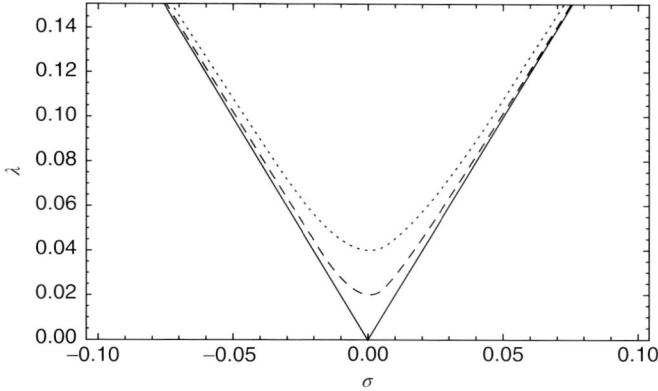

Figure 4.2 Wedge of instability associated with the principal parametric resonance. The solid line depicted here corresponds to $\zeta = 0$, the dashed line to $\zeta = 0.005$, and the dotted line to $\zeta = 0.01$. Inside of this wedge, the solution is unstable, yielding an unbounded response in the absence of non-linearities. Outside of this wedge, the trivial solution is stable. Note that linear damping does not bound the response, but rather changes the shape of the wedge, as evident from the figure.

excitation level necessary for parametric resonance ($\lambda = 4\zeta$ at $\sigma = 0$). Accordingly, nonlinearities arising from dissipative, transduction, or elastic mechanisms must be included to recover a finite, non-trivial response. These nonlinearities are discussed in the subsequent section.

4.6
Sources of Nonlinearity and Nonlinear Steady-State Behaviors

As in directly excited micro/nanoresonators, nonlinearities can arise in parametrically excited systems through a variety of distinct physical mechanisms. Arguably the most common nonlinearities in such systems are those resulting from large mechanical deflections. As at the macroscale, beam-like structures that undergo near-resonant vibrations are prone to both geometric and inertial nonlinearities. Assuming symmetric vibration about the static equilibrium (rest state) of the device, the former leads to the Duffing-like nonlinearity included in Eq. (4.7):

$$F_{geo} = k_3 x^3 \tag{4.18}$$

Here, generally speaking, $k_3 > 0$, which results in a positive value of χ and, as subsequently described, a so-called *hardening* frequency response characteristic. In contrast, inertial nonlinearities typically lead to a force model of the form:

$$F_{inert} = \alpha \left(x\dot{x}^2 + x^2\ddot{x} \right) \tag{4.19}$$

where α is a strong function of the geometry of the system. Though clearly distinct from the geometric nonlinearity as presented here, inertial nonlinearities lead to a similar averaged equation structure to that presented in Eqs. (4.15) and (4.16) and rarely lead to qualitatively distinct behaviors [54].

Another common source of nonlinearities in small-scale, parametrically excited systems are the transduction mechanisms used for actuation. Examples of this include the variable-gap electrostatic actuator and noninterdigitated comb drive actuators introduced above. As highlighted by Eqs. (4.3) and (4.4), these actuators render both time-invariant and time-varying nonlinearities, the latter of which can alter not only the quantitative nature of the system's response, but its qualitative nature as well [42, 50, 51].

Though much more difficult to model accurately from first principles, nonlinearities attributable to material effects (e.g., in piezoelectrically excited systems), viscous damping, and material damping (e.g., in viscoelastic systems) have also been shown to play a role in parametrically excited systems [55, 56].

The scope of nonlinearities considered in this work has been constrained, for the sake of brevity, to include only those associated with geometric effects, as represented in Eq. (4.9) by the term $k_3 x^3$. The impact of this term on the representative parametrically excited system considered herein, can be determined by analyzing the steady-state behaviors predicted by Eqs. (4.15) and (4.16) when $\chi \neq 0$. Doing so reveals that in addition to the trivial solution considered in Section 4.5, the system has two additional steady-state solutions of the form:

$$a_2 = \sqrt{\frac{4\sigma + 2\sqrt{\lambda^2 - 16\zeta^2}}{3\chi}}, \qquad \psi_2 = \frac{1}{2}\arctan\left(\frac{-4\zeta}{\sqrt{\lambda^2 - 16\zeta^2}}\right) \qquad (4.20)$$

$$a_3 = \sqrt{\frac{4\sigma - 2\sqrt{\lambda^2 - 16\zeta^2}}{3\chi}}, \qquad \psi_3 = \frac{1}{2}\arctan\left(\frac{4\zeta}{\sqrt{\lambda^2 - 16\zeta^2}}\right) \qquad (4.21)$$

Of particular note here is that these solutions exist only over specific ranges of detuning (excitation frequency) *and* when the conditions for parametric resonance delineated above have been satisfied. As with the trivial solution, the stability of each of these solutions can be determined by linearizing the system's averaged equations about the steady-state and subsequently examining the eigenvalues associated with the Jacobian of the linearized system.

Figure 4.3 highlights the nonlinear frequency response of a representative system for (a) $\chi > 0$ and (b) $\chi < 0$, respectively. Note that when $\chi > 0$, the system exhibits a so-called *hardening* response, wherein the system undergoes pitchfork bifurcations at $\sigma = \sigma_2$ and $\sigma = \sigma_1$, leading to the creation of a stable and an unstable nontrivial response branch, respectively. This leads to bistability over a wide range of excitation frequencies (i.e., for $\sigma > \sigma_2$), which, in turn, renders a strong dependence on initial conditions. When $\chi < 0$, an effectively mirrored response is recovered – the so-called *softening* response. As with the hardening case, this response also exhibits bistability over a wide range of excitation frequencies (i.e., for $\sigma < \sigma_1$). Softening or hardening nonlinearities akin to this can be introduced by applying a DC voltage across appropriately shaped electrostatic actuators as discussed in Section 4.2.

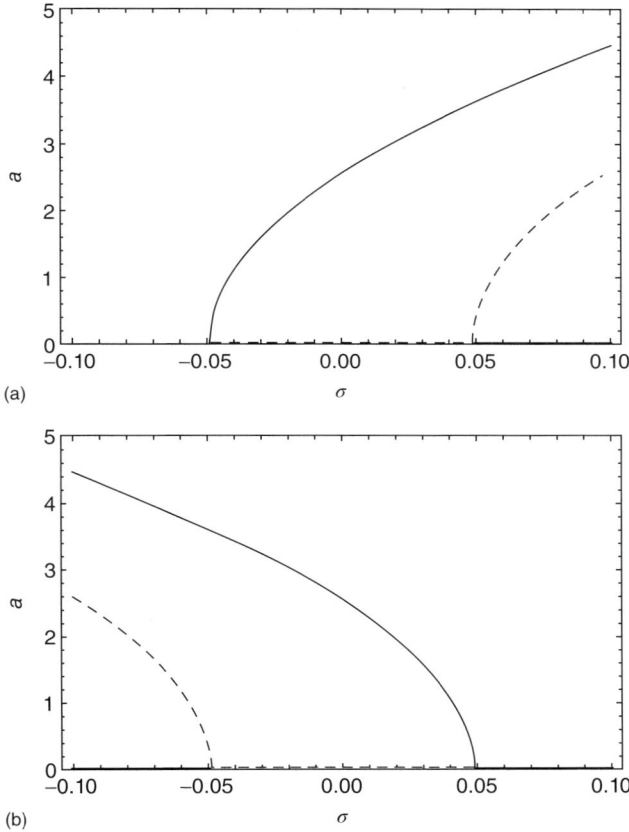

Figure 4.3 The nonlinear frequency response curves associated with a representative parametrically excited system when (a) $\chi = 0.01$ and (b) $\chi = -0.01$. Here, $\zeta = 0.005$ and $\lambda = 0.1$, which renders $\sigma_1 = 0.049$ and $\sigma_2 = -0.049$. Note that in this figure solid lines are used to designate stable steady-state solutions, and dashed lines are used to represent unstable steady-state solutions. In addition, note that in the absence of a nonlinear dissipation mechanism or higher-order nonlinearities, the non-trivial stable and unstable response branches do not intersect for finite values of detuning (frequency).

The nonlinear frequency responses highlighted here differ from those associated with directly excited micro/nanosystems in a few key ways: First, the non-resonant response of the parametrically excited systems is truly zero (or, more accurately, at the noise floor) and the system transitions rapidly to a non-zero state once the conditions for parametric resonance have been satisfied. In addition, the bandwidth of the resonant response is wide and highly controllable through variations in the parametric excitation amplitude. Finally, the recovered response amplitudes are largely insensitive to damping. Collectively, these behavioral features create significant appeal for device designers, as they are well suited for exploitation in signal processing and resonant mass-sensing applications, among many others. Details related to applied studies of parametrically excited micro/nanosystems can be found in Section 4.9.

4.7
Complex Dynamics in Parametrically Excited Micro/Nanosystems

While the steady-state dynamics explored in the preceding sections are representative of those exhibited by many parametrically excited micro/nanosystems, it is important to note that more complex dynamics have been investigated in prior literature as well. For example, a number of works have considered the impact of parametric nonlinearities on the response of parametrically excited microsystems [30, 42, 50–52, 57]. These works have revealed that in the presence of time-varying nonlinearities, parametrically excited microresonators are capable of not only exhibiting hardening and softening nonlinear frequency response characteristics akin to those described above, but also *mixed* response characteristics, wherein the non-trivial steady-state response branches bend toward or away from one another near resonance. Transitions to chaos are also possible in parametrically excited nonlinear systems, and this was investigated in a MEMS resonator by DeMartini *et al.*, both experimentally and by using Melnikov analysis on a system model [57]. This effort built upon the earlier work of Wang and collaborators [58]. Such behaviors, and the ability to rapidly change between them via selective tuning, have been posited to be useful in a variety of signal processing and resonant mass sensing contexts.

Other advanced studies of parametrically excited micro/nanosystems have considered the influence of transient forms of excitation, such as frequency sweeps and noise, on system dynamics. Requa and Turner, for example, considered the influence of frequency sweep rates on the near-resonant response of parametrically excited microbeams actuated via Lorentz forces [41]. Likewise, Cleland explored the influence of thermomechanical noise on parametrically excited nanoresonators suitable for use in mass-sensing applications [59]. Finally, Chan and Stambaugh have conducted analytical and experimental investigations of noise-induced switching in parametrically excited microsystems [60–62]. Though it is difficult to fully characterize the results of these works in this necessarily constrained format, suffice it to note that the influence of noise and transients in parametrically excited systems is markedly different than that associated with their directly excited counterparts.

Also worthy of note, are the recent research efforts that seek to explore the dynamics of multi-degree-of-freedom parametrically excited micro/nanosystems. Included among these works are the efforts of Vyas, van der Avoort, Strachan, and their respective collaborators that explore the dynamics of parametrically excited microstructures with modal coupling [63–67], and the ever-increasing body of literature that explores the dynamics of coupled arrays of parametrically excited micro- and nanoresonators, as well as their associated collective and emergent behaviors [55, 68–73].

4.8
Combined Parametric and Direct Excitations

Though the emphasis of this chapter, thus far, has been on micro/nanosystems that are driven by purely parametric excitations, it is important to note that since the early efforts of Rugar and Grütter [11], there has been considerable interest in micro/nanosystems driven by combined, direct and parametric, excitations, as well. These excitations are embodied by Case 4 in Section 4.3. The interest in combined, parametric and direct, excitations, largely stems from the distinct potential associated with *parametric amplification* – the process of amplifying (or attenuating) a harmonic signal through the use of a time-varying impedance, or *parametric pump* [74–76] – and associated *noise squeezing* [11, 48]. As noted in [56], over the past 20 years parametric amplification has been utilized to achieve low-noise, phase-sensitive response amplification, in some cases with gains in excess of 65 dB, in a wide variety of systems, including torsional microresonators [49, 77], optically excited micromechanical oscillators [43], microgyroscopes [78–81], MEMS diaphragms [82], micromechanical mixers [83], various micro/nanobeams (including those based on carbon nanotube and silicon nanowire structures) [35, 37, 84–91], and coupled microresonators [92–94].

In addition, to these works, there is also a small body of literature focused on the rich nonlinear dynamical behaviors that can arise in resonant micro/nanosystems in the presence of combined excitations. These include the efforts of Zhang, Nayfeh, and their respective collaborators that investigate the nonlinear response of electrostatically-actuated microbeams [14, 15, 95].

4.9
Select Applications

Given their unique, damping-insensitive, near-resonant response characteristics, parametrically excited micro/nanosystems have been seriously considered for use in practical application since their early development. This section seeks to highlight just a few of the application areas where their impact has been noted over the past decade.

4.9.1
Resonant Mass Sensing

As noted in Sections 4.5 and 4.6, parametrically excited systems are capable of exhibiting extremely sharp transitions (in the amplitude-frequency domain) between coexistent stable states; and this fact makes them well suited for use in resonant mass sensing applications. Traditionally, resonant mass sensors leverage the fact that the natural frequency of a given resonator is strongly dependent on the device's effective mass and stiffness. As such, any changes in resonant frequency that occur in a selectively-functionalized device in the presence of a target analyte can be directly correlated with a chemomechanical process, such as adsorption, and thus mass detection. Complementing this approach, is so-called bifurcation-based sensing, wherein abrupt changes in amplitude, resulting from an abrupt transition across a bifurcation point in the frequency domain, are utilized to indicate the presence of a target analyte [54]. Parametrically excited micro/nanoresonators are well suited for use in both of these contexts, as demonstrated in part by prior work. Zhang *et al.* [29, 96], for example, utilized planar comb-driven microresonators to validate the efficacy of parametrically excited mass sensors. Their work revealed that such sensors compared favorably to microsensors based on purely linear dynamical phenomenon, exhibiting markedly higher operating sensitivities in air environments. This result was subsequently reflected, and in many cases extended, in the works of Yu, Zhang, Requa, Cleland, Prakash, Yie, and their respective collaborators [17, 40, 41, 48, 59, 93, 97 – 99] that leveraged alternate form factors and transduction mechanisms, in conjunction with mathematical arguments, to highlight the relative utility of parametrically excited systems in resonant mass sensing applications.

4.9.2
Inertial Sensing

Parametric excitations have also been utilized to enhance the performance of microscale vibratory rate gyroscopes. As detailed in another chapter of this book, traditional resonant gyroscopes exploit two mechanical modes of vibration (often spatially orthogonal) that are coupled through Coriolis accelerations resulting from external rotations of the device. In the most common implementation, one of these vibratory modes is driven into steady-state oscillation and the Coriolis effect couples the motion of the first mode to the second, whose displacement or velocity, in turn, is sensed to yield a measure of rotational rate [100]. Efficient coupling is realized when the modes are matched in resonant frequency; however, this condition is difficult to achieve in practice with high-quality factor systems due to manufacturing variations and the large voltages required for electrostatic compensation. Oropeza-Ramos, et al. demonstrated that the use of properly designed nonlinear parametric resonance in the drive mode gives rise to a much broader frequency response (as compared to a directly-excited gyroscope) that can be overlapped with a linear sensing mode [101 – 103]. Their gyroscope featured a harmonic resonance at $\approx$7800 Hz and a parametrically driven bandwidth of 1 kHz, corresponding to approximately a 13% range in which to overlap the sense mode with no required tuning. The mode overlap range is far higher than what can be achieved with practical electrostatic spring constant tuning. The parametric drive also provided substantial amplification of the effective quality factor (Q) associated with the drive mode along with full Q amplification of the sense mode displacement. Similar results were exploited for practical gain in the work of Krylov and collaborators [104]. Additionally, subsequent analytical studies by Miller *et al.* using perturbation techniques have shown that the nonlinear parametric drive can also be tuned to achieve nearly-linear sense response in the rotation rate [105, 106].

Combined parametric and direct excitations (i.e., parametric amplification) have also been utilized in prior work to improve the performance characteristics of microscale inertial sensors. For example, Gallacher and collaborators utilized this effect in ring gyroscopes seeking to lower the required harmonic drive voltage and thus lower capacitive parasitic feedthrough to the sense mode of the gyroscope [79, 80]. In a subsequent experimental study, a parametric drive gain of 38 dB was achieved while maintaining a linear frequency response characteristic [81]. In contrast to application of parametric amplification in the drive mode, Sharma *et al.* have explored the use of parametric amplification in the Coriolis-induced sense signal [107]. They reported a parametric gain of 23 arising from an AC pump voltage of ≈ 17 V on the parallel-plate sense combs with a corresponding factor of 2.25 reduction in equivalent noise input angular rate. Subsequently, the same group has employed sloped-shaped comb fingers for large-stroke sense-mode parametric amplification [108].

4.9.3
Micromirror Actuation

Resonant scanning micromirrors are of use in a diverse array of applications that includes handheld projection displays, barcode scanners, laser printers, spectroscopy, and endoscopic optical coherence tomography. For practical reasons associated with steering external optical beams through angles exceeding $\pm 10°$, these applications generally dictate that the mirror be torsionally scanned along one or two axes in the plane of the substrate, and thus they require out-of-plane (i.e., vertical) actuation. Many MEMS processes with underlying parallel-plate electrodes or with vertically offset comb fingers add process complexity. Urey *et al.* utilized parametric excitation to drive mirrors with vertically aligned interdigitated comb fingers made in a relatively simple single-layer SOI-MEMS process [31]. In subsequent work from their group, parametric excitation of a 1 mm × 1.5 mm oblong scanning mirror resulted in a 76° total optical scan angle near a 21.8 kHz resonance with a 98 V AC excitation amplitude [32]. While comb drives generally have less damping in air than parallel-plate drives, this mirror design took one step further in lowering damping by coupling a comb-driven outer frame with an inner mirror having no combs. The outer frame moved at a lower speed than the mirror so the damping effect of the comb drive was reduced when compared to a directly excited mirror comb drive. In related work, Baskaran and Turner, demonstrated both degenerate and nondegenerate parametric amplification in a vertical comb driven, dual-torsional-mode resonator [49]. The nondegenerate parametric excitation was driven at the sum of the two modal frequencies, which decoupled feedthrough from the drive and output. More recent work on parametrically excited micromirrors can be found in [109].

4.9.4
Bifurcation Control

To explore the utility of parametric resonance in the design of practical devices, such as ultrasensitive mass sensors or sensitive strain gauges, a drive scheme is needed to excite parametric resonance and reveal the system's underlying bifurcation behavior. To this end, in early research, an open-loop frequency sweep was utilized to locate, in an approximate manner, the bifurcation points associated with various parametrically excited devices [28, 51]. This approach was used in sensing contexts, for example, to establish a link between various bifurcation frequencies and added mass resulting from deposition [97] and chemical absorption [110]. Unfortunately, this open-loop method proved to be comparatively slow due to the slow sweep rates needed for proper bifurcation frequency estimation.

An operating point servo at, or near, a bifurcation point can be designed to facilitate control right at a bifurcation "jump" event, allowing for, among other things, a more accurate identification of a given bifurcation frequency. However, this proves difficult as systems exhibiting nonlinear parametric resonance commonly exhibit latch-up instability and hysteretic behavior. Moreover, these underdamped systems have relatively long ring-down and build-up times. Collectively, these issues have rendered attempts at classical analog control near the bifurcation points largely unsuccessful.

Burgner *et al.* implemented a successful bifurcation-based control scheme by observing the coherency of the signal phase and appropriately changing the parametric excitation [111]. In this work, the resonant amplitude was kept very small (140 pm) in order to detect the onset of phase coherency. This small servo amplitude at the edge of phase coherency avoided the slow-time bifurcation jump event capable of causing latch-up instability.

In contrast, Guo and Fedder, developed an amplitude-state controller that operated at any instant in one of two parametric drive states corresponding to steady-state "on" and "off" operation while driving at a fixed parametric drive frequency [112, 113]. The system rapidly switched between the unstable region within the $\lambda-\sigma$ instability wedge and the stable region outside of the wedge by modulating λ via a DC voltage that modified the b_1 parameter in Eq. (4.9). Pulse-width modulation of the feedback state at 200 kHz avoided the slow-time latch-up that occurred in steady-state operation and facilitated relatively high displacement amplitudes of around 2 μm, with a noise floor of 2Å in the displacement amplitude for a bandwidth of 0.2 Hz.

4.10
Some Parting Thoughts

Near-zero, nonresonant response amplitudes, multi-valued steady-state responses, sharp amplitude transitions near bifurcations, widely tunable bandwidths, and noise squeezing are just some of the interesting dynamical phenomena that arise in parametrically excited systems. Exploration of these effects in the context of MEMS and NEMS has been ongoing since the early 1990s, leaving a blazed trail that may eventually lead toward the purposeful incorporation of parametric excitations, and/or associated nonlinearities, in commercial devices. In particular, several exciting opportunities to improve sensor performance exist, including using parametric excitations to obtain improved sensitivities in air compared to harmonic excitation, employing advanced control techniques to facilitate more rapid analyte detection, using parametric effects to couple various modes and lower feedthrough in multi-degree-of-freedom systems, and leveraging parametric amplification to improve signal to noise ratios.

First-principles modeling and analytical perturbation theory are requisite for characterizing the (often nonlinear) behavior of parametrically excited micro/nanosystems and provide a foundation for future progress. Fortunately, a critical mass of this fundamental knowledge, along with design methodologies and concepts tailored for microsystems, are widely available today, enabling a wide set of MEMS and NEMS researchers and product developers to take advantage of parametric excitations.

Acknowledgment

Effort by G. K. Fedder was financially supported in part by the U.S. National Science Foundation grant CNS 0941497 as part of the Cyber-enabled Discovery and Innovation (CDI) program.

References

1. Faraday, M. (1831) On a peculiar class of acoustical figures and on certain forms assumed by a group of particles upon vibrating elastic surfaces. *Philos. Trans. R. Soc. Lond.*, **121**, 299–340.

2. Melde, F. (1860) Ueber die erregung stehender wellen eines fadenformigen korpers. *Ann. Phys.*, **187** (12), 513–537.

3. Mathieu, E. (1868) Memoire sur le mouvement vibratoire d'une membrane de forme elliptique. *J. Math. Pures Appl.*, **13**, 137–203.

4. Rayleigh, L. (1883) On maintained vibrations. *The London, Edinburgh, and Dublin Philos. Mag. J. Sci. (Fifth Series)*, **15** (94), 229–235.

5. Rayleigh, L. (1887) On the mantenance of vibrations by forces of double frequency, and on the propagation of waves through a medium endowed witha periodic structure. *The London, Edinburgh, and Dublin Philos. Mag. J. Sci. (Fifth Series)*, **24** (147), 145–159.

6. Stoker, J.J. (1950) *Nonlinear Vibrations in Mechanical and Electrical Systems*, John Wiley & Sons, Inc., New York.

7. Campbell, R. (1955) *Theorie Generale De L'Equation De Mathieu*, Masson et Cie, Paris.

8. Nayfeh, A.H. and Mook, D.T. (1979) *Nonlinear Oscillations*, John Wiley and Sons, New York.

9. Cartmell, M. (1990) *Introduction to Linear, Parametric, and Nonlinear Vibrations*, Chapman and Hall, London.

10. Turner, K.L., Miller, S.A., Hartwell, P.G., MacDonald, N.C., Strogatz, S.H., and Adams, S.G. (1998) Five parametric resonances in a microelectromechanical system. *Nature*, **396** (6707), 149–152.

11. Rugar, D. and Grutter, P. (1991) Mechanical parametric amplification and thermomechanical noise squeezing. *Phys. Rev. Lett.*, **67** (6), 699–702.

12. Napoli, M., Baskaran, R., Turner, K., and Bamieh, B. (2003) Understanding mechanical domain parametric resonance in microcantilevers. *Proceedings of the IEEE 16th Annual International Conference on Micro Electro Mechanical Systems*, pp. 169–172.

13. Napoli, M., Olroyd, C., Bamieh, B., and Turner, K. (2005) A novel sensing scheme for the displacement of electrostatically actuated microcantilevers. *Proceedings of the 2005 American Control Conference*, pp 2475–2480.

14. Zhang, W. and Meng, G. (2005) Nonlinear dynamical system of microcantilever under combined parametric and forcing excitations in MEMS. *Sens. Actuators, A*, **119** (2), 291–299.

15. Zhang, W.M. and Meng, G. (2007) Nonlinear dynamic analysis of electrostatically actuated resonant MEMS sensors under parametric excitation. *IEEE Sens. J.*, **7** (3), 370–380.

16. Krylov, S., Harari, I., and Cohen, Y. (2005) Stabilization of electrostatically actuated microstructures using parametric excitation. *J. Micromech. Microeng.*, **15** (6), 1188–1204.

17. Yu, M.F., Wagner, G.J., Ruoff, R.S., and Dyer, M.J. (2002) Realization of parametric resonances in a nanowire mechanical system with nanomanipulation inside a scanning electron microscope. *Phys. Rev. B*, **66** (7), 073406.

18. Ahmad, A. and Tripathi, V.K. (2005) Parametric excitation of higher-order electromechanical vibrations of carbon nanotubes. *Phys. Rev. B*, **72** (19), 193409.

19. Liu, C.S. and Tripathi, V.K. (2004) Observational consequences of parametrically driven vibrations of carbon nanotubes. *Phys. Rev. B*, **70** (15), 115414.

20. Adams, S.G., Bertsch, F., and MacDonald, N.C. (1998) Independent tuning of linear and nonlinear stiffness coefficients. *J. Microelectromech. Syst.*, **7** (2), 172–180.

21. Adams, S.G., Bertsch, F., Shaw, K.A., Hartwell, P.G., Moon, F.C., and MacDonald, N.C. (1998) Capacitance based tunable resonators. *J. Microelectromech. Syst.*, **8** (1), 15–23.

22. Guo, C. and Fedder, G.K. (2013) Behavioral modeling of a CMOS-MEMS nonlinear parametric resonator. *J. Microelectromech. Syst.*, **22** (6), 1447–1457.

23. Guo, C., Tatar, E., and Fedder, G.K. (2013) Large-displacement parametric resonance using a shaped comb drive. *Proceedings of the IEEE 26th International Conference on Micro Electro Mechanical Systems*, pp. 173–176.

24. Guo, C. and Fedder, G.K. (2012) Behavioral modeling and testing of a CMOS-MEMS parametric resonator governed by the nonlinear Mathieu equation. *Proceedings of the IEEE 25th International Conference on Micro Electro Mechanical Systems*, pp. 535–538.

25. Linzon, Y., Ilic, B., Lulinsky, S., and Krylov, S. (2013) Efficient parametric excitation of silicon-on-insulator microcantilever beams by fringing electrostatic fields. *J. Appl. Phys.*, **113** (16), 163508.

26. Guo, C. (2013) *Bi-state control of microelectromechanical nonlinear and parametric resonance*. PhD thesis, Carnegie Mellon University.

27. Guo, C. and Fedder, G.K. (2013) A quadratic-shape-finger comb parametric resonator. *J. Micromech. Microeng.*, **23** (9), 095007.

28. Zhang, W., Baskaran, R., and Turner, K.L. (2003) Tuning the dynamic behavior of parametric resonance in a micromechanical oscillator. *Appl. Phys. Lett.*, **82** (1), 130–132.

29. Zhang, W., Baskaran, R., and Turner, K.L. (2002) Effect of cubic nonlinearity on auto-parametrically amplified resonant MEMS mass sensor. *Sens. Actuators, A*, **102** (1-2), 139–150.

30. Rhoads, J.F., Shaw, S.W., Turner, K.L., and Baskaran, R. (2005) Tunable microelectromechanical filters that exploit parametric resonance. *J. Vib. Acoust.*, **127** (5), 423–430.

31. Ataman, C. and Urey, H. (2006) Modeling and characterization of comb-actuated resonant microscanners. *J. Micromech. Microeng.*, **16** (1), 9–16.

32. Arslan, A., Brown, D., Davis, W.O., Holmstrom, S., Gokce, S.K., and Urey, H. (2010) Comb-actuated resonant torsional microscanner with mechanical amplification. *J. Microelectromech. Syst.*, **19** (4), 936–943.

33. Jensen, B.D., Mutlu, S., Miller, S., Kurabayashi, K., and Allen, J.J. (2003) Shaped comb fingers for tailored electromechanical restoring force. *J. Microelectromech. Syst.*, **12** (3), 373–383.

34. Hirano, T., Furuhata, T., Gabriel, K.J., and Fujita, H. (1992) Design, fabrication, and operation of submicron gap comb-drive microactuators. *J. Microelectromech. Syst.*, **1** (1), 52–59.

35. Mahboob, I. and Yamaguchi, H. (2008) Piezoelectrically pumped parametric amplification and Q enhancement in an electromechanical oscillator. *Appl. Phys. Lett.*, **92** (17), 173109.

36. Mahboob, I. and Yamaguchi, H. (2008) Bit storage and bit flip operations in an electromechanical oscillator. *Nat. Nanotechnol.*, **3** (5), 275–279.

37. Thomas, O., Mathieu, F., Mansfield, W., Huang, C., Trolier-McKinstry, S., and Nicu, L. (2013) Efficient parametric amplification in micro-resonators with integrated piezoelectric actuation and sensing capabilities. *Appl. Phys. Lett.*, **102** (16), 163504.

38. Krylov, S., Gerson, Y., Nachmias, T., and Keren, U. (2010) Excitation of large-amplitude parametric resonance by the mechanical stiffness modulation of a microstructure. *J. Micromech. Microeng.*, **20** (1), 015041.

39. Kaajakari, V. and Lal, A. (2004) Parametric excitation of circular micromachined polycrystalline silicon disks. *Appl. Phys. Lett.*, **85** (17), 3923–3925.

40. Requa, M.V. and Turner, K.L. (2006) Electromechanically driven and sensed parametric resonance in silicon microcantilevers. *Appl. Phys. Lett.*, **88** (26), 263508.

41. Requa, M.V. and Turner, K.L. (2007) Precise frequency estimation in a microelectromechanical parametric resonator. *Appl. Phys. Lett.*, **90** (17), 173508.

42. Rhoads, J.F., Kumar, V., Shaw, S.W., and Turner, K.L. (2013) The non-linear dynamics of electromagnetically actuated microbeam resonators with purely parametric excitations. *Int. J. Non Linear Mech.*, **55**, 79–89.

43. Zalalutdinov, M., Olkhovets, A., Zehnder, A., Ilic, B., Czaplewski, D., Craighead, H.G., and Parpia, J.M. (2001) Optically pumped parametric amplification for micromechanical oscillators. *Appl. Phys. Lett.*, **78** (20), 3142–3144.

44. Sibgatullin, T., Schreiber, D., and Krylov, S. (2013) Excitation of parametric resonance in micro beams by Joule's heating. *Proceedings of COMDYN 2013: The 4th ECCOMAS Thematic Conference on Computational Methods in Structural Dynamics and Earthquake Engineering*.

45. Moreno-Moreno, M., Raman, A., Gomez-Herrero, J., and Reifenberger, R. (2006) Parametric resonance based scanning probe microscopy. *Appl. Phys. Lett.*, **88** (19), 193108.

46. Prakash, G., Hu, S., Raman, A., and Reifenberger, R. (2009) Theoretical basis of parametric-resonance-based atomic force microscopy. *Phys. Rev. E*, **79** (9), 094304.

47. Villanueva, L.G., Karabalin, R.B., Matheny, M.H., Kenig, E., Cross, M.C., and Roukes, M.L. (2011) A nanoscale parametric feedback oscillator. *Nano Lett.*, **11** (11), 5054–5059.

48. Prakash, G., Raman, A., Rhoads, J., and Reifenberger, R.G. (2012) Parametric noise squeezing and parametric resonance of microcantilevers in air and liquid environments. *Rev. Sci. Instrum.*, **83** (6), 065109.

49. Baskaran, R. and Turner, K.L. (2003) Mechanical domain coupled mode parametric resonance and amplification in a torsional mode micro electro mechanical oscillator. *J. Micromech. Microeng.*, **13** (5), 701–707.

50. Rhoads, J.F., Shaw, S.W., and Turner, K.L. (2006) The nonlinear response of resonant microbeam systems with purely-parametric electrostatic actuation. *J. Micromech. Microeng.*, **16** (5), 890–899.

51. Rhoads, J.F., Shaw, S.W., Turner, K.L., Moehlis, J., DeMartini, B.E., and Zhang, W. (2006) Generalized parametric resonance in electrostatically actuated microelectromechanical oscillators. *J. Sound Vib.*, **296** (4-5), 797–829.

52. DeMartini, B.E., Rhoads, J.F., Turner, K.L., Shaw, S.W., and Moehlis, J. (2007) Linear and nonlinear tuning of parametrically excited MEMS oscillators. *J. Microelectromech. Syst.*, **16** (2), 310–318.

53. Sanders, J.A. and Verhulst, F. (1985) *Averaging Methods in Nonlinear Dynamical Systems, Applied Mathematical Sciences*, vol. 59, Springer-Verlag, New York.

54. Kumar, V., Yang, Y., Boley, J.W., Chiu, G.T.-C., and Rhoads, J.F. (2012) Modeling, analysis, and experimental validation of a bifurcation-based microsensor. *J. Microelectromech. Syst.*, **21** (3), 549–558.

55. Lifshitz, R. Cross, M.C. (2008) Nonlinear dynamics of nanomechanical and micromechanical resonators, *Reviews of Nonlinear Dynamics and Complexity*, Vol. 1, Schuster, H.G. (ed.), Wiley-VCH, Weinheim, Germany, pp. 1–52.

56. Rhoads, J.F., Shaw, S.W., and Turner, K.L. (2010) Nonlinear dynamics and its applications in micro- and nanoresonators. *J. Dyn. Syst. Meas. Control*, **132** (3), 034001.

57. DeMartini, B.E., Butterfield, H.E., Moehlis, J., and Turner, K.L. (2007) Chaos for a microelectromechanical oscillator governed by the nonlinear Mathieu equation. *J. Microelectromech. Syst.*, **16** (6), 1314–1323.

58. Wang, Y.C., Adams, S.G., Thorp, J.S., MacDonald, N.C., Hartwell, P., and Bertsch, F. (1998) Chaos in MEMS, parameter estimation and its potential application. *IEEE Trans. Circuits Syst. I: Fundam. Theory Appl.*, **45** (10), 1013–1020.

59. Cleland, A.N. (2005) Thermomechanical noise limits on parametric sensing with nanomechanical resonators. *New J. Phys.*, **7** (235), 1–16.

60. Chan, H.B. and Stambaugh, C. (2007) Activation barrier scaling and crossover for noise-induced switching in a micromechanical parametric oscillator. *Phys. Rev. Lett.*, **99** (6), 060601.

61. Chan, H.B., Dykman, M.I., and Stambaugh, C. (2008) Paths of fluctuation induced switching. *Phys. Rev. Lett.*, **100** (13), 130602.

62. Chan, H.B. and Stambaugh, C. (2009) Activated switching in a parametrically driven micromechanical torsional oscillator, *Applications of Nonlinear Dynamics*, Understanding Complex Systems, In, V. *et al.* (eds.), Spring-Verlag Berlin Heidelberg, pp. 15–23.

63. Vyas, A., Peroulis, D., and Bajaj, A.K. (2008) Dynamics of a nonlinear microresonator based on resonantly interacting flexural-torsional modes. *Nonlinear Dyn.*, **54** (1-2), 31–52.

64. Vyas, A., Peroulis, D., and Bajaj, A.K. (2009) A microresonator design based on nonlinear 1:2 internal resonance in flexural structural modes. *J. Microelectromech. Syst.*, **18** (3), 744–762.

65. van der Avoort, C., van der Hout, R., Bontemps, J.J.M., Steeneken, P.G., Le Phan, K., Fey, R.H.B., Hulshof, J., and van Beek, J.T.M. (2010) Amplitude saturation of MEMS resonators explained by autoparametric resonance. *J. Micromech. Microeng.*, **20** (10), 105012.

66. van der Avoort, C., van der Hout, R., and Hulshof, J. (2011) Parametric resonance and Hopf bifurcation analysis

for a MEMS resonator. *Physica D*, **240** (11), 913–919.

67. Strachan, B.S., Shaw, S.W., and Kogan, O. (2013) Subharmonic resonance cascades in a class of coupled resonators. *J. Comput. Nonlinear Dyn.*, **8** (4), 041015.

68. Napoli, M., Zhang, W., Turner, K., and Bamieh, B. (2005) Characterization of electrostatically coupled microcantilevers. *J. Microelectromech. Syst.*, **14** (2), 295–304.

69. Bromberg, Y., Cross, M.C., and Lifshitz, R. (2006) Response of discrete nonlinear systems with many degrees of freedom. *Phys. Rev. E*, **73** (1), 016214.

70. Lifshitz, R. and Cross, M.C. (2003) Response of parametrically driven nonlinear coupled oscillators with application to micromechanical and nanomechanical resonator arrays. *Phys. Rev. B*, **67** (13), 134302.

71. Buks, E. and Roukes, M.L. (2002) Electrically tunable collective response in a coupled micromechanical array. *J. Microelectromech. Syst.*, **11** (6), 802–807.

72. Zhu, J., Ru, C.Q., and Mioduchowski, A. (2007) High-order subharmonic parametric resonance of nonlinearly coupled micromechanical oscillators. *Eur. Phys. J. B*, **58** (4), 411–421.

73. Gutschmidt, S. and Gottlieb, O. (2012) Nonlinear dynamic behavior of a microbeam array subject to parametric actuation at low, medium, and large DC-voltages. *Nonlinear Dyn.*, **67** (1), 1–36.

74. Howson, D.P. and Smith, R.B. (1970) *Parametric Amplifiers, European Electrical and Electron Engineering Series*, McGraw-Hill, London.

75. Louisell, W.H. (1960) *Coupled Mode and Parametric Electronics*, John Wiley & Sons, Inc., New York.

76. Mumford, W.W. (1960) Some notes on the history of parametric transducers. *Proc. IRE*, **48** (5), 848–853.

77. Carr, D.W., Evoy, S., Sekaric, L., Craighead, H.G., and Parpia, J.M. (2000) Parametric amplification in a torsional microresonator. *Appl. Phys. Lett.*, **77** (10), 1545–1547.

78. Gallacher, B.J., Burdess, J.S., Harris, A.J., and Harish, K.M. (2005) Active damping control in MEMS using parametric pumping. *Proceedings of Nanotech 2005: The 2005 NSTI Nanotechnology Conference and Trade Show*, vol. 7, pp. 383–386.

79. Gallacher, B.J., Burdess, J.S., and Harish, K.M. (2006) A control scheme for a MEMS electrostatic resonant gyroscope excited using combined parametric excitation and harmonic forcing. *J. Micromech. Microeng.*, **16** (2), 320–331.

80. Harish, K.M., Gallacher, B.J., Burdess, J.S., and Neasham, J.A. (2009) Experimental investigation of parametric and externally forced motion in resonant MEMS sensors. *J. Micromech. Microeng.*, **19** (1), 015021.

81. Hu, Z., Gallacher, B.J., Harish, K.M., and Burdess, J.S. (2010) An experimental study of high gain parametric amplification in MEMS. *Sens. Actuators, A*, **162** (2), 145–154.

82. Raskin, J.P., Brown, A.R., Khuri-Yakub, B.T., and Rebeiz, G.M. (2000) A novel parametric-effect MEMS amplifier. *J. Microelectromech. Syst.*, **9** (4), 528–537.

83. Koskenvuori, M. and Tittonen, I. (2008) Parametrically amplified microelectromechanical mixer. *Proceedings of the IEEE 21st International Conference on Micro Electro Mechanical Systems*, pp. 1044–1047.

84. Dana, A., Ho, F., and Yamamoto, Y. (1998) Mechanical parametric amplification in piezoresistive gallium arsenide microcantilevers. *Appl. Phys. Lett.*, **72** (10), 1152–1154.

85. Roukes, M.L., Ekinci, K.L., Yang, Y.T., Huang, X.M.H., Tang, H.X., Harrington, D.A., Casey, J., and Artlett, J.L. (2004) An apparatus and method for two-dimensional electron gas actuation and transduction for GaAs NEMS, International Patent No. WO/2004/041998.

86. Ono, T., Wakamatsu, H., and Esashi, M. (2005) Parametrically amplified thermal resonant sensor with pseudo-cooling effect. *J. Micromech. Microeng.*, **15** (12), 2282–2288.

87. Mahboob, I. and Yamaguchi, H. (2008) Parametrically pumped ultrahigh Q

electromechanical resonator. *Appl. Phys. Lett.*, **92** (25), 253109.

88. Karabalin, R.B., Masmanidis, S.C., and Roukes, M.L. (2010) Efficient parametric amplification in high and very high frequency piezoelectric nanoelectromechanical systems. *Appl. Phys. Lett.*, **97** (18), 183101.

89. Wu, C.C. and Zhong, Z. (2011) Parametric amplification in single-walled carbon nanotube nanoelectromechanical resonators. *Appl. Phys. Lett.*, **99** (8), 083110.

90. Westra, H.J.R., Karabacak, D.M., Brongersma, S.H., Crego-Calama, M., van der Zant, H.S.J., and Venstra, W.J. (2011) Interactions between directly- and parametrically-driven vibration modes in a micromechanical resonator. *Phys. Rev. B*, **84** (13), 134305.

91. Nichol, J.M., Hemesath, E.R., Lauhon, L.J., and Budakian, R. (2009) Controlling the nonlinearity of silicon nanowire resonators using active feedback. *Appl. Phys. Lett.*, **95** (12), 123116.

92. Olkhovets, A., Carr, D.W., Parpia, J.M., and Craighead, H.G. (2001) Non-degenerate nanomechanical parametric amplifier. *Proceedings of the IEEE 14th International Conference on Micro Electro Mechanical Systems*, pp. 298–300.

93. Yie, Z., Miller, N.J., Shaw, S.W., and Turner, K.L. (2012) Parametric amplification in a resonant sensing array. *J. Micromech. Microeng.*, **22** (3), 035004.

94. Miller, N.J. and Shaw, S.W. (2012) Frequency sweeping with concurrent parametric amplification. *J. Dyn. Syst. Meas. Control*, **134** (2), 021007.

95. Nayfeh, A.H. and Younis, M.I. (2005) Dynamics of MEMS resonators under superharmonic and subharmonic excitations. *J. Micromech. Microeng.*, **15** (10), 1840–1847.

96. Zhang, W., Baskaran, R., and Turner, K.L. (2002) Nonlinear dynamics analysis of a parametrically resonant MEMS sensor. *Proceedings of the 2002 SEM Annual Conference and Exposition on Experimental and Applied Mechanics.*

97. Zhang, W. and Turner, K.L. (2005) Application of parametric resonance amplification in a single-crystal silicon micro-oscillator based mass sensor. *Sens. Actuators, A*, **122** (1), 23–30.

98. Requa, M.V. (2006) Parametric resonance in microcantilevers for applications in mass sensing. PhD thesis, University of California, Santa Barbara, CA.

99. Yie, Z., Turner, K.L., Miller, N.J., and Shaw, S.W. (2010) Sensitivity enhancement using parametric amplification in a resonant sensing array. *Proceedings of Hilton Head 2010: A Solid-State Sensors, Actuators, and Microsystems Workshop*, pp. 483–486.

100. Xie, H. and Fedder, G.K. (2003) Integrated microelectromechanical gyroscopes. *J. Aerosp. Eng.*, **16** (2), 65–75.

101. Oropeza-Ramos, L.A. and Turner, K.L. (2005) Parametric resonance amplification in a memgyroscope. *Proceedings of IEEE Sensors 2005: The Fourth IEEE Conference on Sensors*, pp. 660–663.

102. Oropeza-Ramos, L.A. (2007) Investigations on novel platforms of micro electro mechanical inertial sensors: analysis, construction and experimentation. PhD thesis, University of California, Santa Barbara, CA.

103. Oropeza-Ramos, L.A., Burgner, C.B., and Turner, K.L. (2009) Robust micro-rate sensor actuated by parametric resonance. *Sens. Actuators, A*, **152** (1), 80–87.

104. Krylov, S., Lurie, K., and Ya'akobovitz, A. (2011) Compliant structures with time-varying moment of inertia and non-zero averaged momentum and their application in angular rate microsensors. *J. Sound Vib.*, **330** (20), 4875–4895.

105. Miller, N.J., Shaw, S.W., Oropeza-Ramos, L.A., and Turner, K.L. (2008) A MEMS-based rate gyro based on parametric resonance. *Proceedings of ESDA08: The 9th Biennial ASME Conference on Engineering Systems Design and Analysis.*

106. Miller, N.J., Shaw, S.W., Oropeza-Ramos, L.A., and Turner, K.L. (2008) Analysis of a novel MEMS gyroscope actuated by parametric resonance. *Proceedings of ENOC 2008: The*

6th EUROMECH Nonlinear Dynamics Conference.

107. Sharma, M., Sarraf, E.H., Baskaran, R., and Cretu, E. (2012) Parametric resonance: amplification and damping in MEMS gyroscopes. *Sens. Actuators, A,* **177**, 79–86.

108. Sharma, M., Sarraf, E.H., Baskaran, R., and Cretu, E. (2013) Shaped combs and parametric amplification in inertial MEMS sensors. *Proceedings of IEEE Sensors* 2013.

109. Shahid, W., Qiu, Z., Duan, X., Li, H., Wang, T.D., and Oldham, K.R. (2013) Modeling and simulation of a parametrically-excited micro-mirror with duty-cycled square wave excitation. *Proceedings of IDETC/CIE 2013: The ASME 2013 International Design Engineering Technical Conferences and Computers and Information in Engineering Conference,* DETC2013-13036.

110. Yie, Z., Zielke, M.A., Burgner, C.B., and Turner, K.L. (2011) Comparison of parametric and linear mass dection in the presence of detection noise. *J. Micromech. Microeng.,* **21** (2), 025 027.

111. Burgner, C.B., Snyders, W.S., and Turner, K.L. (2011) Control of MEMS on the edge of instability. *Proceedings of Transducers 2011: The 16th International Solid-State Sensors, Actuators, and Microsystems Conference,* pp. 1990–1993.

112. Guo, C. and Fedder, G.K. (2013) Bi-state control of parametric resonance. *Appl. Phys. Lett.,* **103** (18), 183512.

113. Guo, C. and Fedder, G.K. (2013) Bi-state control of a duffing resonator on the falling edge of instability. *Proceedings of Transducers '13: The 17th International Conference on Solid-State Sensors, Actuators, and Microsystems,* pp. 1703–1706.

5
Finite Element Modeling of Resonators

Reza Abdolvand, Jonathan Gonzales, and Gavin Ho

5.1
Introduction to Finite Element Analysis

Finite element analysis is a numerical method for finding an approximate solution to a set of differential equations. These differential equations define the physics of systems, such as heat flow and structural mechanics, and are constrained by a set of boundary conditions. This method, widely known as the finite element method/finite element modeling or FEM, is commonly used in many engineering disciplines to model the behavior of complex systems. FEM is considered one of the most versatile tools at the design stage of developing new products in order to predict and optimize their final performance. FEM can substantially shorten the design cycle and dramatically cut the research and development cost.

The field of MEMS resonators has greatly benefited from the application of FEM in recent years and the intention in this chapter is to briefly cover some of these applications. In order to better understand the process through which FEM is implemented in the study of resonant systems, we will start by introducing the fundamental mathematics behind the method. Next, we will proceed to explain the general procedures for initializing FEM experiments and we will use simple examples to elaborate the concepts. Finally, we will briefly present several specific instances of using FEM to study and improve MEMS resonators and refer the reader to the literature for more details.

5.1.1
Mathematical Fundamentals

In order to avoid the need for excessive prerequisites, a simplified and conceptual introduction to the basic mathematics behind the FEM is provided. We will use examples to describe essential steps in solving a problem using FEM techniques and explain some of the intricacies involved in the process.

Resonant MEMS – Fundamentals, Implementation and Application, First Edition.
Edited by Oliver Brand, Isabelle Dufour, Stephen M. Heinrich and Fabien Josse.
© 2015 Wiley-VCH Verlag GmbH & Co. KGaA. Published 2015 by Wiley-VCH Verlag GmbH & Co. KGaA.

5.1.1.1 Static Problems

Let's start by considering the relatively simple problem of calculating the displacement in an axially loaded solid bar of initial length L (Figure 5.1).

The bar is fixed at one end (fixed boundary), while a force F is applied to the other end normal to the bar cross section. The general differential equation governing the displacement at any distance x along the bar is [1]

$$EA(x)\frac{du(x)}{dx} = F \tag{5.1}$$

where E is the Young's modulus of the solid material, $A(x)$ is the cross-sectional area of the bar at any distance x from the fixed end, and $u(x)$ is the displacement function. Therefore, for a bar with a constant cross section, the differential equation will reduce to

$$EA\frac{du(x)}{dx} = F \tag{5.2}$$

The general solution to this differential equation is a first-order polynomial given as

$$u(x) = a + bx \tag{5.3}$$

The unknown values of a and b are calculated by satisfying the differential equation and the boundary condition (i.e., $u(0) = 0$), resulting in

$$u(x) = \frac{F}{EA}x \tag{5.4}$$

Therefore, the displacement at the end of the bar is

$$u(L) = \frac{F}{EA}L \tag{5.5}$$

From this equation, one can easily extract an equivalent spring constant (i.e., stiffness) for the bar:

$$K = F/u(L) \rightarrow K = \frac{EA}{L} \tag{5.6}$$

Despite the existence of a simple solution for the above problem, it is clear that for an arbitrarily shaped bar, where $A(x)$ is not uniform, the differential equation could quickly become very complicated to solve. This is exactly the motivation for seeking alternative numerical solutions by means of the FEM. To realize this point,

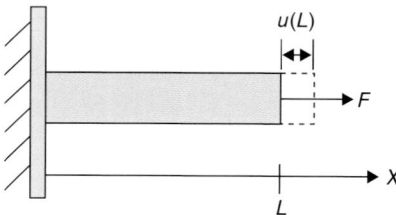

Figure 5.1 An axially loaded solid bar.

let's add one level of complexity and assume that the bar has a linearly tapered shape (see Figure 5.2).

Therefore $A(x)$ is now represented as a linear function of x:

$$A(x) = A_0 + A_1 x/L \tag{5.7}$$

As a result, the new differential equation is

$$E(A_0 + A_1 x/L)\frac{du(x)}{dx} = F \tag{5.8}$$

Although this new differential equation has a closed-from solution, one could approach this problem differently by assuming that the displacement of the tapered bar can be approximated with the displacement in a structure composed of two bars with constant cross sections joined in the middle as shown in Figure 5.3.

The advantage of this approach is that the displacement for the middle and the end point of this structure can be simply calculated using the earlier analysis. The only consideration is that the displacement at the end point should be calculated as the combined individual elongations for the first element and the second element (i.e., the continuity of the displacement at the junction of the two bars):

$$u\left(\frac{L}{2}\right) = \frac{F}{K_1} \tag{5.9}$$

$$u(L) = \frac{F}{K_2} + u\left(\frac{L}{2}\right) \tag{5.10}$$

where K_1 and K_2 are the stiffness coefficients of elements one and two, respectively. The approach described here is the essence of FEM and the overall process could be summarized as follows:

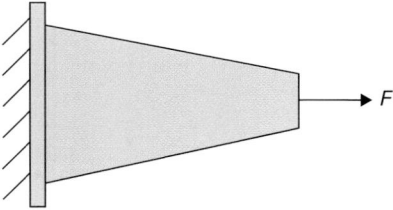

Figure 5.2   An axially loaded linearly tapered bar.

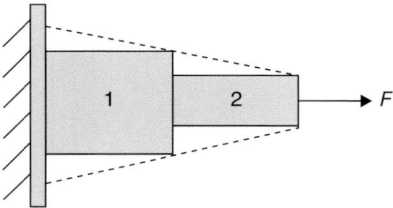

Figure 5.3   An approximate representation of the tapered bar consisting of two connected bars each with uniform cross sections.

1) The domain on which the differential equation is defined will be discretized into a "finite" number of smaller connected domains known as "elements"; hence the name "FEM." The ends of each of these elements are known as nodes.
2) The parameter of interest (e.g., displacement) at all nodes will be calculated using a set of linear equations that include the boundary conditions (e.g., zero displacement at one end of the bar for the above example) and external inputs (e.g., the applied force).
3) The parameter of interest inside the domain of each element is approximated using a polynomial function that is based on assumed "shape functions." Linear and quadratic polynomials are the most commonly used shape functions.

It is of practical value to identify some of the sources of error in solving a problem using the FEM. From Figure 5.3, it is clear that the first source of error arises from the loss of information for some portions of the original structure due to incomplete coverage of the discrete elements with predefined shapes. The second source of error is rooted in using shape functions to approximate the values of the quantity of interest at locations between the nodes. These errors can be reduced by increasing the number of elements (Figure 5.4).

The reduction of error when using a larger number of elements (a "refined mesh") is intuitively comprehensible since by choosing infinitely small elements one could imagine that the finite element solution will approach the original partial differential equation. However, the use of more elements demands more computational resources. Therefore, a suitable number of elements is practically chosen to balance the computational capacity with acceptable levels of error.

5.1.1.2 Dynamic Problems (Modal Analysis)

The process of identifying the natural resonance modes of an elastic body using FEM is often called modal analysis. The differential equation governing the dynamic response of the lumped mass-spring-damper system of Figure 5.5 is formulated as (Newton's laws of motion)

$$m\ddot{u}(t) + c\dot{u}(t) + ku(t) = f(t) \qquad (5.11)$$

where m, k, and c are the lumped mass, spring constant, and damping coefficient, respectively and $u(t)$, $\dot{u}(t)$, and $\ddot{u}(t)$ correspond to displacement, velocity, and acceleration of the lumped mass.

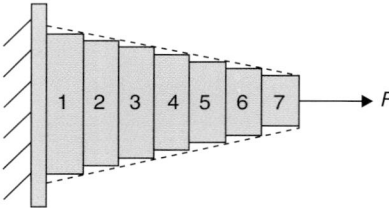

Figure 5.4 The tapered bar of Figure 5.2 approximated with a larger number of discrete elements.

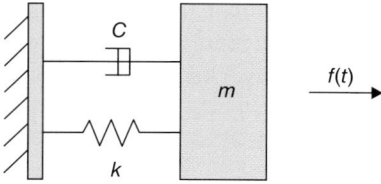

Figure 5.5 The schematic representation of a lumped mass-spring-damper system.

Following the described finite element approach, an elastic body such as the bar in Figure 5.1 is discretized into smaller fractions (i.e., elements). For the resulting discrete system, the dynamics of each element is governed by Eq. (5.11) and thus a generalized system equation is formulated as [1]

$$M\ddot{u}(t) + C\dot{u}(t) + Ku(t) = F(t) \tag{5.12}$$

where M, K, and C are mass, stiffness, and damping matrices, respectively and $u(t)$, $\dot{u}(t)$, and $\ddot{u}(t)$ represent the nodal displacement, velocity, and acceleration vectors.

For modal analysis, any external forces are ignored because the natural resonance modes are associated with free (unforced) vibrations. Also, in many cases the energy loss is either small or not of interest and consequently, the damping term in the equation can be disregarded. Therefore, Eq. (5.12) is reduced to

$$M\ddot{u}(t) + Ku(t) = 0 \tag{5.13}$$

With the assumption that the displacement vector is of constant shape and oscillates harmonically in time, it can be written as

$$u(t) = e^{i\omega t} \tag{5.14}$$

Substituting Eq. (5.14) into Eq. (5.13) results in the following eigenvalue problem:

$$KX = -\omega^2 MX \tag{5.15}$$

The values for ω and the vectors X that satisfy the above eigenvalue equation are called the eigenfrequencies and the eigenmodes, respectively. It is useful to note that the same approach can be extended to systems with loss (e.g., non-zero damping coefficient). In that case, the eigenfrequencies will be complex numbers [1]. Later in this chapter, we will see how this concept can be exploited to model quality factors in resonators.

5.1.2
Practical Implementation

Several commercial software packages are currently available for FEM. However, regardless of the specific software in use, the general steps involved in the analysis are similar and can be classified into three sections: setup (preprocessing), processing (solution), and post-processing. We will explain these steps using an example. Let's consider the problem of finding the first-harmonic lateral-extensional resonance mode (also known as contour mode) of a silicon block resonator. The software package used in this chapter is COMSOL multi-physics [2].

5.1.2.1 Set Up

The FEM usually starts by choosing the dimensional space and the type of analysis that is planned to be executed (e.g., structural modal analysis in our example). Next, the geometry of the resonant body should be defined. Most FEM packages include a graphical user interface (GUI) in which the resonator will be drawn. In our example, the resonator is a 3-D silicon block as seen in Figure 5.6. The dimensions of this block are chosen as 200 µm long, 150 µm wide, and 35 µm thick.

Next, the material of the resonant body is defined. Usually, the software package includes a library of commonly used materials. Otherwise, the physical properties should be manually defined by the user. If the material is anisotropic in nature (e.g., single crystalline silicon in our example) it is critical to properly align the simulated resonator to the correct crystallographic planes.

The next step is to define the boundary conditions. Properly defining the boundary conditions is critical in acquiring accurate results. In case the location and the geometry of the suspension tethers are not identified at the preliminary FEM stage, all boundaries could be left free to move. Note that in some software packages, the user may need to specify support conditions to eliminate rigid-body motion. The final step in setting up the simulation is to break the geometry into several elements. This step is called meshing and is one of the most important stages of the modeling process. Some advanced FEM software packages (e.g., ANSYS) offer a large range of element types that can be selected based on the device geometry and the type of targeted analysis. The type of element will influence the accuracy and the computational time of the analysis.

All software packages will give the user control over the number of the elements that are used and also the shape function(s) that are used for approximating the values of the unknown field quantity (e.g., displacement) inside each element. As mentioned earlier, there is an inherent trade-off between the computational load and the accuracy, which needs to be balanced by the user. One could optimize

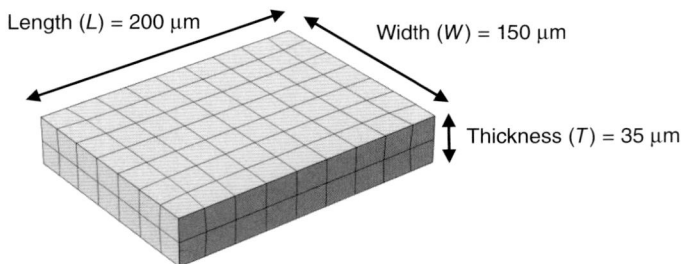

Length (*L*) = 200 µm   Width (*W*) = 150 µm
Thickness (*T*) = 35 µm

Figure 5.6 The silicon block drawn and meshed using the graphic user interface of COMSOL software.

the number of elements by starting with a low-density mesh and evaluating the residual error in an iterative matter. Increasing the mesh density may be stopped when the output convergence meets a targeted threshold.

5.1.2.2 Processing

After setting up the simulation, it is time to run the analysis. For modal analysis, the parameters that should be defined at this stage include the range of frequency and the number of modes that are of interest. In many instances, a closed-form equation exists for predicting the resonance frequency of the targeted mode shape(s) and can be used as a guideline for setting the solver parameters. In our example, we are interested in evaluating the frequency and the mode shape of the fundamental lateral-extensional mode of the silicon block aligned to the <110> crystalline plane. Assuming that the Poisson's ratio is zero, the width-extensional frequency of the block can be approximated as [3]

$$f_n = \frac{n}{2W} \sqrt{\frac{E}{\rho}} \tag{5.16}$$

where n is the extensional mode number, W is the width (frequency-determining dimension) of the block, and E is the Young's modulus of the structural material (silicon) in the primary direction of the vibration (i.e., the <110> direction in our example), and ρ is the material density. Using this equation, the frequency of the first harmonic ($n = 1$) will be approximately

$$f_1 = \frac{1}{2 \times 150e^{-6}} \sqrt{\frac{169e^9}{2330}} \approx 28.4 \, \text{MHz}$$

The frequency range for the modal analysis is therefore chosen according to this estimated value.

5.1.2.3 Post-processing

After the completion of the analysis processing stage, the user can choose among several methods of visualizing the output data. One of the more useful output configurations for modal analysis is to generate the deformed shape (mode shape) of the resonator. The outer surface of the deformed body can be color-coded to exhibit the contours of response quantities such as displacement, stress, or strain. It is important to note that for modal analysis, the magnitudes of these parameters are not informative since the actual magnitude of the system's dynamic response is a function of the external forces that excite the resonator or the initial conditions that set it into a specific free vibration. In Figure 5.7, the resulting deformation of the silicon block at the fundamental extensional mode is pictured at the two ends of one vibration half-cycle. The surface color represents the *relative* total displacement (since the magnitude of a mode shape is arbitrary). The dark blue color represents the minimum displacement and the dark red represents the maximum displacement for this mode shape.

The simulated resonance frequency of this mode as obtained from the COM-SOL analysis is about 26.8 MHz, which is presumed to be a more accurate estimation of the resonance frequency than Eq. (5.16). In many software packages,

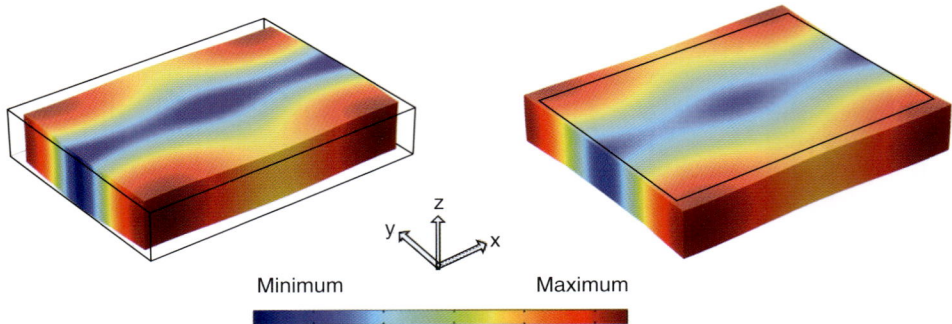

Figure 5.7 The deformed shapes of the silicon block at the two ends of a half-cycle of vibration in the first lateral-extensional mode. The total displacement is color-coded on the surface.

the user can generate an animation that captures the dynamics of a complete harmonic cycle.

5.2
Application of FEA in MEMS Resonator Design

In this section, we will introduce some of the most common use of FEM in studying MEMS resonators and we will refer to examples from the literature when appropriate. The applications are classified into three categories: modal analysis, loss analysis, and frequency response analysis.

5.2.1
Modal Analysis

The most widely used application of FEM in MEMS resonators is modal analysis. As discussed earlier, the mode shapes and the resonance frequencies of resonators with arbitrarily shaped structures can be determined with great accuracy. Therefore, the effect of geometry on the behavior of a resonator can be studied and the design of the resonator can be optimized to suit the application.

5.2.1.1 Mode Shape Analysis for Design Optimization
Modal analysis at the design stage of an MEMS resonator can be utilized to optimize the location of the suspension element and the transduction electrodes. Let's refer to the block resonator discussed in the last section and learn through practice.

An important aspect of a resonator design is the placement and the dimensions of the features that suspend the resonant structure (i.e., anchors or tethers). The design criterion for such suspension features is commonly to reduce their interference with the targeted mode shape, while guaranteeing the structural integrity of the whole system. Therefore, the optimized locations for placing the tethers are at the boundary location(s) which undergo minimal displacement (i.e., the "modal nodes" or "vibrational nodes"). The modal nodes for the width-extensional resonance mode in the silicon block example are located on the symmetry plane of the mode shape ($x-z$ plane in the middle of the resonator of Figure 5.7); therefore, the tethers can be ideally located at this location as seen in Figure 5.8.

To minimize the effect of the anchors on the mode shape, the width of the anchor beams should be as narrow as possible. However, the final beam width is usually constrained by the fabrication process and the shock survivability requirement. The boundary condition for the end of the tether is commonly chosen to be a fixed boundary. This is a reasonable assumption given that the tether is practically terminated to a large body of the silicon substrate. The mode shape of the revised block resonator (including the tethers) is shown in Figure 5.8 and upon close inspection one can identify the slight differences in the total displacement field plotted on the surface compared to the free (untethered) silicon block of Figure 5.7.

Alternatively, the tether ends can be left free to move. Accurately speaking, the displacement at the end of the tether is never zero and this is exactly the mechanism through which elastic energy will radiate out of the resonator. This radiation is one of the important mechanisms that limits the quality factor of the resonator and is commonly referred to as the anchor/support loss in the literature [4].

One can study the effect of the tether and the device geometry on the anchor loss by monitoring the pressure developed at the tether/substrate interface. For example, it is intuitively perceived that designing the length of the tether to be equal to a quarter of the resonance wavelength will reduce the anchor loss [5]; however, this perception is valid only if the displacement at the substrate end of the tether is zero (i.e., fixed boundary). Using our block resonator, we can expand our model and put this idea to the test. The new structure will include silicon semi circles (300 μm radius) of the same thickness and the far edges are set to fixed boundaries (Figure 5.9). Modal analysis is used to generate the eigenfrequencies and the pressure data is integrated for the tether/substrate boundary in

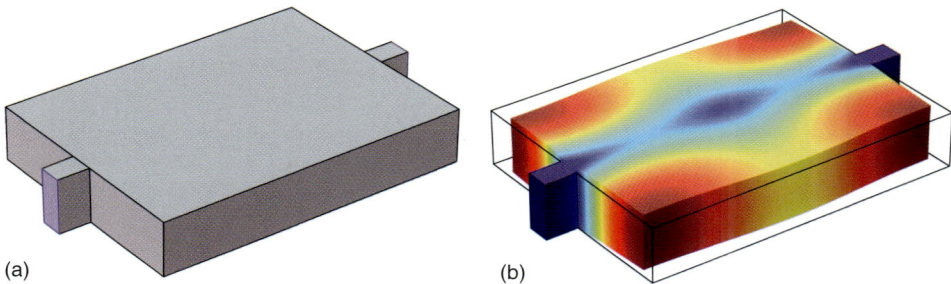

(a)   (b)

Figure 5.8 (a, b) Modal analysis of the revised silicon block resonator with tethers and the resulting mode shape.

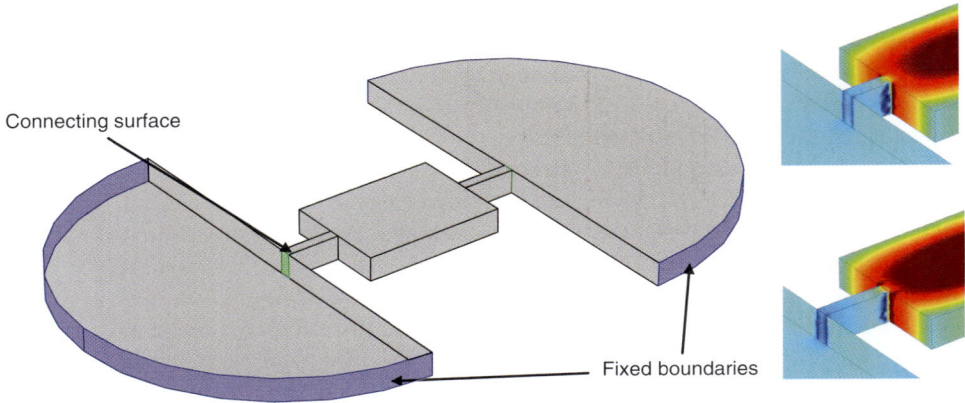

Figure 5.9 The geometry used for simulating the pressure created at the substrate/tether boundary. The existing resonator is connected to silicon slabs and the far sides of the substrate are set to a fixed boundary condition (blue). The top of the left substrate piece is hidden to show the interior tether/substrate boundary (green), which is used to gather pressure data.

post-processing. Although the pressure difference for tether lengths of 45 μm versus 75 μm ($\lambda/4$) is visibly slight, the integration shows the pressure to be 1.5 times larger for the 75 μm tether. This suggests that the anchor loss for such tether length could be more dominant in contrast to the common intuitive perception. Later in this chapter, we will confirm this hypothesis using a different FEM approach.

The design of a resonator is not complete without considering the means to excite and sense the targeted resonance mode. Another important use of the modal analysis is in designing the transduction mechanism to offer the most efficient electromechanical energy coupling. There are several transduction mechanisms that could be used to excite the previously discussed extensional resonance mode and two of the most common are capacitive [6] and piezoelectric [7]. In the following, we will explain the process of transducer design for both cases. The objective for the design of both transducers is to maximize the electromechanical coupling. The electromechanical coupling is a measure of efficiency of the energy conversion between the electrical and mechanical domains.

Let's start by discussing the case of a capacitive transducer. It is intuitive to see that for an efficient energy conversion, the applied input signal should generate a force component in the same direction as the main displacement component of the resonance mode. Therefore, the capacitive electrodes should certainly be placed on the sides of the resonator as shown in of Figure 5.10a. The coverage area of the electrodes can be optimized by carefully studying the mode shape. The coupling will increase with the coverage area as long as the polarity of displacement for every point covered by that area is identical in the mode shape. Therefore, if there is no limitation imposed by the fabrication process, all the area on the sides should be covered with the electrodes in our example (Figure 5.10). However, this

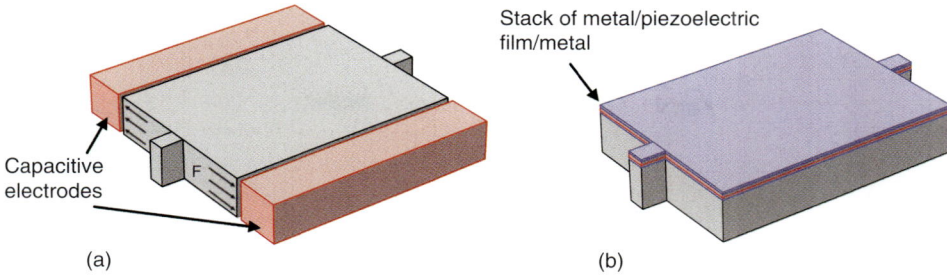

Figure 5.10 The design of (a) capacitive and (b) piezoelectric electrodes for maximum coupling in the silicon block resonator.

is not necessarily true for an arbitrarily sized silicon block. For example, increasing the length of the block will introduce nonuniform displacement across the length. Therefore, it is likely that for a specific block dimensions, the full coverage of the sidewalls results in charge cancelation and reduction of the coupling as suggested in [6].

In the case of piezoelectric transduction, an effective approach is to add a stack of a thin piezoelectric film sandwiched between two metal layers on top of the resonant structure (Figure 5.10b). The metallic electrodes are patterned to selectively apply an electrical potential across the film [7]. Unlike the capacitive transducer, the piezoelectric film generates force components in several directions as a function of the piezoelectric coefficients and the film crystalline alignment. In the case of in-plane extensional modes in a silicon block covered with sputtered Wurtzite piezoelectric material such as AlN and ZnO, d_{31} is generally the dominant piezoelectric coefficient responsible for the efficiency of the transduction.

Since the piezoelectric film (e.g., AlN) may influence the natural frequencies and mode shapes of the resonator, we need to include the film in our finite element model and re-run the modal analysis in order to optimize the electrode pattern. Also, we will choose the strain in the primary direction of vibration (Y direction) as the output quantity to be color-coded in the contour plot on the deformed shape of the resonator (Figure 5.11). If there are no other constraints, all the areas with the same polarity of strain should be covered with the electrode in order to maximize the coupling. Otherwise, the areas with the maximum strain should preferably be covered (areas in the middle of the block). Therefore, for the fundamental extensional mode in our block resonator (Figure 5.11a), the whole surface area of the piezoelectric film can be connected to a single piece of metal. However, in the case of the third-order extensional mode (Figure 5.11b), the situation is different.

In this case, covering the outer two-thirds of the device as shown in Figure 5.12 results in a maximum coupling as the two covered areas are under the same strain polarity (which is opposite the strain polarity in the middle third of the block). Moreover, the area in the middle of the block can be covered by a different set of electrodes and used as the second port in a two-port resonator configuration.

It is of practical value to mention that by increasing the geometrical complexity of the resonant structure, the number of elements in the corresponding finite

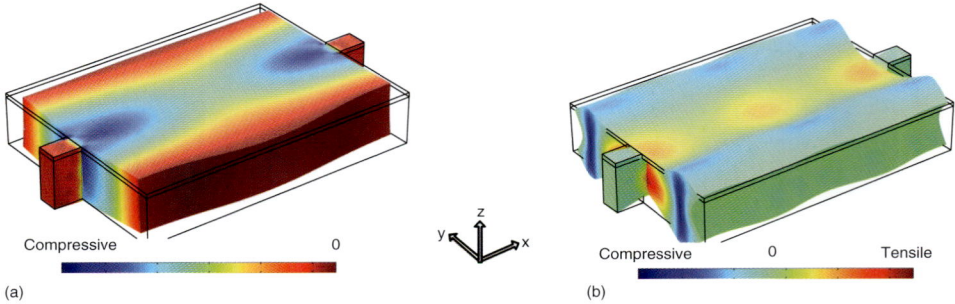

(a)   (b)

Figure 5.11 The simulated fundamental (a) and third-order (b) extensional modes of the silicon block resonator covered by a thin layer of AlN. The strain in Y direction is color-coded on the surface.

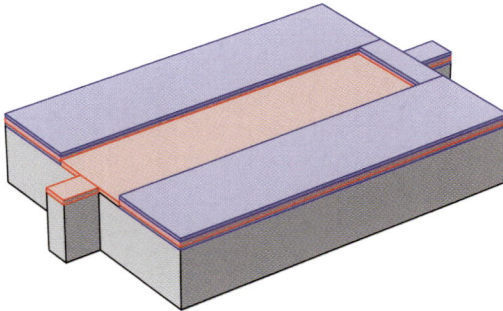

Figure 5.12 Electrode shapes designed for piezoelectric excitation of the third extensional mode in the silicon block resonator.

element model can grow rapidly to the point at which computational capacities are exceeded. In that case, symmetry planes can be used to reduce the number of elements without compromising the accuracy. For example, the targeted fundamental lateral-extensional mode shape of the silicon block covered with an AlN film is symmetric with respect to the XZ and YZ mid-planes. Therefore, one quarter of the device can be simulated provided that the boundary condition on the symmetry planes are properly defined (i.e., the displacement component normal to the symmetry planes and the shear stress components on the symmetry planes are set to zero). In this case, the number of finite elements is potentially one fourth of what is required for modeling the complete resonator (assuming the same element size), while the calculated mode shape and frequency are almost identical to the full model (compare Figure 5.13 and Figure 5.11a).

5.2.1.2 Modeling Process-Induced Variation

Another useful application of FEM in the design of MEMS resonators is to study the effect of structural variation due to the fabrication process on the performance. The final shape and dimensions of the fabricated resonator always differs from the

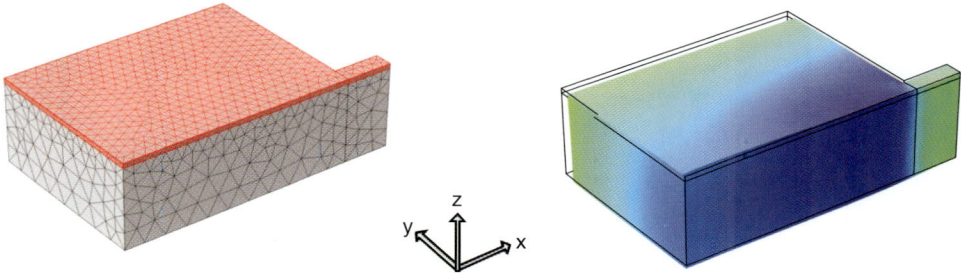

Figure 5.13 The mesh and the mode shape for a quarter of the silicon block constrained with symmetry plane boundary conditions to reduce the computational load. The simulated frequency is less than 2 kHz and different from the frequency obtained from the full model.

design values due to the inevitable inaccuracies and nonuniformities involved in all fabrication steps. Let's revisit our silicon block resonator example and assume that the resonant structure is built from the device layer of a silicon-on-insulator (SOI) substrate. The silicon block should be etched using anisotropic deep reactive ion etching tools (DRIE) [8]. The exact lateral dimensions of the silicon block across a wafer are not uniform due to both lithography and DRIE bias. Consequently, the frequency won't be uniform across the wafer. Clever design features have been used to compensate for such process-induced frequency variations [9]. However, the thickness of an SOI wafer varies across the wafer and the frequency variation due to thickness variation is inevitable even if the resonator design compensates for variation in lateral dimensions. In Figure 5.14 the simulated resonance frequency of the block resonator is shown as a function of the thickness.

From this plot, it can be concluded that to minimize the effect of thickness variations on the frequency of lateral-extensional resonators, the thickness of the silicon block should be decreased. This result may initially sound counterintuitive but upon careful consideration, one can realize that for smaller thickness values, the coupling between the in-plane and out-of-plane (i.e., thickness) motion

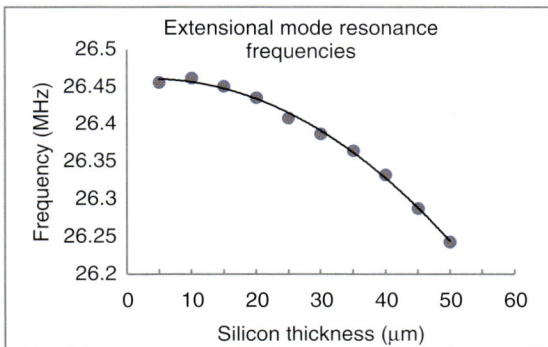

Figure 5.14 The simulated first-harmonic, lateral-extensional resonance frequency of the block resonator versus the thickness of silicon.

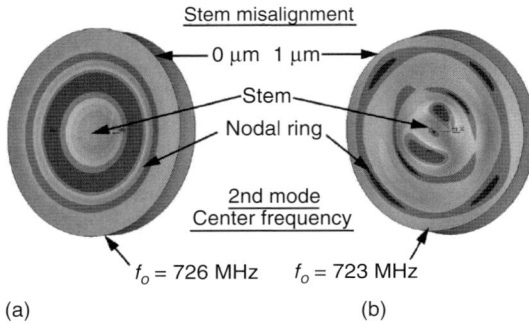

Figure 5.15 (a, b) The simulated effect of support post misalignment on the radial mode shape of a polysilicon disk resonator [10].

through the Poisson ratio is less and hence the effect of thickness variation on the frequency is lower.

Another great example of process-induced variation modeling is presented in a work by Wang *et al.* [10]. In this work, the effect of a misaligned support post from the optimal central position in a capacitive disk resonator is studied. The mode shapes for the two disk resonators with and without misalignment highlight the considerable effect of such misalignment (Figure 5.15). The authors offer a solution to completely avoid such an issue by altering the original process flow.

5.2.2
Loss Analysis

The quality factor (Q) of a mechanical resonator is a very important parameter that affects the performance in any application perceived for the resonator. For example, the resonator Q directly affects the short-term stability of oscillators and the roll-off skirt of the filters that are built based on a specific type of resonator. Therefore, the ability to predict and design for appropriate Q values is very empowering at the design stage. However, Q in a resonator is affected by several mechanisms, some of which are very difficult to control and model. Among the better-known and more-studied loss mechanisms that affect Q are anchor loss [11] and thermoelastic damping (TED) [12]. Fortunately, these two loss mechanisms can be modeled using FEM software. The following section will briefly explain the basics of modeling anchor loss and TED in MEMS resonators and refer the interested readers to some of the work published on these topics.

5.2.2.1 **Anchor Loss**
Anchor loss, or clamping loss, is the quantification of the elastic energy radiation from the resonator through the structural elements connecting it to a frame. Even though the anchors/tethers are optimally positioned at the resonator's modal nodes for a target mode shape, some periodic displacement still occurs at these points as seen earlier in this chapter. As the resonator vibrates, it exerts pressure

on the tether, causing acoustic energy to travel down the tether and to subsequently be radiated into the substrate. This energy is commonly considered lost since any reflection from the substrate boundaries are scattered and the effect is insignificant. Attempts to model such effects have been made from the early years of developing advanced numerical methods by introducing concepts such as absorbing boundaries [13]. The more recent work on the simulation of anchor loss in MEMS resonators using FEM was pioneered by Bindel and Govindjee [14]. The proposed method is based on the application of a layer of artificial material with finite length that models the behavior of a perfectly matched semi-infinite media. This layer quickly absorbs elastic waves that impinge upon it at a wide range of incidence angle and reflects near zero energy; hence, these layers are called perfectly matched layers (PMLs). To use PML in modeling a resonant system, the surrounding substrate to which the resonator tethers are attached will be terminated with the PML and the complex eigenfrequencies of the system are simulated using finite element analysis (FEA). Finally, to calculate Q, the real part of the eigenfrequency is divided by twice the imaginary part [14]. To design the PML, a complex-valued change of coordinates will be applied to the anisotropic silicon's material properties to create a material that will attenuate any wave from that medium using

$$C_{PML} = \widetilde{C}_{Si} \det(\Lambda) \tag{5.17}$$

$$\rho_{PML} = \widetilde{\rho}_{Si} \det(\Lambda) \tag{5.18}$$

where C_{PML} and ρ_{PML} are the stiffness matrix and density of the PML, respectively, $\widetilde{C}_{Si}$ and $\widetilde{\rho}_{Si}$ are coordinate transformations of the original silicon properties, and Λ is the Jacobian of the transformed coordinate.

Since defining a PML is relatively complicated, a simplified matched layer (ML) has been introduced by Steeneken *et al.* [15]. This ML behaves as a perfectly ML only for waves with an incident angle of 90°. Therefore, to improve the accuracy of the simulation the geometry of the ML should be chosen to achieve a near-normal angle of incidence for a large portion of the waves radiated from the tether. Going back to our silicon block example, assuming that the tether behaves like a point source, the ML geometry should be chosen as a ring centered at the tethers end (Figure 5.16).

With the inclusion of the ML, the eigenfrequencies calculated using the FEM will be complex numbers and the Q_{anchor} can be estimated as explained earlier. We use this method to confirm our earlier study where we concluded that the anchor loss for a tether length of 75 μm ($\lambda/4$) could be higher than the case with 45 μm long anchors. It should be noted that the relative magnitude of the quality factor calculated using this method is of interest as opposed to the absolute value. The complex eigenfrequency for the 45 μm and the 75 μm tether lengths are 26.3e6 + 2874i and 26.3e6 + 4109i resulting in estimated Q's of 4583 and 3204, respectively. Therefore, this study predicts an increase in Q of ∼1.4 times when the tether length is varied from 75 to 45 μm, which is in agreement with our earlier observation based on induced pressure at the substrate/tether interface.

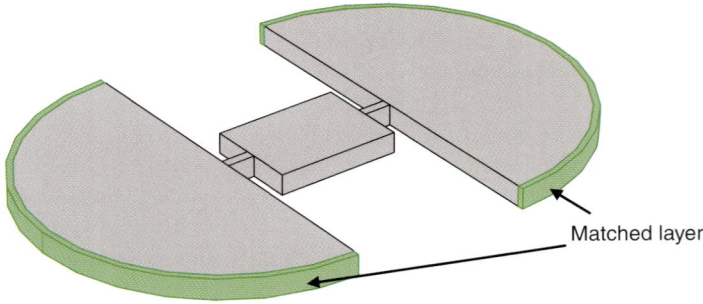

Figure 5.16 The inclusion of circular-matched layers around the silicon block resonator for simulating anchor loss.

The same approach explained above has been used to confirm the efficiency of in-plane acoustic reflectors strategically etched into the substrate to reflect some of the elastic energy back into the resonator [16]. A similar approach using "infinite elements" has been also proposed to predict improved quality factors of dome resonators fabricated on a mesa structure [17].

5.2.2.2 Thermoelastic Damping

The coupling between the strain and temperature fields, known as thermoelasticity, exists in all materials with a nonzero linear thermal expansion coefficient. TED is the relaxation in an oscillating structure, which is caused by irreversible thermal diffusion. The thermal currents are driven by local temperature gradients as a result of local normal strain fields. Therefore, typical MEMS resonators, which are usually designed to operate in modes that are flexural or extensional, experience TED. The coupling between the strain and temperature field can be formulated by the following partial differential equations [18]:

$$\rho \ddot{u}_i - \frac{E}{2(a+\sigma)} \frac{\partial^2 u_i}{\partial x_k^2} - \frac{E}{2(1+\sigma)(1-2\sigma)} \frac{\partial^2 u_l}{\partial x_i \partial x_l} = -\frac{\alpha E}{3(1-2\sigma)} \frac{\partial T}{\partial x_i} \quad (5.19)$$

$$C_\mathrm{p} \frac{\partial T}{\partial t} - \nabla \cdot (\kappa \nabla T) = -\frac{\alpha E T_0}{3(1-2\sigma)} \frac{\partial}{\partial t} \left(\frac{\partial u_i}{\partial x_i} \right) \quad (5.20)$$

where ρ is density, σ is Poisson's ratio, E is Young's modulus, α is the linear thermal expansion coefficient, C_p is specific heat capacity, κ is thermal conductivity, and T_0 is the equilibrium temperature. Departure of the resonator variables from equilibrium are described by a temperature field T and displacement field, u_i, for $i = x, y, z$. (The summation convention is to be employed for repeated indices within an individual term.) Exact closed form solutions to TED are available only for simple geometries and FEA has been suggested to model TED in complex structures [19]. To implement the above-coupled partial differential equations in FEM software such as COMSOL, the temperature, and displacement are assumed to be separable functions of time and position

$$u_i(x, y, z, t) = \hat{u}_i(x, y, z) f(t) \quad (5.21)$$

$$T(x, y, z, t) = \widehat{T}(x, y, z)f(t) \tag{5.22}$$

where $f(t)$ is a decaying exponential function of time,

$$f(t) = e^{(i\omega - \delta)t} = e^{\lambda t} \tag{5.23}$$

With such assumptions, the problem is reduced to a general eigenvalue problem that can be solved by the COMSOL eigenvalue solver. (Note that these eigenvalues are different from the complex eigen-frequencies that resulted from running a modal analysis.) After calculating λ, the imaginary part of λ (i.e., ω) defines the angular resonance frequency, and the magnitude of the real part (δ) is the damping parameter. Therefore, the quality factor of a resonating structure can be calculated as

$$Q = \frac{Im\,\lambda}{-2.Re\,\lambda} \tag{5.24}$$

This method has been utilized to model Q_{TED} in a variety of resonant structures including solid beam resonators [20], slotted beam resonators [21], ring resonators [22, 23], and trench-refilled polysilicon beam resonators [24].

5.2.3
Frequency Response Analysis

The experimental evaluation of a resonator performance often involves observing the electrical output of the device in response to a sinusoidal electrical input signal over a range of input frequencies (i.e., a frequency response analysis). Through such an analysis, the actual performance of the device is studied and important parameters such as quality factor (Q) and the equivalent electrical resistance of the resonator at resonance are measured. Also, unwanted resonance peaks (i.e., spurious peaks) inside the frequency range of interest are identified. Advancement in FEM theory has enabled the frequency response of a complete resonant system (including the transduction mechanism) to be analyzed. The coupled field equations that describe the conversion of energy from the electrical to the mechanical domain are now included in FEM software packages such as COMSOL. Combining such capabilities along with methods to capture and model energy loss (such as those described in the last section for anchor loss) can be utilized to predict the performance of a resonator more accurately than ever before. In the next two sections, we briefly review some specific applications of such an analysis.

5.2.3.1 Spurious Mode Identification and Rejection
As mentioned earlier in this chapter, the resonance modes that are excited in a resonant system depend, among other things, on the coupling coefficient of the transducer. For the piezoelectrically coupled resonator of our example, the force components resulting from the application of an electric field across the piezo-electric film are spread in all directions; therefore, such resonators are prone to spurious mode excitation close to the target mode. Such modes can be detected

through FEM and, should the strength of such peaks be considered detrimental to an application, they can be mitigated through the modification of the resonant structure's geometry and the design of the electrodes. The plot shown in Figure 5.17 is the simulated frequency response of the piezoelectrically coupled block resonator discussed throughout this chapter for a 3-MHz span around the target resonance frequency.

The resonator electrodes for this simulation are modified and the top electrode is patterned into an interdigitated transducer to form a two-port resonator [25]. The input signal is then applied to the electrode in the middle and the output is picked up from the two electrodes on the side while the bottom of the piezoelectric film is grounded. As seen in the plot, a second resonance peak is detected and upon investigation, the mode shape for that resonance peak is identified. The spurious mode appears to be a high-order flexural mode with a thickness-dependent frequency. Therefore, to push this mode out of this frequency span, one can simply modify the thickness. The plot in Figure 5.18 is the simulated frequency response

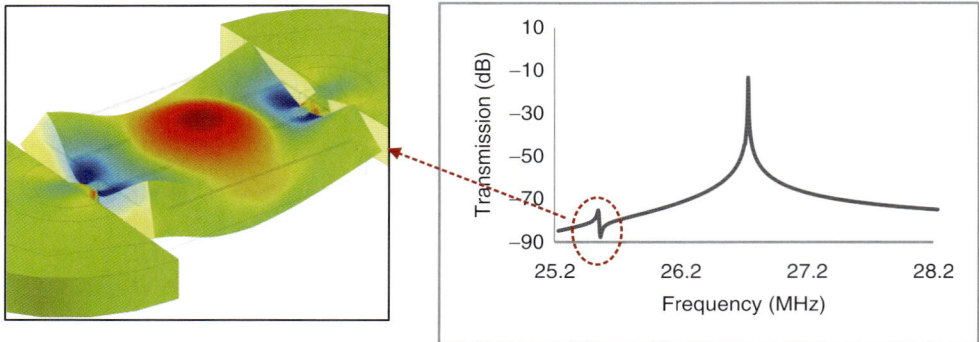

Figure 5.17 The simulated frequency response of the 35-μm-thick silicon block resonator with interdigitated piezoelectric transducer on top. The mode shape for a spurious peak is also shown.

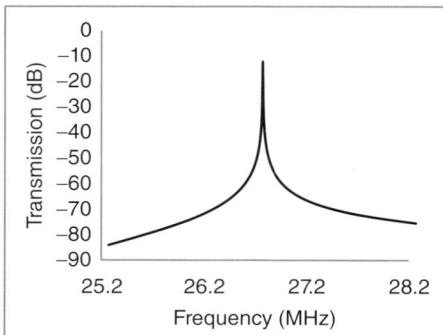

Figure 5.18 The simulated frequency response of a 33-μm-thick silicon block resonator with interdigitated piezoelectric transducer on top.

of the same resonator with 33-μm thickness. As seen the spurious mode is successfully moved out of the frequency range of interest.

The frequency response modeling technique discussed above has been used to
identify and mitigate the spurious modes for piezoelectric resonators in the past
[26 – 28].

5.2.3.2   Filter Design

An important application of micromechanical resonators is in signal filtering
when very high selectivity is required [29]. Among the different techniques to
build coupled resonant systems for filtering applications, the mechanical/acoustic
coupling [30, 31] possesses significant advantages compared to the more common
electrical coupling techniques. For example, mechanically coupled resonator
filters are usually smaller in size and offer higher out-of-band rejection for
similar filter orders. However, the performance of such filters is much harder to
predict and modeling techniques for the design of such filters are not extensively
developed. The finite element frequency response analysis can be of significant
help in providing a better understanding of the geometrical effects on the performance of such coupled systems. However, due to the inherent complexity of
coupled-resonator systems, the frequency response analysis for the actual devices
demands significant computational resources, forcing researchers to implement
modeling simplifications to date [32, 33]. The plot shown in Figure 5.19 is the
simulated frequency response of a second-order monolithic AlN-on-diamond
filter simulated in COMSOL [30].

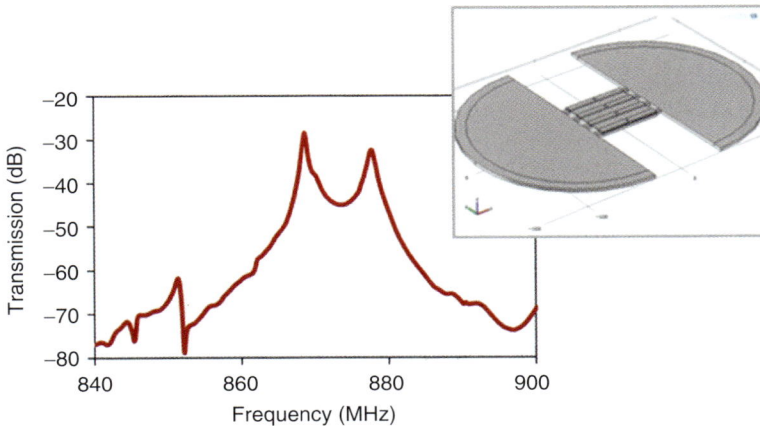

Figure 5.19   The simulated frequency response of a second-order, monolithic piezoelectric-
on-diamond filter.

5.3
Summary

In this chapter, a brief introduction to the fundamentals of the FEM and its application in the design of MEMS resonators was presented. It was shown that to capture the complexities associated with an MEMS resonator at the design stage, the only viable approach is to utilize FEM. Significant advances continue to be made in FEM software and computational power so that the use of FEM in resonant MEMS applications will undoubtedly become more prevalent in the decades to come.

References

1. Öchsner, A. and Merkel, M. (2013) *One-dimensional Finite Elements: An Introduction to the FE Method*, Springer.

2. COMSOL *http://www.comsol.com/shared/downloads/IntroductionToCOMSOLMultiphysics.pdf* (accessed 1 September 2014).

3. Ginsberg, J.H. (2001) *Mechanical and Structural Vibrations: Theory and Applications*, John Wiley & Sons, Inc., New York.

4. Park, Y.-H. and Park, K.C. (2004) High-fidelity modeling of MEMS resonators. part I. anchor loss mechanisms through substrate. *J. Microelectromech. Syst.*, **13** (2), 238–247.

5. Piazza, G., Stephanou, P.J., and Pisano, A.P. (2006) Piezoelectric aluminum nitride vibrating contour-mode MEMS resonators. *J. Microelectromech. Syst.*, **15** (6), 1406–1418.

6. Pourkamali, S., Ho, G.K., and Ayazi, F. (2007) Low-impedance VHF and UHF capacitive silicon bulk acoustic wave resonators—part I: concept and fabrication. *IEEE Trans. Electron Devices*, **54** (8), 2017–2023.

7. Ho, G.K., Abdolvand, R., Sivapurapu, A., Humad, S., and Ayazi, F. (2008) Piezoelectric-on-silicon lateral bulk acoustic wave micromechanical resonators. *J. Microelectromech. Syst.*, **17** (2), 512–520.

8. Laermer, F. and Urban, A. (2003) Challenges, developments and applications of silicon deep reactive ion etching. *Microelectron. Eng.*, **67–68**, 349–355.

9. Ho, G.K., Perng, J.K.C., and Ayazi, F. (2007) Process compensated micromechanical resonators. IEEE 20th International Conference on Micro Electro Mechanical Systems, MEMS, January 21–25, 2007, pp. 183–186.

10. Wang, J., Ren, Z., and Nguyen, C.T.-C. (2004) 1.156-GHz self-aligned vibrating micromechanical disk resonator. *IEEE Trans. Ultrason. Ferroelectr. Freq. Control*, **51** (12), 1607–1628.

11. Lee, J.E.Y., Yan, J., and Seshia, A.A. (2011) Study of lateral mode SOI-MEMS resonators for reduced anchor loss. *J. Micromech. Microeng.*, **21** (4), 045010.

12. Lifshitz, R. and Roukes, M.L. (2000) Thermoelastic damping in micro-and nanomechanical systems. *Phys. Rev. B*, **61** (8), 5600.

13. Engquist, B. and Majda, A. (1977) Absorbing boundary conditions for numerical simulation of waves. *Proc. Natl. Acad. Sci. U.S.A.*, **74** (5), 1765–1766.

14. Bindel, D.S. and Govindjee, S. (2005) Elastic PMLs for resonator anchor loss simulation. *Int. J. Numer. Methods Eng.*, **64** (6), 789–818.

15. Steeneken, P.G., Ruigrok, J.J.M., Kang, S., Van Beek, J.T.M., Bontemps, J., and Koning, J.J. (2013) Parameter Extraction and Support-loss in MEMS Resonators, arXiv preprint arXiv:1304.7953.

16. Harrington, B.P. and Abdolvand, R. (2011) In-plane acoustic reflectors for reducing effective anchor loss in lateral–extensional MEMS resonators. *J. Micromech. Microeng.*, **21** (8), 085021.

17. Pandey, M. *et al.* (2009) Reducing anchor loss in MEMS resonators using mesa isolation. *J. Microelectromech. Syst.*, **18** (4), 836–844.

18. Nayfeh, A. and Nemat-Nasser, S. (1971) Thermoelastic waves in solids with thermal relaxation. *Acta Mech.*, **12** (1-2), 53–69.

19. Gorman, J.P. (2002) Finite element analysis of thermoelastic damping in MEMS. MS Thesis. Massachusetts Institute of Technology.

20. Yi, Y.B. and Matin, M.A. (2007) Eigenvalue solution of thermoelastic damping in beam resonators using a finite element analysis. *J. Vib. Acoust. Trans.-Am. Soc. Mech. Eng.*, **129** (4), 478.

21. Prabhakar, S. and Vengallatore, S. (2009) Thermoelastic damping in hollow and slotted microresonators. *J. Microelectromech. Syst.*, **18** (3), 725–735.

22. Koyama, T., Bindel, D.S., Wei, He, Quevy, E.P., Govindjee, S., Demmel, J.W., and Howe, R.T. (2005) Simulation tools for damping in high frequency resonators. Sensors, 2005 IEEE, October 30 2005–November 3 2005, 4 pp.

23. Yi, Y.B. (2008) Geometric effects on thermoelastic damping in MEMS resonators. *J. Sound Vib.*, **309** (3-5), 588–599.

24. Abdolvand, R., Johari, H., Ho, G.K., Erbil, A., and Ayazi, F. (2006) Quality factor in trench-refilled polysilicon beam resonators. *J. Microelectromech. Syst.*, **15** (3), 471–478.

25. Reza, A., Lavasani, H., Ho, G., and Ayazi, F. (2008) Thin-film piezoelectric-on-silicon resonators for high-frequency reference oscillator applications. *IEEE Trans. Ultrason. Ferroelectr. Freq. Control*, **55** (12), 2596–2606.

26. Fatemi, H., Zeng, H., Carlisle, J.A., and Abdolvand, R. (2013) High-frequency thin-film AlN-on-diamond lateral–extensional resonators. *J. Microelectromech. Syst.*, **22** (3).

27. Olsson, R.H., Wojciechowski, K.E., and Branch, D.W. (2010) Origins and mitigation of spurious modes in aluminum nitride microresonators. Ultrasonics Symposium (IUS), 2010 IEEE.

28. Yazici, S., Giovannini, M., Kuo, N.-K., and Piazza, G. (2012) Suppression of spurious modes via dummy electrodes and 2% frequency shift via cavity size selection for 1 GHz AlN MEMS contour-mode resonators. Frequency Control Symposium (FCS), 2012 IEEE International IEEE, 2012, pp. 1–5.

29. Nguyen, C.T.-C. (2007) MEMS technology for timing and frequency control. *IEEE Trans. Ultrason. Ferroelectr. Freq. Control*, **54** (2), 251–270.

30. Abdolvand, R. and Ayazi, F. (2009) High-frequency monolithic thin-film piezoelectric-on-substrate filters. *Int. J. Microwave Wireless Technolog.*, **1** (01), 29–35.

31. Stephanou, P.J., Piazza, G., White, C.D., Wijesundara, M.B.J., and Pisano, A.P. (2007) Piezoelectric aluminum nitride MEMS annular dual contour mode filter. *Sens. Actuators, A*, **134** (1), 152–160.

32. Fatemi, H. and Reza, A. (2013) Low-loss lateral-extensional piezoelectric filters on ultra-nano-crystalline diamond. *IEEE Trans. Ultrason. Ferroelectr. Freq. Control*, **60** (9), 1978–1988.

33. Thakar, V.A., Pan, W., Ayazi, F., and Rais-Zadeh, M. (2012) Acoustically coupled thickness-mode AlN-on-Si band-pass filters-part II: simulation and analysis. *IEEE Trans. Ultrason. Ferroelectr. Freq. Control*, **59** (10).

Part II
Implementation

Resonant MEMS – Fundamentals, Implementation and Application, First Edition.
Edited by Oliver Brand, Isabelle Dufour, Stephen M. Heinrich and Fabien Josse.
© 2015 Wiley-VCH Verlag GmbH & Co. KGaA. Published 2015 by Wiley-VCH Verlag GmbH & Co. KGaA.

6
Capacitive Resonators

Gary K. Fedder

6.1
Introduction

Microresonators that employ capacitive transduction are used in electrometers [1], gravimetric sensors [2], gyroscopes [3], optical scanners [4], oscillators [5], signal filters [6], strain sensors [7], and vibromotors [8]. The earliest capacitive microresonator device made with lithographic techniques was the resonant body transducer, which comprised a cantilever acting as the movable gate capacitor electrode of a field-effect transistor [9, 10]. A comprehensive overview of the application literature is not attempted here; instead, this chapter conveys some of the fundamental principles in design of capacitive transducers for resonant devices.

Capacitive sensing and actuation is an attractive design option for resonant micro-electromechanical systems (MEMS) devices for a number of reasons. Any conductive material will function as a capacitive electrode so that no specialty materials are required for implementation. In most cases, specific degrees of freedom for actuation and for sensing in a micromechanical device are set solely by appropriate geometrical design of flexures and electromechanical capacitors. Capacitor geometry is scalable, being limited by mechanical integrity and chip cost. No DC current is drawn in capacitive systems enabling low power integration with chip-scale electronics. Electric field interactions allow for voltage-based tuning of transducer gain and nonlinear capacitive effects can provide voltage-based tuning of resonant frequency. An ideal capacitor has no intrinsic noise and so is well suited for low-noise motion sensing. Field interactions are extremely fast, generally limited by electrical time constants that are relatively small compared with most mechanical and other physical behavior.

The main weakness of capacitive transduction is the weak electrostatic force efficiency relative to electrothermal and piezoelectric actuation. Higher forces require higher voltages and smaller gaps; the former is generally limited by practical system and safety constraints, while the latter is limited by process technology and stability. This chapter provides an overview of design considerations for capacitive resonators, including basic modeling equations and a small sampling of some exemplary device instantiations.

Resonant MEMS – Fundamentals, Implementation and Application, First Edition.
Edited by Oliver Brand, Isabelle Dufour, Stephen M. Heinrich and Fabien Josse.
© 2015 Wiley-VCH Verlag GmbH & Co. KGaA. Published 2015 by Wiley-VCH Verlag GmbH & Co. KGaA.

6.2
Capacitive Transduction

Capacitance is a quantity relating the charge between two conducting structures to the electric potential (i.e., voltage) between the structures. As such, capacitance is dependent on a structure's geometry and its position relative to other electrically conductive and insulating structures. For a conducting electrode pair carrying a charge $\pm q$ on each electrode with potential $v = v_+ - v_-$ between the electrodes, the capacitance is defined as

$$C = \frac{q}{v} \tag{6.1}$$

The simplest case, and one often used in MEMS analyses and approximated in capacitive resonant MEMS devices, is that of the parallel-plate capacitor, shown in Figure 6.1, that for generality is shown to be free to move in three orthogonal directions. In the ideal parallel-plate approximation, the electric field, **E** is confined to the region of overlap between the two plates, determined by x_o and y_o, and "fringing fields" located in other regions are assumed small and are neglected. With this approximation, the capacitance after a small arbitrary relative displacement $(x = x_b - x_a, y = y_b - y_a, z = z_b - z_a)$ of the two plates is

$$C = \frac{\varepsilon_0 \varepsilon_r (x_o + x)(y_o + y)}{z_o + z} \tag{6.2}$$

where z_o is the gap between the plates, measured normal to the plates' opposing surfaces. From Eq. (6.2), the capacitance scales linearly with size if all dimensions are scaled equally. Note, the relationship between capacitance and the displacements depends on overlap orientation of the plates and Eq. (6.2) assumes the orientation in Figure 6.1.

Micromechanical capacitance values are typically femtofarads to picofarads with capacitor gaps that are most often micron or sub-micron in scale. The macro-scale electric field breakdown strength of air is around $3\,\mathrm{MV\,m^{-1}}$. However, because micromechanical gaps are relatively close to the mean free path of air (67 nm at 1 atm), breakdown occurs at much larger field strengths [11]. Gas composition, humidity, pressure, electrode material, and surface roughness affect the breakdown voltage. In dry air at 1 atm, a measured minimum breakdown voltage between Si surfaces is 380 V at a gap of $2\,\mu\mathrm{m}$ ($190\,\mathrm{MV\,m^{-1}}$) and 300 V at

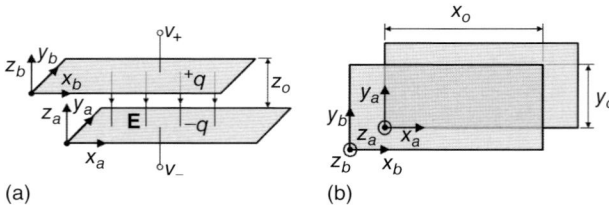

Figure 6.1 Parallel-plate capacitor with plates free to move in three orthogonal directions. (a) Three-quarter view. (b) Plan view.

a gap of $5\,\mu m$ $(60\,MV\,m^{-1})$ [12]. For metal electrodes, a greater amount of field emission current arises at micron-scale gaps, leading to the breakdown being roughly limited by a constant minimum breakdown field for gaps below around $7\,\mu m$ [13]. The ability to apply large electric fields in MEMS devices leads to high capacitive sensitivity and electrostatic force relative to macro-scale counterparts.

6.3
Electromechanical Actuation

Capacitive forces generally arise from charge interaction between electrodes. For MEMS devices smaller than $1\,mm$, the electrostatic field approximation (sometimes called an "electroquasistatic" approximation) applies for electromagnetic wavelengths above $100\,mm$, corresponding to frequencies below $3\,GHz$. Mechanical devices resonating at gigahertz frequencies are much smaller than $1\,mm$, so electrostatic analysis of the force generally always applies for these MEMS devices as well. A partial exception to this rule can be for system-level analysis of radio-frequency (RF) MEMS that includes interconnections longer than about $1/10$ of a wavelength. Therefore, capacitive forces in MEMS are generally called "electrostatic forces" even when they are time varying.

6.3.1
Electromechanical Force Derivation

Electromechanical force is derived from power conservation equating the total electrical power into a system to the rate of change in electrical potential energy stored, W_e, within the system plus the rate of mechanical work accomplished [14]. For a single-mass resonator,

$$\sum_{j=1}^{n} v_j \frac{dq_j}{dt} = \frac{d}{dt} W_e(q_1, \ldots, q_n, \mathbf{r}) + \mathbf{F}_e \cdot \frac{d\mathbf{r}}{dt} \tag{6.3}$$

where v_j and q_j are the voltage and charge, respectively, on the jth port, $\mathbf{F}_e$ is the electrostatic force vector and $\mathbf{r}$ is the displacement vector of the mass. While Eq. (6.3) may be solved to find forces as functions of electrode charge and plate displacement, it is generally more appropriate to solve forces in terms of port voltages and displacements. Toward this end, a "co-energy," $\widetilde{W}_e$, of the system is defined such that

$$W_e(q_1, \ldots, q_n, \mathbf{r}) + \widetilde{W}_e(v_1, \ldots, v_n, \mathbf{r}) = \sum_{j=1}^{n} q_j v_j \tag{6.4}$$

Integrating Eq. (6.3) over time, taking the differential of Eq. (6.4) and substituting, a conservative relationship between co-energy and electrostatic force arises:

$$d\widetilde{W}_e(v_1, \ldots, v_n, \mathbf{r}) = \sum_{j=1}^{n} q_j \, dv_j + \mathbf{F}_e \cdot d\mathbf{r} \tag{6.5}$$

The total co-energy, $\widetilde{W}_e$, must be found through the path integration of Eq. (6.5), which yields multiple terms for multi-port systems. For a MEMS air-gap capacitor, the $q-v$ relation (Eq. (6.1)) is always linear, that is, the capacitance is only a function of displacement and not of voltage. For an electrically linear single-port system ($j = 1; q_1 = q = Cv$), the co-energy is

$$\widetilde{W}_e(v_1, \mathbf{r}) = \frac{1}{2} C(\mathbf{r}) v^2 \tag{6.6}$$

Substituting Eq. (6.6) into Eq. (6.4), the total electrical potential energy and the total co-energy are numerically equal for an electrically linear system. The electrostatic force is found by taking the gradient of Eq. (6.6) while holding all voltages in the system constant. In the case of a single-port (i.e., one motional capacitor), the electrostatic force is

$$\mathbf{F}_e = \nabla \widetilde{W}_e|_v = \frac{1}{2} v^2 \left(\frac{dC}{dx}\hat{\mathbf{x}} + \frac{dC}{dy}\hat{\mathbf{y}} + \frac{dC}{dz}\hat{\mathbf{z}} \right) \tag{6.7}$$

where $\hat{\mathbf{x}}$, $\hat{\mathbf{y}}$, and $\hat{\mathbf{z}}$ are unit vectors. Equation (6.7) illustrates that a single capacitor can produce multiple components of force. The suspension type, dimensions, and layout of a capacitor meant for actuation is generally designed to select, or accentuate, a single desired force component.

Substituting Eq. (6.2) into Eq. (6.7) gives a general expression for the force on a parallel-plate actuator with nonaligned plates as in Figure 6.1:

$$\mathbf{F}_e = \frac{1}{2} \frac{\varepsilon_0 \varepsilon_r v^2}{(z_o + z)} \left((y_o + y)\,\hat{\mathbf{x}} + (x_o + x)\hat{\mathbf{y}} - \frac{(x_o + x)(y_o + y)}{(z_o + z)}\hat{\mathbf{z}} \right) \tag{6.8}$$

Each force component is increased by reducing the gap between the two plates. This is a typical means for increasing transduction efficiency but is limited by the process technology, by the desired stroke, and by snap-in phenomena as discussed in Section 6.4.3.

In the special case where the plate electrodes maintain a fixed overlap area (e.g., when one plate is larger than the other), the force has only a z-directed component:

$$\mathbf{F}_e = -\frac{1}{2} \frac{\varepsilon_0 \varepsilon_r x_o y_o}{(z_o + z)^2} v^2 \hat{\mathbf{z}} \tag{6.9}$$

which is a highly nonlinear function of the separation of the two plates. Note that the parallel-plate design may be oriented along any axis, not just out of the wafer plane as implied by Figure 6.1, as will be explored in Section 6.4 on motional capacitor topologies.

6.3.2
Voltage Dependent Force Components

The v^2 nonlinearity on the electrostatic force (Eq. (6.7)) enables several scenarios that may be exploited for actuation. One of the most common cases is application of a sinusoidal drive signal, v_d, of amplitude V_d and frequency ω_d to one electrode *a* of the motional capacitor and application of a DC voltage V_p, on the opposing electrode *b*. Alternatively, the two voltages can be applied in series to one of the electrodes. The result in the former case is that the voltage from electrode *b* to *a* is

$$v = V_p - V_d \cos \omega_d t \tag{6.10}$$

Substituting in Eq. (6.7) yields the resulting electrostatic force,

$$\mathbf{F}_e = \frac{1}{2} \left(\frac{dC}{dx}\hat{\mathbf{x}} + \frac{dC}{dy}\hat{\mathbf{y}} + \frac{dC}{dz}\hat{\mathbf{z}} \right) \left(V_p^2 + \frac{V_d^2}{2} - 2V_p V_d \cos \omega_d t + \frac{V_d^2}{2} \cos 2\omega_d t \right) \tag{6.11}$$

which has a pair of DC terms, a term at the drive frequency, ω_d, and a second drive term at twice the drive frequency, $2\omega_d$. The sign of V_p sets the sign of the ω_d term and so V_p is often called a "polarization voltage." For the case where $V_p \gg V_d$, the ω_d drive term is much larger than the $2\omega_d$ term, and this double frequency term may usually be neglected if ω_d is set to the system resonance frequency.

If $V_p = 0$, then the force in Eq. (6.11) has a DC term and a $2\omega_d$ term, but no ω_d term. In this case, the system may be driven into resonance by setting ω_d to half of the resonant frequency. One advantage of this latter approach is that any electrical feedthrough capacitive current from v_d is at half of the frequency of the electromechanical motional displacement current at resonance and so can be separated by filtering. However, high-frequency modulation techniques, visited at the end of this section and discussed further in Section 6.7, provide superior sensing results.

In many resonant devices, the large DC force is undesirable as it may result in a large DC operating-point displacement. Excessive values of polarization voltage can even lead to the capacitor plates snapping together, as discussed in Section 6.4.3. A common solution that eliminates the DC force is a symmetric placement of drive capacitors attached to the resonator's mass, as exemplified in the physical schematic of Figure 6.2. The capacitor electrodes attached to the mass, *m*, are connected to the polarization voltage. The opposing electrodes are driven with AC voltage sources that are 180° out of phase and with the same

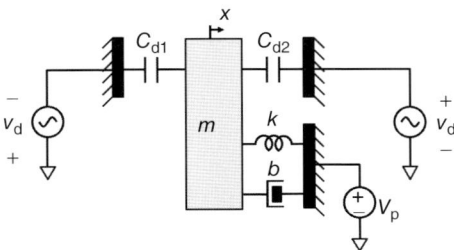

Figure 6.2 Physical schematic of a differentially driven mass-spring-damper (*m-k-b*) resonator with symmetrically placed motional drive capacitors, C_{d1} and C_{d2}.

amplitude V_d and frequency ω_d. The total force acting on the mass is given by the superposition of the individual capacitive forces, governed by Eq. (6.11). This total force is

$$\mathbf{F}_e = V_p V_d \left(\frac{dC}{dx}\hat{\mathbf{x}} + \frac{dC}{dy}\hat{\mathbf{y}} + \frac{dC}{dz}\hat{\mathbf{z}} \right) \cos \omega_d t \qquad (6.12)$$

Exploiting the symmetry of the capacitive placement, the DC and $2\omega_d$ terms of each capacitor are common-mode and cancel, leaving only the ω_d term.

For a single degree-of-freedom (DoF) mass-spring-damper system that is driven and moves in the direction x, the sum of the electrostatic force and the reaction forces leads to the canonical second-order differential equation:

$$F_{e,x} = m\frac{d^2x}{dt^2} + b\frac{dx}{dt} + kx = m\left(\frac{d^2x}{dt^2} + \frac{\omega_r}{Q}\frac{dx}{dt} + \omega_r^2 x \right) \qquad (6.13)$$

where m is the effective mass, b is the effective damping coefficient, k is the system spring constant, $\omega_r = \sqrt{k/m}$ is the angular resonant frequency and $Q = \sqrt{km}/b$ is the quality factor. Substituting Eq. (6.12) into Eq. (6.13), the resulting x displacement at resonance (i.e., when $\omega_d = \omega_r$) is

$$x = \frac{Q}{k}V_p V_d \frac{dC}{dx} \sin \omega_r t \qquad (6.14)$$

The displacement amplitude is directly proportional to the polarization voltage and to the drive amplitude and is 90° lagging the AC drive voltage.

Various modulation components, beyond the $2\omega_d$ modulation described earlier, can be created by applying various combinations of sinusoidal voltages across motional capacitors. One particularly useful example of modulation is a modification of the setup in Figure 6.2 where a sinusoidal voltage with amplitude V_a and frequency ω_a substitutes for the polarization voltage V_p that was connected to the mass. The resulting electromechanical force, assuming only x-directed force, is

$$F_{e,x} = V_a V_d \frac{dC}{dx} (\cos[(\omega_d - \omega_a)t] + \cos[(\omega_d + \omega_a)t]) \qquad (6.15)$$

For the case where $|\omega_d - \omega_a| = \omega_r$, the displacement is given by Eq. (6.14) with V_a replacing V_p. If furthermore both $\omega_d \gg \omega_r$ and $\omega_a \gg \omega_r$, then the second $\omega_d + \omega_a$ term in Eq. (6.15) is well above resonance and creates negligible displacement. The advantage of this driving scheme is that any electrical feedthrough capacitive current from the ω_d and ω_a drive voltages that may enter into capacitive detection circuits can be very easily filtered since the frequency content of the feedthrough is far from the resonant displacement signal [15]. This kind of drive modulation is also exploited in RF MEMS mixer-filters [16].

More complex modulation arises when coupled with motional capacitors having displacement dependence. Some aspects of such displacement modulation will be discussed in Section 6.7.

6.4
Capacitive Sensing and Motional Capacitor Topologies

The current–voltage relation for a single-port (i.e., one electrode pair) microme-chanical capacitive device is

$$i = \frac{dq}{dt} = C\frac{dv}{dt} + v\frac{dC}{dt} \tag{6.16}$$

where the first term on the right-hand side is the conventional electrical displace-ment current and the second term on the right-hand side is an electromechani-cal motional displacement current that is present when the capacitance changes with time.

Capacitive sensing is generally confined by appropriate flexure design to one of three dynamic configurations of parallel-plate capacitors: parallel-moving plates, perpendicularly moving plates, and angularly moving plates. Some design devi-ations from these basic configurations are discussed in Chapter 4 on parametric resonance.

6.4.1
Parallel-Moving Plates

Parallel-moving plates in capacitive microresonators are most often configured as a set of interdigitated fingers, known as a "lateral comb" sensor or "comb drive" [17] and shown schematically in Figure 6.3a, b. Note, the axes are modified from those in Figure 6.1 to correspond to a typical layout design, where the $x-y$ axes are defined parallel to the substrate plane and the z axis is defined to point out of plane. It is assumed that the flexure compliances are designed so that the motion is only

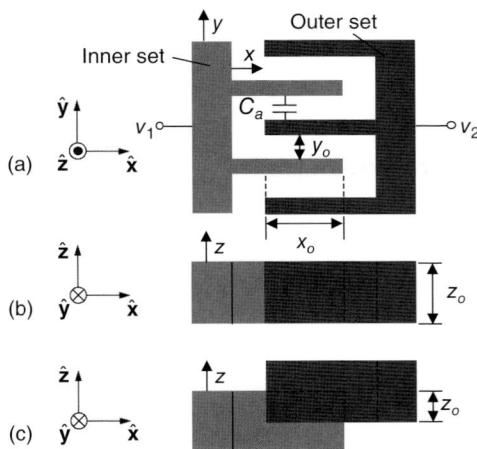

Figure 6.3 Lateral comb schematic. (a) Plan view. (b) Side view with aligned fingers. (c) Side view with offset fingers. The different gray color fill distinguishes the two sets of fin-gers and does not denote a difference in materials.

along the x-direction, that is, only the displacement difference x between the finger sets is non-zero in Eq. (6.2). Most comb capacitors are designed with all fingers of one set (the "inner" set) being overlapped on both sides by the fingers of the other set (the "outer" set) in order to cancel electrostatic forces in y by symmetry as long as $y = 0$ through flexure constraints. Sufficiently high voltages can cause electrostatic instability, which is discussed in Section 6.4.3.

An example double-ended tuning-fork resonator with two masses and four lateral comb capacitors is shown in Figure 6.4a, where the combs as shown are oriented 90° relative to Figure 6.3a [7]. The outer two combs drive the masses into *anti*-phase motion, while the inner two combs transduce the motion into an electrical signal. The fixed-fixed flexure constrains motion to the x direction (using the axes in Figure 6.3a). Combs can also follow curved geometries. The parallel-moving interdigitated plate design in Figure 6.4b enables capacitive sensing or actuation of in-plane rotational displacement around the z-axis [18].

For N inner fingers in a single lateral comb, there are $2N$ sidewall capacitors in parallel leading to a total capacitance,

$$C(x) = 2N C_a(x) = 2N \frac{\varepsilon_0 (x_o + x) z_o}{y_o} \tag{6.17}$$

where C_a is the capacitance between finger sidewalls and the geometry is defined in Figure 6.3 with finger thickness z_o, finger gap y_o, and finger overlap x_o. The motional sensitivity is

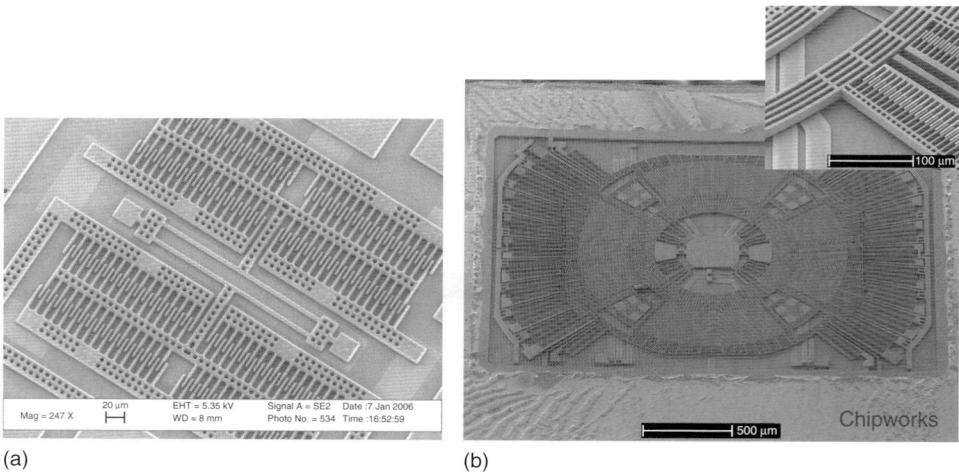

(a)

(b)

Figure 6.4 Plan view schematics of lateral comb capacitors with motion parallel to the plate surfaces. (a) Polycrystalline silicon carbide resonant strain gage with two fixed-fixed flexures and four lateral comb capacitors [7]. (Photo provided courtesy of Muthu B.J. Wijesundara.) (b) A z-axis rotary comb resonator in the ST LYPR540AH tri-axis gyroscope. Inset: Close-up of the rotary drive comb [18]. (Photos provided courtesy of Chipworks.)

$$S_{C,x} = \frac{dC}{dx} = 2N\frac{\varepsilon_0 z_o}{y_o} \tag{6.18}$$

and is not dependent on the displacement x nor on the finger overlap x_o.

By substituting Eq. (6.17) into Eq. (6.7), the x-directed electrostatic force is

$$\mathbf{F}_e = \frac{N\varepsilon_0 z_o}{y_o} v^2 \hat{\mathbf{x}} \tag{6.19}$$

where $v = v_1 - v_2$. A key attribute of the lateral comb drive is that its force is independent of the x displacement. The force scales linearly with the number of inner fingers and with the finger thickness and inversely with the finger gap.

Parallel-motion combs may also be used to detect motion or to actuate in the z direction. If the fingers are aligned in the z direction, as in Figure 6.3b, then the overlap in the z direction is $z = z_o - |z|$ and the capacitive sensitivity changes sign when z passes through zero. This nonlinearity is useful for parametrically driven resonance (see Chapter 4), but does not allow for a direct harmonic drive. Several versions of linear combs for z-directed transduction have been designed with the fingers offset in z, as in Figure 6.3c. The silicon vertical comb drives in Figure 6.5a are made through a self-aligned wafer bonding and deep reactive ion etch (DRIE) process [19] and actuate a torsional-scanning micromirror [20]. Vertical comb fingers with an angled offset may be formed using intrinsic-stress-based self-assembly performed after processing, as evidenced by the CMOS-MEMS (complementary metal-oxide-semiconductor) micromirror design shown in Figure 6.5b [21]. Angular offset has also been achieved through self-assembly by photoresist reflow [4].

6.4.2
Perpendicular Moving Plates

Resonator designs having a relatively small displacement often benefit from the incorporation of perpendicular moving plates (i.e., motion normal to the plate surface) because of the resulting high-capacitive motional sensitivity per unit area. Conventional planar MEMS processing enables ready implementation of

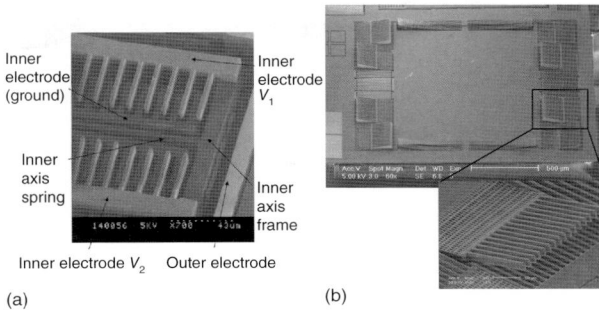

Inner electrode (ground)

Inner axis spring

Inner electrode V_2 Outer electrode

Inner electrode V_1

Inner axis frame

(a)

(b)

Figure 6.5 (a) Silicon vertical comb drives that are offset in the z direction [20]. (b) Vertical comb drive angularly offset in z through intrinsic-stress-based self-assembly [21].

the single parallel-plate capacitor oriented for *z*-axis motion with capacitance given by Eq. (6.2) and force given by Eq. (6.9). An example is the Draper Laboratory tuning-fork gyroscope shown in Figure 6.6 [22]. The two proof masses are perforated polysilicon plates that are driven *anti*-phase in the *x* direction with lateral comb drives. When an input rotational rate, Ω, occurs around the y axis, the *anti*-phase motion of the two proof masses in the *z* direction due to Coriolis force is sensed by the parallel-plate capacitors.

Lateral-axis motion in MEMS resonators is commonly detected using the parallel-plate interdigitated-finger topology shown in Figure 6.7a. This topology is an efficient means for large parallel-plate capacitance per layout area and is sometimes used in MEMS gyroscopes (Figure 6.7b) [23] and other resonant

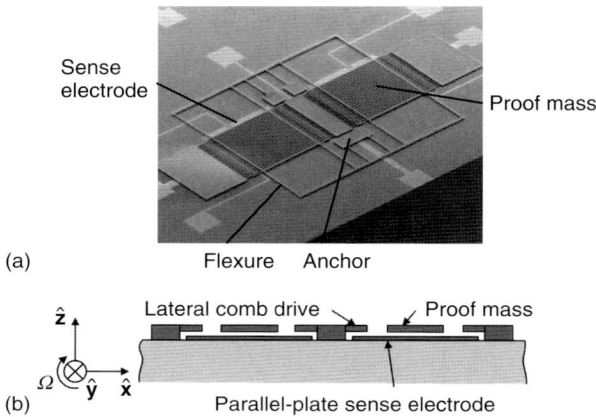

Figure 6.6 (a) Draper Laboratory dual-mass tuning-fork gyroscope with parallel-plate capacitive transducers in the z direction (out of the substrate plane) [22]. (b) Cross section view. (Photo provided courtesy of The Charles Stark Draper Laboratory, Inc.)

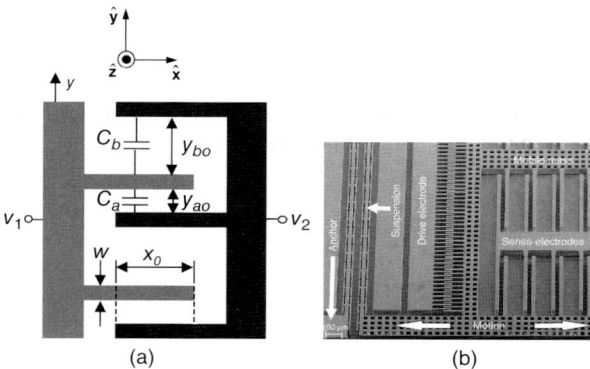

Figure 6.7 (a) Plan view of an interdigitated-finger schematic intended for motion in the *y* direction that is perpendicular to the parallel-plate surfaces. (b) A portion of a *z*-axis gyroscope with a parallel-plate comb (labeled "sense electrodes") and a lateral comb drive (labeled "drive electrode") [23].

devices. The primary design constraint is that the output displacement must be smaller than the capacitor gap.

The total capacitance for such a parallel-plate capacitor with N inner fingers is

$$C(y) = N[C_a(y) + C_b(y)] = N\varepsilon_0 x_o z_o \left(\frac{1}{y_{ao} + y} + \frac{1}{y_{bo} - y} \right) \tag{6.20}$$

and the motional sensitivity is

$$S_{C,y} = \frac{dC}{dy} \approx N\varepsilon_0 x_o z_o \left(\frac{1}{y_{bo}^2} - \frac{1}{y_{ao}^2} \right) \tag{6.21}$$

where it is assumed that $y \ll y_{ao}$ and $y \ll y_{bo}$. A useful design metric is sensitivity per unit layout area

$$\frac{S_{C,y}}{A} \approx \frac{\varepsilon_0 z_o}{2w + y_{ao} + y_{bo}} \left(\frac{1}{y_{bo}^2} - \frac{1}{y_{ao}^2} \right) \tag{6.22}$$

where the layout area of the comb trusses and nonoverlapping parts of the fingers is omitted. The maximum sensitivity per unit area is found by taking the derivative of Eq. (6.22) with respect to y_{bo} and setting it to zero. This maximum occurs at a gap ratio of

$$\frac{y_{bo}}{y_{ao}} = \sqrt[3]{1 + 2\beta + 2\sqrt{\beta(1+\beta)}} + \sqrt[3]{\frac{1}{1 + 2\beta + 2\sqrt{\beta(1+\beta)}}} \tag{6.23}$$

where $\beta = w/y_{ao}$. If the finger width $w = y_{ao}$, the maximum sensitivity per unit area occurs when $y_{bo} \approx 2.35 y_{ao}$.

Substituting Eq. (6.21) into Eq. (6.7), the corresponding first-order electrostatic force acting on the left-hand side of the parallel-plate capacitor in Figure 6.7a is

$$\mathbf{F}_e \approx \frac{N\varepsilon_0 x_o z_o}{2} \left(\frac{1}{y_{bo}^2} - \frac{1}{y_{ao}^2} \right) v^2 \hat{\mathbf{y}} \tag{6.24}$$

once again assuming y is small compared to the gaps. The force acting on the right-side of the comb is of equal and opposite sign. The maximum force per unit area follows Eq. (6.23). Note that if $y_{ao} < y_{bo}$ as indicated in the example drawing in Figure 6.7a, the force in Eq. (6.24) is negative valued and acts to close the gap y_{ao}.

Lateral parallel-plate capacitors intended to sense or actuate motion perpendicular to the plate surfaces are sometimes designed in curved geometries, particularly radial geometries [24–27]. For example, the disk resonator design in Figure 6.8b with two semi-circular sidewall parallel-plate capacitors enables drive and sense of the disk's radial contour mode [25]. The polysilicon disk has a 34 μm diameter and a first contour-mode resonance at 156 MHz. The electroplated gold outer electrodes are self-aligned to the disk by a 0.1 μm-thick sacrificial oxide gap. A different approach to stimulating and detecting high-frequency bulk-mode resonance is "internal dielectric transduction," where the resulting transduction is effectively electrostriction [28]. The method, where parallel-plate capacitors are formed with gaps filled with dielectric, instead of air, has been used to make "resonant body transistors" that resonate at gigahertz frequencies [29]. The nanoscale-sized gaps enable detection of sub-nm resonant deformation of the gap.

Another example of a curved parallel-plate topology is shown in Figure 6.8b. The threefold-symmetric z-axis gyroscope design includes 16 sidewall capacitive electrodes around its perimeter to allow drive, sense, and tuning of its radial "wine-glass" mode [24, 30].

6.4.3
Electrostatic Spring Softening and Snap-In

The first-order electrostatic force in Eq. (6.24) leads to the conclusion that the force is constant with displacement; however, this is an approximation that holds only for small displacements. Even a small amount of force dependent on displacement can have significance in resonators as it affects the overall system spring constant and hence the resonant frequency. For the lateral parallel-plate design of Figure 6.7a, inclusion of non-zero displacement leads to the force

$$\mathbf{F_e} = \frac{N\varepsilon_0 x_o z_o}{2} \left(\frac{1}{\left(y_{bo} - y\right)^2} - \frac{1}{(y_{ao} + y)^2} \right) v^2 \, \hat{\mathbf{y}} \tag{6.25}$$

Taking the Taylor's series expansion:

$$\mathbf{F_e} = \frac{N\varepsilon_0 x_o z_o}{2} v^2 \left[\sum_{j=1}^{\infty} j \left(\frac{1}{y_{bo}^{j+1}} + \frac{(-1)^j}{y_{ao}^{j+1}} \right) y^{j-1} \right] \hat{\mathbf{y}} \tag{6.26}$$

In the general case, the force comprises an infinite set of polynomials terms, y^j for $j = 0$ to ∞. The special case of the comb capacitor where the fingers are centered (i.e., where $y_{ao} = y_{bo}$), only the odd displacement terms ($y, y^3, y^5, \dots$) are non-zero. The coefficient of the first, y, term may be interpreted a linear electrostatic spring

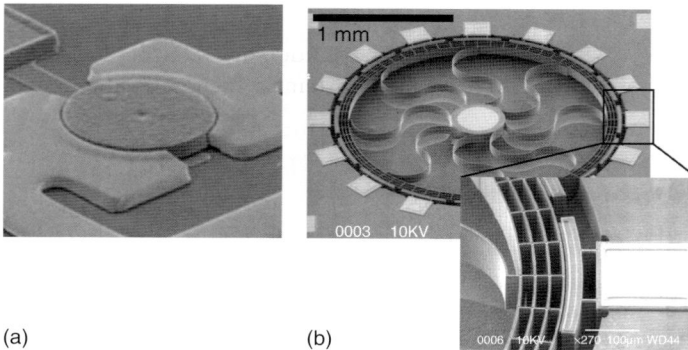

(a)   (b)

Figure 6.8 Examples of curved resonant capacitive structures. (a) A silicon z-axis disk resonator for timing applications [24, 30]. (b) A silicon ring gyroscope with 16 polysilicon electrodes. Inset – a sidewall capacitor [25]. (Photos in (b) provided courtesy of Farrokh Ayazi.)

constant,

$$k_{e,y} = -N\varepsilon_0 x_o z_o \left(\frac{1}{y_{bo}^3} + \frac{1}{y_{ao}^3} \right) v^2 \tag{6.27}$$

within a mass-spring-damper system. The negative sign in Eq. (6.27) arises by moving the linear displacement force term to the reaction side of a canonical mass-spring-damper system equation. The parallel-plate electrostatic spring effect "softens" the system by reducing the overall system spring constant,

$$k_y = k_{m,y} + k_{e,y} \tag{6.28}$$

where $k_{m,y}$ is the mechanical spring constant of the system. The resulting radian resonant frequency in a canonical spring-mass-damper system is

$$\omega_{r,y} = \sqrt{\frac{k_{m,y} + k_{e,y}}{m}} \tag{6.29}$$

One of the most useful features of Eq. (6.29) is that the resonant frequency can be tuned electrically by changing the voltage, v, applied to the parallel-plate comb.

In static (e.g., non-resonant) operation, the force in Eq. (6.25) is equal to the restoring force from the mechanical spring:

$$\mathbf{F}_e = k_{m,y} y \, \hat{\mathbf{y}} \tag{6.30}$$

This relation has a physically realizable static displacement solution that is stable for a sufficiently small applied voltage, v. An instability phenomenon called electrostatic "snap-in" or "pull-in" occurs when the curvature (i.e., the second derivative) of the total system potential, including electrical and mechanical potential energy, is less than or equal to zero [31]. For the lateral parallel-plate design, the onset of the snap-in condition is equivalent to having the first derivative of the left- and right-hand sides of Eq. (6.30),

$$\frac{dF_{e,y}}{dy} = k_{m,y} \tag{6.31}$$

hold simultaneously at the static displacement solution derived from Eq. (6.30). The total stiffness of the electromechanical system is then zero; the system becomes unstable and the plates will snap together. For example, in the special case where $y_{ao} = y_{bo}$, the static displacement solution remains at $y = 0$ for all voltages. In the further special case of a balanced differential comb resonator design such as in Figure 6.2, the finger overlap x_o is not a function of the DC applied voltage. Substituting $y = 0$ in Eq. (6.31) and solving for the static "snap-in" voltage yields

$$v_{\text{snap-in}} = \sqrt{\frac{k_{m,y} y_{ao}^3}{2N\varepsilon_0 x_o z_o}} \tag{6.32}$$

The dynamic snap-in voltage while the system is moving in resonance will be lower than this static case, and requires more complex calculations [31–34]. Above the snap-in voltage, the plates will move in the y direction until they come into contact. Due to the geometric symmetry of this case, one cannot predict which way the plates will move to pull in. In a perfectly manufactured comb, the snap-in direction is determined by thermomechanical noise; however, in practice, the preferred snap-in direction will be determined by manufacturing imperfections.

While Eq. (6.32) is specific to the interdigitated comb, the preceding analysis for softening and snap-in behavior holds for any parallel-plate transducer with perpendicular moving plates. The snap-in displacement and voltage is highly dependent on the particular motional capacitor topology. For example, for the lateral comb drive, the snap-in occurs at $y = 0$ rather than at the well-known solution of one-third of the initial gap for a spring connected to a single-sided parallel-plate capacitor. The electrostatic tuning and snap-in effects must be analyzed, taking into account the cumulative effects of all the motional capacitors connected to the system, as their combined behavior determines the displacement operating point of the system.

6.4.4
Angular Moving Plates

Torsional resonators that rotate out of plane, for example, in scanning mirrors (e.g., Figure 6.9a) [35] and in certain lateral-axis gyroscopes, often rely on angular moving plates for capacitive detection or electrostatic actuation. The canonical configuration shown in Figure 6.9b is constrained by its flexure to only rotate out of plane around its lateral y axis with angle, θ. The left- and right-hand capacitances, C_+ and C_- respectively, are

$$C_\pm(\theta) = \frac{\varepsilon_0 W}{\theta} \ln \left[\frac{h_o + \left(\pm x_c + \frac{L}{2}\mathrm{sgn}\theta \right) \tan \theta}{h_o + \left(\pm x_c - \frac{L}{2}\mathrm{sgn}\theta \right) \tan \theta} \right] \tag{6.33}$$

where W is the width of the rectangular stator plates in the y direction, L is the length of the stator plates in the x direction, and h_o is the gap between the moving and stator plates when $\theta = 0$.

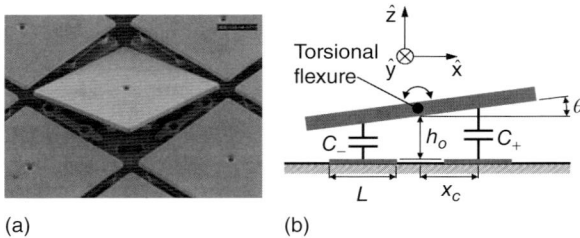

(a)　　　　　　　　　　　(b)

Figure 6.9 Torsional resonator. (a) Texas instruments micromirror [35]. (b) Side view schematic of a common topology for differential angular moving plate capacitors.

6.5
Electrical Isolation

Electrical connection to capacitive plates may be limited by process constraints. In systems made with a single conducting MEMS material, a single device must share a common capacitive electrode that is connected to the moving structure. Multiple capacitors associated with the device may be formed across gaps to adjacent stator or rotor structures.

Some processes include one or more insulating materials that can electrically isolate parts of a moving structure. For example, wafer-bonded structures such as the micromirror in Figure 6.5a include a silicon oxide layer between the structural silicon layers [19]. Some silicon inertial sensors employ oxide-filled trench isolation, as shown in the gyroscope in Figure 6.10a [36]. Mechanically, the device acts as a rigid body, but electrically, it forms two circuits that are coupled by a fixed capacitance. This arrangement allows distinct DC and AC voltages to be driven at these nodes.

A second example of isolation is by wiring electrodes in a CMOS-MEMS process, as indicated in Figure 6.10b. An example of this design in practice is the CMOS MEMS square-frame resonator in Figure 6.10c [37]. The CMOS back-end-of-line metal layers are made part of the moving structure and can be interconnected to distinct voltage sources or interface circuits.

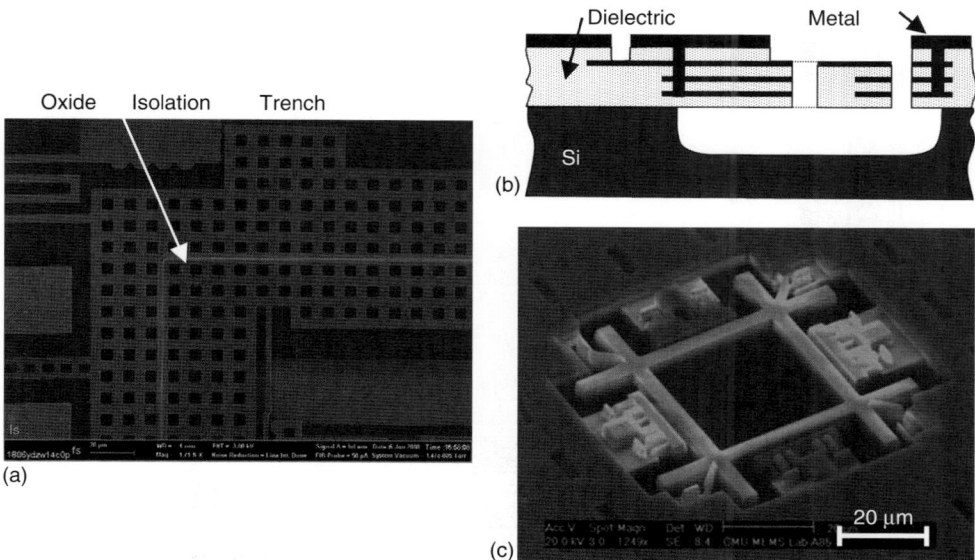

Figure 6.10   Electrically isolated MEMS examples. (a) Oxide-filled trench isolation. (Photo provided courtesy of B. Vigna, ST Microelectronics [36].) (b) Cross section CMOS MEMS schematic showing an example of isolation. The dotted line indicates the structure is connected. (c) CMOS MEMS square-frame resonator with four capacitor electrodes on the resonant structure [37].

6.6
Capacitive Resonator Circuit Models

A lumped model for an electromechanical capacitive transducer when operating at resonance is found by linearizing and combining the current–voltage relation (Eq. (6.16)) and the force-voltage relation (Eq. (6.7)). For a single degree-of-freedom (1-DoF, in x) capacitive resonator drive having a DC polarization voltage, V_p, in series with an AC voltage source, v_d, the corresponding linearized relations are:

$$i = C_o \frac{dv_d}{dt} + V_p \frac{dC}{dx} \frac{dx}{dt} \tag{6.34}$$

$$F_{e,x} = V_p \frac{dC}{dx} v_d \tag{6.35}$$

where C_o is the transducer capacitance at the operating point $x = 0$. For systems where the capacitive sensitivity, dC/dx, is constant, the linear relations are represented in a circuit form by the electromechanical transformer element and fixed capacitance in Figure 6.11a. The electromechanical transformer ratio is

$$\eta = V_p \frac{dC}{dx} \tag{6.36}$$

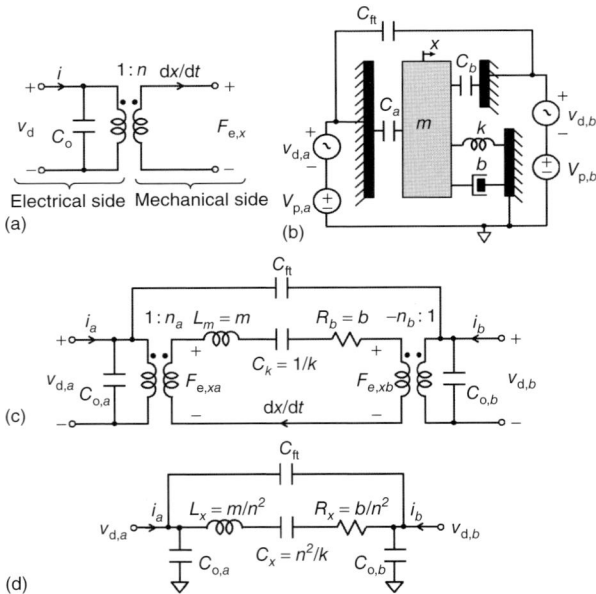

Figure 6.11 Linear lumped-parameter capacitive resonator equivalent circuit models. (a) Electromechanical transformer model of a capacitive transducer. (b) Conceptual schematic for a two-port 1-DoF capacitive resonator. (c) Corresponding equivalent circuit model. (d) Simplified circuit when $\eta = \eta_a = -\eta_b$.

An example 1-DoF resonator equivalent circuit that uses the electromechanical transformer model is shown in Figure 6.11c, corresponding to the conceptual schematic in Figure 6.11b. The circuit model is valid for analyzing the fundamental harmonic signal. The DC behavior, second-harmonic behavior, and other nonlinearities are not captured in this model. The equivalence between the differential equations for the mass-spring-damper and a series inductor-capacitor-resistor circuit is specified by the coefficients on the mechanical side of the circuit in Figure 6.11c. Equivalent "voltages" and "currents" on the mechanical side correspond to forces and velocities, respectively. This example has two capacitive transducers, labeled a and b, connected to the opposing sides of the effective mass of the resonator. The port voltages are small signal voltages and so do not directly correspond to the exact voltages impressed across the physical transducers. Specifically, the dc polarization voltages are not impressed on the ports but instead are present in the electromechanical transformer ratios. Any electrical circuits can be connected to the two ports; the connections do not have to be the voltage sources depicted in Figure 6.11b.

One important effect in many MEMS resonators is the presence of fixed electrical feedthrough capacitance, C_{ft}, that bridges between the electrical ports of the two capacitive transducers. Feedthrough capacitance arises from fringing electric field lines spanning between proximal motional capacitances. A much larger feedthrough capacitance is created when interconnect lines connecting to motional capacitances are crossed or are closely adjacent on the device chip. Careful attention to design eliminates this issue. An example resonant response with feedthrough is addressed in Section 6.7.1.

The linear circuit in Figure 6.11c is general in that it models any 1-DoF resonator where the electromechanical transformer ratios of the two capacitive transducers may be different. In microresonators with a solid electrically conducting mass, one terminal from each of the electrical ports in the equivalent circuit model is connected to the AC ground. If the two transformer ratios are the same, that is, $\eta = \eta_a = -\eta_b$, then an impedance transformation may be applied to the circuit elements on the mechanical side, resulting in the simpler circuit of Figure 6.11d. The $L-C-R$ circuit provides a compact two-port electrical model at frequencies around resonance and illustrates why such MEMS resonators can be considered series tank circuits. The transformed damping resistance is called the "motional resistance."

$$R_{\rm m} = \frac{b}{\eta^2} \tag{6.37}$$

Lower damping, higher polarization voltage, and higher capacitive sensitivity give rise to lower motional resistance.

In some resonator designs, more capacitive transducers are connected to the mass, for example, two for differential actuation and two for differential sensing. In these cases, correspondingly more electromechanical transformers would be added in series to the mechanical circuit in Figure 6.11c. Additional series $L-C-R$ circuits are required to model each additional degree of mechanical freedom. Spring coupling, with spring constant k_c, between two mass-spring-damper systems may be modeled by adding a shunt spring capacitance element, $C_{k_c} = 1/k_c$, to ground between the two equivalent $L-C-R$ circuits.

6.7
Capacitive Interfaces

Capacitive sensing interfaces may be categorized as measuring the voltage across, the current through or the charge on the capacitor. The choice of circuit topology is often constrained by the MEMS structure. For example, if the resonator is made of a homogeneously conducting material, such as silicon, then all drive and sense capacitors share an electrode corresponding to the motional node. Topologies must be chosen to avoid direct feedthrough and minimize capacitive feedthrough between the drive output and sense circuit input electrodes.

6.7.1
Transimpedance Amplifier

Motional current detection, such as the single-ended scheme in Figure 6.12a, is a very common sensing approach for resonant microsystems. A DC polarization voltage, V_p, is applied to one electrode of the motional capacitance, C_s. A transimpedance amplifier then keeps the other electrode at a fixed potential (e.g., electrical ground) through use of feedback and converts the motional current, i_x,

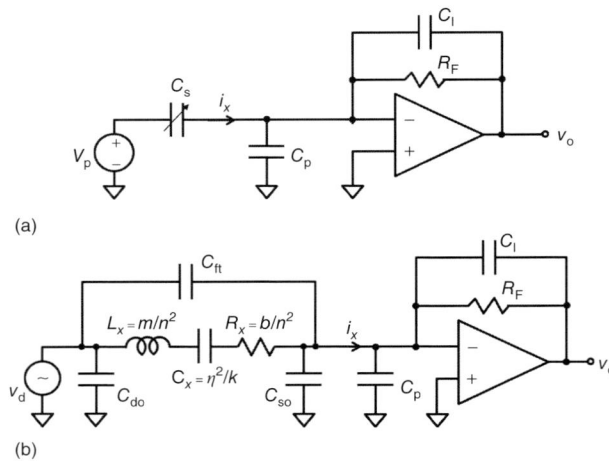

(a)

(b)

Figure 6.12 (a) Circuit of a transimpedance amplifier sensing motional current, i_x, generated by a time-varying motional capacitance, C_s. (b) Equivalent linear small-signal circuit of a transimpedance sense amplifier coupled to a two-port micromechanical resonator that is driven by an AC drive voltage, v_d. The sense capacitor model is embedded within the electromechanical transformer ratio, η.

to an output voltage, v_o. Advantages of this configuration are that the DC bias across the motional capacitance is fixed and that the parasitic capacitance (C_p) to ground does not reduce the sensitivity. The AC transfer function is

$$\frac{V_o(j\omega)}{X(j\omega)} = \frac{-j\omega R_F}{1 + j\omega R_F C_I} \frac{dC_s}{dx} V_P \tag{6.38}$$

Usually, the circuit is designed to operate in one of two modes. At frequencies where $\omega \ll 1/(R_F C_I)$, the output is proportional to velocity, dx/dt, and the gain is set by the feedback resistor, R_F. At frequencies where $\omega \gg 1/(R_F C_I)$, the output is proportional to displacement, x, and the gain is set by inverse of the integration capacitance, $1/C_I$.

The equivalent mixed electrical-micromechanical small-signal circuit in Figure 6.12b models the particular case where a two-port resonator is electrostatically driven with one motional capacitance, C_d, and sensed with a second motional capacitance, C_s. The drive and sense capacitances have corresponding quiescent capacitances at zero displacement of C_{do} and C_{so}, respectively. The linear coupled mechanical and electrical behavior assumes that the signal of interest is at the electrical drive frequency around resonance. The equivalent motional circuit elements and feedthrough capacitance were introduced previously in Section 6.6.

Figure 6.13 illustrates the effect of the feedthrough capacitance, C_{ft}, on the frequency response of the resonator in Figure 6.12b for the case where $\omega \ll 1/(R_F C_I)$. The polarization voltage is set to zero and thus $\eta = 0$ for the frequency response indicated by $|v_{o,ft}|$ in Figure 6.13. The micromechanical elements, L_x, C_x, and R_x then have infinite impedance and do not enter into the circuit analysis. The output response increases linearly with frequency as is expected for current through the fixed feedthrough capacitance. The polarization

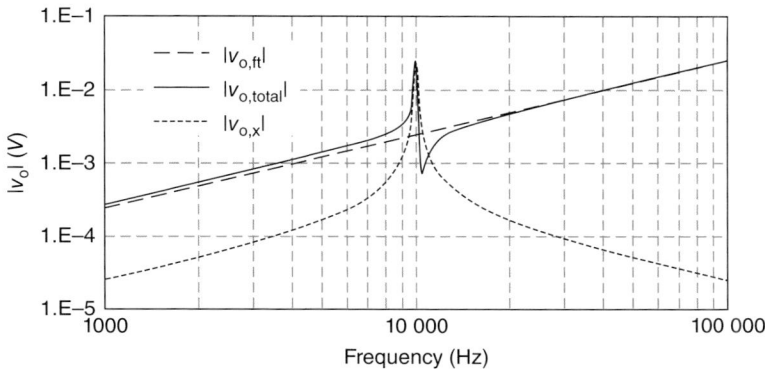

Figure 6.13 Frequency responses from the resonator equivalent circuit in Figure 6.12b with $|v_d| = 0\,V$. $|v_{o,ft}|$ is the response with polarization voltage set to zero, corresponding to feedthrough current. $|v_{o,total}|$ is the response with non-zero polarization voltage. $|v_{o,x}|$ is the difference of the responses, corresponding to motional current.

voltage and η are non-zero for the frequency response indicated by $|v_{\text{o,total}}|$. The equivalent micromechanical impedance becomes small at frequencies around the mechanical resonance, which is 10 kHz in this example, and a substantial motional current flows through the R_x branch. The phase of the motional current changes rapidly as the drive frequency advances through resonance. An *anti*-resonance null occurs at the frequency where the phase of the motional current is 180° out of phase with the feedthrough current. This null always occurs after the resonant peak. A common practice in microresonator characterization is to perform two measurement sweeps of the frequency response – one with the polarization voltage off, the other with the polarization voltage on – and then subtract the responses to obtain the motional current signal without the feedthrough. This subtracted response is indicated by $|v_{\text{o},x}|$.

It is common for circuit flicker noise to be larger than thermal noise sources even into megahertz frequencies and especially with CMOS circuits. MEMS resonance frequencies are often well below megahertz values and in those cases, the noise performance of the circuit in Figure 6.12b is affected by flicker noise.

Sufficiently high frequency electronic modulation will shift the resonant displacement signal out of the low-frequency band where flicker noise is dominant and achieve superior noise performance. The motional signal also becomes separated in frequency from the capacitive feedthrough current, which remains at the drive frequency. This system-level approach to reducing low frequency interference is called "chopper stabilization." In the MEMS literature, this technique is sometimes called "electromechanical amplitude modulation" [38, 39].

A conceptual schematic of a single-ended capacitive resonant device indicating a basic implementation of high-frequency modulation is shown in Figure 6.14. The modulation voltage, v_{m}, can be a sinusoidal signal or a square wave signal. In contrast to the models in Figure 6.12, the modulation component of the motional current through the sense capacitor, C_{s}, is decoupled from the polarization voltage. The appropriate circuit model for high-frequency modulation is identical to Figure 6.12a, but with V_{p} replaced by v_{m}.

The example differential transimpedance amplifier in Figure 6.15a illustrates the principle of chopper stabilization [40–42]. The feedback resistors are dashed

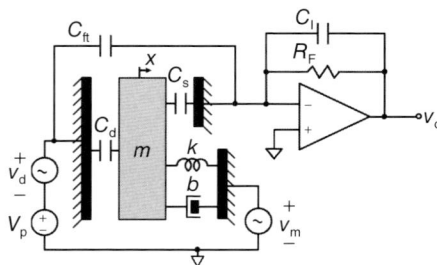

Figure 6.14 Conceptual schematic for a basic capacitive resonator driven electrostatically with AC drive v_{d} and with displacement x sensed using a high-frequency modulation source v_{m} that implements chopper stabilization.

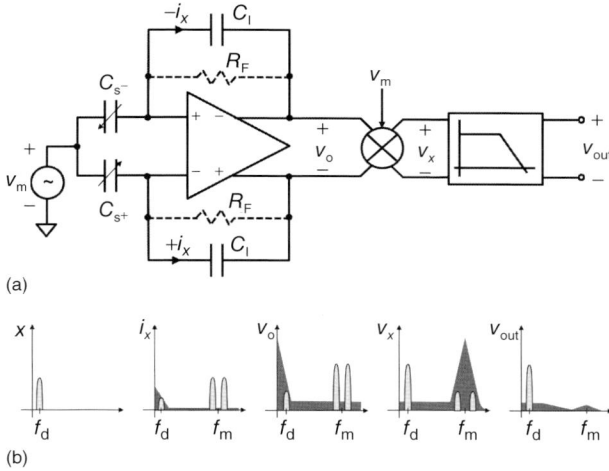

Figure 6.15 (a) Chopper-stabilized differential transimpedance amplifier for sensing motional current, i_x [40]. (b) Corresponding signals in the frequency-domain. The signals are depicted as peaks and the noise is shaded.

to indicate operation at high modulation frequency, $\omega_m \gg 1/(R_F C_I)$, where the motional current mainly passes through the integration capacitors. The pre-amplifier differential output is then

$$v_o = \frac{C_{s+} - C_{s-}}{C_I} v_m \tag{6.39}$$

Subsequent demodulation and low-pass filtering after the pre-amplifier reject dc bias and low-frequency drift and disturbance components. Figure 6.15b provides a visual description of the signal chain in the frequency domain to illustrate how chopper stabilization removes the flicker noise and feedthrough from the signal band.

Lateral comb capacitors with parallel-moving plates, have a highly linear characteristic and when used as the motional capacitances, C_{s+} and C_{s-}, they provide distinct modulation sidebands at $f_m \pm f_d$. Parallel-plate comb capacitors with perpendicularly moving plates can be approximated as linear for small displacements. As displacement becomes larger, approximately $p\%$ higher-order signal harmonics will exist at a motional amplitude of $p\%$ of the quiescent gap; this approximation holds up to around $p = 43\%$ of the gap for 5% accuracy. The nonlinear nature of the parallel-plate sense capacitors creates additional modulation sidebands at $f_m \pm k f_d$ where k is a positive non-zero integer. The ratio of adjacent sideband amplitudes (i.e., at k and $k + 1$) relate directly to the ratio of the motional amplitude to the quiescent gap, x/g, and can be used to compensate the gain to account for large-signal nonlinearity [23, 43].

6.7.2
High-Impedance Voltage Detection

Voltage detection is the second primary capacitive sensing scheme and is commonly implemented with a capacitive divider, as exemplified in Figure 6.16, where a differential capacitive divider is driven by balanced AC modulation sources, v_m. The high-impedance node, v_i, will inevitably have parasitic capacitance, C_{p1} and C_{p2}, to AC ground arising from interconnect, bond pads, and the pre-amplifier input stage. The portion of this capacitance (C_{p2}) that can be shielded can be eliminated from the sensitivity calculation by feeding back the unity-gain pre-amplifier output in a "bootstrapping" configuration. The DC bias, V_b, on the high-impedance node is usually set by a subthreshold-biased transistor or by a transistor switch in order to keep additional parasitic capacitance to a minimum. The former is usually incorporated into the amplifier circuit, while the latter requires clocking. Ignoring the DC bias, the output signal is

$$v_o = \frac{C_{s+} - C_{s-}}{C_{s+} + C_{s-} + C_{p1}} v_m \tag{6.40}$$

6.7.3
Switched-Capacitor Detection

The third primary scheme for capacitive motion detection is the switched capacitor circuit, which can take on many variations. One canonical form is shown in Figure 6.17 [40]. The capacitors are switched in three nonoverlapping clock phases. The first phase, Φ_z, resets all nodal voltages to create a zero differential output. The second phase, Φ_A, applies a DC voltage, V_p, to the input sense capacitors resulting in a corresponding differential signal at the output of the pre-amplifier. The hold capacitors, C_H, hold that signal as charge once Φ_A is switched off. The second and final phase, Φ_B, applies a negative DC voltage, $-V_p$, to the input sense capacitors resulting in a corresponding negative differential signal at the output of the pre-amplifier. The resulting hold capacitor charge is the sum of the signals from the Φ_A and Φ_B phases. The output voltage during the final clock phase Φ_B is

$$v_o = 2 \left(\frac{C_{s+} - C_{s-}}{C_{so} + C_I} \right) V_p \tag{6.41}$$

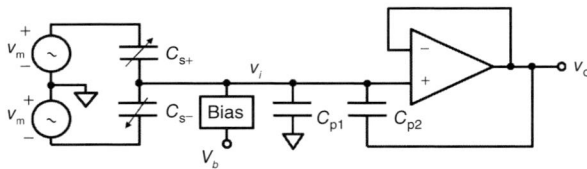

Figure 6.16 Voltage sensing circuit interfacing to a differential motional capacitance divider.

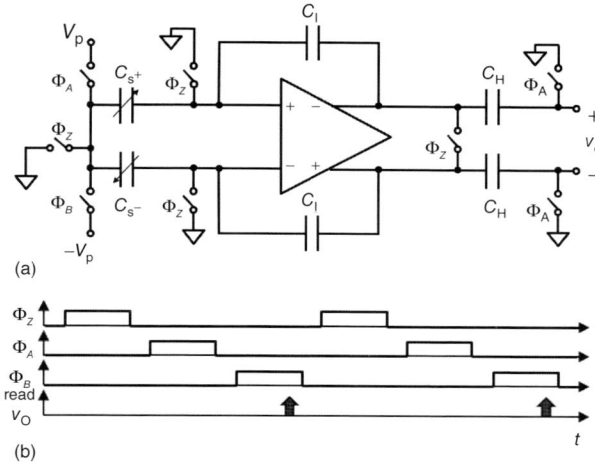

Figure 6.17 (a) A differential switched capacitor circuit that implements double correlated sampling [40]. (b) Corresponding clock signals.

This two-phase technique is called "double correlated sampling" or CDS [41, 42]. The two-phase signals are correlated and sum at the output. Given a clock that is faster than a few megahertz, any DC bias or low-frequency flicker noise and disturbances are also correlated between the two phases, but are subtracted at the output during phase Φ_B. Thus, the CDS approach is similar to chopper stabilization in that it provides the rejection of DC bias and low-frequency noise and disturbances.

6.8
Conclusion

Fundamentals of electrostatic actuation and capacitive sensing provide a toolbox to analyze the plethora of potential resonant devices that may be designed. Common motional capacitor topologies in resonant MEMS include lateral comb drives, lateral interdigitated parallel-plate fingers, vertical parallel plates and rotational-mode parallel plates. These topologies are all modeled to first order with various derivatives of parallel-plate approximations. Application of modulation frequencies with polarization voltages can give rise to multiple forcing terms, though the significant terms are generally only those stimulating resonance. Modulation techniques, including chopper stabilization and correlated double sampling, aid in removing DC offsets, low-frequency disturbances, and noise from the output of detection circuits. Equivalent circuit models of the second-order differential equations governing the micromechanical dynamics enable compact co-simulation and analysis of microresonators connected to these kinds of interface circuits.

Acknowledgment

The author gratefully acknowledges discussions with Dr. Peter J. Gilgunn in the formation of this book chapter. Effort by G. K. Fedder was financially supported in part by the U.S. National Science Foundation grant CNS 0941497 as part of the Cyber-enabled Discovery and Innovation (CDI) program.

References

1. Lee, J.E.-Y., Zhu, Y., and Seshia, A.A. (2008) A micromechanical electrometer approaching single-electron charge resolution at room temperature. Proceeding of the 21st International Conference on Micro Electro Mechanical Systems (MEMS 2008), pp. 948–951, doi: 10.1109/MEMSYS.2008.4443814

2. Bedair, S.S. and Fedder, G.K. (2007) Polymer mass loading of CMOS/MEMS microslot cantilever for gravimetric sensing. Technical Digest of the IEEE Sensors 2007 Conference, pp. 1164–1167, doi: 10.1109/ICSENS.2007.4388614

3. Xie, H. and Fedder, G.K. (2003) Integrated microelectromechanical gyroscopes. *J. Aerosp. Eng.*, **16** (2), 65–75. doi: 10.1061/(ASCE)0893-1321(2003)16:2(65)

4. Hah, D., Patterson, P.R., Nguyen, H.D., Toshiyoshi, H., and Wu, M.C. (2004) Theory and experiments of angular vertical comb-drive actuators for scanning micromirrors. *IEEE J. Sel. Top. Quantum Electron.*, **10**, 505–513. doi: 10.1109/JSTQE.2004.829200

5. Nguyen, C.T.-C. (2007) MEMS technology for timing and frequency control. *IEEE Trans. Ultrason. Ferroelectr. Freq. Control*, **54** (2), 251–270. doi: 10.1109/TUFFC.2007.240

6. Nguyen, C.T.-C. (2013) MEMS-based RF channel selection for true software-defined cognitive radio and low-power sensor communications. *IEEE Commun. Mag.*, **51** (4), 110–119. doi: 10.1109/MCOM.2013.6495769

7. Azevedo, R.G., Jones, D.G., Jog, A.V., Jamshidi, B., Myers, D.R., Chen, L., Fu, X., Mehregany, M., Wijesundara, M.B.J., and Pisano, A.P. (2007) A SiC MEMS resonant strain sensor for harsh environment applications. *IEEE Sens. J.*, **7** (4), 568–576. doi: 10.1109/JSEN.2007.891997

8. Daneman, M.J., Tien, N.C., Solgaard, O., Pisano, A.P., Lau, K.Y., and Muller, R.S. (1996) Linear microvibromotor for positioning optical components. *J. Microelectromech. Syst.*, **5** (3), 159–165. doi: 10.1109/84.536622

9. Nathanson, H.C. and Wickstrom, R.A. (1965) A resonant-gate silicon surface transistor with high-Q band-pass properties. *Appl. Phys. Lett.*, **7**, 84–86. doi: 10.1063/1.1754323

10. Nathanson, H.C., Newell, W.E., Wickstrom, R.A., and Davis, J.R. Jr., (1967) The resonant-gate transistor. *IEEE Trans. Electron Devices*, **ED-14** (3), 117–133. doi: 10.1109/T-ED.1967.15912

11. Paschen, F. (1889) Ueber die zum Funkenübergang in Luft, Wasserstoff und Kohlensäure bei verschiedenen Drucken erforderliche Potentialdifferenz. (On the potential difference required for spark initiation in air, hydrogen, and carbon dioxide at different pressures). *Ann. Phys.*, **273** (5), 69–75. doi: 10.1002/andp.18892730505

12. Ono, T., Sim, D.Y., and Esashi, M. (2000) Micro-discharge and electric breakdown in a micro-gap. *J. Micromech. Microeng.*, **10**, 445–451. doi: 10.1088/0960-1317/10/3/321

13. Tirumala, R. and Go, D.B. (2010) An analytical formulation for the modified Paschen's curve. *Appl. Phys. Lett.*, **97**, 151502. doi: 10.1063/1.3497231

14. Woodson, H.H. and Melcher, J.R. (1968) *Electromechanical Dynamics Part I: Discrete Systems*, Chapter 3, John Wiley & Sons, Inc., New York, pp. 60–102, (Massachusetts Institute of Technology: MIT OpenCourseWare,

License: Creative Commons BY-NC-SA), *http://ocw.mit.edu/resources/res-6-003-electromechanical-dynamics-spring-2009* (accessed 5 April 2014).

15. Clark, J.R., Hsu, W.-T., Abdelmoneum, M.A., and Nguyen, C.T.-C. (2005) High-Q UHF micromechanical radial-contour mode disk resonators. *J. Microelectromech. Syst.*, **14** (6), 1298–1310. doi: 10.1109/JMEMS.2005.856675

16. Wong, A.-C. and Nguyen, C.T.-C. (2004) Micromechanical mixer-filters ("mixlers"). *J. Microelectromech. Syst.*, **13** (1), 100–112. doi: 10.1109/JMEMS.2003.823218

17. Tang, W.C., Nguyen, T.-C.H., Judy, M.W., and Howe, R.T. (1990) Electrostatic-comb drive of lateral polysilicon actuators. *Sens. Actuators, A*, **21** (1-3), 328–331. doi: 10.1016/0924-4247(90)85065-C

18. iPhone 4 gyroscope teardown, *http://www.ifixit.com/Teardown/iPhone+4+Gyroscope+Teardown/3156*. File of main image uploaded by Miroslav Djuric; File in inset uploaded by Andrew Bookholt and cropped. Used with permission, license Creative Commons BY-NC-SA. Accessed 29 March 2014. Photos provided courtesy of Chipworks.

19. Krishnamoorthy, U., Lee, S., and Solgaard, O. (2003) Self-aligned vertical electrostatic combdrives for micromirror actuation. *J. Microelectromech. Syst.*, **12** (4), 458–464. doi: 10.1109/JMEMS.2003.811728

20. Jung, I.W., Krishnamoorthy, U., and Solgaard, O. (2006) High fill-factor two-axis gimbaled tip-tilt-piston micromirror array actuated by self-Aligned vertical electrostatic combdrives. *J. Microelectromech. Syst.*, **15** (3), 563–571. doi: 10.1109/JMEMS.2006.876666

21. Xie, H., Pan, Y., and Fedder, G.K. (2003) A CMOS-MEMS mirror with curled-hinge comb drives. *J. Microelectromech. Syst.*, **12**, 450–457. doi: 10.1109/JMEMS.2003.815839

22. Weinberg, M.S. and Kourepenis, A. (2006) Error sources in in-plane silicon tuning-fork MEMS gyroscopes. *J. Microelectromech. Syst.*, **15** (3), 479–491. doi: 10.1109/JMEMS.2006.876779

23. Trusov, A.A. and Shkel, A.M. (2007) A novel capacitive detection scheme with inherent self-calibration. *J. Microelectromech. Syst.*, **16** (6), 1324–1333. doi: 10.1109/JMEMS.2007.906077

24. Hsu, W.-T., Clark, J.R., and Nguyen, C.T.-C. (2001) A sub-micron capacitive gap process for multiple-metal-electrode lateral micromechanical resonators. Proceeding of the 14th International Conference on Micro Electro Mechanical Systems (MEMS 2001), pp. 349–352, doi: 10.1109/MEMSYS.2001.906550

25. Ayazi, F. and Najafi, K. (2001) A HARPSS polysilicon vibrating ring gyroscope. *J. Microelectromech. Syst.*, **10** (2), 169–179. doi: 10.1109/84.925732

26. Kubena, R.L. and Chang, D.T. (2009) Disc resonator gyroscopes. US Patent 7,581,443 B2, Assignee: The Boeing Company.

27. Stewart, R.E. (2012) Micro hemispheric resonator gyro. US Patent 8109145 B2. Assignee: Northrup Grumman.

28. Weinstein, D. and Bhave, S.A. (2009) Internal dielectric transduction in bulk-mode resonators. *J. Microelectromech. Syst.*, **18** (6), 1401–1408. doi: 10.1109/JMEMS.2009.2032480

29. Weinstein, D. and Bhave, S.A. (2010) The resonant body transistor. *Nano Lett.*, **10**, 1234–1237. doi: 10.1021/nl9037517

30. Discera *http://www.discera.com/* (accessed 3 October 2014).

31. Elata, D. and Bamberger, H. (2006) On the dynamic pull-in of electrostatic actuators with multiple degrees of freedom and multiple voltage sources. *J. Microelectromech. Syst.*, **15** (1), 131–140. doi: 10.1109/JMEMS.2005.864148

32. Nayfeh, A.H., Younis, M.I., and Abdel-Rahman, E.M. (2007) Dynamic pull-in phenomenon in MEMS resonators. *Nonlinear Dyn.*, **48** (1–2), 153–163. doi: 10.1007/s11071-006-9079-z

33. Leus, V. and Elata, D. (2008) On the dynamic response of electrostatic MEMS switches. *J. Microelectromech. Syst.*, **17** (1), 236–243. doi: 10.1109/JMEMS.2007.908752

34. Fargas-Marques, A., Casals-Terre, J., and Shkel, A.M. (2007) Resonant pull-in condition in parallel-plate

electrostatic actuators. *J. Microelectromech. Syst.*, **16** (5), 1044–1053. doi: 10.1109/JMEMS.2007.900893

35. Texas Instruments DLP Technology *http://www.dlp.com/technology/how-dlp-works/default.aspx* (accessed 3 October 2014).

36. Vigna, B. (2009) MEMS epiphany. Proceeding of the 22nd International Conference on Micro Electro Mechanical Systems (MEMS 2009), pp. 1–6, doi: 10.1109/MEMSYS.2009.4805304.

37. Lo, C.-C., Chen, F., and Fedder, G.K. (2005) Integrated HF CMOS-MEMS square-frame resonators with on-chip electronics and self-assembling narrow gap mechanism. Technical Digest of the 13th International Conference on Solid-State Sensors, Actuators, and Microsystems (Transducers '05), vol. 2, pp. 2074–2077; doi 10.1109/SENSOR.2005.1497511

38. Nguyen, C.T.-C. (1994) Micromechanical signal processors. PhD Dissertation, Department of Electrical Engineering and Computer Sciences, University of California at Berkeley, December 1994.

39. Cao, J. and Nguyen, C.T.-C. (1999) Drive amplitude dependence of micromechanical resonator series motional resistance.

Techinical Digest of the 10th International Conference on Solid-State Sensors and Actuators (TRANSDUCERS '99), Sendai, Japan, pp. 1826–1829.

40. Petkov, V.P. and Boser, B.E. (2004) *Enabling Technologies for MEMS and Nanodevices*, Advanced Micro and Nanosystems Book Series, vol. **1**, Chapter 3, Wiley-VCH Verlag GmbH, Weinheim, pp. 49–92. doi: 10.1002/9783527616701

41. Enz, C.C. and Temes, G.C. (1996) Circuit techniques for reducing the effects of op-amp imperfections: autozeroing, correlated double sampling, and chopper stabilization. *Proc. IEEE*, **84** (11), 1584–1614. doi: 10.1109/5.542410

42. Schreier, R., Silva, J., Steensgaard, J., and Temes, G.C. (2005) Design-oriented estimation of thermal noise in switched-capacitor circuits. *IEEE Trans. Circuits Syst. I*, **52** (11), 2358–2368. doi: 10.1109/TCSI.2005.853909

43. Trusov, A.A. and Shkel, A.M. (2007) Capacitive detection with in resonant MEMS with arbitrary amplitude of motion. *J. Micromech. Microeng.*, **17**, 1583–1592. doi: 10.1088/0960-1317/17/8/022

7
Piezoelectric Resonant MEMS

Gianluca Piazza

7.1
Introduction to Piezoelectric Resonant MEMS

Historically, the term microelectromechanical systems (MEMSs) has been used to identify sensors and actuators that are microfabricated using manufacturing processes based on those employed for the making of integrated circuits (ICs). In most cases, MEMS devices are fabricated on or from silicon wafers and made out of the same type of materials that are conventionally used for the synthesis of complementary metal oxide semiconductor (CMOS) technologies, such as polysilicon, silicon dioxide, silicon nitride, and aluminum. In the early stages of MEMS development (starting around 1960), piezoelectric materials were rarely used and mostly restricted to ZnO, as it was hard to access deposition techniques that would guarantee repeatable results. Furthermore, the most common piezoelectric used for the making of macroscale transducers was lead zirconate titanate (PZT), a material that was more difficult to process and integrate as a thin film on silicon, and was of concern to IC fabrication facilities because of contamination risks in CMOS lines. Nonetheless, the development of inkjet printing devices based on PZT transducers and more repeatable and controllable deposition methods of PZT films nurtured a continuous interest in thin piezoelectric films for MEMS technologies. With the introduction of aluminum nitride (AlN) and especially the commercial success of the thin-film bulk acoustic resonator (TFBAR or FBAR), interest in piezoelectrics for MEMS blossomed. The consequent investments in the development of repeatable and IC-compatible physical vapor deposition (PVD) techniques for the growth of AlN films on silicon have spurred a great deal of activities in the field of piezoelectric and AlN MEMS. Practically, most conventional MEMS devices that used to be made out of silicon have recently been reproduced (in most cases with enhanced performance) by using thin-film piezoelectric technology. For example, resonators [1–8], filters [9–12], switches [13–16], energy harvesters [17, 18], ultrasonic transducers [19, 20], microphones [21, 22], resonant chemical sensors [23], accelerometers [24, 25], and gyroscopes [26] have been demonstrated using piezoelectric thin films. Another interesting and relatively recent trend in the field of resonant piezoelectric MEMS has been

Resonant MEMS – Fundamentals, Implementation and Application, First Edition.
Edited by Oliver Brand, Isabelle Dufour, Stephen M. Heinrich and Fabien Josse.
© 2015 Wiley-VCH Verlag GmbH & Co. KGaA. Published 2015 by Wiley-VCH Verlag GmbH & Co. KGaA.

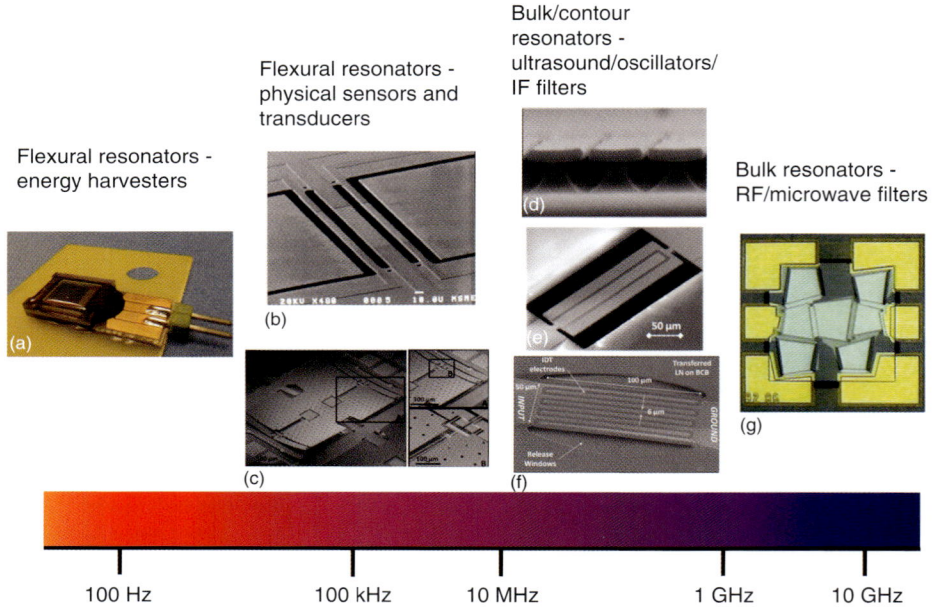

Figure 7.1 Plot of piezoelectric devices versus frequency with example of possible applications. (a) AlN piezoelectric harvester from Ref. [18]. (b) ZnO resonant beam from Ref. [31]. (c) AlN resonant accelerometer from Ref. [25]. (d) PZT xylophone ultrasound transducers from Ref. [19]. (e) AlN width-extensional contour-mode resonators from Ref. [32]. (f) Lithium niobate thin film laterally vibrating resonator from Ref. [30]. (g) AlN FBAR from Ref. [10].

the use of suspended thin films made out of single crystal piezoelectric substrates (such as Quartz [27, 28] and Lithium Niobate (LN) [29, 30]). This approach takes advantage of either the intrinsically low damping or the high electromechanical coupling of the single crystal bulk piezoelectric material and transfers it to the microscale.

To offer an idea of the extent to which the field of piezoelectric resonant MEMS has grown, several examples of some of the devices fabricated using thin film piezoelectric technology are shown in Figure 7.1. The scalability of the thin-film technology is particularly evident if we consider that these resonant devices can cover a broad range of the frequency spectrum going from a few kilohertz to several gigahertz. This translates to the broad applicability of resonant piezoelectric thin film technology to various areas such as ultrasonic imaging, physical sensing, filtering, and frequency generation.

This chapter will provide a brief introduction to the concept of piezoelectricity and review the most relevant properties of thin-film piezoelectric materials used for the making of resonant MEMS devices. In an attempt to provide a relatively broad description of resonant piezoelectric MEMS, a simplified and generalized equivalent model for a piezoelectric resonator is derived and used to describe the most important parameters that affect the device response. Finally, some practical examples of vibrating piezoelectric beams and plates are presented. Simplified analytical models of these structures are derived to describe the device behavior. Experimental data of demonstrated devices based on these conceptual models and using different classes of piezoelectric materials are also introduced to highlight the broad applicability of the piezoelectric thin-film technology.

7.2
Fundamentals of Piezoelectricity and Piezoelectric Resonators

Piezoelectricity is a physical phenomenon that was discovered in 1880 by the Curie brothers. It expresses a linear interaction between mechanical and electrical domains that happens within a single elastic body. It is an inherent property of a material and it is more precisely described as electric polarization induced by mechanical strain in crystals belonging to certain classes, the polarization being proportional to the strain and changing sign with it. The converse effect, whereby a piezoelectric crystal becomes strained when an electric field is applied to it, also exists and shows the reversibility of such a phenomenon. The materials that exhibit this particular property have a crystallographic structure that is not symmetrical with respect to a point (non centrosymmetrical). This configuration is the only one that permits mechanical strain to induce a net electrical polarization in a crystal. Materials that belong to the wurtzite family, such as AlN and ZnO, have piezoelectric coefficients that depend on the orientation of the crystals as deposited. Ferroelectric materials, such as PZT, need to be poled in order to exhibit strong piezoelectricity.

Depending on the kind of independent variables that are chosen (stress or strain; electric field or electric displacement), the fundamental equations of piezoelectricity relating mechanical and electrical variables can be expressed, for example, using the IEEE standard notation (ANSI/IEEE Std 176 – 1987) in the following way:

$$S = sT + eE \tag{7.1a}$$

$$D = e^{\mathrm{T}}S + \varepsilon E \tag{7.1b}$$

Equations (7.1a) and (7.1b) are known as the strain-charge form (also known as the *e*-form) and render in a matrix form the complex description of piezoelectricity. In this case, *S* represents the strain (6 by 1 matrix), *T* the stress (6 by 1 matrix), *E* the electric field (3 by 1 matrix), and *D* the electric displacement (3 by 1 matrix); *s* is the compliance matrix (6 by 6 matrix), *e* the direct piezoelectric matrix (6 by 3 matrix), e^{T} is the transpose of the piezoelectric matrix, and ε is the dielectric constant matrix (3 by 3 matrix). The *e*-form is widely used because of its suitability to solve problems that involve the dynamics of elastic bodies. Another very effective way of reducing the piezoelectric phenomenon into equations is

$$T = cS + dE \tag{7.2a}$$

$$D = d^{\mathrm{T}} T + \varepsilon E \tag{7.2b}$$

This set of equations is known as the stress-charge form (also known as the *d*-form), in which *d* is the piezoelectric matrix and *c* the stiffness matrix. To offer an intuitive understanding of these equations, we should consider the main differences with respect to the equations that would describe a simple elastic material or the electrical characteristics of a dielectric, that is, materials in which the electrical and mechanical fields are not coupled. In the case of the mechanical properties of the film, the piezoelectric effect adds to the conventional stress–strain relationship (through either the compliance or the stiffness matrices), a strain or stress term that depends on the magnitude of the electric field applied in a specific direction. In the case of the electrical characteristics of the film, the piezoelectric effect introduces a term that represents how a stress or strain is converted into an infinitesimal charge per unit area. This chapter is meant to offer just a brief introduction to the fundamentals of piezoelectricity. For more in-depth information on the topic, the reader is referred to other sources in the literature [33, 34].

From a practical standpoint, it is important to note that the matrices representing the piezoelectric coefficients are not full and generally only three or four independent terms are used. For example, in the case of thin film piezoelectrics such as AlN, ZnO, and PZT, the most common coefficients used for MEMS devices are d_{33} and d_{31}. These piezoelectric coefficients relate an electric field in the "3 direction" (i.e., generally applied across the film thickness) to a strain either in the same (3) direction or in an orthogonal direction (1 or 2 as in most cases the films are isotropic in the plane). This is schematically represented in Figure 7.2, which shows that an electric field in the 3 direction produces equal in-plane strain in a hexagonal unit lattice. By taking advantage of these coefficients and properly designing their compound (integral) effect on a mechanical structure, it is possible to excite most of the piezoelectric MEMS structures into vibrations. The anisotropic nature of materials such as LN renders the analysis a bit more complex as several piezoelectric coefficients are available depending on the particular material cut and orientation.

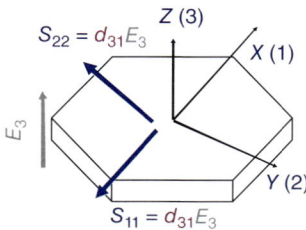

Figure 7.2 Main axis and directions used to describe the piezoelectric effect in a hexagonal crystal. Superimposed is an example of how an electric field in the 3 direction generates strain in the 1 and 2 directions of the same crystal through the d_{31} piezoelectric coefficient (in-plane symmetry was assumed).

A piezoelectric resonator is an elastic body consisting at least partly of a piezo-electric material that can be excited into vibrations when an alternating electric field at the proper frequency is applied to it. Basically, the electric field drives the piezoelectric body into motion through the converse piezoelectric effect; vice versa, when it vibrates, the deformations induce periodic piezoelectric charges on the surface of the electrodes through the direct effect. The mechanical vibrations in the neighborhood of the resonant frequency depend on the inertia and the elastic compliance (or stiffness) of the resonant element. In a similar manner, the electrical reaction of the vibrating body in any electronic circuit can be described by an equivalent network, consisting of inductance, capacitance, and resistance, corresponding respectively to inertia, compliance, and damping of the mechanical system. (More details on how to derive these equivalent components are provided in the section about the equivalent circuit of piezoelectric resonant MEMS.)

Depending on which piezoelectric coefficient is used to excite a structure into resonance, how the piezoelectric media are arranged, and which geometrical dimension sets the resonance frequency, the piezoelectric transducer can be actuated primarily in a flexural, thickness-extensional, shear, or contour mode shape to quote some of the most used modes of operation of piezoelectric crystals (Figure 7.3). Piezoelectric crystals have been used for about 100 years to manufacture resonators because of their extremely sharp resonance, high quality factors, and minimal losses. Quartz tuning forks, Quartz crystals, ceramic resonators, and Surface Acoustic Wave (SAW) devices are some examples of commercially available piezoelectric resonators. Piezoelectric MEMS resonators have emerged as alternatives to these existing devices because their miniaturized dimensions and the use of IC-compatible methods for their fabrication will yield less-expensive components, easier arrangement into large-scale systems, and direct integration with electronics. The general principle of operation and specific examples of MEMS resonators are described in the following sections.

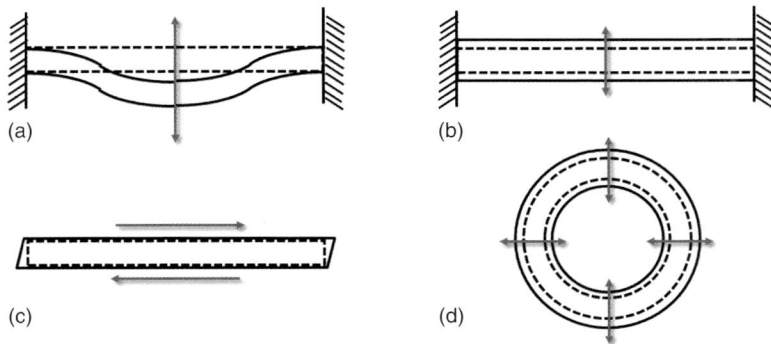

Figure 7.3 Common examples of different types of vibrations induced in piezoelectric MEMS resonators: (a) flexural vibrations in a beam, (b) thickness-extensional vibrations in a piezoelectric plate, (c) shear vibrations in a piezoelectric plate, and (d) contour vibrations in an annulus. The undeformed shape is shown by a dashed line. The deformed shape is in bold. Arrows indicate the direction of vibration.

7.3
Thin Film Piezoelectric Materials for Resonant MEMS

The piezoelectric material plays a key role in defining the characteristics and applicability of a resonant device. In fact, the values of the piezoelectric coefficient, the dielectric constant, and stiffness of a material determine the maximum electromechanical coupling coefficient, k_t^2, (intuitively seen as the ratio of the output mechanical to the input electrical energy) that a transducer can provide. In a given material, the k_t^2 of a transducer varies depending on the film orientation and the selected mode of vibration (i.e., the selected piezoelectric coefficient used to excite the resonator into vibrations). In some of the most commonly available thin film piezoelectrics, such as AlN and ZnO, the film orientation is constrained by the deposition method with the c-axis (3 direction in Figure 7.2) generally orthogonal to the substrate. PZT films have been deposited with different orientations, (001) and (111) being the most common. The orientation of piezoelectric thin films is generally characterized by X-ray diffraction, which also permits one to quantify the degree of crystallinity of a sample. The degree of orientation (how well the grains are directed with respect to the main c-axis) of the thin film depends on several deposition parameters and the quality of the underlying layer. Generally, smooth and well-oriented seed layers yield highly oriented films. In the case of AlN sputtering, the pre-sputter etch step and gas mixture ratio play a key role in defining the ultimate film orientation [35]. Beyond affecting the piezoelectric properties of the material, the degree of orientation also has an effect on the quality factor (Q) of the resonator, whenever the majority of the losses are to be attributed to the intrinsic material damping. It is important to note that a higher degree of orientation will have a greater impact on the device Q rather than its piezoelectric properties. For example, a full width half maximum value of 4° obtained from X-ray diffraction measurements of sputter-deposited AlN films is sufficient to obtain almost nominal piezoelectric coefficients, but the resonators built with such films are likely to have a poor Q.

Ensuring reproducible and highly oriented thin films is of paramount importance and one of the major bottlenecks to the widespread utilization of thin film piezoelectric MEMS. Rapid adoption by industry and transfer to production requires that the thin film deposition and resonator manufacturing process is also compatible with IC foundries and processes. Metal diffusion in ZnO and the risk of contamination posed by Zn in an IC line have made ZnO less attractive than AlN, a wider band gap thin film piezoelectric that can be readily deposited at low temperature ($<400\,°\mathrm{C}$) by sputtering [36] and integrated with CMOS processes [37, 38]. Ferroelectric PZT films have been extensively developed for memory applications creating a process know-how that has transferred to piezoelectric films. The high piezoelectric coupling of PZT makes it very attractive for thin film devices despite the lead content and explains why it has been considered for the development of certain products such as ultrasonic transducers, switches, and inkjet printers.

Whether sputtering (see AlN and PZT) or sol–gel (see PZT) deposition techniques are used, there is limited control available on the type of orientation of the deposited thin film materials. Examples of tilted AlN grains have shown limited repeatability. An intrinsic design advantage for devices such as SAW and Quartz Crystals is that they use bulk piezoelectric materials available in different cuts and orientations, and therefore have access to a broader range of material properties. Recent research efforts have shown that thin films of materials such as quartz or LN can be transferred to another substrate and micromachined to form suspended MEMS resonators. Methods based either on etching and polishing [28] or ion-slicing [39, 40] have been developed to transfer high-quality, thin films of a piezoelectric crystal onto a carrier wafer. Despite potential drawbacks associated with the cost of the transfer process (note, though, that the same was thought of silicon on insulator (SOI) substrates), the use of these thin films significantly broadens the range of available materials for piezoelectric resonant MEMS.

As briefly mentioned, another important parameter that influences the performance of a resonator and is affected by the material selection is the intrinsic damping, a direct measure of energy loss and an inverse measure of the ultimate achievable quality factor. For example, ferroelectric materials such as PZT suffer from domain wall motion at high frequency, which results in reduced Qs in the gigahertz range [41]. Materials such as AlN, LN, and Quartz Crystal have instead shown low losses at high frequency. Phonon–phonon dissipation [42] can be considered the ultimate limit for most insulating piezoelectric materials. In this case, the material quality factor is inversely proportional to the frequency of operation of the resonator. Measured Qs for some of the most common thin film piezoelectric materials are reported in Table 7.1.

The method of deposition, the presence of defects, and the overall crystal structure of the thin film also affect the intrinsic material damping. Additionally, the interface between the resonant piezoelectric body and the metal electrode used to transduce it into vibrations is a considerable source of damping, especially at high frequency [44]. Although this is an ongoing topic of investigation and no conclusive statements can yet be made, it is important to note that the design of a piezoelectric MEMS resonator involves careful consideration of the piezoelectric material properties, the metal electrodes, and the fabrication equipment and steps used to manufacture the device.

7.4
Equivalent Electrical Circuit of Piezoelectric Resonant MEMS

The most general equivalent circuit representation of a piezoelectric transducer is schematically shown in Figure 7.4.

In this simplified lumped model (please refer to [45] for more device-specific models), C_O represents the physical capacitance of the electrode part of the piezoelectric device. The transformer with a turn ratio, η, represents the conversion between electrical and mechanical variables at a specific location on the device

Table 7.1 Characteristic piezoelectric (k_t^2) and elastic properties (phase velocity) of most commonly used thin film piezoelectrics for resonant MEMS used in different modes of vibration.

Material (mode of vibration)	Phase velocity (m s^{-1})	k_t^2 (%)	f-Q_{max} (Hz)
AlN (TE)	9 000–10 000	6–7	1×10^{13}
AlN (WE)	9 000–10 000	1–2.5	1.1×10^{13}
ZnO (TE)	~6 000	8–9	3.5×10^{12}
PZT (TE)	1 500–2 000	9–35	4.4×10^{11}
PZT (WE)	~3 000	5–9	6.4×10^{10}
Quartz (S)	~3 000	0.04–0.1	2.1×10^{13}
Lithium niobate (WE)	6 000–7 000	21	1.5×10^{12}

TE, thickness-extensional; WE, width-extensional; S, shear.
The value of the *f-Q* (frequency-quality factor) product is based on reported experiments and should be taken as a general indication of the film quality. *Q* measured in mechanical resonators is affected by several damping mechanisms; hence, the reported *Q*s represent an ensemble of various effects.
Source: Data for AlN (TE), ZnO (TE), and PZT (TE) are from Ref. [41], for AlN (WE) are from Ref. [43], for PZT (WE) are from Ref. [6], for quartz (S) are from Ref. [27], and Lithium niobate (WE) are from Ref. [30].

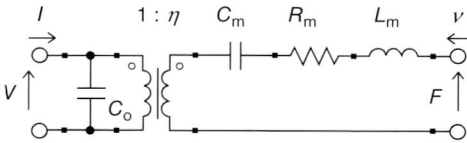

Figure 7.4 Simplified lumped circuit model for a piezoelectric transducer.

(generally the point of maximum displacement). η is a function of the piezoelectric material and its mode of vibration. Examples on how to calculate η depending on the resonator and electrode geometry will be presented in the next section. The motional capacitance, C_m, resistance, R_m, and inductance, L_m, represent the mechanical variables of the MEMS resonator, each being respectively associated with the resonator equivalent compliance ($1/k_{eq}$), damping (c_{eq}), and mass (m_{eq}). It is important to note that the variables on the right-hand side of the transformer are effectively mechanical elements relating force, F and velocity, v, for a specific point on the resonator (which has been reduced from a continuum vibrating 3D body to a 1D lumped element). The transformer permits the conversion of these mechanical quantities into current, I and voltage, V.

The transducer static capacitance and the transformer ratio are generally expressed as

$$C_O = \varepsilon_{ij} \frac{\text{Electroded Area}}{\text{Equivalent thickness of electroded Area}} \qquad \eta = \frac{F}{V} = \frac{I}{v} \qquad (7.3)$$

where ε_{ij} is the permittivity of the piezoelectric material in the specific direction along which the electric field is applied. Note that the transformer ratio can represent the conversion of either the voltage into force (generally used for actuation) or the velocity into current (generally used for sensing).

The lumped mechanical quantities, m_{eq} and k_{eq}, are derived by equating the energy (either kinetic or elastic) stored in the resonator body to the energy of one point on the resonator (generally selected as the point of maximum displacement). For example, the kinetic energy of the vibrating system, $E_{kinetic}$, is obtained by integrating the product of the mass density, ρ and square of the velocity, v, of each point of the resonator over its entire volume, V_{vol}:

$$E_{kinetic} = \frac{1}{2} \int \rho v^2 dV_{vol} \tag{7.4}$$

The kinetic energy of the system, which depends on the particular mode of vibration as determined by its velocity field, is then equated to the kinetic energy at one point moving at a known velocity. This equality is used to define the equivalent lumped mass of the resonator:

$$m_{eq} v^2_{\text{fixed-point}} = \int \rho v^2 dV_{vol} \quad m_{eq} = L_m \tag{7.5}$$

Once m_{eq} is known, k_{eq} can be readily computed using the following equation:

$$k_{eq} = m_{eq} \omega_n^2 = \frac{1}{C_m} \tag{7.6}$$

The natural frequency of the resonator, ω_n, depends on the geometry and material properties of the resonator, the mode of vibration, and the associated phase velocity. ω_n is the eigenvalue of the differential equation that describes the vibrations in the resonator. Examples of how ω_n is derived for specific devices are described in the following sections.

The determination of the equivalent damping, c_{eq}, requires knowing the resonator quality factor:

$$c_{eq} = \frac{m_{eq}}{Q} = R_m \tag{7.7}$$

As different factors can impact a device Q, it is difficult to calculate an exact value prior to device demonstration. Generally a value based on prior experiments or simplified predictive models is used to calculate c_{eq}. It is clear that the equivalent parameters will change depending on the material properties, the geometry of the resonator and more specifically on the mode of vibration of the resonator (as described by its velocity field). More details on how to compute the lumped mechanical variables of a resonator can be found, for example, in Johnson [46] or in Tilmans [47, 48]. The schematic model of Figure 7.4 represents a general transducer. How the transducer is then mechanically terminated (boundary conditions for F and v) determines whether we have a one or two-port resonator.

7.4.1

One-Port Piezoelectric Resonators

A one-port piezoelectric resonator is a device for which the output mechanical variables (F, v) are terminated in a short. This condition represents, in the acoustic domain, a perfectly reflective boundary such as air. This is true of most MEMS resonators, whose mechanical boundaries are well defined by an etch step into the structural material. Effectively, in a one-port configuration, the resonator only has two accessible electrical terminals and is preferentially tested by grounding one of them and applying a known signal to the other. In this configuration, the static capacitance of the device, C_O, appears in parallel with the motional components of the resonator. The equivalent electrical circuit can be reduced to the conventional Butterworth Van Dyke (BVD) model. At the MEMS scale, this model is generally modified to include the effect of parasitic components (R_S and R_O in Figure 7.5), which tend to play a larger role given the miniaturized size of the resonator. The parasitic element, R_S, represents the equivalent series resistance due to the resonator electrodes and via contacts (if present), whereas R_O is a function of the losses in the dielectric material and the substrate over which the pads of the resonant device are fabricated. This is the primary reason why, at high frequencies, a high resistivity silicon substrate is generally used. R_S and R_O are generally extracted from experimental measurements. The other parameters, R_M, L_M, and C_M represent the equivalent motional parameters of the resonator in the electrical domain and are obtained by simply taking the equivalent mechanical impedance and dividing it by the square of the transformer ratio (as conventionally done with an electrical transformer). Note that these parameters are electrical equivalent elements and not the same as those that were presented in the previous section in

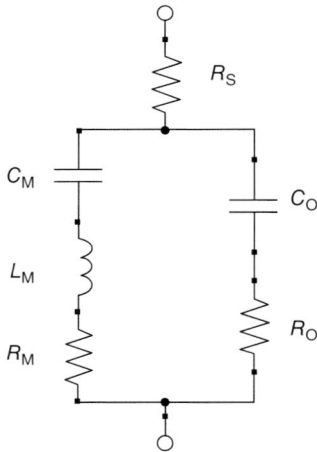

Figure 7.5 Equivalent electrical model of a one-port piezoelectric resonator. This circuit is known as the modified BVD model [49].

which the same variables appeared with a lower case m (indicating the mechanical domain).

7.4.2
Two-Port Piezoelectric Resonators

A two-port piezoelectric resonator is formed by two of the transducers in Figure 7.4 placed back to back. This implies that the two transducers are mechanically coupled via an infinitely stiff element (as the two transducers would have the same force and velocity at the coupling point). In this case, the model can be simplified to the equivalent circuit of Figure 7.6. As shown in the figure, the resonator is a three electrical terminal device (effectively four of which two are generally shorted together to ground), two representing the input and output terminals and a third generally used to set the reference ground. The transformer, not to be confused with the transformer of the individual transducer of Figure 7.4, represents any difference in size between the input and output ports and its turn ratio, N, is equivalent to the ratio of the transducer's individual turn ratios ($\eta_{input}/\eta_{output}$). To avoid confusion, the parasitic components, R_S and R_O, of Figure 7.5 were neglected, although they could be included, R_O being in series with C_O, and R_S in series with each side of the input and output terminals. A parasitic capacitance, C_f, is introduced to account for substrate and device feedthrough between the input and output ports of the resonator. This capacitance is generally determined experimentally and, if the transducer is properly designed, can be neglected.

The two-port configuration enables greater electrical isolation of the input and output terminals, therefore facilitating the detection of the resonant signal. Additionally, this isolation permits the use of the device as a voltage transformer. The main drawback of a two-port topology (assuming that the two ports are defined side to side [50]) is that the equivalent value of the motional impedance is 4 times that of a one-port resonator for a given transduction area. Also, just half the resonator body is excited into the desired mode of vibration, hence other spurious vibrations could be present. If the two ports were stacked on top of each other [51], then one- and two-port devices would yield the same motional impedance, although the stacked device would be more difficult and costlier to manufacture.

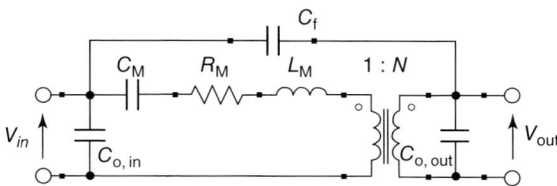

Figure 7.6   Equivalent electrical model of a two-port piezoelectric resonator. This model is generally used to describe the behavior of macroscale resonant piezoelectric transformers.

Both kinds of topologies have been demonstrated and their use depends on the end applications. Examples of one- and two-port resonators will be presented in the following sections.

7.4.3
Resonator Figure of Merit

A different description of the equivalent electrical model of a piezoelectric resonator resorts to the use of the electromechanical coupling coefficient, k_t^2, of the transducer rather than its transformer ratio, η. This representation is very effective (and the most commonly used in the acoustics community) as it permits one to directly define all of the resonator's equivalent parameters as functions of the device static capacitance, C_O, its center frequency, ω_n, the k_t^2, and the Q of the resonator. For a simple one-port resonator, the most general set of equivalent parameters can be approximately expressed as

$$R_M = \frac{\pi^2}{8}\frac{1}{\omega_n C_O}\frac{1}{k_t^2 Q} \qquad L_M = \frac{\pi^2}{8}\frac{1}{\omega_n^2 C_O}\frac{1}{k_t^2} \qquad C_M = \frac{8}{\pi^2}C_O k_t^2 \qquad (7.8)$$

This set of equations is very important as they clearly state that for a given material and mode of vibration (fixed k_t^2, Q, and ω_n) a circuit designer can only modify the device impedance by altering its geometrical characteristics (C_O).

Most importantly, these equations clearly spell out how the equivalent motional resistance of the piezoelectric resonator is inversely proportional to the product of k_t^2 and Q. In fact, the product of these two quantities is defined as the Figure of Merit (FoM) of a resonator. The higher the FoM, the lower the equivalent R_M for a given impedance, $1/j\omega C_O$. A low value of R_M is extremely important in setting the performance of any device in which the resonator will be used. For example, R_M impacts the insertion loss of a filter [32], the power consumption in an oscillator [52], and the efficiency of a transformer [53]. $k_t^2 - Q$ products around 10 are considered acceptable. The best piezoelectric resonators exhibit $k_t^2 - Q$ values around 300. This does not necessarily mean that the resonator with the highest FoM is best suited for all applications. In fact, depending on how the resonator will be utilized, k_t^2 can be traded for Q or vice versa as each of them could independently affect other system-level performance.

7.5
Examples of Piezoelectric Resonant MEMS: Vibrations in Beams, Membranes, and Plates

The following sections describe the principle of operation of some of the most common thin film MEMS piezoelectric resonators. The resonators are classified according to their mode of vibration, as they represent a distinct class of devices for which specific design considerations have to be taken into account. Each section provides a simplified description of the general device and then delves into the details of deriving the equivalent electromechanical parameters for a specific resonator that has been experimentally demonstrated and reported in the literature.

7.5.1
Flexural Vibrations

Resonators excited to exhibit flexural vibrations are formed either by a thin film piezoelectric material deposited on top of another structural material or by a stack of piezoelectric layers (Figure 7.7). The reason for having a stacked structure is to ensure that the generated piezoelectric stress is offset with respect to the neutral axis of the formed beam or membrane. This offset is essential in generating a bending moment in the structure and hence exciting flexural vibrations. The need for this offset is dictated by the intrinsic nature of the piezoelectric material, which in all available thin films can only develop uniform stress in one direction and cannot directly form stress gradients. The use of an external layer, which could be the metal electrode itself, to break the symmetry in the piezoelectric layer enables the generation of a bending moment. The generated flexural vibrations correspond to the excitation of an asymmetric wave and generally result in out-of-plane displacement. In most piezoelectric thin films, such as AlN, ZnO, and PZT, such motion is obtained by means of the e_{31} piezoelectric coefficient.

The classification of the resonator as a beam or plate depends exclusively on the aspect ratio of the structure. When $L \gg W \gg H$, then the structure is classified

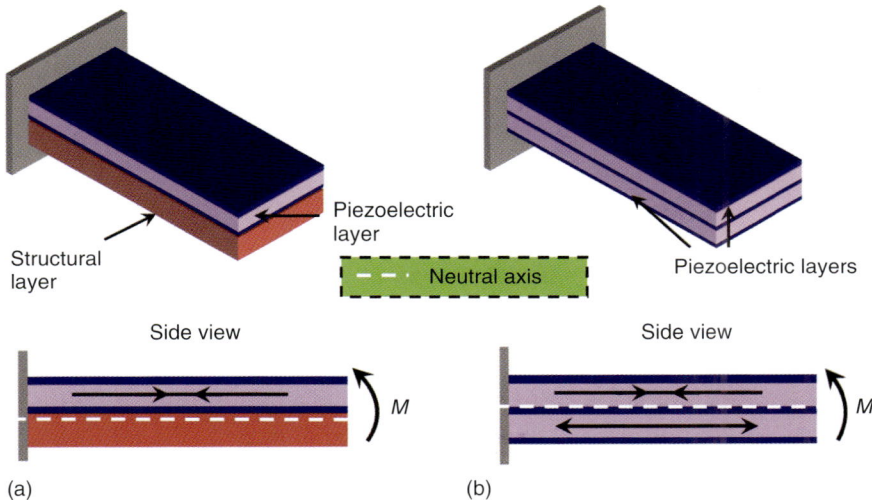

Figure 7.7 Schematic representation of flexural resonators formed by either (a) a piezoelectric thin film on top of another structural layer or (b) a stack of piezoelectric layers. These configurations enable the excitation of flexural vibration in the suspended beam or plate. The in-plane stress generated via the e31 piezoelectric coefficient (indicated by arrows) is converted into a bending moment because the piezoelectric layer is offset with respect to the neutral axis of the beam/plate.

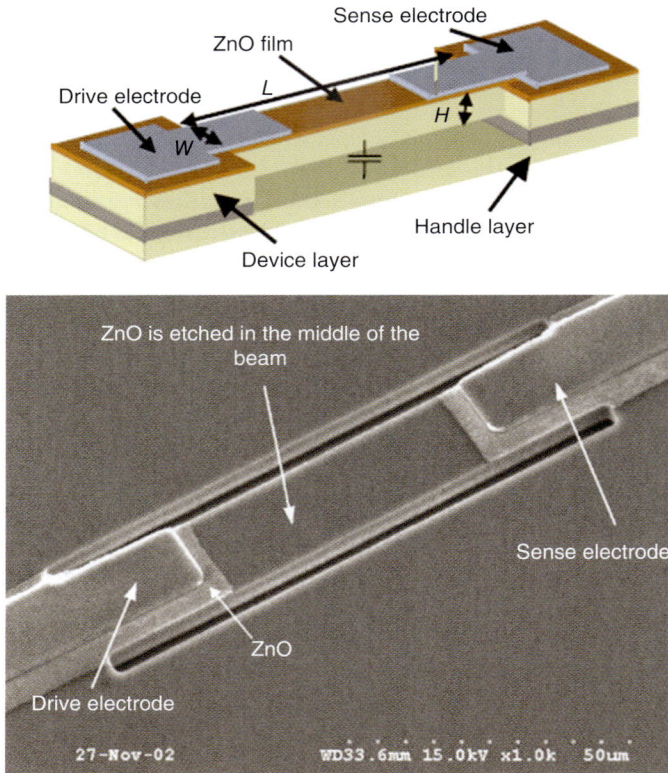

Figure 7.8 Schematic and scanning electron micrograph images of an example of a clamped–clamped beam excited into flexural vibrations by a thin piezoelectric layer deposited on top of silicon [18].

as a beam. As the in-plane dimensions (L and W) of the structure become comparable, but still much greater than the thickness, 2D effects need to be considered and the device is better classified as a plate. Nonetheless, the general principle of operation is similar and, for simplicity, only beams will be considered.

As an example, this section focuses on the derivation of the equivalent parameters for a clamped–clamped beam, formed by a thin layer of ZnO on silicon or another structural layer [8, 31], excited to vibrate in its first mode of vibration (Figure 7.8).

The resonance frequency, f_o, of the structure shown in Figure 7.8 is determined by the length, L, the thickness, H, of the beam, and equivalent Young's modulus and mass density of the material stack:

$$f_o = 1.03 \frac{H}{L^2} \sqrt{\frac{E_{eq}}{\rho_{eq}}}$$

(7.9)

The equivalent value of the mass density can be obtained by computing the average of the mass density of each material in the stack weighted by each layer thickness, while the equivalent Young's modulus can be determined by classical composite beam theory. (The assumption that all layers cover the entire extent of the beam is made for simplicity.) From Eq. (7.9), it is clear that most MEMS flexural resonators operate in the few kilohertz to megahertz range. This is the main reason why flexural vibrations are used for the making of structures that operate as ultrasound devices or low-frequency references and sensors. Aggressive scaling to reach gigahertz frequencies has been attempted, but generally results in reduced transduction efficiency given the very small device dimensions.

In order to couple into the desired first mode of vibration, the piezoelectric film is actuated in specific regions of the beam that are consistent with the mode shape to be excited for the clamped–clamped beam (Figure 7.8). To clarify, the piezoelectric film is excited with similar polarity in the regions between the inflexion point of the mode shape and the anchors (i.e., the outer quarters of the beam span). In these two regions, the stress profile is in the same direction and matches the natural behavior of the piezoelectric film (both portions undergoing expansion or compression). Generally, most of the flexural resonators are arranged in a two-port configuration and use one set of electrodes for excitation and a second for sensing the charge generated by the piezoelectrically induced strain. Hence, one electrode could be used to excite the beam into motion by covering the regions close to the anchors and a second electrode could be used in the center of the beam for sensing the produced charge. For practical reasons and difficulty in routing of electrodes, in most cases, only half of the beam area is used and the actuation electrode is placed by one anchor with the sensing electrode close to the second anchor. Electrode shaping to maximize coupling into other and higher-order flexural modes is described in more detail in [54].

The previous paragraphs provided an intuitive understanding of the operation and design of a piezoelectrically excited, clamped–clamped beam. To more quantitatively and rigorously define the resonator, it is important to derive its equivalent electromechanical parameters.

The equivalent mass and stiffness of the resonator can be quantified according to Eqs. (7.5) and (7.6), assuming that the mode of vibration has the following form [55], $u(x)$, from which the corresponding velocity can be derived by differentiation:

$$u(x) = A \sin(kx) + B \cos(kx) + C \sin h(kx) + D \cos h(kx) \tag{7.10}$$

A, B, C, and D are constants that depend on the boundary conditions. k is the wavenumber (related to the eigenvalue of the mode) associated with the particular mode of vibration.

For the first mode of vibration of the clamped–clamped beam, the resulting equivalent mass is approximately given by:

$$M_{eq} = 0.38 \, \rho_{eq} \, WLH \tag{7.11}$$

To find the equivalent damping of the resonator, a value of Q has to be assumed. Depending on the set of materials used, the resonator geometry, and the surrounding gas and its pressure, Q can vary from a few hundred to a few thousand. Pressure has the largest impact on Q for these low-frequency devices, and when high Q is required, flexural resonators are generally operated under vacuum.

In order to fully define the resonator, an expression for the transformer ratio, η, in the model of Figure 7.4 needs to be derived [18]:

$$\eta = \eta_{\text{in}} = \frac{d_{31}E_p T_s}{2} \int_0^L W''(x)\frac{u(x)}{u_{\text{max}}}dx = 2.49\frac{d_{31}E_p W T_s}{2L} \tag{7.12}$$

where W'' represents the second derivative of the electrode width, and u_{max} is the maximum displacement of the beam. Equation (7.12) quantitatively defines what was intuitively described at the beginning of this section and explains why the location and polarity of the piezoelectric layers is fundamental in maximizing the transduction efficiency. Assuming symmetrical placement of the electrodes $\eta_{\text{in}} = \eta_{\text{out}}$, hence $N = 1$ in Figure 7.6.

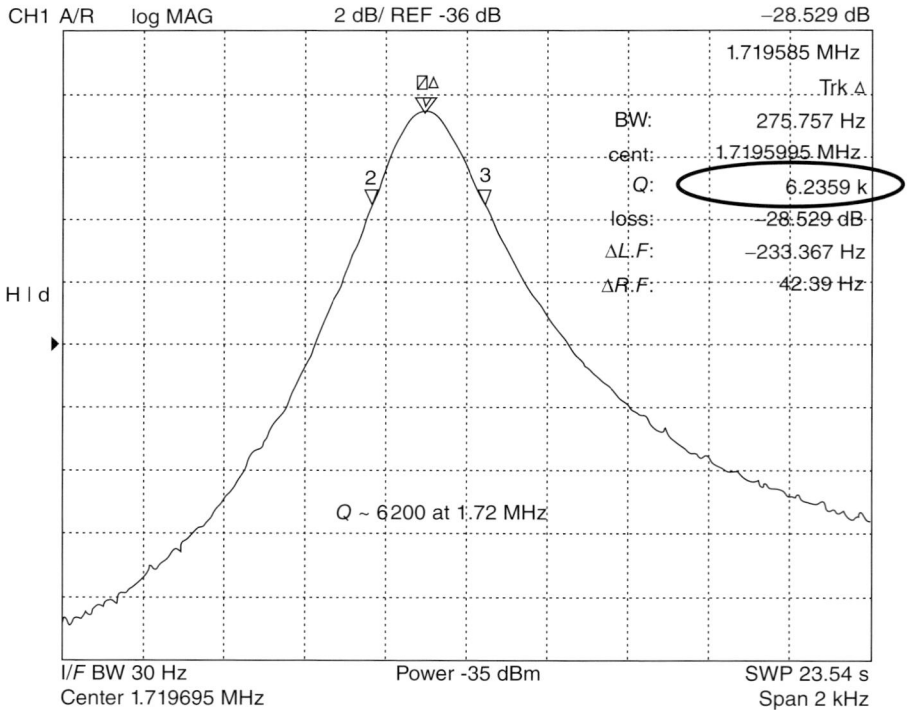

Figure 7.9 Characteristic transmission response of a 100 µm long and 20 µm wide ZnO-on-silicon clamped–clamped beam resonator operating at 1.72 MHz. The *x*-axis represents the frequency span over which the device was measured. The *y*-axis represents the magnitude of the transmission response. A high *Q* of 6236 was obtained. Its value is circled on the same plot.

The characteristic response of a two-port flexural resonator such as the one shown in Figure 7.8 is shown in Figure 7.9.

7.5.2
Width-Extensional Vibrations

Width-extensional vibrations are commonly excited in thin film piezoelectric plates to achieve high frequencies of operation. Width- or length-extensional vibrations, depending on the primary dimension of the plate along which vibrations are excited, are naturally induced in a piezoelectric film by patterning electrodes on the top surface of the plate and alternating the polarity of the voltage applied to each of them (Figure 7.10). In this way, adjacent elements alternatively expand and contract. The resonance frequency of the plate is set by the pitch of the electrodes and the acoustic velocity of the material stack forming the resonator. The number of electrode pairs (2 to n) defines the mode number excited in the plate. When n is large, instead of viewing the vibrating plate as formed by n adjacent width-extensional resonators, the vibration can be analyzed as the propagation of a Lamb wave in an infinite plate[45].

A single electrode (Figure 7.11) can also be used as long as a bottom electrode is present to confine the electric field in the vertical direction. In this case, the fundamental mode of vibration of the plate is excited. Even when more than one electrode is present, a bottom electrode is also used to better confine the electric field within the body of the piezoelectric film and generally obtain a higher electromechanical coupling. Most importantly, in thin films such as AlN, ZnO, and PZT, the presence of a bottom electrode ensures that the electromechanical coupling, k_t^2, is non-dispersive, and enables the use of the same film thickness at different frequencies [56]. It is because of this particular feature in conjunction with a high and constant acoustic phase velocity that the development of these resonators has become particularly attractive for high-frequency applications. In fact, the ability to manufacture multiple frequency resonators on the

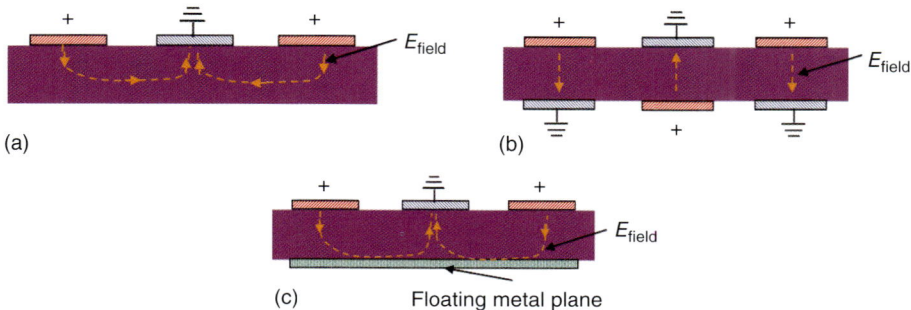

Figure 7.10 Schematic cross-sectional view of a piezoelectric plate excited into width-extensional vibrations by (a) top electrodes of alternating polarity, (b) patterned top and bottom electrodes, and (c) top patterned electrodes and continuous bottom floating plate.

(a)

(b)

(c)

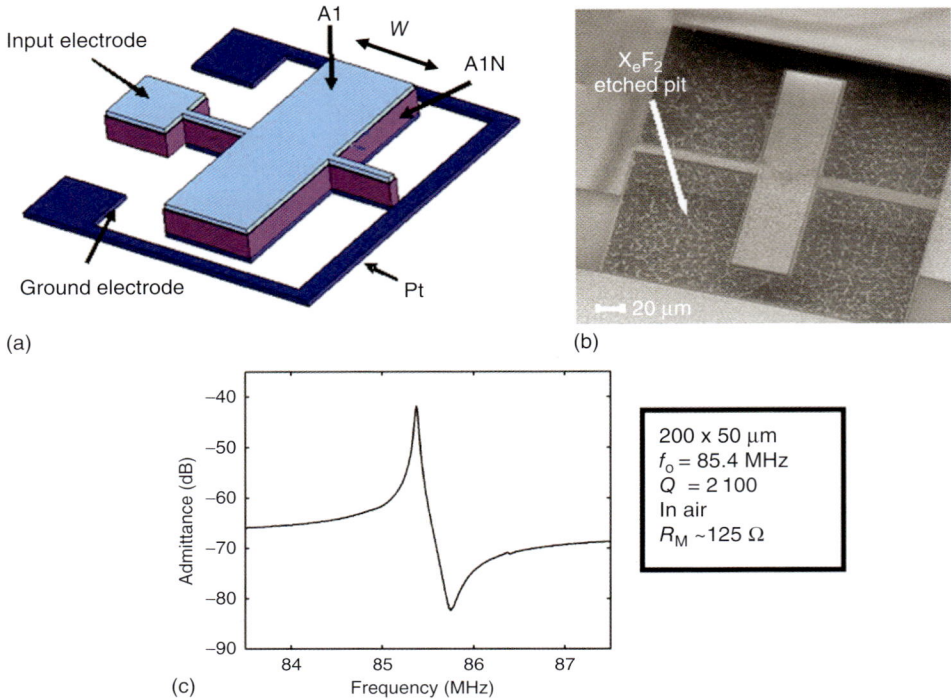

Figure 7.11 (a) Schematic representation, (b) Scanning Electron Microscope (SEM) image, and (c) electrical response of an AlN plate vibrating in its fundamental width extensional mode.

same chip would ensure covering various bands without requiring multiple packages, for example, enabling monolithic and fully integrated filters and oscillators. Devices operating between a few MHz and 10 GHz have been built and demonstrated using width-extensional vibrations in plates [57]. These devices have been made mostly out of the piezoelectric material and the metal electrodes sandwiching it. In some configurations, a layer of silicon dioxide or another semiconductor material (for example, silicon [7], silicon carbide [58], diamond [59], or sapphire [60]) has been added under the piezoelectric film to provide for temperature compensation or enhance the quality factor of the structure. The main piezoelectric materials used for the demonstration of this class of devices have been AlN, PZT, and more recently LN [30]. AlN-based devices exhibit the highest quality factor when the resonator is made without any other supporting layer. LN offers the highest coupling coefficient with a relatively high Q. LN devices have been recently explored and their ultimate performance still needs to be unveiled.

To provide a simple and intuitive understanding of the principle of operation of width- (or length)-extensional resonators, and to derive their equivalent electrical parameters, the case of a rectangular AlN plate sandwiched by top and bottom electrodes is presented in detail (Figure 7.11).

In this case, the equivalent electromechanical parameters of the one-port resonant circuit are derived using a different method than the one presented in the previous section. This method uses the fundamental equations of piezoelectricity and rigid body dynamics. It is selected to offer further insights into the fundamental physics of the device. A main limitation of this method is that it becomes more laborious and lengthy for 2D problems. For this reason, the piezoelectric plate is assumed to vibrate exclusively along its width, W (x direction). Effectively, the plate will be analyzed as equivalent to a bar expanding and contracting along its width. The fundamental equation of motion without damping is given by

$$\rho \frac{\partial^2 u}{\partial t^2} - E_p \frac{\partial^2 u}{\partial x^2} = 0 \tag{7.13}$$

where ρ is the effective material density, u the bar displacement in the x direction, and t a variable representing time. To further simplify the derivation, given the in-plane symmetry of AlN, an equivalent in-plane elastic modulus, E_p, for the metal and piezoelectric layer is used. Assuming a sinusoidally time-varying solution, we can focus only on the space dependence of u. Hence the particular solution is given by:

$$u(x) = C_1 \sin(kx) \tag{7.14}$$

where $k = \frac{\omega W}{2 v_{ac}}$ is the wavenumber (v_{ac} being the acoustic velocity of the stack, $\sqrt{E_p/\rho}$); ω is the frequency of excitation, and W is the width of the bar. The constant C_1 can be evaluated by using the constitutive equations of piezoelectricity. We assume that the origin of the axis of vibration is at the center of the bar. At the free surface of the bar in the direction of vibration, the stress is equal to zero and the strain is purely due to the piezoelectric term. Therefore:

$$\frac{\partial u}{\partial x} = d_{31} \frac{V}{H} \quad - \rightarrow \quad C_1 = \frac{d_{31} V}{kH \cos\left(\frac{kW}{2}\right)} \tag{7.15}$$

where V is the voltage applied across the thickness, H, of the film. Using the e-form for the direct piezoelectric effect, the total current, i, flowing through the device can be obtained as the time derivative (product by $j\omega$ for a sinusoidally time-varying variable) of the integral of the electric displacement (surface charge):

$$i = j\omega \iint_A D_3 dA = j\omega \int_{-W/2}^{W/2} L \left(e_{31} \frac{\partial u}{\partial x} + \varepsilon_3 E_3 \right) dx$$

$$= j\omega \left[2 \frac{L}{H} \frac{d_{31} e_{31}}{k} \tan\left(\frac{kW}{2}\right) + \varepsilon_3 \frac{WL}{H} \right] V \tag{7.16}$$

where A is the electrode area of the bar (width, W and length, L), D_3 is the electric flux in the 3 direction (film thickness), E_3 is the electric field in the 3 direction, e_{31} and d_{31} are the piezoelectric coefficients in the e and d-forms, and ε_3 is the dielectric constant of AlN in the 3 direction. The overall device admittance, Y, is then derived as

$$Y = j\omega \left[2 \frac{L}{H} \frac{d_{31} e_{31}}{k} \tan\left(\frac{kW}{2}\right) + \varepsilon_3 \frac{WL}{H} \right] \tag{7.17}$$

The second term between brackets represents the physical capacitance due to the dielectric property of the piezoelectric material sandwiched between the top and bottom electrodes. This term corresponds to C_o in the equivalent circuit of the resonator. The first term represents the motional part of the resonator. As no losses were included, this term only describes L_M and C_M. By introducing a complex expression for k, damping can be included and a full representation of the resonator equivalent parameters can be attained. Given $= k_r + jk_i = \frac{\omega W}{2c}\left(1 + j\frac{1}{2Q}\right)$, where Q is the quality factor of the resonator, it can be shown that the first term between brackets in Eq. (7.17) corresponds to an electrical circuit formed by the series combination of a resistance, a capacitance, and an inductance. Using a 1D approximation, $e_{31} = d_{31}E_p$ and the equivalent circuit parameters are given by

$$R_M = \frac{\pi}{8}\frac{H}{L}\frac{\rho^{1/2}}{E_p^{3/2}d_{31}^2} \qquad L_M = \frac{1}{8}\frac{WH}{L}\frac{\rho}{E_p^2 d_{31}^2} \qquad C_M = \frac{8}{\pi^2}\frac{WL}{H}E_p d_{31}^2 \qquad (7.18)$$

which can be further reduced to the format of Eq. (7.8) if $k_t^2 = \frac{d_{31}^2 E_p}{\epsilon_3}$ is substituted into Eq. (7.18). It is interesting to note that the resonator motional resistance is controlled by the FoM of the device and the static capacitance. By altering the value of the capacitance (which can be controlled independently of the frequency via H and L), the resonator equivalent impedance can be modified. The characteristic response of an AlN plate sandwiched by two metal electrodes is show in Figure 7.11 next to an SEM image of a fabricated device.

Although limited to a single electrode, the derivation of the equivalent electrical parameters for this one-port piezoelectric resonator can be generalized to higher order modes of vibration. If n electrodes were to be present, the same model can be applied by viewing the resonator as formed by the parallel combination of n individual electrode devices. If the electrodes are patterned only on the top surface of the plate, particular considerations have to be made when computing the electromechanical coupling and the static capacitance of the resonator. More details on this can be found in [30].

7.5.3
Thickness-Extensional and Shear Vibrations

Piezoelectric plates (circular or rectangular) can also be excited to vibrate across the thickness of the film. This is the main mode of vibration for the AlN FBAR, a commercially available MEMS resonator that is used for the synthesis of acoustic filters for mobile applications, and for larger scale and lower-frequency PZT-based ultrasonic transducers. The resonator is generally arranged in a one-port configuration and is equivalent to a parallel plate capacitor formed by a thin piezoelectric layer sandwiched by two metal electrodes. For mode selection and elimination of spurious vibrations, the FBAR top electrode shape is generally apodized (see Figure 7.12). The resonator topology is rather simple and easy to fabricate and thanks to the use of a thin film technology, it can operate in the high-frequency

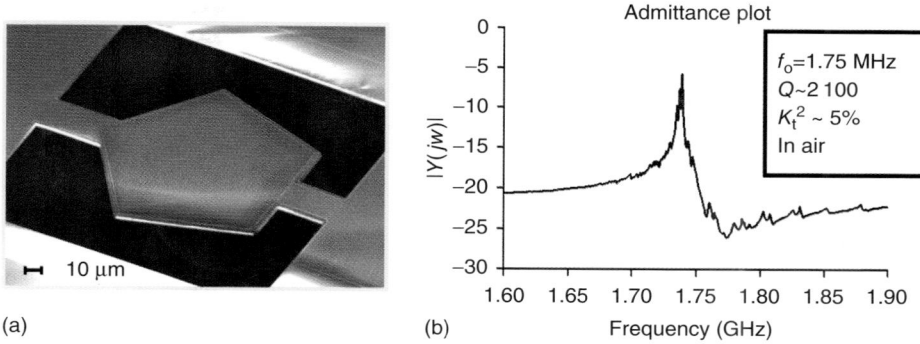

Figure 7.12 (a) SEM image and (b) electrical response of an apodized thickness-extensional AlN resonator.

range (500 MHz to 40 GHz). In fact, the film thickness sets the resonance frequency of the device for a given acoustic velocity of the material stack, v_{ac}:

$$f_o = \frac{1}{2H} v_{ac} \tag{7.19}$$

Commercially available FBAR mostly operate between 800 MHz and 2.4 GHz (corresponding to a piezoelectric film thickness between 1 and 2 µm), but demonstrations of 40 GHz devices exist. It is also interesting to note that a solidly mounted resonator (SMR) based on a thickness-extensional mode in the piezoelectric film have been developed and are competing against FBARs in the wireless communication space. The SMR relies on the use of acoustic reflectors on the side of the device, which is directly anchored to the substrate. Because of the high level of acoustic energy confinement provided by vertical Bragg reflectors, respectable Qs comparable to those of fully suspended membranes have been attained.

The main piezoelectric material used for the implementation of FBARs and SMRs is AlN, although demonstration of devices with PZT, ZnO, and LN exist (mostly research/development efforts). The main piezoelectric coefficient used to excite thickness-extensional vibrations is d_{33}, which is generally higher than d_{31} in all thin film materials. For this reason, thickness-extensional vibrations result in a high electromechanical coupling (>6% for AlN), which makes the technology amenable to wideband filtering.

The equivalent electromechanical parameters of this device can be derived in a way similar to what was presented in the previous section for width-extensional modes in a plate. Effectively, using a 1-D approximation, a thickness-extensional vibration can be treated as a width-extensional mode in which the width is exchanged by the thickness of the plate. Using this approximation, the electromechanical parameters of the device can be calculated to be the following:

$$R_M = \frac{\pi}{8} \frac{H^2}{WL} \frac{\rho^{1/2}}{E_p^{3/2} d_{33}^2} \qquad L_M = \frac{1}{8} \frac{H^3}{WL} \frac{\rho}{E_p^2 d_{33}^2} \qquad C_M = \frac{8}{\pi^2} \frac{WL}{H} E_p d_{33}^2 \tag{7.20}$$

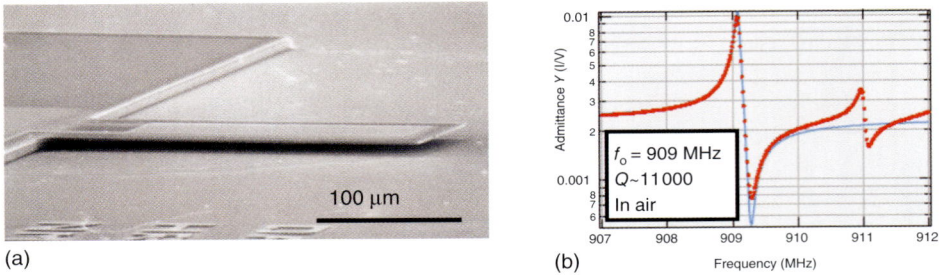

Figure 7.13 (a) Photograph and (b) electrical response of a thickness-shear quartz crystal [28].

It is interesting to note that for this class of devices, the motional resistance scales directly with the square of the film thickness making the structure truly appropriate for high-frequency operation. It is believed that the FBAR and SMR devices will expand the capabilities of existing communication devices by providing high performance at a significantly reduced form factor over very large range of frequency bands.

Quartz crystal products used for reference oscillators utilize thickness-shear vibrations. Miniaturized versions of these devices have been realized by means of MEMS fabrication techniques [28]. Device miniaturization enables operation in the hundreds of megahertz to gigahertz. Most interestingly, the same or even a higher *f-Q* product typical of quartz crystals was achieved also in these miniaturized prototypes. An example of these devices in the form of a patterned mesa is shown in Figure 7.13. These devices are effectively a miniaturized version of quartz crystals and their principle of operation and electromechanical characteristics can be analyzed in the same way [28].

7.6
Conclusions

This chapter has offered a general introduction to piezoelectric MEMS resonators, reviewed the fundamentals of piezoelectricity, and discussed in detail a few examples of some of the most established piezoelectric MEMS devices. For over one century, piezoelectric resonators have served as the core elements for frequency control applications, and have been used for the making of oscillators, filters, and sensors. With the advent of the semiconductor era and the development of MEMS processes, the miniaturization of piezoelectric resonators has been heavily pursued because of the potential gains in size, power consumption, and direct integration with electronics. The AlN FBAR probably represents the greatest commercial success for piezoelectric MEMS resonators. It is easy to imagine, though, that the innovative and entrepreneurial spirit of MEMS researchers and new commercial opportunities will drive the deployment of other products based on piezoelectric resonant MEMS devices. The need for higher performance will also support research activities aimed at the development of materials and resonators with a higher Q and k_t^2. Finally, scaling of the piezoelectric technology is deemed to continue into nanoelectromechanical systems (NEMS) devices. This will happen in conjunction with the demonstration of large arrays of resonant systems directly integrated with advanced electronics, hence enabling frequency selection and sensing capabilities that can barely be envisioned.

References

1. Piazza, G., Stephanou, P.J., and Pisano, A.P. (2006) Piezoelectric aluminum nitride vibrating contour-mode MEMS resonators. *J. Microelectromech. Syst.*, **15** (6), 1406–1418.

2. Ruby, R.C., Bradley, P., Oshmyansky, Y., Chien, A., and Larson, J.D. III, (2001) Thin film bulk wave acoustic resonators (FBAR) for wireless applications. 2001 IEEE Ultrasonics Symposium: Proceedings: An International Symposium, October 7–10, 2001, pp. 813–821.

3. Bjurstrom, J., Katardjiev, I., and Yantchev, V. (2005) Lateral-field-excited thin-film Lamb wave resonator. *Appl. Phys. Lett.*, **86** (15), 154103 (3 pp.).

4. Moreira, M., Bjurstrom, J., Katardjev, I., and Yantchev, V. (2011) Aluminum scandium nitride thin-film bulk acoustic resonators for wide band applications. *Vacuum*, **86** (1), 23–26.

5. Yantchev, V. and Katardjiev, I. (2013) Thin film Lamb wave resonators in frequency control and sensing applications: a review. *J. Micromech. Microeng.*, **23** (4), 043001(14 pp).

6. Chandrahalim, H., Bhave, S.A., Polcawich, R.G., Pulskamp, J.S., and Kaul, R. (2010) PZT transduction of high-overtone contour-mode resonators. *IEEE Trans. Ultrason. Ferroelectr. Freq. Control*, **57** (9), 2035–2041.

7. Humad, S., Abdolvand, R., Ho, G.K., Piazza, G., and Ayazi, F. (2003) High frequency micromechanical piezo-on-silicon block resonators. IEEE International Electron Devices Meeting, December 8–10, 2003, pp. 957–960.

8. Piazza, G., Abdolvand, R., Ho, G.K., and Ayazi, F. (2004) Voltage-tunable piezoelectrically-transduced single-crystal silicon micromechanical resonators. *Sens. Actuators, A*, **A111** (1), 71–78.

9. Ruby, R. (1996) Micromachined cellular filters. 1996 IEEE MTT-S International Microwave Symposium Digest, June 17–21, 1996, pp. 1149–1152.

10. Ruby, R., Bradley, P., Clark, D., Feld, D., Jamneala, T., and Kun, W. (2004) Acoustic FBAR for filters, duplexers and front end modules. 2004 IEEE MTT-S International Microwave Symposium Digest, June 6–11, 2004, pp. 931–934.

11. Chengjie, Z., Sinha, N., and Piazza, G. (2010) Very high frequency channel-select MEMS filters based on self-coupled piezoelectric AlN contour-mode resonators. *Sens. Actuators, A*, **160** (1-2), 132–140.

12. Bongsang, K., Olsson, R.H. III, and Wojciechowski, K.E. (2013) AlN microresonator-based filters with multiple bandwidths at low intermediate frequencies. *J. Microelectromech. Syst.*, **22** (4), 949–961.

13. Mahameed, R., Sinha, N., Pisani, M.B., and Piazza, G. (2008) Dual-beam actuation of piezoelectric AlN RF MEMS switches monolithically integrated with AlN contour-mode resonators. *J. Micromech. Microeng.*, **18** (10), 105011, (11 pp.).

14. Piazza, G., Rinaldi, M., and Zuniga, C. (2004) Nanoscaled piezoelectric aluminum nitride contour mode resonant sensors. 2010 Ninth IEEE Sensors Conference (SENSORS 2010), November 1–4, 2010, pp. 2202–2207.

15. Ivanov, T.G., Pulskamp, J.S., Polcawich, R.G., and Proie, R.M. (2012) Shunt RF MEMS contact switch based on PZT-on-SOI technology. 2012 IEEE/MTT-S

International Microwave Symposium – MTT 2012, June 17–22 2012, 3 pp.

16. Proie, R.M., Polcawich, R.G., Pulskamp, J.S., Ivanov, T., and Zaghloul, M.E. (2011) Development of a PZT MEMS switch architecture for low-power digital applications. *J. Microelectromech. Syst.*, **20** (4), 1032–1042.

17. Ting-Ta, Y., Hirasawa, T., Wright, P.K., Pisano, A.P., and Liwei, L. (2011) Corrugated aluminum nitride energy harvesters for high energy conversion effectiveness. *J. Micromech. Microeng.*, **21** (8), 085037, (9 pp.).

18. Elfrink, R., Kamel, T.M., Goedbloed, M., Matova, S., Hohlfeld, D., van Andel, Y., and van Schaijk, R. (2009) Vibration energy harvesting with aluminum nitride-based piezoelectric devices. *J. Micromech. Microeng.*, **19** (9), 094005, (8 pp.).

19. Mina, I.G., Kim, H., Kim, I., Park, K.S., Choi, K., Jackson, T.N., Tutwiler, R.L., and Trolier-McKinstry, S. (2007) High frequency piezoelectric MEMS ultrasound transducers. *IEEE Trans. Ultrason. Ferroelectr. Freq. Control*, **54** (12), 2422–2430.

20. Guedes, A., Shelton, S., Przybyla, R., Izyumin, I., Boser, B., and Horsley, D.A. (2011) Aluminum nitride pMUT based on a flexurally-suspended membrane. Transducers 2011–2011 16th International Solid-State Sensors, Actuators and Microsystems Conference, June 5–9, 2011, pp. 2062–2065.

21. Littrell, R. and Grosh, K. (2012) Modeling and characterization of cantilever-based MEMS piezoelectric sensors and actuators. *J. Microelectromech. Syst.*, **21** (2), 406–413.

22. Williams, M.D., Griffin, B.A., Reagan, T.N., Underbrink, J.R., and Sheplak, M. (2012) An AlN MEMS piezoelectric microphone for aeroacoustic applications. *J. Microelectromech. Syst.*, **21** (2), 270–283.

23. Zuniga, C., Rinaldi, M., Khamis, S.M., Johnson, A.T., and Piazza, G. (2009) Nanoenabled microelectromechanical sensor for volatile organic chemical detection. *Appl. Phys. Lett.*, **94** (22), 223122, (3 pp.).

24. Olsson, R.H. III, Wojciechowski, K.E., Baker, M.S., Tuck, M.R., and Fleming, J.G. (2009) Post-CMOS-compatible aluminum nitride resonant MEMS accelerometers. *J. Microelectromech. Syst.*, **18** (3), 671–678.

25. Vigevani, G., Goericke, F.T., Pisano, A.P., Izyumin, I.I., and Boser, B.E. (2012) Microleverage DETF aluminum nitride resonating accelerometer. 2012 IEEE International Frequency Control Symposium (FCS), May 21–24, 2012, 4 pp.

26. Goericke, F.T., Vigevani, G., Izyumin, I.I., Boser, B.E., and Pisano, A.P. Novel thin-film piezoelectric aluminum nitride rate gyroscope. 2012 IEEE International Ultrasonics Symposium, IUS 2012, October 7–10, 2012, pp. 1067–1070.

27. Kubena, R.L., Stratton, F.P., Chang, D.T., Joyce, R.J., Hsu, T.Y., Lim, M.K., and M'Closkey, R.T. (2006) Next generation quartz oscillators and filters for VHF-UHF systems. 2006 IEEE MTT-S International Microwave Symposium Digest, June 11–16 2006, 4 pp.

28. Stratton, F.P., Chang, D.T., Kirby, D.J., Joyce, R.J., Tsung-Yuan, H., Kubena, R.L., and Yook-Kong, Y. (2004) A MEMS-based quartz resonator technology for GHz applications. Proceedings of the 2004 IEEE International Frequency Control Symposium And Exhibition, August 23–27 2004, pp. 27–34.

29. Gong, S. and Piazza, G. (2012) Weighted electrode configuration for electromechanical coupling enhancement in a new class of micromachined lithium niobate laterally vibrating resonators. 2012 IEEE International Electron Devices Meeting, IEDM 2012, December 10–13, 2012, pp. 15.6.1–15.6.4.

30. Gong, S. and Piazza, G. (2013) Design and analysis of lithium-niobate-based high electromechanical coupling RF-MEMS resonators for wideband filtering. *IEEE Transa. Microwave Theory Tech.*, **61** (1), 403–414.

31. DeVoe, D.L. (2001) Piezoelectric thin film micromechanical beam resonators. *Sens. Actuators, A*, **A88** (3), 263–272.

32. Zuo, C., Sinha, N., and Piazza, G. (2010) Very high frequency channel-select MEMS filters based on self-coupled

piezoelectric AlN contour-mode resonators. *Sens. Actuators, A*, **160** (1-2), 132–140.

33. Cady, W. (1946–1947) *Piezoelectricity*, McGraw-Hill, New York.

34. Ikeda, T. (1990) *Fundamentals of Piezoelectricity*, Vol. 2, Oxford University Press, Oxford.

35. Piazza, G., Felmetsger, V., Muralt, P., Olsson, R.H. III, and Ruby, R. (2012) Piezoelectric aluminum nitride thin films for microelectromechanical systems. *MRS Bull.*, **37** (11), 1051–1061.

36. Mishin, S., Marx, D.R., Sylvia, B., Lughi, V., Turner, K.L., and Clarke, D.R. (2003) Sputtered AlN thin films on Si and electrodes for MEMS resonators: Relationship between surface quality microstructure and film properties. 2003 IEEE Ultrasonics Symposium – Proceedings, October 5–8, 2003, pp. 2028–2032.

37. Olsson, R.H., III, Fleming, J.G., Wojciechowski, K.E., Baker, M.S., and Tuck, M.R. (2007) Post-CMOS compatible aluminum nitride MEMS filters and resonant sensors. 2007 IEEE International Frequency Control Symposium Jointly with the 21st European Frequency and Time Forum, May 29–June 1 2007, pp. 412–419.

38. Dubois, M.A., Carpentier, J.F., Vincent, P., Billard, C., Parat, G., Muller, C., Ancey, P., and Conti, P. (2006) Monolithic above-IC resonator technology for integrated architectures in mobile and wireless communication. *IEEE J. Solid-State Circuits*, **41** (1), 7–16.

39. Rabiei, P. and Gunter, P. (2005) Submicron thin films of lithium niobate single crystals prepared by crystal ion slicing and wafer bonding. 2005 Conference on Lasers and Electro-Optics (CLEO), May 22–27 2005, pp. 235–237.

40. Levy, M., Osgood, R.M. Jr., Liu, R., Cross, L.E., Cargill, G.S. III, Kumar, A., and Bakhru, H. (1998) Fabrication of single-crystal lithium niobate films by crystal ion slicing. *Appl. Phys. Lett.*, **73** (16), 2293–2295.

41. Muralt, P., Antifakos, J., Cantoni, M., Lanz, R., and Martin, F. (2005) Is there a better material for thin film BAW applications than AlN? 2005 IEEE Ultrasonics Symposium, September 18–21, 2005, pp. 315–320.

42. Wauk, M.T. (1969) *Attenuation in Microwave Acoustic Transducers and Resonators*, Stanford University.

43. Gong, S., Kuo, N.-K., and Piazza, G. (2011) GHz AlN lateral overmoded bulk acoustic wave resonators with a f. Q of 1.17 1013. 2011 Joint Conference of the IEEE International Frequency Control and the European Frequency and Time Forum (FCS), May 2–5 2011, 5 pp.

44. Frangi, A., Cremonesi, M., Jaakkola, A., and Pensala, T. (2013) Analysis of anchor and interface losses in piezoelectric MEMS resonators. *Sens. Actuators, A*, **190**, 127–135.

45. Royer, D. and Dieulesaint, E. (2000) *Elastic Waves in Solids II: Generation, Acousto-Optic Interaction, Applications*, Vol. 2, Springer.

46. Johnson, R.A. (1983) *Mechanical Filters in Electronics*, John Wiley & Sons, Inc, New York.

47. Tilmans, H.A. (1996) Equivalent circuit representation of electromechanical transducers: I. Lumped-parameter systems. *J. Micromech. Microeng.*, **6** (1), 157.

48. Tilmans, H.A. (1997) Equivalent circuit representation of electromechanical transducers: II. Distributed-parameter systems. *J. Micromech. Microeng.*, **7** (4), 285.

49. Larson, J.D., III, Bradley, P.D., Wartenberg, S., and Ruby, R.C. (2000) Modified butterworth-van dyke circuit for FBAR resonators and automated measurement system. 2000 IEEE Ultrasonics Symposium. Proceedings. An International Symposium, October 22–25, 2000, pp. 863–868.

50. Piazza, G., Stephanou, P.J., and Pisano, A.P. (2007) One and two port piezoelectric higher order contour-mode MEMS resonators for mechanical signal processing. *Solid-State Electron.*, **51** (11-12), 1596–1608.

51. Piazza, G. and Pisano, A.P. (2007) Two-port stacked piezoelectric aluminum nitride contour-mode resonant MEMS. *Sens. Actuators, A*, **136** (2), 638–645.

52. Zuo, C., Sinha, N., Van Der Spiegel, J., and Piazza, G. (2010) Multifrequency

pierce oscillators based on piezoelectric AlN contour-mode MEMS technology. *J. Microelectromech. Syst.*, **19** (3), 570–580.

53. Bedair, S.S., Pulskamp, J.S., Polcawich, R.G., Morgan, B., Martin, J.L., and Power, B. (2013) Thin-film piezoelectric-on-silicon resonant transformers. *J. Microelectromech. Syst.*, **22** (6), 1383–1394.

54. Sanchez-Rojas, J., Hernando, J., Donoso, A., Bellido, J., Manzaneque, T., Ababneh, A., Seidel, H., and Schmid, U. (2010) Modal optimization and filtering in piezoelectric microplate resonators. *J. Micromech. Microeng.*, **20** (5), 055027.

55. Weaver, W. Jr., Timoshenko, S.P., and Young, D.H. (1990) *Vibration Problems in Engineering*, John Wiley & Sons, Inc., New York.

56. Kuypers, J.H., Lin, C.-M., Vigevani, G., and Pisano, A.P. (2008) Intrinsic temperature compensation of aluminum nitride Lamb wave resonators for multiple-frequency references. 2008 IEEE International Frequency Control Symposium, pp. 240–249.

57. Rinaldi, M., Zuniga, C., and Piazza, G. (2009) 5-10 GHz AlN contour-mode

nanoelectromechanical resonators. IEEE 22nd International Conference on Micro Electro Mechanical Systems, MEMS, 2009, pp. 916–919.

58. Gong, S., Kuo, N.-K., and Piazza, G. (2012) GHz high-Q lateral overmoded bulk acoustic-wave resonators using epitaxial sic thin film. *J. Microelectromech. Syst.*, **21** (2), 253–255.

59. Abdolvand, R., Ho, G.K., Butler, J., and Ayazi, F. (2007) ZnO-on-nanocrystalline diamond lateral bulk acoustic resonators. IEEE 20th International Conference on Micro Electro Mechanical Systems, MEMS, 2007, p. 795–798.

60. Kuo, N.-K., Songbin, G., Hartman, J., Kelliher, J., Miller, W., Parke, J., Krishnaswamy, S.V., Adam, J.D., and Piazza, G. (2012) Micromachined sapphire GHz lateral overtone bulk acoustic resonators transduced by aluminum nitride. 2012 IEEE 25th International Conference on Micro Electro Mechanical Systems (MEMS), January 29–February 2, 2012, pp. 27–30.

8
Electrothermal Excitation of Resonant MEMS

Oliver Brand and Siavash Pourkamali

8.1
Basic Principles

Thermal actuation is often associated with bimetal strips, which consist of two joined metal strips with different thermal expansion coefficients that bend when subjected to a temperature change. Originally invented by John Harrison (1693–1776) for temperature compensation in marine clocks [1], bimetal strips and coils have been widely used in thermometers, but are more and more replaced by electronic thermometers. While the environmental temperature induces a mechanical actuation in case of bimetal thermometers, this chapter focuses on *electrothermal* actuation, where power dissipation in a resistor R causes a temperature increase, which results in material expansion and, ultimately, mechanical actuation. Thus, electrothermal actuation as a transduction mechanism involves three energy domains, and the conversion of energy from the electrical to the thermal, and from the thermal to the mechanical domains.

The main advantage of electrothermal actuation in (resonant) microelectromechanical systems (MEMS) is the straightforward implementation, as resistive elements are readily available in many microfabrication processes. Moreover, electrothermal actuators can generate large actuation forces while operating at low voltages. However, due to the involvement of the thermal energy domain, they are often plagued with relatively high power consumption. Traditionally, thermal actuation is considered to be slow, but recent research has demonstrated efficient actuation up to the hundreds of megahertz [2] and potentially even the gigahertz range [3]. While this chapter focuses on electrothermal actuation in resonant MEMS, a thorough overview on the modeling of electrothermal actuators has been published in a recent chapter of this series [4].

8.1.1
Fundamental Equations for Electro-Thermo-Mechanical Transduction

The basic transduction element in an electrothermal actuator is a resistive heater, which is either embedded on the surface or in the bulk of the (resonant)

Resonant MEMS – Fundamentals, Implementation and Application, First Edition.
Edited by Oliver Brand, Isabelle Dufour, Stephen M. Heinrich and Fabien Josse.
© 2015 Wiley-VCH Verlag GmbH & Co. KGaA. Published 2015 by Wiley-VCH Verlag GmbH & Co. KGaA.

MEMS. If the heating resistor with resistance R is driven by an AC voltage V_{ac} $\cos(\omega t)$ superimposed on a DC voltage V_{dc}, the thermal power dissipation P is given by [5, 6]:

$$P = P_{stat} + P_{dyn1} + P_{dyn2} = \frac{V_{dc}^2}{R} + \frac{V_{ac}^2}{2R} + \frac{2V_{dc}V_{ac}}{R}\cos(\omega t) + \frac{V_{ac}^2}{2R}\cos(2\omega t) \quad (8.1)$$

The thermal power consists of one static component P_{stat} and two dynamic components P_{dyn1} and P_{dyn2} with frequencies ω and 2ω, respectively. The DC voltage V_{dc} is needed to obtain a frequency component with the same frequency as the applied AC voltage V_{ac} and, thus, to ensure device actuation at the same frequency as the applied driving signal. With $V_{dc} \gg V_{ac}$, one can make $P_{dyn1} \gg P_{dyn2}$.

The power dissipation results in a static and dynamic temperature distribution in the microstructure. Assuming that heat is transported by conduction only, the thermal transport is described by the heat flow equation [7]:

$$\frac{\partial Q}{\partial t} = \nabla \kappa \nabla T + P \quad (8.2)$$

Here, Q represents the stored thermal energy per volume, κ the thermal conductivity, T the temperature, P the power dissipation per volume, and ∇ the nabla operator. If we further assume a temperature-independent specific heat (per mass) C and a spatially constant thermal conductivity, Eq. (8.2) can be rewritten as

$$\frac{\partial T}{\partial t} = \frac{\kappa}{\rho C}\nabla^2 T + \frac{P}{\rho C} \quad (8.3)$$

with the mass density ρ and the Laplace operator ∇^2. The ratio $\kappa(\rho C)^{-1}$ is often referred to as the thermal diffusivity. Since Eq. (8.3) is linear in temperature, the effects of the static and dynamic power dissipations in Eq. (8.1) can be treated separately.

The *static heating component* P_{stat} causes a static temperature elevation ΔT_{stat} in the microstructure. In constrained microstructures, such as clamped–clamped beams or clamped plates, compressive stress will develop as a result of the static temperature elevation, which can significantly affect the resonator characteristics, such as the resonance frequency. In case of clamped-free beams, however, the (static) thermal expansion does not cause any axial thermal stress and the resonance frequency is expected to be largely independent of P_{stat} (it should be noted that changes in the device dimensions and material properties, for example, the Young's modulus E, caused by the static temperature increase will of course still affect the resonance frequency). Although a thermally induced frequency shift seems to be a disadvantage at first sight, it allows tuning of the resonance frequency in a controlled, reproducible way. In case of complex microstructures with multiple materials and heat transport in three dimensions, the temperature across the microstructure is typically modeled using finite element modeling (FEM) software tools. However, for simple device geometries, such as cantilever beams, Eq. (8.3) can often be solved analytically.

The *dynamic heating power* P_{dyn1} produces a dynamic temperature change (thermal wave) in the vicinity of the heating resistors. In case of resonant MEMS, this dynamic temperature distribution is the most important for the resonator actuation. The temperature increase $\Delta T = T - T_0$ stemming from the dynamic heating power and the static heating power strain the resonator material [7]

$$\varepsilon(T) = \varepsilon(T_0) + \alpha_T(T - T_0) \tag{8.4}$$

with ε being the material strain, α_T the linear coefficient of thermal expansion, and T_0 the reference temperature. The periodic strain stemming from the thermal wave in the vicinity of the heating resistors will ultimately generate the mechanical forces or moments, which actuate the resonant structure.

8.1.2
Time Constants and Frequency Dependencies

The thermal domain is in many ways similar to the electrical domain and thermal phenomena and systems can usually be modeled as low-pass electrical RC circuits. Just as an electrical potential gradient (i.e., voltage difference between two points) leads to an electrical current flow in an electrical conductor, a temperature gradient leads to a heat flow in a thermal conductor. Accurate analysis of thermal systems is, however, much more challenging than the analysis of electrical systems due to the distributed nature of the different elements, a variety of heat-transfer mechanisms (conduction, convection, and radiation), and the lack of perfect thermal insulators (heat can flow by conduction, convection, or radiation through any medium, while electricity generally only flows via conduction in electrical conductors).

Still, lumped-element modeling is widely used to develop reduced-order models that describe, for example, the maximum temperature elevation T_{max} as a function of the effective thermal resistance R_{th} (in units of K/W) and thermal capacitance C_{th} (in units of J/K) of the microstructure. The resulting equivalent circuit (see Figure 8.1) can be analyzed easily, resulting in a complex transfer function

$$H_{th}(\omega) = \frac{T_{max}}{P} = \frac{R_{th}}{(1 + j\omega R_{th} C_{th})} \tag{8.5}$$

with the thermal time constant

$$\tau_{th} = R_{th} C_{th} \tag{8.6}$$

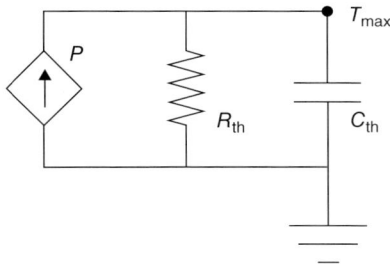

Figure 8.1 Equivalent electrical circuit for a thermal actuator in the thermal domain.

The challenge often lies in extracting proper values for R_{th} and C_{th} for a particular microstructure because of the distributed nature of the thermal circuit elements and the different heat transfer mechanisms that need to be considered. Moreover, R_{th} and C_{th} are generally frequency dependent themselves and, thus, are not the same for the static heating power P_{stat} and the dynamic heating power P_{dyn1}.

As demonstrated by Eq. (8.5), the temperature to heating power transfer function has the form of a low-pass transfer function (similar to the response of electrical low pass RC circuits). Just as it takes time for an electrical capacitor to settle to its steady state voltage when a resistor limits the current flow, it takes time for an object to reach thermal equilibrium due to heat flows being limited by the thermal resistances in the heat paths. Due to this low-pass behavior, as well as our experience in the macro world where thermal phenomena typically have relatively large time constants, thermal actuators are generally considered to be slow actuators suitable only for DC or very low frequency applications. It is hard to imagine a thermal system with small enough thermal time constants to respond to high-frequency excitations in the gigahertz range. As we will see, this is however possible due to the reduced length scales of micro- and nano-scale resonators.

Figure 8.2 shows a qualitative thermal analysis of a dog-bone structure that is electro-thermally actuated by an alternating electrical current flowing through the structure [3]. The current flow enters and exits the structure through the two short anchoring arms of the structure as shown in the figure. The same anchoring points also act as heat sinks. The structure has a length, width, and thickness of 48, 22, and 15 μm and conduction is the only heat dissipation/transfer mechanism considered in the analysis. Figure 8.2a shows the resulting dynamic temperature distribution

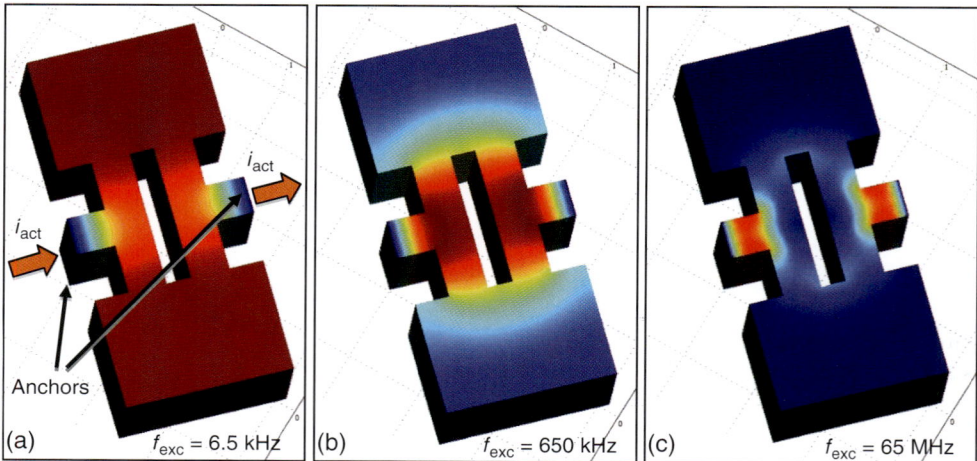

Figure 8.2 Qualitative transient thermal analysis of a dog-bone micro-structure showing the dynamic (AC) temperature distribution when thermally actuated by an alternating electrical current with frequency of: (a) 6.5 kHz, (b) 650 kHz, and (c) 65 MHz.

(ΔT_{dyn}) for a relatively low excitation frequency of 6.5 kHz. In this case, the electrical excitation is slow enough for the dynamic temperature across the structure to follow the excitation (i.e., the excitation frequency is small compared to the inverse of the thermal time constant of the whole structure). Results from a 100x faster excitation with frequency of 650 kHz, however, show a completely different distribution of the temperature fluctuation amplitude (Figure 8.2b). In this case, the temperature of locations farther away from the anchoring points barely changes and the maximum temperature variation occurs in the thermal actuation beams that have a lower thermal capacitance as well as lower thermal resistance (due to their vicinity to the anchoring points). At another 100x faster excitation of 65 MHz (Figure 8.2c), which is way beyond the cut-off frequency of the low pass thermal system in this case, very small temperature fluctuations occur only at areas close to the anchoring points. It should be noted that the temperature scales in the three figures are not the same and the ratio of the temperature difference between the dark red and dark blue regions are approximately 80 : 15 : 1 for Figure 8.2a : b : c. In summary, the alternating electrical excitation current leads to a heat wave in the mechanical structure where the heat wave becomes more confined and also smaller in amplitude as the excitation frequency increases.

Although, according to the discussion above, the response of a thermal actuator appears to deteriorate as the excitation frequency increases, a closer look at the scaling behavior of a thermal actuator reveals an interesting trend. Similar to electrical RC circuits, the thermal time constant of a thermal actuator is the product of its effective thermal capacitance and thermal resistance (see Eq. (8.6)): $\tau_{th} = R_{th}C_{th}$. The thermal resistance of an element can be calculated using a very similar equation to that of the electrical resistance, that is, $R_{th} = \kappa^{-1}L/A$, where κ is the thermal conductivity of the structural material and L and A are the length and cross-sectional area of the element, respectively. Similar to electrical resistance, the thermal resistance increases proportionally as all the dimensions of the element are scaled down. In other words, if a thermal actuator is scaled down by X times, its thermal resistance increases by a factor of X. On the other hand, the thermal capacitance is proportional to the mass and consequently the volume of the element and therefore decreases by a factor of X^3 if the dimensions are scaled by a factor of X. Hence, the thermal time constant of the element reduces by a factor of X^2 upon shrinking the dimensions by X times [3].

This may not seem significant as it is a general trend and somewhat expected for physical phenomena to occur faster in smaller systems. The significance of the scaling behavior of the thermal response time is more clearly revealed when compared to the scaling behavior of mechanical response. If a mechanical structure is scaled down by a factor of X, its mechanical resonant frequency increases only by a factor of X, that is, the mechanical time constant decreases by a factor of X. The overall conclusion is that if all the dimensions of a thermally actuated resonant structure are scaled down, its thermal time constant decreases faster than its mechanical time constant. To be more specific, both thermal and mechanical responses of the system become faster upon scaling the dimensions down, however, the increase in the speed of the thermal response is sharper (proportional to X^2) than the increase in the mechanical resonant frequency (proportional to X). Therefore, as the structural dimensions are scaled down, the performance of thermally actuated mechanical resonators is expected to improve (more actuation force for the same actuation power) as the thermal response of the system can catch up with the mechanical vibrations more effectively. This can present an interesting opportunity for development of high-performance, high-frequency, thermally actuated resonators with extremely small dimensions comparable to those of state-of-the-art transistors. Both piezoelectric and electrostatic transduction require relatively large device dimensions in the tens to hundreds of microns to provide a large enough electrode area for transduction strength. Thermally actuated resonators could therefore be the solution for high-frequency mechanical resonant components that can be integrated on-chip in a size-efficient manner.

8.2
Actuator Implementations

Arguably, the two most common electrothermal actuator structures are the U-shape and bent-beam actuators. The *U-shape electrothermal actuator* [8, 9] typically comprises a surface-micromachined "U" featuring two legs with different widths (see Figure 8.3a). This way, the electrical resistance of the narrow leg is large compared to the wider leg. Thus, if a voltage is applied between the two anchor points, more power is dissipated in the narrow leg, it heats up and expands more, and the actuator bends toward the wider leg. The narrow suspension beam close to the clamped edge of the wide beams reduces the

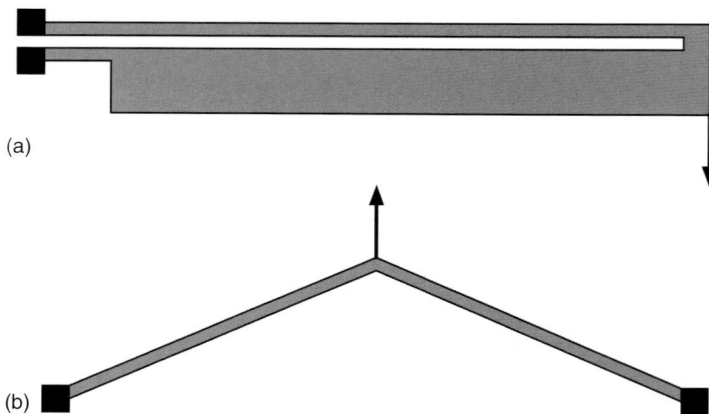

(a)

(b)

Figure 8.3 Schematic view of (a) U-shaped and (b) bent-beam electrothermal actuators. The microstructures are anchored to the substrate at the location of the black squares. The actuators are operated by applying a voltage between the two anchor points. The actuation direction is indicated by the arrow.

stiffness of the actuator structure and allows for larger deflections. Characteristic for such actuators is relatively large actuation distances.

The *bent-beam electrothermal actuator* [10, 11] is basically a clamped–clamped beam structure consisting of two angled beams (see Figure 8.3b). By applying a voltage between the anchor points, power is dissipated in the angled clamped-clamped beam and thermal expansion pushes the apex of the V-shaped beam outward. Bent-beam actuators are especially suited if large actuation forces are needed.

Both types of electrothermal actuators have been extensively modeled, but are typically operated at low frequencies and are not widely used in resonant MEMS. It should be noted, however, that the dimensions of both U-shaped and bent-beam actuators can be easily adjusted to reach higher operating frequencies (typically at the expense of a reduced maximum deflection); an example along this line is the U-shaped micromechanical resonator/oscillator developed at NXP Semiconductor, which operates at 1.26 MHz ([12], see also Section 8.3.3).

It is interesting that the use of electrothermal actuators to excite microfabricated resonators by far predates the U-shaped and bent-beam actuator development: Wilfinger, Bardell, and Chhabra devised and analyzed a thermally excited resonant cantilever beam in the late 1960s [13, 14]. While the silicon cantilevers used were still fairly large, these so-called "resonistors" had many of the features of today's thermally actuated resonators, such as diffused resistors for electrothermal excitation and piezoresistive detection of transverse beam vibrations. After this initial work, it took until the early 1990s for electrothermal actuation to become more widely investigated in (resonant) MEMS [5, 6].

Today, two fundamental approaches can be distinguished: (i) electrothermal actuators with heaters located on the surface of the MEMS structure (thin-film/surface actuators) [15–17] and (ii) electrothermal actuators, where the heater comprises most of the cross-section of the MEMS structure (bulk actuators) [3, 18]. In the following section, typical representatives of these actuator classes will be analyzed in more detail.

8.2.1
Thin-Film/Surface Actuators

Figure 8.4a, b show a scanning electron microscope (SEM) micrograph and a schematic of a silicon cantilever with two diffused heating resistors at its clamped edge, respectively [15]. The cantilever also features a U-shaped Wheatstone bridge with four diffused piezoresistors that was designed to particularly detect in-plane vibration modes ([15], see Section 8.3 for details). The out-of-plane component of the vibration amplitude (measured using an optical vibrometer) of a thermally actuated 400 μm long, 45 μm wide, and 20 μm thick silicon cantilever as a function of frequency is shown in Figure 8.4c. Three resonance frequencies are clearly visible in the tested frequency range and the associated mode shapes have been identified using FEM. The second mode at 338 kHz is actually the fundamental in-plane vibration mode of the cantilever; it has a strong in-plane

(a)   (b)

(c)

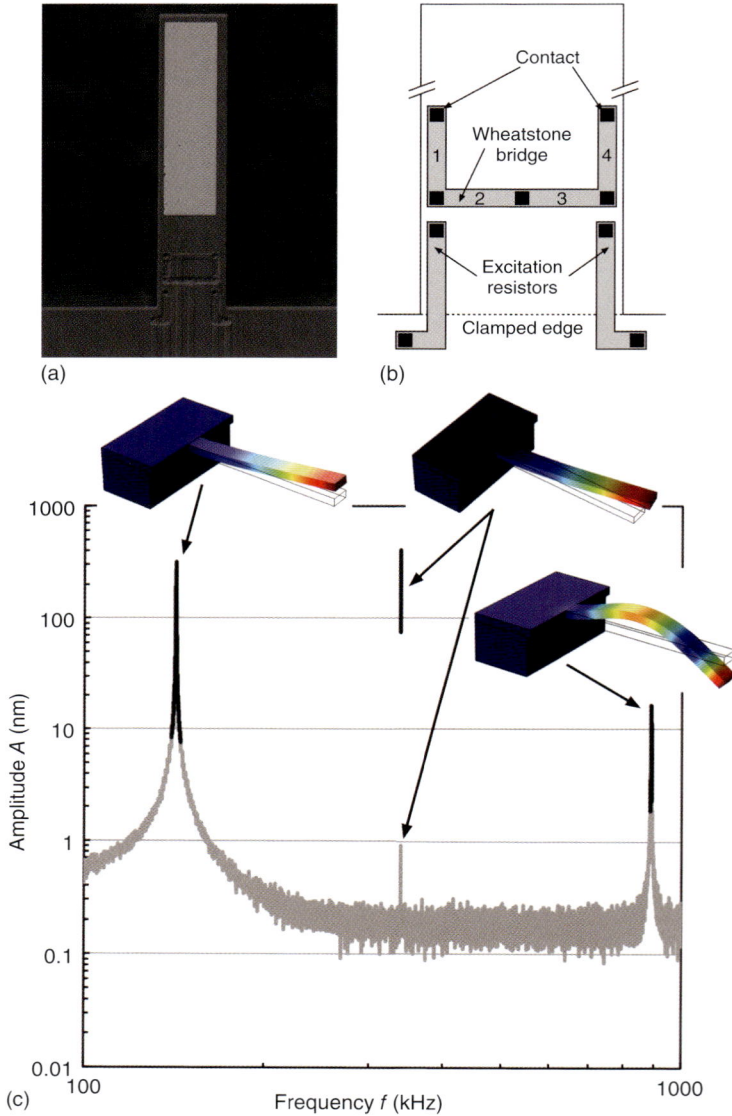

Figure 8.4 (a) SEM photograph and (b) schematic of resonant silicon cantilever with two diffused silicon resistors for electrothermal excitation and four piezoresistors connected in a Wheatstone bridge for vibration sensing. (c) Out-of-plane vibration amplitude (measured using an optical vibrometer) at the tip of a 400 μm long, 45 μm wide, and 20 μm thick cantilever driven with a dynamic heating power of $P_{dyn1} = 4$ mW as a function of the excitation frequency; the gray curve represents a coarse scan over the frequency range 100 kHz to 1 MHz, while the solid black lines correspond to detailed measurements around the resonance modes; in case of the in-plane mode at 338 kHz, the black line represents the in-plane vibration amplitude. The individual modes have been identified by FEM.

vibration component (solid line), but also a non-zero out-of-plane vibration component (gray line). The following section briefly discusses how the bending moments responsible for the out-of-plane flexural vibrations of the cantilever are created, even in the case of single material structures.

In the case of diffused heating resistors, the thermal wave and associated temperature gradients in the vicinity of the heaters, especially in the thickness direction, ensure the actuation. If the surface area A covered by the resistor is much larger than the thickness h of the resonator (see schematic in Figure 8.5), the dynamic temperature distribution can be assumed to be constant in the xy-plane, and the temperature distribution in the thickness direction z can be approximated by solving the one-dimensional heat-flow equation [7]

$$\frac{\partial T(z,t)}{\partial t} = \frac{\kappa}{\rho C} \frac{\partial^2 T(z,t)}{\partial z^2} \tag{8.7}$$

with the boundary conditions:

$$J_z(0,t) = -\kappa \left. \frac{\partial T(z,t)}{\partial z} \right|_{z=0} = \frac{P}{A} \tag{8.8}$$

$$J_z(h,t) = 0 \tag{8.9}$$

These boundary conditions assume that the heating power $P = P_{dyn1}e^{j\omega t}$ is dissipated across a resonator surface area A, defining an incoming heat flux J_z, and that no heat can escape from the lower surface of the resonator. The resulting temperature distribution in the z-direction becomes [6]

$$T_{dyn1}(z,t) = Re \left\{ \frac{P_{dyn1}}{A\sigma\kappa} \left[\frac{e^{-\sigma h}e^{\sigma z}}{2\sinh(\sigma h)} + \frac{e^{\sigma h}e^{-\sigma z}}{2\sinh(\sigma h)} \right] e^{j\omega t} \right\} \tag{8.10}$$

with $\sigma = (1+j)/\delta$ and the penetration depth of the thermal wave $\delta = \sqrt{2\kappa/(\rho C\omega)}$. Equation (8.10) describes a thermal wave with exponentially decreasing amplitude that penetrates through the beam in the z-direction and is

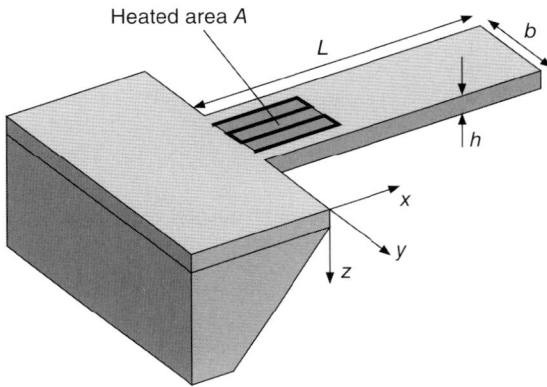

Figure 8.5 Schematic of cantilever with surface heater at its clamped edge.

reflected at the surfaces [6]. The amplitude of this dynamic temperature profile depends on the dissipated power and on the excitation frequency. The penetration depth δ of the thermal wave depends on the excitation frequency and the thermal properties of the cross-section of the beam structure. Assuming a silicon resonator ($\kappa = 156\,\mathrm{Wm^{-1}\,K^{-1}}$, $\rho = 2330\,\mathrm{kg\,m^{-3}}$, $C = 0.7\,\mathrm{Jg^{-1}\,K^{-1}}$), Figure 8.6 shows the resulting temperature profile through the resonator thickness ($h = 20\,\mu\mathrm{m}$) for two frequencies, $f = 50\,\mathrm{kHz}$ and $5\,\mathrm{MHz}$, and for $P_{\mathrm{dyn1}}/A = 10\,\mu\mathrm{W}\,\mu\mathrm{m^{-2}}$. At these frequencies, the corresponding penetration depths of the thermal wave are 25 and 2.5 μm, respectively. As the frequency increases, the thermal wave becomes more localized and its amplitude decreases (which is similar to the trend demonstrated in Figure 8.2).

The resulting temperature gradients in the z-direction cause an oscillating effective bending moment (around the y-axis) even in single material cantilevers. In multilayer sandwich structures, the dynamic heating produces an additional bending moment due to the different thermal expansion coefficients of the materials involved. In the case of a homogeneous beam, the resulting effective bending moment M generated along the length of the resistor can be calculated from the first moment of the temperature distribution as [6]

$$M(t) = Eb\alpha_T \int_0^h T_{\mathrm{dyn1}}(z,t)[h/2 - z]\mathrm{d}z \tag{8.11}$$

with the Young's modulus E and thermal expansion coefficient α_T of the beam material and the width of the beam b. Equation (8.11) assumes that the heater spans the whole width of the beam; if this is not the case, b should be replaced

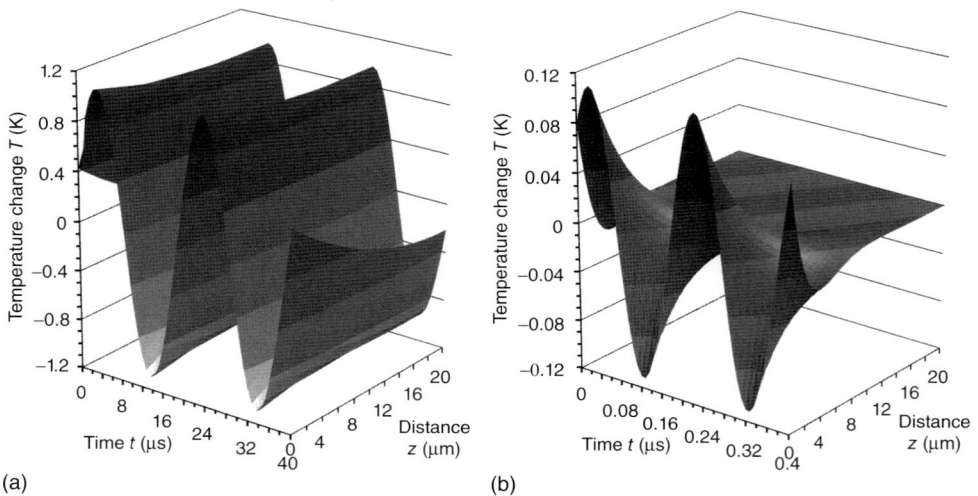

(a)

(b)

Figure 8.6 Temperature profile as a function of distance z from the heater (in the thickness direction) and time according to Eq. (8.10) for a silicon cantilever with $h =$ 20 μm, a heating power per area $P_{\mathrm{dyn1}}/A =$ 10 μW μm^{-2} and two excitation frequencies of (a) 50 kHz and (b) 5 MHz.

by the effective width of the resistor. The resulting effective moment exhibits a low-pass characteristic with a characteristic turnover frequency of [6]

$$\omega_t = 2\pi f_t = 12\frac{\kappa}{\rho C}\frac{1}{h^2} \tag{8.12}$$

For $\omega \ll \omega_t$, the amplitude of the bending moment becomes frequency independent [6]

$$M(t) = P_{dyn1}\frac{\alpha_T}{2\kappa}EIe^{j\omega t} \tag{8.13}$$

with the moment of inertia $I = \frac{1}{12}bh^3$ for the cantilever beam with rectangular cross-section. For the silicon beam with 20 μm thickness, the turnover frequency is $f_t = 460$ kHz. Decreasing the cantilever thickness will further increase this frequency limit; as was found for the simple scaling considerations of the thermal time constant in Section 8.1.2, the cut-off frequency for the thermal actuation increases by a factor of X^2 when reducing the dimension (thickness) by a factor of X. Again, it should be noted that the turnover frequency associated with this electrothermal excitation is generally orders of magnitude larger than the frequency associated with the thermal RC time constant of the full micromechanical structure (see Eq. (8.6)), because the thermal wave only exists over a characteristic length scale defined by δ, while the temperature increase associated with the static heating power heats the complete microstructure. This way the thermal capacitance associated with the dynamic heating power is greatly reduced and the cut-off frequency for electrothermal excitation is greatly increased. Finally, it should be noted that an alternative modeling approach is to treat the thermal loading as a discontinuity in the beam's rotation, which has been used to simulate the response of electrothermally excited cantilever beams vibrating in the in-plane modes [19].

For the cantilever beam shown in Figure 8.4, the heaters do not span the cantilever width and the lateral heater extensions (especially the heater width) are also not large compared to the cantilever thickness. As a result, there will not only be a thermal wave in z-direction but also in x- and y-directions. However, the characteristic penetration depth of the thermal wave will be similar in all directions. If both heating resistors in Figure 8.4a are excited in parallel, preferentially out-of-plane vibration modes will be excited; if only one heating resistor is used, also in-plane vibration modes can be excited effectively. Because $b > h$, the cut-off frequency for the excitation of in-plane modes will be lower than for out-of-plane modes. In the case of the 400 μm long, 45 μm wide, and 20 μm thick cantilever (see Figure 8.4c) driven by one heating resistor ($P_{stat} = 3$ mW and $P_{dyn1} = 4$ mW), the measured tip vibration amplitude was 315 nm (with a quality factor of 1100 in air) at the fundamental out-of-plane resonance frequency $f = 142$ kHz and 350 nm (with a quality factor of 3400 in air) at the fundamental in-plane resonance frequency $f = 339$ kHz.

8.2.2
Bulk Actuators

In contrast to the thin-film/surface actuators, bulk actuators are uniformly heated throughout the device thickness and, thus, the temperature in the thickness direction is constant assuming that there is no (or negligible) heat flow from the top and bottom surfaces. In the in-plane direction, temperature fluctuations will extend beyond the heater area; similar to the surface heater, the temperature swing will exponentially decrease with the distance from the heater. Assuming that there is no reflecting boundary in the lateral direction, the temperature distribution is described by the first term of Eq. (8.10), with P_{dyn1}/A being the dissipated power per cross-sectional area A.

As an example for a bulk actuator, Figure 8.7 shows an SEM photograph of an I-shaped bulk acoustic wave resonator or IBAR (also known as a dog-bone resonator) [3]. In the in-plane extensional resonance mode, the two blocks attached on the two sides of the structure move back and forth in opposite directions subjecting the two beams connecting them to periodic tensile and compressive stress. This resonance mode can simply be actuated by applying the actuation voltage (or current) between the two support pads on the two sides of the structure. The ohmic loss caused by the current flow is maximized in the narrower parts of its path, that is, along the extensional beams in the middle of the structure (the beams connecting the two blocks to each other). Upon application of a fluctuating actuation voltage, the fluctuating ohmic power generation results in a fluctuating temperature gradient and therefore periodic thermal expansion of the extensional actuator beams. The alternating extensional force resulting from the fluctuating temperature in the beams can actuate the resonator in its in-plane extensional resonance mode. If the ohmic power and consequently, the resulting temperature fluctuations have the same frequency as the mechanical resonance frequency

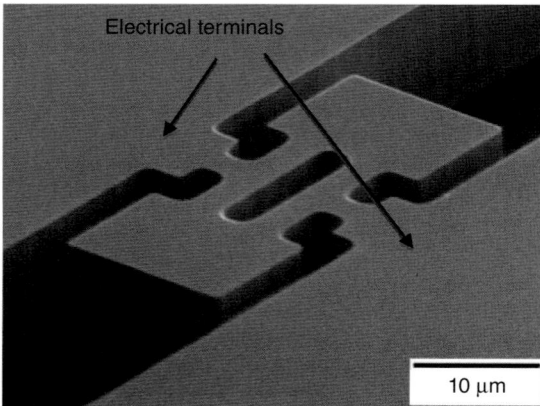

Figure 8.7 SEM view of a 61 MHz 15 μm thick single crystalline silicon thermal-piezoresistive IBAR (dog-bone resonator) fabricated on a low resistivity *p*-type SOI substrate.

of the structure, the vibration amplitude is amplified by the mechanical quality factor (Q) of the resonator. The amplified alternating stress in the pillars leads to considerable fluctuations in their electrical resistance due to the piezoresistive effect. When biased with a DC voltage, resistance fluctuations modulate the current passing through the structure resulting in an AC current component known as the motional current. A more extensive discussion and analysis of operation of such bulk mode thermally actuated resonators is presented in Section 8.4.

8.3
Piezoresistive Sensing

Because electrothermal actuation is not a reversible transduction mechanism, it is often paired in resonant MEMS with piezoresistive detection of vibrations. Thereby, either separate resistors can be used for sensing or the same resistor is employed for thermal actuation and piezoresistive sensing. In the following section, the fundamentals of piezoresistive sensing are briefly discussed, implementations of piezoresistors in resonant MEMS are highlighted, and the implementation of mechanical thermal-piezoresistive oscillators is presented.

8.3.1
Fundamental Equations for Piezoresistive Sensing

Stress and strain applied to a resistor affect its resistance in two ways: (i) through a change of the resistor geometry and (ii) through a change of its resistivity. In metals, the geometrical effect dominates and the change of resistivity in iron and copper wires subject to elongation has been studied by William Thomson (Lord Kelvin) more than 150 years ago [20]. In semiconductors, in particular silicon, the change in resistivity due to applied stress, the so-called piezoresistivity, is far larger than the geometric effect. The "exceptionally large" piezoresistive effect in silicon and germanium has been studied by C.S. Smith at Bell Laboratories and published in a seminal paper in 1954 [21]. In 1962, the first pressure-sensing diaphragm based on diffused piezoresistors was demonstrated [22]. Today, piezoresistive pressure sensors are a mature technology and piezoresistive transduction is explored for various sensors, including resonant MEMS. An excellent review on semiconductor piezoresistance for microsystems was published in 2009 by Barlian *et al.* [23]. The following text will briefly highlight the fundamental equations for piezoresistive transduction.

A resistor with length L and cross-sectional area A subject to an applied strain ε in length direction experiences a relative resistance change [23]

$$\frac{\Delta R}{R} = (1 + 2\nu)\varepsilon + \frac{\Delta\rho}{\rho} \tag{8.14}$$

with the Poisson ratio ν and the resistivity of the material ρ. The base resistance is simply given by $R = \rho L/A$. The first term in Eq. (8.14) describes the geometric effect, the second term the piezoresistive effect. For small stresses, the relative resistance change is assumed to be a linear function of the applied stress, with the piezoresistive coefficients π being the proportionality factor. Thus, in the presence of applied stress, Ohm's law can be written as [24]

$$E_i = \rho_0 \sum_{j=1}^{3} \left[\delta_{ij} + \sum_{m,n=1}^{3} \pi_{ijmn} \sigma_{mn} \right] J_j \tag{8.15}$$

with the zero-stress resistivity $\rho_0 = (\rho_{11} + \rho_{22} + \rho_{33})/3$ and Kronecker's delta δ_{ij}. σ_{mn} is the stress tensor of rank two and π_{ijmn} is the piezoresistive tensor of rank four. Due to the symmetry of the resistivity and stress tensors, a simplified notation is widely used in which the resistance change and stress are described as six-component vectors, yielding [25]

$$\frac{\Delta\rho_I}{\rho_0} = \sum_{J=1}^{6} \pi_{IJ}\sigma_J \tag{8.16}$$

In this notation, the tensor of piezoresistive coefficients for a material with cubic crystal symmetry, such as silicon, is given by [24, 25]

$$[\pi_{IJ}] = \begin{bmatrix} \pi_{11} & \pi_{12} & \pi_{12} & 0 & 0 & 0 \\ \pi_{12} & \pi_{11} & \pi_{12} & 0 & 0 & 0 \\ \pi_{12} & \pi_{12} & \pi_{11} & 0 & 0 & 0 \\ 0 & 0 & 0 & \pi_{44} & 0 & 0 \\ 0 & 0 & 0 & 0 & \pi_{44} & 0 \\ 0 & 0 & 0 & 0 & 0 & \pi_{44} \end{bmatrix} \tag{8.17}$$

with the coefficients π for low-doped *p*- and *n*-type silicon given in Table 8.1. Using appropriate coordinate transformations, the piezoresistive coefficients in other directions can be obtained from Eq. (8.17).

In most applications, long and narrow resistors are used as piezoresistors, so that the resistor geometry confines the current and electric field. Under this assumption, the relative resistance change can be written as [7]

$$\frac{\Delta R}{R} = \pi_l \sigma_l + \pi_t \sigma_t \tag{8.18}$$

with the longitudinal (in the length direction of the resistor) and transverse stresses σ_l and σ_t, respectively, and the associated piezoresistive coefficients π_l and π_t. For the most common case of piezoresistors implemented in the surface

Table 8.1 Piezoresistive coefficients for low-doped single-crystalline silicon [7, 21, 24, 25].

	ρ_0 (Ω cm)	π_{11} (Pa^{-1})	π_{12} (Pa^{-1})	π_{44} (Pa^{-1})
n-type Si	11.7	-102.2×10^{-11}	53.4×10^{-11}	-13.6×10^{-11}
p-type Si	7.8	6.6×10^{-11}	-1.1×10^{-11}	138.1×10^{-11}

Table 8.2 Longitudinal and transverse piezoresistive coefficients for piezoresistors located on the surface of low-doped (100) silicon wafers.

	π_l (Pa^{-1})	π_t (Pa^{-1})	π_l (Pa^{-1})	π_t (Pa^{-1})
	<100> direction ($\phi = 0°$, i.e., 45° to primary flat)		<110> direction ($\phi = 45°$, i.e., parallel to primary flat)	
n-type Si	-102.2×10^{-11}	53.4×10^{-11}	-31.2×10^{-11}	-17.6×10^{-11}
p-type Si	6.6×10^{-11}	-1.1×10^{-11}	71.8×10^{-11}	-66.3×10^{-11}

of a (100) silicon wafer, one can derive

$$\pi_l = \pi_{11} - 2(\pi_{11} - \pi_{12} - \pi_{44})\cos^2\phi\sin^2\phi$$
$$\pi_t = \pi_{12} + 2(\pi_{11} - \pi_{12} - \pi_{44})\cos^2\phi\sin^2\phi \tag{8.19}$$

with ϕ being the angle between the resistor length direction and the <100> direction in the wafer surface (see Table 8.2 for values in <100> and <110> directions). Typically, p-type silicon piezoresistors are aligned parallel or perpendicular with respect to the primary flat on (100) wafers, while n-type silicon piezoresistors are rotated 45° with respect to the primary flat. (Of course, a fourfold symmetry holds on the (100) surface because of the cubic lattice structure.)

The piezoresistive coefficients decrease as the silicon doping concentration increases [23, 25]. At the same time, the temperature coefficient of the piezoresistive coefficients also decreases with increasing doping concentration [23, 25], making higher doped piezoresistors an interesting choice for applications requiring a wider temperature range.

8.3.2
Piezoresistor Implementations

To suppress common mode signals stemming from, for example, temperature changes, four piezoresistors are often combined in a Wheatstone bridge. By cleverly arranging the resistors on the (resonant) microstructure, the designer can benefit from opposite signs of the longitudinal and transverse piezoresistive coefficients to boost the output signal and can suppress signals from unwanted vibration modes, while amplifying signals of desired modes. To this end, the characteristic stress distributions of the vibrating microstructure in the different resonance modes need to be considered. As an example, Figure 8.8 shows the characteristic (surface) stress distributions of the first out-of-plane, in-plane and torsional modes of a 200 µm long, 45 µm wide, and 20 µm thick silicon cantilever that can be explored to sense only the mode of interest. By arranging four p-type piezoresistors in one of the two bridge layouts, either out-of-plane or in-plane flexural modes are preferentially detected, assuming that the cantilever is released on the surface of a (100) wafer and aligned with respect to the primary flat (i.e., in <110> direction). Considering for the moment only the U-shape bridge layout (Figure 8.8b) with Eq. (8.18), the coefficients $\pi_l = 71.8 \cdot 10^{-11}\,\mathrm{Pa}^{-1}$

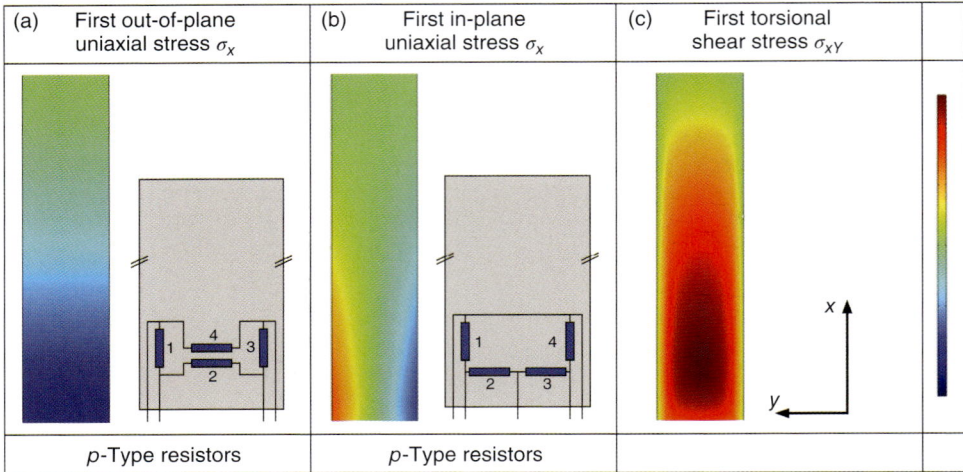

(a) First out-of-plane uniaxial stress σ_x	(b) First in-plane uniaxial stress σ_x	(c) First torsional shear stress σ_{xY}
p-Type resistors	*p*-Type resistors	

Figure 8.8 Simulated stress distribution in 200 μm long, 45 μm wide, and 20 μm thick silicon cantilever for fundamental (a) out-of-plane, (b) in-plane, and (c) torsional flexural modes (red = positive stress, blue = negative stress). The cantilever is rigidly clamped at the bottom. The cantilever is aligned with respect to the primary flat of a (100) silicon wafer, that is, the cantilever length is parallel to <110>. The schematics show piezoresistor layouts that sense the particular flexural mode shape and suppress signals from other mode shapes.

and $\pi_t = -66.3 \cdot 10^{-11}\,\mathrm{Pa}^{-1}$ for *p*-type silicon and a <110> resistor alignment (see Table 8.2), and the stress distribution for the in-plane mode, resistors 1 and 3 exhibit a positive resistance change while resistors 2 and 4 exhibit a negative change, that is, an output signal is created by the Wheatstone bridge. On the other hand, the same bridge layout yields positive resistance changes for resistors 1 and 4 and negative changes for resistors 2 and 3 in case of the out-of-plane mode and – to first order – no signal is generated. To demonstrate how well this works, Figure 8.9 shows the amplitude of the Wheatstone bridge output signal as a function of frequency for a 400 μm long, 45 μm wide, and 20 μm thick cantilever. In comparison to the optical measurement of the vibration amplitude (see Figure 8.4c), the desired in-plane mode at 338 kHz shows a strong signal, while the other modes are well suppressed. This simplifies the application of the resonator as frequency determining element in an amplifying feedback loop [26]. Similarly, the bridge design in Figure 8.8a preferably detects flexural out-of-plane modes. The torsional mode shows a strong shear stress distribution and thus can be detected particularly well with *n*-type piezoresistors aligned in the (100) direction [7].

While the above bridge layouts separate thermal excitation resistors and piezoresistive sensing resistors, each thermal actuation resistor also experiences a resistance change due to a generated stress. Therefore, the same resistors can simultaneously act as both thermal actuators and piezoresistive sensors. This approach is more popular in devices utilizing bulk actuators since providing

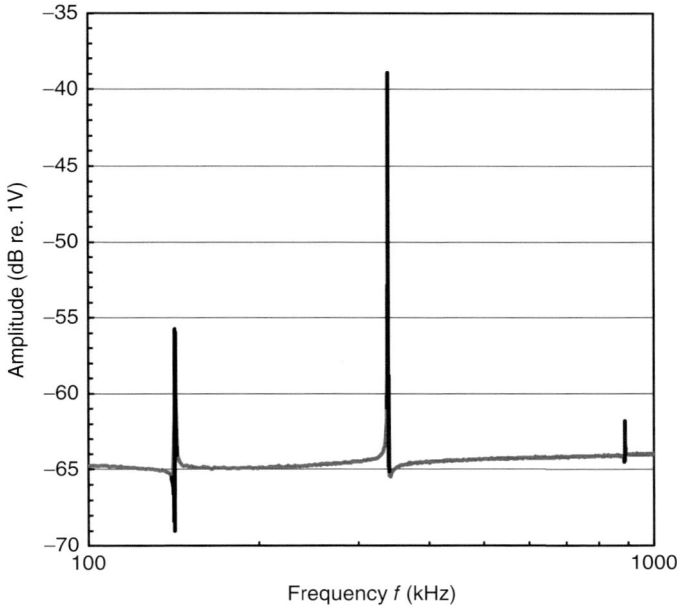

Figure 8.9 Piezoresistive output signal as a function of frequency for the 400 μm long, 45 μm wide, and 20 μm thick silicon cantilever of Figure 8.4. The embedded Wheatstone bridge (according to the layout in Figure 8.8b) senses in-plane flexural modes and suppresses out-of-plane modes; the gray curve represents a coarse scan over the frequency range 100 kHz to 1 MHz, while the solid black lines correspond to detailed measurements around the resonance peaks.

electrical insulation within the bulk of the structure between sensing and actuating resistors is challenging (although possible [27]). Such devices also referred to as "single-port" thermal-piezoresistive resonators are two terminal devices (as opposed to four terminals for two-port devices with separated actuation and sensing resistors), where the electrical resistance between the two terminals shows an abrupt negative or positive spike at the resonance frequency of the structure.

8.3.3
Self-Sustained Thermal-Piezoresistive Oscillators

Generally, in order to implement an electronic oscillator with a mechanical resonator as its frequency reference, the resonator needs to be engaged in a positive feedback loop consisting of amplifying circuitry with appropriate phase shift. One of the interesting aspects of the single-port thermal-piezoresistive mechanical resonators that use the same elements as both thermal actuators and piezoresistive sensors is the internal feedback mechanism formed between thermal actuation force and piezoresistivity. This feedback results from the fact that the piezoresistive motional current of the structure passing through the thermal actuators, also acts as an excitation for the resonator by affecting the ohmic loss in the actuators. Depending on the sign of the structural material piezoresistive coefficient, electrical terminations of the resonator, and the phase shift resulting from thermal and mechanical delays, this could act as a positive feedback stimulating mechanical vibrations or a negative feedback limiting the vibration amplitude of the structure or reducing the Brownian motion of the structure at that specific frequency (refrigeration [12]). For example, ohmic power generation in a thermal actuator with negative piezoresistive coefficient biased with a constant current (terminated with a large electrical resistance) decreases if the actuator is subjected to tensile stress, due to the decreased electrical resistance. As part of a resonant structure, when such an actuator is fully stretched and is about to start to compress, its decreased ohmic power and resulting cooling assist the actuator going into compression. In the same manner, after compression of the actuator, increased ohmic power generation helps its expansion by raising its temperature. Mechanical and thermal time delays play a key role here in determining whether the feedback has an amplifying or attenuating effect. Figure 8.10 schematically shows the feedback loop and the sequence of phenomena that can potentially lead to self-sustained oscillation [28].

It was demonstrated by researchers at NXP semiconductors that such positive feedback can be strong enough to initiate and maintain mechanical oscillations upon providing a DC electrical power source to the structure without the need for any external AC stimulation [12]. Figure 8.11 shows the scanning electron micrograph of the suspended n-type single crystalline silicon structure for which such an effect was demonstrated for the first time [12]. The structure is made of *n*-type

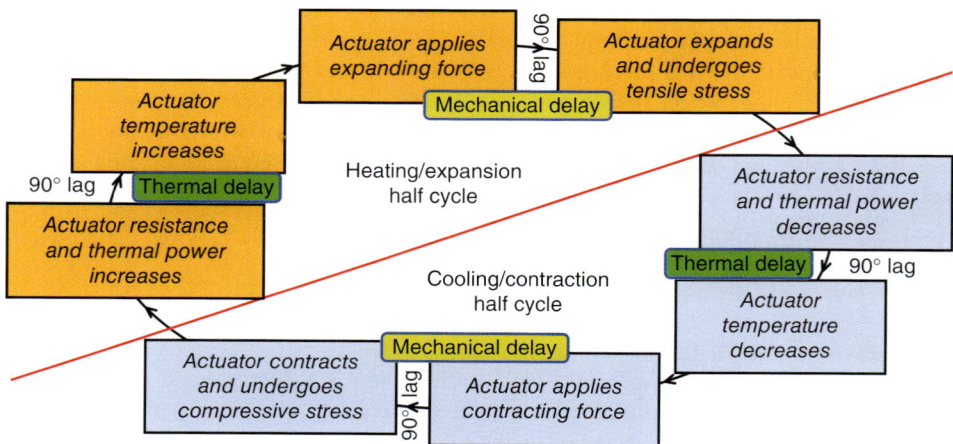

Figure 8.10 Sequence of phenomena causing an internal positive feedback loop in thermal-piezoresistive resonators biased with a constant current that can lead to self-sustained oscillation.

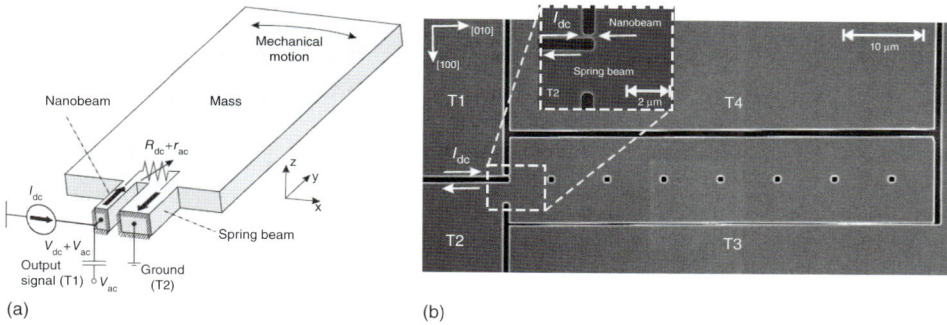

(a)

(b)

Figure 8.11 (a) Schematic view and (b) SEM top view of a fully micromechanical crystalline silicon oscillator. To operate the oscillator, a DC current is applied between the terminals of the device and the output AC voltage resulting from the resonator's spontaneous mechanical vibrations is measured. The inset shows a zoomed-in view of the nanobeam actuator and the wide beam providing the flexural stiffness of the structure. The output electrical signal amplitude of 27 mV with frequency of 1.26 MHz was measured for this resonator when biased with a current of 1.2 mA (1.2 mW power consumption) under vacuum (Reprinted with permission [12] from Macmillan Publisher Ltd: Nature Physics, Copyright 2011).

silicon because n-type silicon has a negative longitudinal piezoresistive coefficient, which is required for such an oscillation under constant bias current.

Figure 8.12 shows the SEM view and time response of an extensional mode 6.6 MHz self-sustained thermal piezoresistive oscillator operating based on the same principle [28]. Such a structure is a derivative of the IBAR structures discussed in the previous sections with relatively larger plates. It operates in an

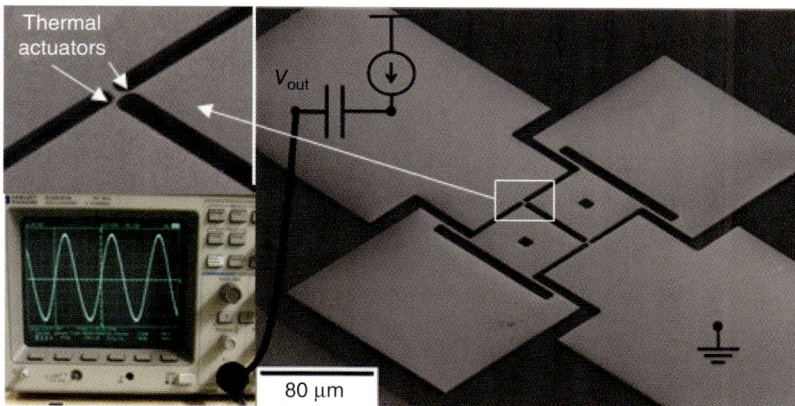

Figure 8.12 SEM view and output signal of a 6.6 MHz single crystalline silicon thermal-piezoresistive dual-plate resonator capable of self-sustained oscillation. Only a DC bias current of 3.0 mA is being applied to the resonator and the resonator is operating in air. The power consumption of the oscillator is about 20 mW and its output voltage amplitude is 70 mV.

in-plane resonant mode with its two plates moving back and forth in opposite directions and the extensional actuator beams being stretched and compressed periodically. Extensional resonant modes can provide higher frequencies as well as higher quality factors both under vacuum and in air. Dual plate oscillators with frequencies in the few megahertz range, power consumption of a few to tens of milliwatts and operating under both vacuum and atmospheric pressure have been demonstrated [28]. The output voltage amplitudes range from tens to hundreds of millivolts.

It can be shown that the requirement for self-sustained oscillation in such devices is that the absolute value of the motional conductance of the resonator becomes larger than the physical static conductance of its actuators [28]. The minimum DC power required to achieve this condition is

$$P_{dc_{Min}} = \frac{KLC_{th}\omega_m}{4\alpha E^2 |\pi_l| QA} \tag{8.20}$$

where L, A, and C_{th} are the length, cross-sectional area, and effective thermal capacitance of the actuator, K, ω_m, and Q are the mechanical stiffness, angular resonance frequency, and quality factor of the resonator, and E and α are the Young's modulus and thermal expansion coefficient of the structural material. Again, we can see that the key to having low power consumption is scaling down the resonator dimensions. Scaling the dimensions down reduces the power consumption of a self-sustained oscillator with a square relationship leading to significantly lower power consumption while having higher oscillation frequencies.

Another area where such devices need significant improvement is the output voltage amplitude, which is a small fraction of the DC voltage across the resonator (typically $0.01-0.1$). What limits the oscillation amplitude of the resonators are nonlinearities (most likely mechanical nonlinearities) limiting the feedback loop gain to unity after reaching a certain amplitude. Therefore, a thorough study on the minimization of the mechanical nonlinearities can possibly lead to fully micromechanical oscillators with higher output amplitudes. The piezoresistive coefficient of the structural material is another parameter that has a direct effect on the output voltage amplitude of such oscillators. With the same mechanical vibration amplitude and bias current, a higher piezoresistive coefficient leads to proportionally higher output voltage amplitude. This further shows the significance of fabricating such devices with sub-100 nm actuator width taking advantage of the large piezoresistivity observed in silicon nanowires [29] to increase the output amplitude.

Elimination of the sustaining amplifier by the self-sustained fully micromechanical oscillators described here might not be a significant improvement as electronics can be cheap and small and readily necessary to support other functions within the systems employing the electromechanical oscillators. However, it is possible that much higher performance and superior phase noise characteristics may be achieved for such devices because the electronic amplifier noise, which is the main source of phase noise in conventional oscillators, has been eliminated in them. Furthermore, for applications where large oscillator arrays are required, for example, sensors arrays, elimination of the amplifiers, and the associated inter-connections could be highly beneficial.

8.4
Modeling and Optimization of Single-Port Thermal-Piezoresistive Resonators

8.4.1
Thermo-Electro-Mechanical Modeling

Operation of a thermal-piezoresistive resonator involves phenomena in three different domains: thermal, mechanical, and electrical. Ohmic loss turns the electrical input voltage into a temperature variation, thermal expansion turns the temperature variation into a mechanical force that leads to mechanical deformation, and finally the piezoresistive effect turns the mechanical deformation back into an electrical signal (motional current).

Figure 8.13 shows the equivalent electrical circuit models for the three domains. In the thermal domain, as discussed in Section 8.1.2, the actuator beams act as a low-pass RC combination. The equivalent circuit consists of a current source representing the dynamic heating power along with a capacitance (C_{th}) and a resistance (R_{th}) representing the effective thermal capacitance and thermal resistance of the thermal actuators respectively (Figure 8.13a). The resulting voltage across the parallel RC combination (T_{ac}) represents the temperature fluctuation amplitude.

Due to the square relationship between ohmic power generation and electrical voltage (or current), application of only an AC actuation voltage with frequency of f_a results in an ohmic power component with frequency of $2f_a$. Therefore, as discussed in Section 8.1.1, in order to have the same frequency as the input AC actuation voltage for the thermal actuation force, a combination of AC and DC

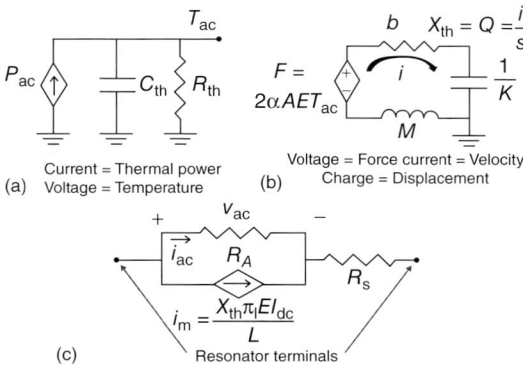

(a) Current = Thermal power / Voltage = Temperature

(b) Voltage = Force / current = Velocity / Charge = Displacement

(c) Resonator terminals

Figure 8.13 Equivalent electrical circuit for a thermal-piezoresistive resonator: (a) the thermal domain; (b) the mechanical domain; and (c) the electrical domain.

voltages (V_{dc} and v_{ac}) needs to be applied between the two terminals of the resonators. The resulting power loss component at the same frequency as v_{ac} in this case will be $P_{ac} = \frac{2V_{dc}v_{ac}}{R_A}$, where R_A is the electrical resistance of the actuator elements and V_{dc} and v_{ac} are the applied DC and AC actuation voltages respectively.

The RLC combination of Figure 8.13b represents the mechanical resonance behavior of the structure. The voltage source represents the thermally generated mechanical force, which is the product of thermal expansion coefficient (α) of the actuator structural material, actuator cross sectional area (A), and Young's modulus (E) of the structural material along the actuator length. The coefficient "2" has been added since in many cases, such as the dog-bone structures of Figure 8.7, there are two actuator beams in each resonator contributing to the actuation force. The inductor, capacitor, and resistor represent the effective mechanical mass (M), stiffness (K), and damping (b) of the resonator, respectively. Consequently, current represents velocity and electrical charge ($q = \int i \cdot dt$) represents displacement (X_{th}), which is the elongation of the thermal actuators (resonator vibration amplitude). The vibration amplitude, which is the output of the mechanical subsystem, acts as an input to the electrical sub-system (Figure 8.13c) causing a change in the resistance of the piezoresistive sensor elements (same as actuator elements for single-port devices). Resistance fluctuations in turn modulate the current flow through the resonator, which is shown as a current source in the equivalent circuit. R_s in Figure 8.13c represents all the parasitic electrical resistances in series with the actuator resistance, which can include resistance of non-actuating parts in the current path to the actuators, wirebonds, and so on.

The overall transfer function that relates the AC motional current (output current) to the AC input voltage is the product of the three transfer functions from the thermal, mechanical, and electrical domains. The simplified transfer function for a dog-bone structure at resonance frequency is [3]:

$$H_T|_{s=j\omega_0} = g_m = 4\alpha E^2 \pi_1 Q \frac{AI_{dc}^2}{KLC_{th}\omega_0} \tag{8.21}$$

where π_1 is the longitudinal piezoresistive coefficient of the structural material, L is the thermal actuator length, and Q and ω_0 are the quality factor and angular resonant frequency of the resonator.

For dog-bone structures, knowing that $K = 2EA/L$, Eq. (8.21) can be further simplified to:

$$g_m = 2\alpha E \pi_1 Q \frac{I_{dc}^2}{C_{th}\omega_0} \tag{8.22}$$

A more detailed version of the modeling and derivations above is available in [3]. The transfer function in Eqs. (8.21) and (8.22) can be referred to as small signal voltage to current gain or motional conductance (g_m) of the resonator at its resonance frequency, which is one of the most important parameters of a thermal-piezoresistive resonator when utilized as an electronic circuit component.

8.4.2
Resonator Equivalent Electrical Circuit and Optimization

For two terminal (single-port) resonators, the physical resistance of the actuator/piezoresistors connects the input and output of the device. The equivalent electrical circuit in this case therefore includes an overall resistance of $R_A + R_s$ connected between the two terminals of the resonator, where R_A is the resistance of the thermal actuators and R_s is the sum of all parasitic resistances present in the actuator path (with support beam resistances typically being the main component). In parallel to the physical resistance of the actuators, there is a series RLC combination that represents the mechanical resonant behavior of the structure. The value of R_m in the RLC has to be set so that at resonance, a motional current of $i_m = g_m \cdot v_{ac} = v_{ac}/R_m$ is added to (or subtracted from) the feed through current passing through R_A. Therefore, $R_m = g_m^{-1}$ and L_m and C_m values can be calculated based on the value of R_m as shown in Figure 8.14 according to the resonance frequency and quality factor of the resonator.

To calculate the thermal conductance (g_m) of a thermal-piezoresistive IBAR, which is the most important parameter to be extracted from the model using Eq. (8.22), all the parameters except the effective thermal capacitance of the actuators (C_{th}) are known for every set of resonator dimensions and structural material. Due to the distributed nature of the thermal parameters including thermal generator, thermal capacitance and thermal conductance, analytical derivations, or finite element analysis should be used to find an accurate value for C_{th}. In Ref. [3], finite element analysis has been used to simulate the transient thermal response of the actuators and calculate the fluctuating temperature amplitude (T_{ac}) at different points along the length of the actuator beams. The mean value of the small signal temperature amplitudes at different points along the actuators is the effective temperature fluctuation amplitude for the actuator from which the effective thermal capacitance value can be extracted. As shown in [3], the effective thermal capacitance for a thermal actuator beam is close to 0.92 of the static lumped element thermal capacitance of the same beam.

- *Resonator optimization*: As an electronic component, it is desirable to maximize the resonator motional conductance (g_m) (for improved signal to noise ratio), while minimizing power consumption. Therefore, a figure of merit for

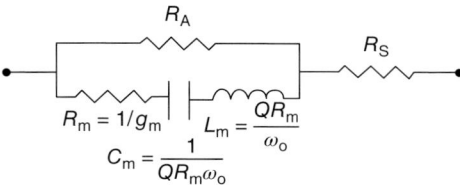

Figure 8.14 Compact equivalent electrical circuit for single-port thermally actuated resonators with piezo-resistive readout.

the resonators can be defined as the ratio of the resonator g_m to the overall DC power consumption:

$$F.M. = \frac{g_m}{P_{dc}} = \frac{2\alpha E \pi_l Q}{C_{th}\omega_0(R_A + R_s)} \tag{8.23}$$

where $P_{dc} = I_{dc}^2 \cdot (R_A + R_s)$ is the static power consumption of the resonator. Different parameters that can be used to maximize *F.M.* are as follows.

- *Actuator beam dimensions*: Generally smaller actuator dimensions lead to smaller C_{th} improving resonator figure of merit. However, the effect of actuator dimensions on its electrical resistance (R_A) should also be taken into account. Since the extensional stiffness of the actuator beams define the resonance frequency of the IBAR, in order to maintain the same resonance frequency for an IBAR while reducing its actuator thermal capacitance, both the length and width of the actuator should be scaled down simultaneously. Such scaling does not affect the actuator electrical resistance (R_A). Therefore, scaling down both the length and width of the actuator beams by a ratio S_a, results in an improvement in the resonator *F.M.* by a factor of S_a^2 while maintaining the resonance frequency for the device.
- *Resonance frequency and resonator scaling*: At a first glance at Eq. (8.23), higher resonance frequencies seem to have a deteriorating effect on the resonator figure of merit. However, if higher resonant frequencies are achieved by shrinking the resonator size, at the same time C_{th} will be shrinking sharply. If all resonator dimensions are scaled down proportionally by a factor S, $C_{th} \propto S^{-3}$, $\omega_0 \propto S$, and $(R_A + R_s) \propto S$. Therefore, $F.M. \propto S$, that is, *F.M.* increases proportionally if the resonator dimensions are scaled down, while its resonance frequency increases at the same time. This confirms the validity of the earlier claim regarding suitability of thermal actuation for high-frequency applications. It can be concluded from the last two discussions that the capability to fabricate thermal actuator elements with very small (nanoscale) dimensions is the key to improved performance of thermal-piezoresistive resonators, both at low and high frequencies.
- *Electrical resistivity and other material properties*: One of the important parameters that can be controlled for some of the most popular resonator structural materials (e.g., Si or SiC), is the electrical resistivity of the structural material. Lower electrical resistivity of the structural material improves the transduction figure of merit by lowering $R_A + R_s$. This can be explained by the fact that, with lower electrical resistivity, the same amount of DC bias current and therefore the same motional conductance can be maintained while burning less ohmic power in the structure.

A number of other structural material properties also affect the figure of merit for thermal-piezoresistive resonators. A higher thermal expansion coefficient and a lower specific heat capacity improve the thermal actuation resulting in proportionally higher *F.M.* A higher piezoresistive coefficient improves the piezoresistive readout having the same effect on *F.M.* One interesting and promising fact is the giant piezoresistive effect found in single crystalline silicon nanowires that has been observed and reported by different researchers [29]. Therefore, as the resonators are scaled down to reach higher resonant frequencies in the hundreds of megahertz and gigahertz range, it is expected that the actuator beams with deep submicron width exhibit such strong piezoresistivity significantly improving resonator performances. Finally, using structural materials with higher Young's modulus (E) can also improve the resonator figure of merit while increasing its resonant frequency.

8.5
Examples of Thermally Actuated Resonant MEMS

Examples of thermally actuated MEMS resonators were shown in Figures 8.4, 8.7, 8.10, and 8.12. Figure 8.15a shows a 20.6 MHz single-port thermal-piezoresistive IBAR utilized as an airborne microparticle detector [30]. The structure is carved out of the thin ($\sim$2 μm thick) device layer of a silicon-on-insulator (SOI) substrate and therefore made of single crystalline silicon. The measured resonance frequency of the resonator during exposure to a flow of artificially generated airborne particles of 1 μm diameter shows how the resonator can resolve the added mass by every individual particle landing on the resonator surface. In general, thermal-piezoresistive resonators are very suitable for sensory applications due to the robustness of their simple and monolithic structure.

Figure 8.16 shows the SEM view of two different thermally actuated cantilever structures utilizing thin film actuators. Figure 8.16a shows a silicon tuning fork structure with embedded diffused thermal excitation elements and piezoresistors [31]. Four piezoresistors connected in a Wheatstone bridge are placed on the tuning fork structure so that an output signal is generated – to a first order – only for the desired tuning-fork mode at 418 kHz. This allows embedding the resonator

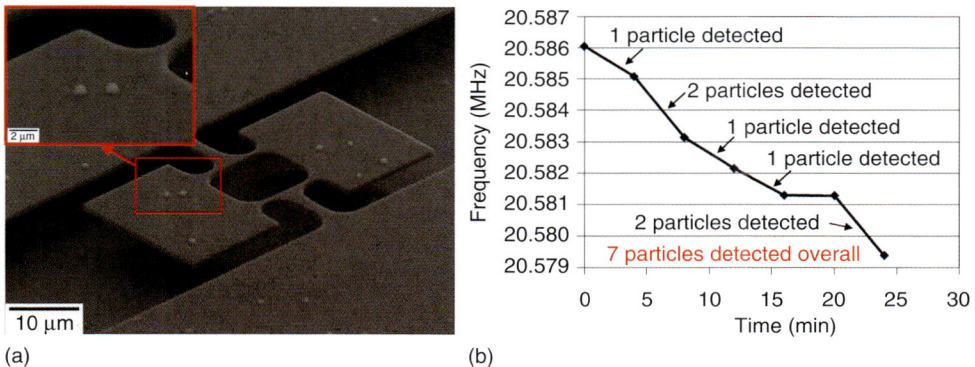

(a)   (b)

Figure 8.15 (a, b) SEM view of a 20.6 MHz IBAR with seven artificially generated 1 μm diameter airborne particles deposited on it. Each added particle reduces the frequency of the resonator by 900 Hz. The changes in the measured resonance frequency over time confirm deposition of seven particles.

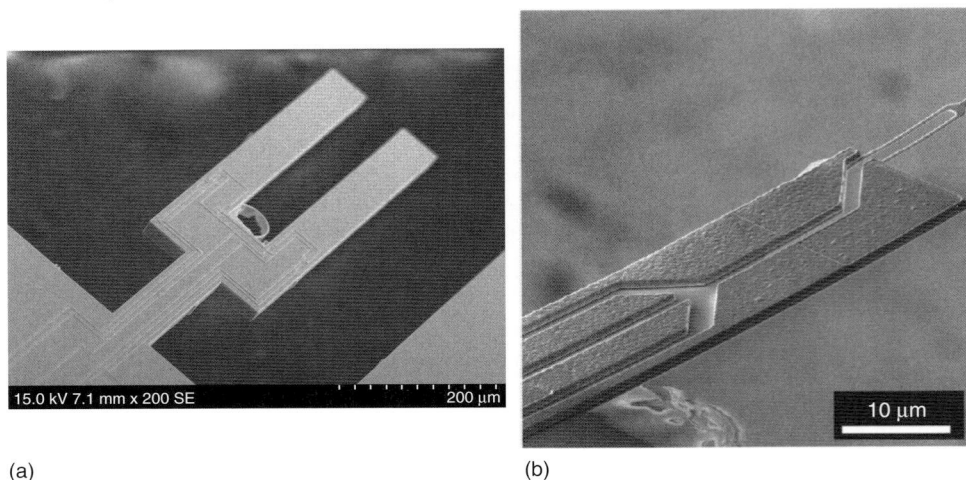

(a)   (b)

Figure 8.16 SEM images of (a) a thermally actuated tuning fork resonator with diffused excitation resistors and piezoresistors [31] and (b) a thermally actuated high-speed force probe with embedded heating resistors and piezoresistors (Reproduced from [32] Copyright 2012 with permission from IOP publishing. All rights reserved).

in an amplifying feedback loop achieving short-term frequency stabilities in air as low as 0.02 ppm [31]. The tuning fork exhibits a higher Q-factor in air compared to a cantilever with the same dimension as the tuning fork tine because of the reduced anchor loss in the tuning fork. The tuning forks have been coated with chemically sensitive polymer films for chemical sensing applications [31]. The cantilever in Figure 8.16b has been designed as a high-speed force probe with pN-scale force resolution [32]. Shallow doped silicon piezoresistors act as force/displacement sensors, while the actuation is performed by silicon heaters inducing a temperature increase and thermal expansion in an overlaid Al film.

Thermal actuation has also been applied to resonant MEMS operating in liquids, providing a robust excitation scheme even in harsh environments. In this case, the damping by the surrounding liquid dominates the damping behavior and other loss mechanisms such as support loss or thermoelectric damping can generally be neglected. Losses to the surrounding liquid are typically reduced by shearing the liquid rather than compressing it; as a result, disk-type resonator designs have shown great promise for liquid operation. The thermal actuation is not compromised in fluids, even considering their high thermal conductivity compared to air. This can be understood by considering the localized thermal wave, which generates the thermal actuation and is not affected by adding more thermally conductive material around the resonator. The reduced Q-factors in liquid demand, however, larger actuation forces/moments to obtain similar deflections as in air. As an example, Figure 8.17 shows two disk resonators that have been successfully operated in water. The device shown in Figure 8.17a comprises two half disks supported from a central beam; diffused thermal resistors located in the

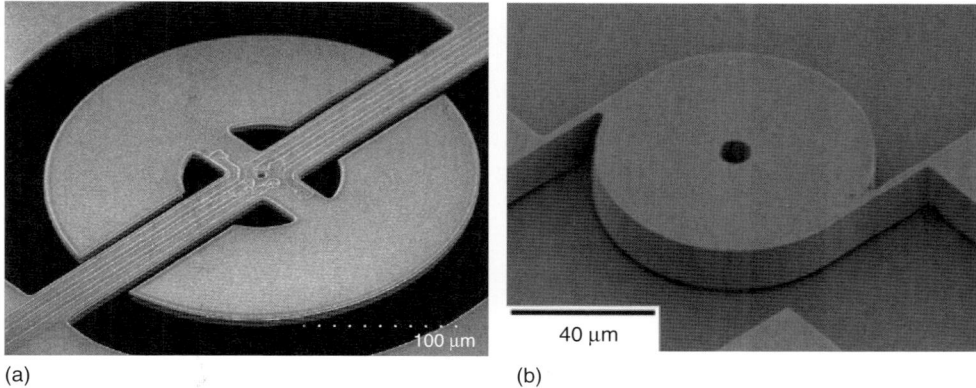

(a)   (b)

Figure 8.17 SEM images of thermally actuated (a) disk resonator with a 120 μm outer radius [33] and (b) a disk resonator with 100 μm diameter and 20 μm thickness [34]. The thermal actuator legs in (b) are aligned along the <100> direction.

short support beams excite the disk into a rotational eigenmode [33]. *Q*-factors in water up to 94 have been measured at the resonance frequency of approx. 580 kHz [33]. In contrast, the design shown in Figure 8.17b comprises a disk supported by two tangential heating beams from the perimeter [34]. This design has yielded exceptional *Q*-factors of up to 280 in water at a rotational frequency of 5.7 MHz [34]. Recent analytical modeling of this device indicates that this device type may be capable of achieving even higher *Q*-factors in liquids [35].

References

1. Betts, J. John Harrison (1693–1776) and Lt. Cdr Rupert T. Gould R.N. (1890–1948), *http://www.rmg.co.uk/harrison%5D* (accessed 2 September 2014).
2. Hall, H.J., Rahafrooz, A., Brown, J.J., Bright, V.M., and Pourkamali, S. (2013) I-shaped thermally actuated VHF resonators with submicron components. *Sens. Actuators, A*, **195**, 160–166.
3. Rahafrooz, A. and Pourkamali, S. (2011) High-frequency thermally actuated electromechanical resonators with piezoresistive readout. *IEEE Trans. Electron Devices*, **58**, 1205–1214.
4. Li, R.-G. and Huang, Q.-A. (2012) in *Advanced Micro and Nanosystems: System-Level Modeling of MEMS* (eds O. Brand, G. Fedder, C. Hierold, J. Korvink, and O. Tabata), Wiley-VCH Verlag GmbH, Weinheim, pp. 125–146.
5. Lammerink, T.S.J., Elwenspoek, M., van Ouwerkerk, R.H., Bouwstra, S., and Fluitman, J.H.J. (1990) Performance of thermally excited resonators. *Sens. Actuators, A*, **A21-A23**, 352–356.
6. Lammerink, T.S.J., Elwenspoek, M., and Fluitman, J.H.J. (1991) Frequency dependence of thermal excitation of micromechanical resonators. *Sens. Actuators, A*, **25–27**, 685–689.
7. Senturia, S.D. (2000) *Microsystem Design*, Springer.
8. Guckel, H., Klein, J., Christenson, T., Skrobis, K., Laudon, M., and Lovell, E.G. (1992) Thermo-magnetic metal flexure actuators. Proceedings of the 5th IEEE Solid-State Sensor and Actuator Workshop, June 22–25, 1992, Hilton Head Island, SC, 1992, pp. 73–75.

9. Li, L. and Uttamchandani, D. (2004) Modified asymmetric microelectrothermal actuator: analysis and experimentation. *J. Micromech. Microeng.*, **14**, 1734–1741.

10. Long, Q., Jae-Sung, P., and Gianchandani, Y.B. (2001) Bent-beam electrothermal actuators-part I: single beam and cascaded devices. *J. Microelectromech. Syst.*, **10**, 247–254.

11. Jae-Sung, P., Chu, L.L., Oliver, A.D., and Gianchandani, Y.B. (2001) Bent-beam electrothermal actuators-part II: linear and rotary microengines. *J. Microelectromech. Syst.*, **10**, 255–262.

12. Steeneken, P.G., Le Phan, K., Goossens, M.J., Koops, G.E.J., Brom, G.J.A.M., van der Avoort, C., and van Beek, J.T.M. (2011) Piezoresistive heat engine and refrigerator. *Nat. Phys.*, **7**, 354–359.

13. Wilfinger, R.J., Bardell, P.H., and Chhabra, D.S. (1968) The resonistor: a frequency selective device utilizing the mechanical resonance of a silicon substrate. *IBM J. Res. Dev.*, **12**, 113–118.

14. Wilfinger, R.J., Chhabra, D.S., and Bardell, P.H. (1966) A frequency selective device utilizing the mechanical resonance of a silicon substrate. *Proc. IEEE*, **54**, 1589–1591.

15. Beardslee, L.A., Addous, A.M., Heinrich, S., Josse, F., Dufour, I., and Brand, O. (2010) Thermal excitation and piezoresistive detection of cantilever in-plane resonance modes for sensing applications. *J. Microelectromech. Syst.*, **19**, 1015–1017.

16. Brand, O., Hornung, M., Baltes, H., and Hafner, C. (1997) Ultrasound barrier microsystem for object detection based on micromachined transducer elements. *J. Microelectromech. Syst.*, **6**, 151–160.

17. Lange, D., Hagleitner, C., Hierlemann, A., Brand, O., and Baltes, H. (2002) Complementary metal oxide semiconductor cantilever arrays on a single chip: mass-sensitive detection of volatile organic compounds. *Anal. Chem.*, **74**, 3084–3095.

18. Rahafrooz, A. and Pourkamali, S. (2010) Fabrication and characterization of thermally actuated micromechanical resonators for airborne particle mass sensing: I. Resonator design and modeling. *J. Micromech. Microeng.*, **20**, 125018, (10 pp.).

19. Heinrich, S., Maharjan, R., Dufour, I., Josse, F., Beardslee, L., and Brand, O. (2010) An analytical model of a thermally excited microcantilever vibrating laterally in a viscous fluid. IEEE Sensors, 2010, pp. 1399–1404.

20. Thomson, W. (1856) On the electrodynamic qualities of metals:–effects of magnetization on the electric conductivity of nickel and of iron. *Proc. R. Soc. London*, **8**, 546–550.

21. Smith, C.S. (1954) Piezoresistance effect in germanium and silicon. *Phys. Rev.*, **94**, 42–49.

22. Tufte, O.N., Chapman, P.W., and Long, D. (1962) Silicon diffused-element piezoresistive diaphragms. *J. Appl. Phys.*, **33**, 3322–3327.

23. Barlian, A.A., Park, W.T., Mallon, J.R., Rastegar, A.J., and Pruitt, B.L. (2009) Review: semiconductor piezoresistance for microsystems. *Proc. IEEE*, **97**, 513–552.

24. Nathan, A. and Baltes, H. (1999) *Microtransducer CAD: Physical and Computational Aspects*, Springer.

25. Kanda, Y. (1982) A graphical representation of the piezoresistance coefficients in silicon. *IEEE Trans. Electron Devices*, **ED-29**, 64–70.

26. Demirci, K.S., Beardslee, L.A., Truax, S., Su, J.J., and Brand, O. (2012) Integrated silicon-based chemical microsystem for portable sensing applications. *Sens. Actuators B*, **180**, 50–59.

27. Hajjam, A., Rahafrooz, A., and Pourkamali, S. (2012) Input–output insulation in thermal-piezoresistive resonant microstructures using embedded oxide beams. 66th IEEE International Frequency Control Symposium, IFCS 2012, Baltimore, MD, 2012, pp. 425–428.

28. Rahafrooz, A. and Pourkamali, S. (2012) Thermal-piezoresistive energy pumps in micromechanical resonant structures. *IEEE Trans. Electron Devices*, **59**, 3587–3593.

29. He, R. and Yang, P. (2006) Giant piezoresistance effect in silicon nanowires. *Nat. Nanotechnol.*, **1**, 42–46.

30. Hajjam, A., Wilson, J.C., and Pourkamali, S. (2011) Individual air-borne particle mass measurement using high-frequency micromechanical resonators. *IEEE Sens. J.*, **11**, 2883–2890.

31. Beardslee, L.A., Lehmann, J., Carron, C., Su, J.J., Josse, F., Dufour, I., and Brand, O. (2012) Thermally actuated silicon tuning fork resonators for sensing applications in air. IEEE 25th International Conference on Micro Electro Mechanical Systems (MEMS), Piscataway, NJ, January 29–February 2, 2012, pp. 607–610.

32. Doll, J.C. and Pruitt, B.L. (2012) High-bandwidth piezoresistive force probes with integrated thermal actuation. *J. Micromech. Microeng.*, **22**, 095012, (14 pp.).

33. Seo, J.H. and Brand, O. (2008) High Q-factor in-plane-mode resonant microsensor platform for gaseous/liquid environment. *J. Microelectromech. Syst.*, **17**, 483–493.

34. Rahafrooz, A. and Pourkamali, S. (2010) Rotational mode disk resonators for high-Q operation in liquid. 2010 Ninth IEEE Sensors Conference (SENSORS 2010), Piscataway, NJ, November 1–4 2010, pp. 1071–1074.

35. Sotoudegan, M.S., Heinrich, S.M., Josse, F., Nigro, N.J., Dufour, I., and Brand, O. (2013) Effect of design parameters on the rotational response of a novel disk resonator for liquid-phase sensing: analytical results. Proceeding of IEEE Sensors, 2013, pp. 1164–1167.

9
Nanoelectromechanical Systems (NEMS)

Liviu Nicu, Vaida Auzelyte, Luis Guillermo Villanueva, Nuria Barniol, Francesc Perez-Murano, Warner J. Venstra, Herre S. J. van der Zant, Gabriel Abadal, Veronica Savu, and Jürgen Brugger

9.1
Introduction

The vibrant activities in nanoscale research for producing finer nanostructures of various novel 1D and 2D materials, as well as advances in on-chip integration and signal transduction strongly support the development of nanoelectromechanical systems (NEMS). Several years of work by the highly interdisciplinary NEMS community at this stage has yielded extensive and important findings. This chapter offers an update of the activities in the research and development of NEMS over the past several years. The following sections will describe some of these achievements, highlighting the work done on fundamental studies using NEMS, their transduction schemes, nonlinear behavior, NEMS fabrication, and incorporation of novel materials, including 1D and 2D carbon-based structures. Applications in the fields of electronics using linear and nonlinear NEMS, bio-NEMS as well as energy scavenging will be covered.

9.1.1
Fundamental Studies

Fundamental and characteristic properties of NEMS, for example, high surface-to-volume ratios, extremely small masses, and small onsets of nonlinearity, make NEMS an outstanding scientific tool to study different physical phenomena that would otherwise be not accessible. In particular, over the years NEMS have extensively been used as tools to probe quantum physics. Originally aimed at detecting individual quanta of electrical [1] and thermal conductance [2–5], we have experienced in the last few years an exciting race to cool down systems to their mechanical ground state. Even though this was finally attained by using micron-sized devices [6, 7], smaller devices in the sub-micron range and even below 100 nm, are still of the greatest interest because they are more susceptible to being affected by back-action of the surroundings, like detection or actuation

Resonant MEMS – Fundamentals, Implementation and Application, First Edition.
Edited by Oliver Brand, Isabelle Dufour, Stephen M. Heinrich and Fabien Josse.
© 2015 Wiley-VCH Verlag GmbH & Co. KGaA. Published 2015 by Wiley-VCH Verlag GmbH & Co. KGaA.

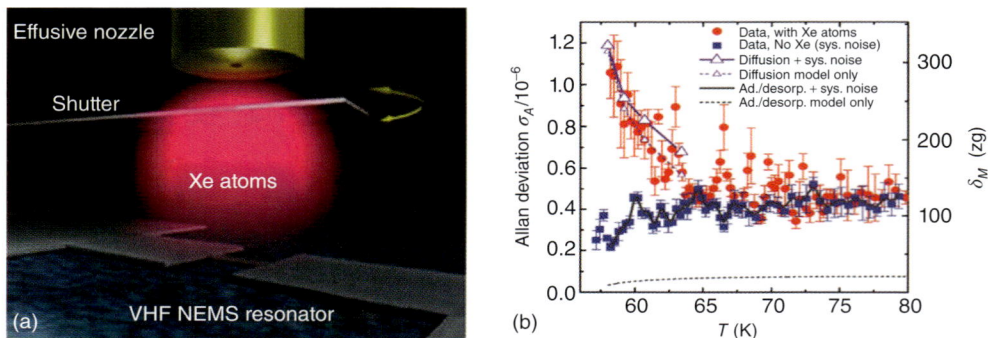

Figure 9.1 Experimental determination of the nature of frequency noise in NEMS [12]. (a) Photo illustration showing the system utilized for the experiment with a clamped–clamped silicon carbide beam in a cryostat with a nozzle that ejects Xe atoms onto the sample. (b) Frequency stability as a function of temperature with and without Xe atoms landing onto the device. The noise cannot be interpreted as being caused by only adsorption/desorption of particles, but by the inclusion of diffusion of particles. (Reprinted with permission from Ref. [12]. Copyright 2011 American Chemical Society.)

techniques. Interaction with a superconducting qubit [8] and coupling to electronic conduction are some of the examples that have been recently proved [9, 10].

The small mass of NEMS makes them ideal for mass-sensing experiments. Usually, mass-detection experiments seek the detection of mass landing onto the device [11], but they can also be used to study adsorption/desorption and diffusion of particles on the device surface [12], leading to deeper understanding of the microscopic dynamics of deposited materials and an understanding of the limitations of mass sensing with NEMS [13]. In [12], the authors used a system to perform mass spectroscopy measurements in order to monitor the landing of Xe atoms onto the NEMS. The landing of atoms can be switched on/off via a shutter (Figure 9.1a). By continuously running a phase-locked loop, the frequency is monitored over time for different temperatures of the device. The frequency stability can therefore be plotted as a function of temperature and different models for adsorption/desorption can be checked. The data indicates that diffusion along the beam, an effect neglected up to date, is the dominant effect at lower temperatures (Figure 9.1b).

An additional topic of interest is the study of nonlinear and complex dynamics with NEMS. Their reduced dimensions result in a very low threshold for the onset of nonlinearity so that the corresponding nonlinear phenomena are readily accessible and easily predicted theoretically. In addition, their high-quality factors and even higher frequencies make them ideally suited to analyze quasi-Hamiltonian systems and also to experimentally measure stationary states reached in very short times, that is, after a few milliseconds. Finally, their small size facilitates array fabrication, which increases the complexity and the interest of the systems to be studied [14]. Numerous theoretical studies have been undertaken and published

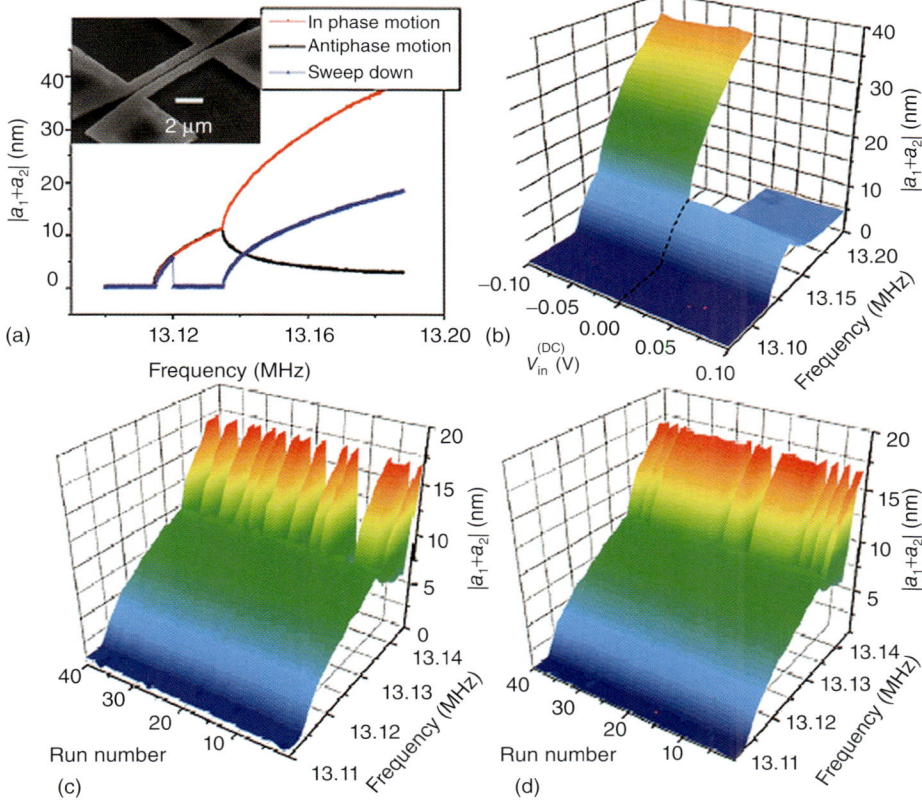

Figure 9.2 Bifurcation topology amplifier [18]. (a) Experimental measurement of the parametric response of two uncoupled NEMS beams, confirming the predicted pitchfork bifurcation. Inset: pair of coupled clamped–clamped beams to perform the experiment. (b–d) Amplitude response of the coupled system as a function of frequency. When the voltage difference between the beams is zero (b) the system presents a perfectly symmetric behavior and the upper branch is chosen 50% of the time. If the voltage is slightly negative (c) or positive (d) (±3 mV), the symmetry gets broken, as predicted by theory. This symmetry breaking can be interpreted as a very effective signal amplifying sensor. (Reprinted with permission from Ref. [18]. Copyright 2011 American Physical Society.)

in recent years dealing with nonlinear behavior and its implications for certain applications. Euler instability in clamped–clamped beams [15] and diffusion-induced bistability [16] are some of those examples.

Experimentally, the verification of the predicted behavior of a system of two coupled resonators has been shown, screening rich nonlinear dynamics and even chaos [17]. A novel detection system has been proposed, based on symmetry breaking close to a Hopf bifurcation (Figure 9.2) [18]. But it has also been proven that it is not necessary to have arrays of individual beams to observe such behavior. Due to the nature of the nonlinearity in NEMS, it is also possible to observe such effects on one single beam using the existing coupling between several vibrational modes, which is of great importance when considering frequency-based sensing or similar applications [19]. Finally, NEMS have also been used to observe nonlinear damping in mechanical resonators, something that had not been observed before and whose origin is not fully understood, making theoretical predictions or model development difficult [20].

9.1.2
Transduction at the Nanoscale

The reduced size is what makes NEMS appealing from both the fundamental and applications points of view. But everything comes at a cost, and the trade-off that must be paid in order to achieve performance advantages (e.g., sensing capabilities) is very low transduction efficiencies.

In order to use NEMS, it is necessary to make it move (actuation) and to detect such motion (detection). The combination of actuation and detection is what is usually referred to as transduction. Finding an optimal transduction technique that works universally for a wide variety of NEMS has been pursued by many different research groups.

We can divide transduction mechanisms into two large groups: those based on (or using) optics and those based on (or using) electronics. The first group had been traditionally overlooked for NEMS, as diffraction effects were assumed to be detrimental when going deep into the sub-micron regime. However, recent experiments have shown that it is possible to not only detect motion [21–23], but also to actuate NEMS devices using optics (Figure 9.3) [24–26]. Another point against optical transduction methods was that they could not be integrated on-chip, which limited the future applicability of such methods. However, this has also been disproved, and integrated solutions for optical transduction have been proposed [24, 27].

But even though optics seems to be catching up, electronic transduction is usually preferred, mainly due to on-chip integration possibilities. A plethora of methods have been used over the years: magnetomotive [28], thermoelastic [29], piezoelectric [30, 31], capacitive [32], ferromagnetic [33], Kelvin polarization force [34], and so on. A common problem for any of these methods is the fact that the motional signal produced by the NEMS is minute when compared to the parasitic cross-talk associated with the actuation signal. This is mainly caused by the great size difference between the NEMS and the necessary elements to connect it to the macroscopic world (metal pads, wire bonds, etc.). This undesired effect creates a background on the response of the device that hides the actual mechanical signal and makes it very difficult to be distinguished, which in turn affects the stability of the closed-loop systems used for frequency-based sensing. A number of solutions have been proposed to solve this issue, most of them based on the use of some mixing mechanism that moves the frequency of the detected signal away from the frequency of the actuating signal.

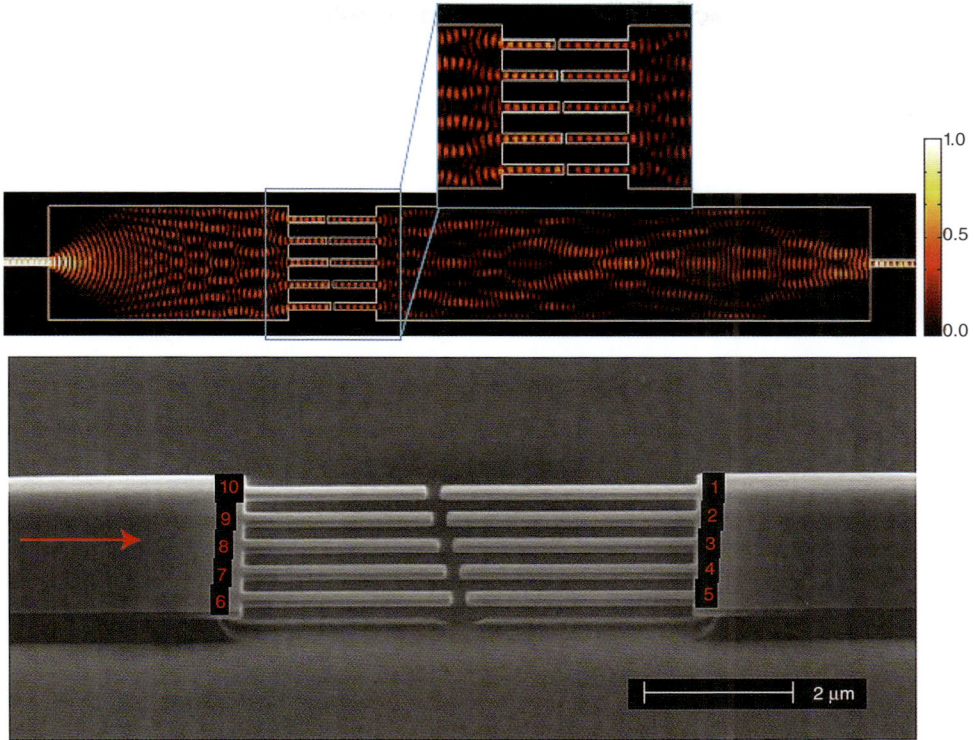

Figure 9.3 Novel transduction mechanisms for the nanoscale-photonic circuit. Simulation of the light propagation through an array of NEMS, shown in SEM image below, when no motion is present. As soon as the cantilevers start moving, misalignment occurs, and therefore the output intensity becomes modulated [24]. (Reprinted with permission from Ref. [24]. Copyright 2009 Nature Publishing Group.)

Down-mixing [35], amplitude modulation [36, 37], frequency modulation [20, 38], use of superior harmonics actuation [39], and parametric actuation [40] are some of the mentioned techniques used to cancel the effect of the background and give much more stability and superior performance than more traditional approaches such as direct bridging of the signal [41, 42].

However, even though these techniques have proven very useful from a research point of view, the most promising option for future integration and with wider applicability is the use of an on-chip amplifier located very close to the NEMS [43], or even within the mechanical device itself [44–47]. The amplifier boosts the motional signal, making it stronger and less susceptible to the parasitic effect mentioned above. At this point, the problem is no longer associated with NEMS *per se*, but becomes one of electronics (how to best lay-out a series of transistors to obtain the biggest gain with the smallest noise) and/or technology (how to integrate on-chip the mechanical device with the adjacent electronic circuitry).

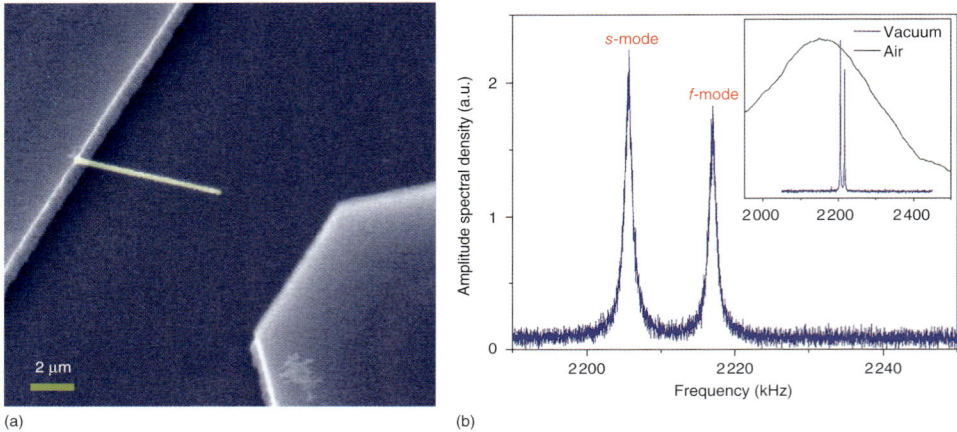

Figure 9.4 (a, b) NEMS resonator made of a single crystalline silicon nanowire provides good mechanical properties, high-resonance frequency, and high-quality factor, that can be exploited to build-up ultra-high sensitivity mass sensors, including extra functionality due to the existence of several resonance modes. The frequency response of the nanowire can be detected by optical methods [57]. (Reprinted with permission from Ref. [57]. Copyright 2010 Nature Publishing Group.)

9.1.3
Materials, Fabrication, and System Integration

Most of the fabrication methods used for investigating the ultimate performance of NEMS devices still rely on silicon-based technology. The reason is that the combination of advantageous mechanical properties and well-established processing methods. Mechanical resonators require materials that provide high-quality factors and high-resonance frequency. For this reason, the most relevant progress in the exploitation of the functional properties of NEMS devices has been obtained with devices made of silicon [48, 49], silicon nitride [50, 51], silicon carbide [52–55], or related materials like SiCN [56], addressing a range of diverse aspects like ultra-high frequency operation [49], information processing [48], parametric amplification [51, 52], or sensing [53, 55]. Remarkably, silicon is also present in the promising approach of building-up devices based on bottom–up fabrication methods. For example, a silicon NEMS resonator made of a single silicon nanowire grown by CVD methods combines good mechanical properties, ultra-high mass sensitivity, and additional functionality given by the possibility of exciting flexural modes in two dimensions [57] (Figure 9.4).

In parallel to this research on silicon-based technology, increased attention is being paid to piezoelectric materials due to the possibility of easy implementation of self-transduction (sensing and actuation) [31, 44, 58, 59]. Realization of SiN/AlN piezoelectric cantilevers was made (Figure 9.5). 50-nm-thick AlN films present high piezoelectric coefficients that enable electrical transduction

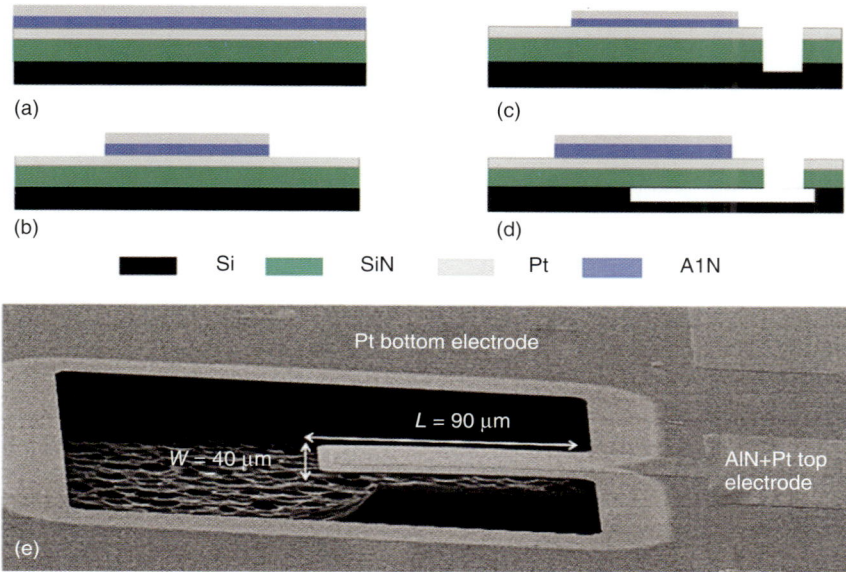

(a)

(b)

(c)

(d)

| ■ Si | ■ SiN | ▨ Pt | ■ A1N |

Pt bottom electrode

$L = 90 \, \mu m$

$W = 40 \, \mu m$

AlN+Pt top electrode

(e)

Figure 9.5 Fabrication process and SEM picture of cantilevers with 50 nm-thick AlN films that provide ultra-high sensitivity mass sensing with electrical transduction [58]. (Reprinted with permission from Ref. [58]. Copyright 2011 Institute Of Physics.)

with excellent frequency stability, while SiN provides the desirable mechanical behavior, demonstrating an achievable limit of detection of $53 \, zg/\mu m^2$.

Electron beam lithography is still the most used method to define NEMS devices from an experimental point of view, as it is a simple and proven method, although not convenient for massive fabrication and future industrial application. New prototyping approaches have been recently reported, including the use of focused ion beams [60–64]. However, little progress has been made in finding processes that would allow scaling up the fabrication of NEMS. Some activities include the technology being developed by the *Nanosystems alliance* between Leti and Caltech [65], optical-based technologies for fabricating single carbon-nanotube devices [66], and the extension of CMOS technology to integrated NEMS resonators in microelectronic circuits taking advantage of the high resolution provided by deep ultraviolet (DUV) optical lithography [67–71] (Figure 9.6).

Besides the top–down fabrication, new approaches have been explored for parallel integration of 1D nanostructures (carbon nanotubes (CNTs), Si nanowires (NWs), ZnO NWs) into functional devices. Usually a bottom–up technique is necessary for growing from small amounts of catalysts wire-like nanostructures in a specific chemical and thermal environment. Blank or micrometer-size "forests" of such 1D nanostructures can be easily grown from lithography-defined catalysts. But reaching a position-controlled, single-digit number of wires using parallel fabrication techniques is still a worth-while goal. The localized synthesis can be done

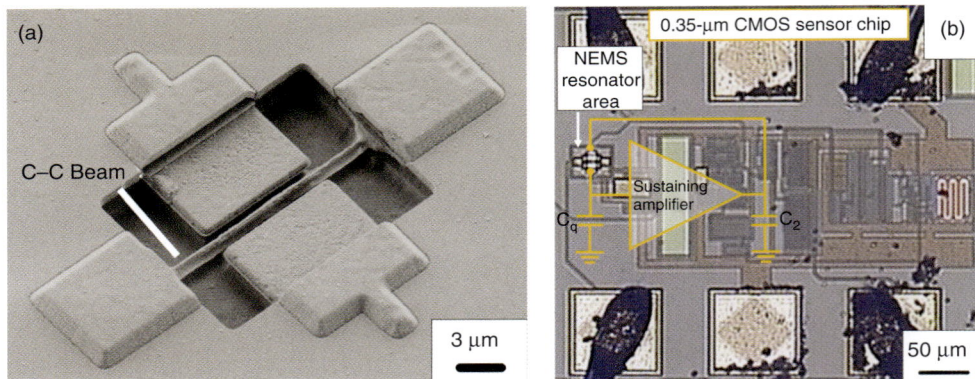

Figure 9.6 Integration of NEMS into CMOS using conventional DUV optical lithography. (a) Shows a SEM image of a clamped-clamped metallic beam: the closely located electrodes serve for electrostatic actuation and capacitive detection. The resonator is monolithically integrated into the CMOS circuit shown in (b) for building-up a self-oscillator system [68]. (Reprinted with permission from Ref. [68]. Copyright 2009 Wiley-VCH Verlag GmbH & Co.)

by having control either over the thermal environment or over the catalyst position. Blank deposition of the catalyst material combined with localized heating during the growth has been demonstrated [72–79]. Both methods are consistent with horizontal growth on vertical sidewalls, enabling the nanostructures to be positioned between two electrodes.

An alternative method is the localized deposition of the catalyst via parallel techniques such as stencil lithography. The advantage here is the possibility of reaching the single nanostructure limit with high-resolution positioning accuracy [80, 81]. Self-limiting deposition of CNTs from suspension gains control on the number and simultaneously on the orientation of nanostructures positioned between electrodes, allowing for piezoresistive pressure sensors based on single-walled CNTs [82].

Combination of NEMS devices with CMOS circuits also provides a route toward system integration, as it incorporates on-chip the functionalities of signal read-out and amplification. Alternative directions to integration have been proposed in order to reduce the parasitic signals in capacitive detection [83, 84]. In addition, many efforts are directed toward combination with diverse devices and structures: integration of field effect transistor for electrical read-out [85, 86], integration of a Schotky diode for optical detection [87], integration of waveguides [88, 89], and integration of nanofluidic channels [90]. In the latter approach, the degradation of the performance that mechanical resonators experience when operated in solution (due to viscous drag) is overcome by defining a suspended nanochannel (Figure 9.7). Using this approach, the best mass resolution in solution using NEMS sensing (27 ag) has been demonstrated. A step further in the use of NEMS for system integration is the proposal of using piezoelectric NEMS resonators for extracting signals from biosensors [91].

Figure 9.7 Suspended nanochannels used for the realization of NEMS mass sensors, providing a mass resolution below 30 ag [90]. (Reprinted with permission from Ref. [90]. Copyright 2010 American Chemical Society.)

9.1.4
Electronics

NEMS devices may play a role in future electronics both in the analog and digital areas. For several years, these devices have been proposed as basic building blocks for telecommunication systems by replacing components that cannot be integrated using conventional technologies. More recently, the application of NEMS in information processing is gaining more attention, mainly because of their lower power consumption and improved resistance to harsh environments compared to pure electronic processing in miniaturized systems. Application of NEMS for developing high-efficiency telecommunication systems is being addressed by several groups [43, 66, 67, 69, 70, 92–99]. In particular, NEMS resonators offer the potential to replace quartz crystals in the field of RF communications due to their capability of being fabricated with a standard IC process, the higher frequencies that they can achieve, and the small area they require.

The primary building blocks for any telecommunication system are oscillators. A self-sustaining oscillator with feedback can be implemented by employing a NEMS resonator as the frequency-determining element for the feedback oscillator. Such NEMS oscillators are active systems that are self-regenerative; thus, they are distinct from the more readily available NEMS resonators, which are passive devices that require external signal sources to provide periodic stimuli and driving forces to sustain the desirable stable oscillations. Therefore, NEMS oscillators clearly have important potential for a number of emerging applications. An important drawback associated with the reduced dimensions of NEMS is the resulting increase of the motional resistance, especially for the use of capacitive detection. In this case, piezoresistive sensing is viewed as a promising alternative [99]. Some other relevant examples of the use of NEMS for telecommunications are the mechanical implementation of filters [69] and frequency convertors [92].

NEMS switches also present great potential for application in other areas, like logic computation and memories [48, 100]. The main advantages are not only the expected lower power consumption [101], but also the possibility for operation at high temperatures [55]. The feasibility of building up several computational and memory blocks by using only passive components and mechanical switches has been recently demonstrated [102]. Although NEMS-based switches cannot compete with MOSFETs in terms of switching speed, they can provide alternative paths for reducing the power consumption of electronic circuits, with possibilities for high-density integration [103]. A paradigmatic example is the realization of the analogous of a semiconductor transistor by means of a three-terminal NEMS switch [104]. NEMS switches provide zero leakage current, almost infinite sharp on/off transitions, and a square hysteresis window.

9.1.5
Nonlinear MEMS/NEMS Applications

Mechanical sensors and actuators are usually operated as linear transducers. It is worthwhile to examine the behavior of mechanical resonators several tens of microns in size to learn and to adapt to the sub-micron scale, if possible. At large amplitudes, nonlinear effects dominate the response and reduce the range over which mechanical resonators can be applied as linear transducers. Nonlinear effects can be balanced in order to restore a linear response [84, 105–107]. On the other hand, there is a growing interest in MEMS/NEMS operating in the nonlinear regime. Characteristic phenomena were observed in the nonlinear regime, including multi-stability, hysteresis, and chaotic motion [14, 19, 108, 109]. Several new applications based upon nonlinear MEMS/NEMS have been demonstrated. This is an important realm in sensor applications, and to enhance it several concepts have been proposed.

Sensors and actuators usually operate as linear transducers, meaning that the amplitude of the motion is proportional to the driving force [1–3]. Nanoscale devices are extremely susceptible to external forces and large motion amplitudes are readily attained. At large amplitudes, nonlinear effects dominate the response. The resonance frequency, often the parameter of interest, is not well defined because it depends on the amplitude of the motion. The high susceptibility also results in a high noise floor: thermal fluctuations give rise to large mechanical motion. Combined with the early onset of nonlinearity, this limits the (linear) dynamic range of NEMS. Several concepts have been introduced to enlarge the dynamic range, for instance by balancing counteracting nonlinear effects [84, 105–107]; however, there is also a growing interest in NEMS operating in the nonlinear regime, and several applications have been demonstrated. NEMS also present an experimental platform to study the fundamental aspects of nonlinear dynamics.

Figure 9.8 summarizes four nonlinearities in doubly clamped bridges and singly clamped cantilevers, geometries that can be fabricated by top–down or bottom–up technology. Perhaps the best-known instability in mechanics is the Euler buckling (Figure 9.8a) in axially compressed structures. A buckled beam is bistable and does not require energy to remain in a bistable state. This enables applications in micromechanical relays, switches, and nonvolatile memory [100, 110–112].

An example of a nonlinear resonance is the stiffening of doubly clamped bridges by a displacement-induced tension (Figure 9.8b). This nonlinear effect causes the resonance frequency to increase with the vibration amplitude. Beyond a critical point, two vibration amplitudes are stable at the same driving conditions. Transitions between the states are induced by applying excitation pulses [118]. Frequency pulling leads to interesting behavior, such as new detection schemes and synchronized motion in arrays [107, 114, 118].

Strongly driven cantilevers exhibit a similar response, as is shown in Figure 9.8c. In a cantilever, the nonlinearity results from geometric and inertial effects [19, 87, 119]. Instead of driving a resonator directly with a force, one can also periodically modulate one of its parameters, for example, the spring constant. When this parametric drive exceeds a threshold, oscillations occur and two phases of the resonator are stable [120] (see Figure 9.8d). A small change in the driving signal alters the symmetry of the system and can be detected very accurately by measuring the probability of the resonator being in either phase [121]. Small signals can also be detected by dynamically changing the coupling between two parametric oscillators [18].

A bistable mechanical resonator can be used to represent digital information and several groups have demonstrated mechanical memory elements. Elementary mechanical computing algorithms have been implemented by coupling resonators in a tuneable way [30, 122]. The coupling can also be formed by properties intrinsic to the resonator. In this case, no connections are needed as the algorithm is executed in a single resonator. This concept was demonstrated by exciting a parametric resonator at multiple frequencies, where each excitation signal represents a logic input with the resulting motion being the output (Figure 9.9) [119].

The high Q-factor of mechanical systems compared to electronic circuits and the low "on"/high "off" resistance of MEMS/NEMS switches could provide a key to low-dissipation signal processors, and it is believed that mechanics holds the promise of low-power computing in the distant future [120]. Mechanical computing is also applicable in harsh conditions where electronics fail, for example, at high-temperatures or in high-radiation environments (oil industry, space, and defense).

In a bistable MEMS/NEMS, otherwise detrimental noise can be employed to amplify the response to weak signals. This counter-intuitive process in which noise enables the detection of weak signals is called stochastic resonance and has been demonstrated in doubly clamped beams at high noise levels [121, 123–125]. Nonlinearity and noise-induced switching can also improve the figure of merit in energy harvesting applications [126]. A nonlinear harvester

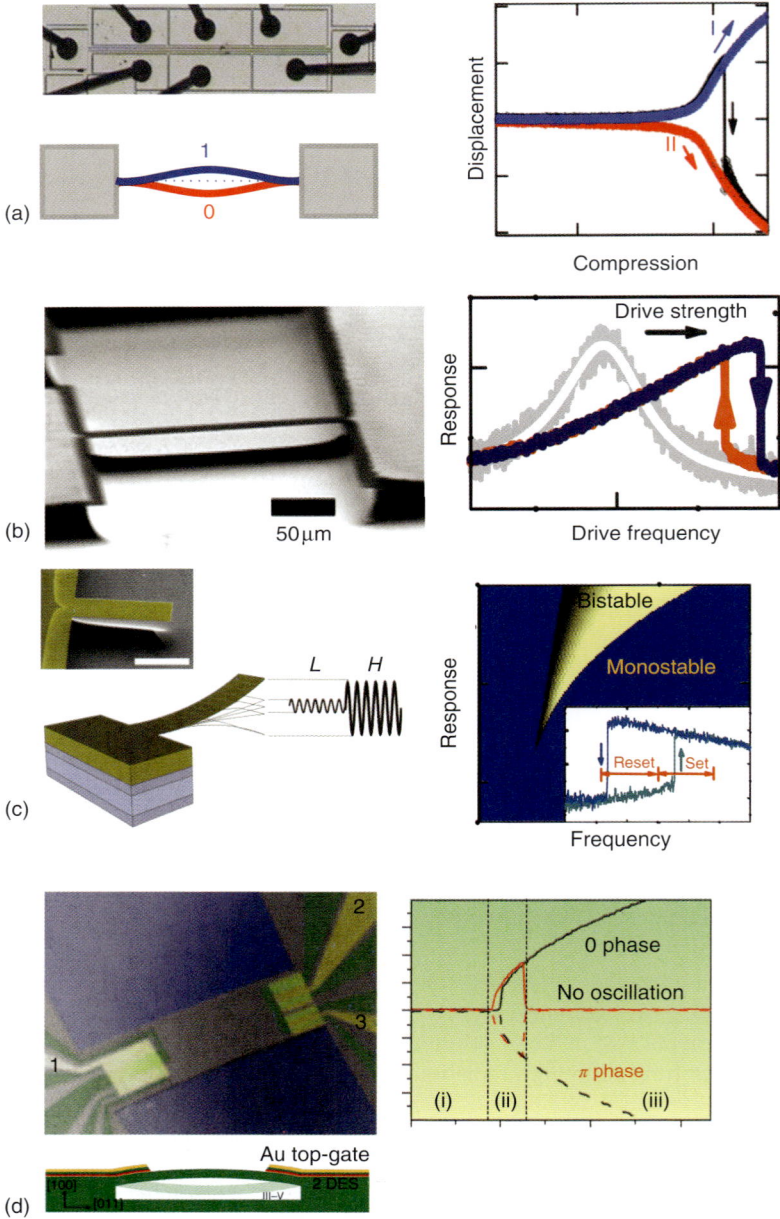

Figure 9.8 Nonlinearities in microresonators: the designs and responses of doubly clamped beams, wires, and singly clamped cantilevers. (a) *Euler buckling:* A doubly clamped beam buckles when the compressive stress exceeds a critical limit. The post-buckled state, up(1) or down(0), represents one bit of information [100, 110–112]. (b) *Duffing nonlinearity:* In doubly-clamped bridges, the displacement-induced tension results in bistability. At the same drive conditions, vibrations with a high and a low amplitude are stable [19, 113]. (c) *Geometric nonlinearity:* In cantilever beams vibrating at large amplitudes, the geometric nonlinearity causes the resonance frequency to depend on the amplitude. The amplitude of a strongly driven cantilever is bistable [84, 114, 115]. (d) *Parametric instability:* A parametric oscillator is driven by modulating the spring constant at twice the resonance frequency [116, 117]. Information can be encoded in the oscillator phase: $\Phi = 0(0)$ or $\Phi = \pi$ (1) is stable. (Reprinted with permission from Ref. [116]. Copyright 2008 Nature Publishing Group.)

responds to a wider spectrum of vibrations, far exceeding the bandwidth of a linear transducer. Implementing this scheme in MEMS/NEMS could lead to efficient power generators for stand-alone devices.

9.2
Carbon-Based NEMS

Despite most NEMS devices being based on silicon technology, in recent years, carbon-based NEMS devices have been gaining more and more interest, mostly because of their high Young's modulus and small diameter. Carbon-based mechanical resonators have large tunable frequencies and exhibit large amplitudes. Due to their low mass, they operate in a different regime from their silicon-based counterparts, as evidenced by their very strong nonlinear response. Moreover, they generally exhibit an extremely high sensitivity to external stimuli, making them interesting candidates for various sensing applications for fundamental studies as well as applications. Fabrication processes for carbon nanotube-based and graphene-based NEMS resonators are now well established for prototyping and demonstration activities. From an applied point of view, the challenge is to fabricate devices suitable for large-scale applications that operate at room temperature. Future research is still necessary to elucidate which of the two carbon forms, CNTs or (few-layer) graphene, due to its larger area, is easier to fabricate on an industry-scale [127].

An example of a suspended CNT device is shown in Figure 9.10. These bottom–up devices are expected not to suffer from excessive damping, as their surface can be defect-free at the atomic scale. Combined with their low mass, the expected low damping makes them ideal building blocks. Moreover, because of their small size, carbon-based resonators typically have frequencies in the megahertz-to-gigahertz range. All these properties (low mass, high Q, high frequencies) are advantageous for sensing applications and the study of quantum properties of resonating objects [128]. For example, when cooled to dilution refrigerator temperatures, carbon-based resonators can be in the quantum

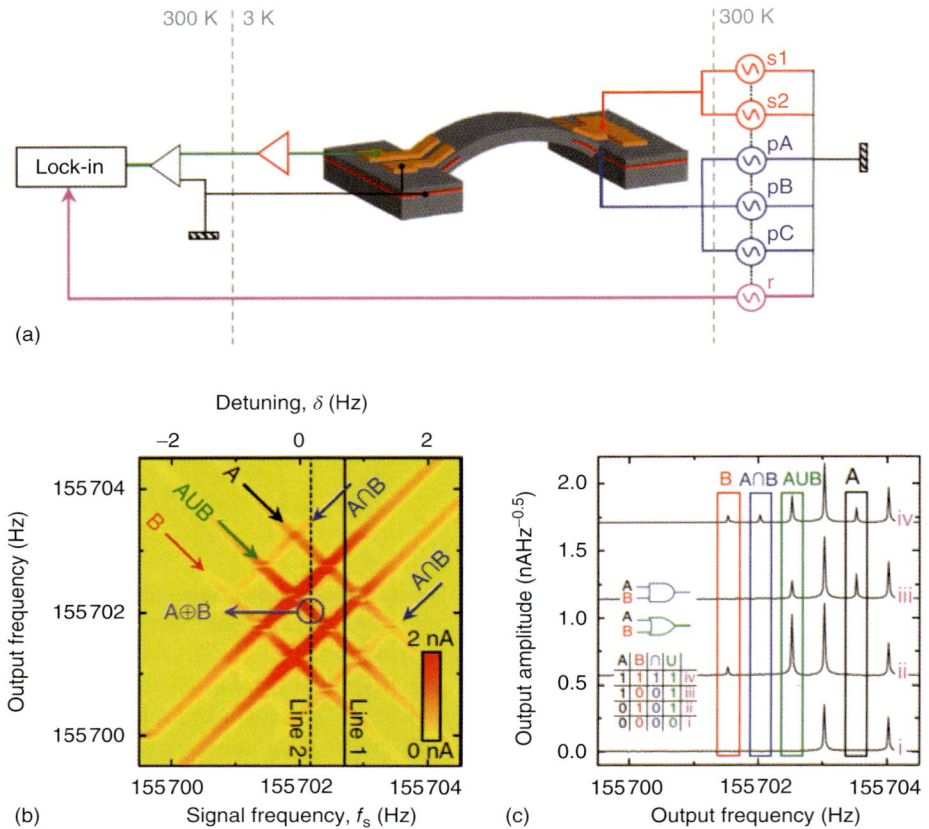

Figure 9.9 Multi-bit logic in a doubly-clamped resonator. (a) Setup and resonator with integrated piezoelectric transducers for direct and parametric driving and on-chip motion detection. The response is measured close to the resonance frequency f_0 by directly driving the resonator at $f_s + \delta$, while parametrically exciting it at $f_{pA} = 2f_0 + \Delta$ and $f_{pB} = 2f_0 - \Delta$. (b) Mixing between the parametric and direct drive results in splitting of the mode, where higher-order mixing frequencies occur when $\Delta \neq 0$. When the parametric driving signals are considered logic inputs, where A (B) denotes the presence of the parametric excitation at f_{pA} (f_{pB}), logic functions can be implemented. Depending on the drive frequency (horizontal axis) and the detection frequency (vertical axis), the resonator functions as an AND ($\cap$), OR ($\cup$), and XOR ($\oplus$) gate. (c) Shows cross-sections of (b), to demonstrate the resonator logic response when the parametric signals are switched on and off. In these experiments, $\Delta = 0.5$ Hz. More complex logic functions are possible by generating more mixing frequencies by driving the resonator at multiple frequencies. (Reprinted with permission from Ref. [119]. Copyright 2011 Nature Publishing Group.)

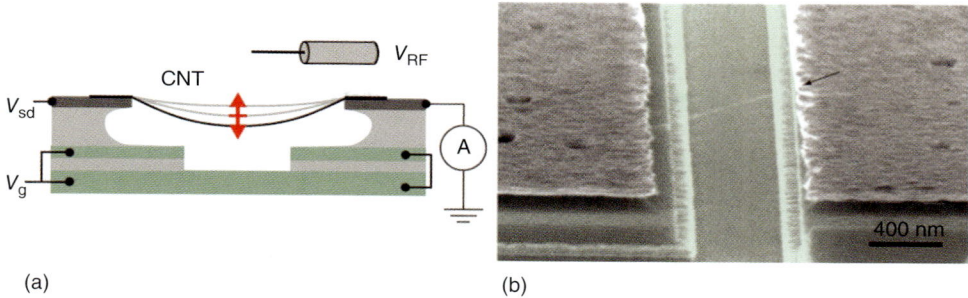

(a)   (b)

Figure 9.10 (a) A high Q mechanical resonator layout with suspended carbon nanotube; A suspended carbon nanotube is
excited into mechanical motion by applying an AC voltage to a nearby antenna. (b)
SEM image of a suspended carbon nanotube
clamped between two metal electrodes. A
bottom gate can be used to tune the frequency of the resonator [10, 129]. (Reprinted
with permission from Ref. [129]. Copyright
2009 American Chemical Society.)

mechanical ground-state while exhibiting relatively large amplitude zero-point
fluctuations.

Position detectors of carbon-based bottom–up NEMS, however, are not yet
as sophisticated as those for the larger top–down silicon-based counterparts.
Consequently, neither non-driven motion at cryogenic temperatures (either
Brownian or zero-point motion) nor active cooling have been reported for
carbon-based NEMS. Nevertheless impressive progress in understanding the
electromechanical properties of bottom–up resonators has been made in recent
years using so-called self-detecting schemes. In these schemes, the nanotube or
graphene resonator acts both as the actuator and detector of its own motion.

Various device geometries with carbon-based materials exist. The motion of
singly-clamped CNTs has been visualized in scanning [130] and transmission electron microscopy [131]. Another method to detect motion of singly-clamped carbon resonators is based on field emission of electrons from a vibrating tip [132].
When a large voltage is applied between a multi-walled CNT and an observation screen, it lights up at the position where electrons accelerated by the electric
field are impinging. The vibration amplitude is enlarged by applying an RF driving signal, and the spot blurring becomes even more pronounced on resonance.
Furthermore, the electric field also pulls on the nanotube, thereby increasing the
resonance frequency. The method has also been used to build a nanotube radio
[133] and a mass sensor approaching and achieving atomic resolution [134].

For doubly clamped resonator geometries, it is also advantageous to use
the suspended device also as a detector of motion. Using current rectification
and frequency mixing, information about the driven motion of the suspended
nanotube and graphene has been obtained. Sazonova *et al.* [36] were the first to
apply frequency mixing to suspended CNT resonators. They observed multiple
gate-tunable resonances with Q-factors on the order of 100 at room temperature.
Subsequently, the bending-mode vibrations of a CNT were also identified [135].

At present, the technique is being employed by many groups, not only restricted to CNTs, but also to suspended doubly clamped graphene sheets [37]. Furthermore, several variations to the original mixing scheme have been implemented, including frequency [136] and amplitude modulations [137].

At low temperatures, a Coulomb blockade can be used to drastically enhance the displacement sensitivity. For example, the change in equilibrium position of a suspended nanotube quantum dot after adding a single electron easily surpasses the zero-point motion. A strong coupling results between mechanical motion and the charge on the nanotube, leading to frequency shifts and changes in damping as a function of gate voltage [10, 138]. The readout using current rectification is employed instead of frequency mixing [10, 129]. While the nanotube motion is actuated by an RF signal on a nearby antenna, the detected signal is at DC (Figure 9.10). The key to understanding this behavior is the notion that nanotube

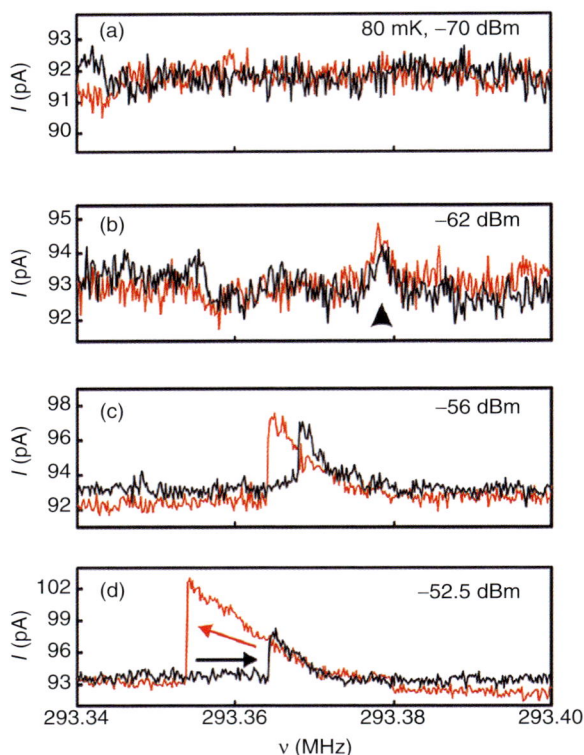

Figure 9.11 Evolution of the resonance peak with increasing driving power (a–d) at a temperature of 80 mK [10]. Black (red) traces are upward (downward) frequency sweeps. At low powers, the peak is not visible, but upon increasing power, a resonance peak with $Q = 128\,627$ appears. As the power is increased further, the line shape of the resonance resembles that of a Duffing oscillator exhibiting hysteresis between the upward and downward sweep [129]. (Reprinted with permission from Ref. [129]. Copyright 2009 American Chemical Society.)

motion effectively translates into an oscillating gate voltage, leading to changes in the DC current, which are maximal at resonance.

The technique is of special interest as it allows for motion detection with small currents, enabling the observation of ultra-high Q-factors, exceeding 100 000 at millikelvin temperatures [129]. Furthermore, experiments show that the dynamic range is small, that is, carbon-based resonators are easily driven into the nonlinear regime as illustrated in Figure 9.11. This can be understood from the small tube diameters or the extremely thin membrane-like shapes of the graphene flakes: with increasing driving power the amplitude of flexural motion rapidly grows to their characteristic sizes inducing sizable tension in the resonator. In addition, non-linear damping effects have recently been reported using mixing techniques in nanotube and graphene resonators [139]. Finally, electron tunneling and mechanical motion are found to be strongly coupled [129, 138], resulting in single-electron tuning oscillations of the mechanical frequency and in energy transfer to the electrons causing additional mechanical damping.

9.3
Toward Functional Bio-NEMS

High-frequency NEMS are attracting more and more interest as a new class of sensors and actuators for potential applications to single (bio)molecule sensing [53, 140]. For NEMS to be considered a viable alternative to their actual biosensing macro counterparts, they must simultaneously meet three major requirements: high mass responsivity (MR), low minimum detectable mass (MDM) and low response time (RT).

Without any doubt, as emphasized by theoretical studies [141, 142], the first two specifications (MR and MDM) can be successfully addressed by NEMS devices. Such predictions have already been validated in the case of virus sensing [143], enumeration of DNA molecules [144] or even single-molecule nanomechanical mass spectrometry [53]. Unfortunately, nanometer-scale sensors have been proved in theory [145, 146] to be inadequate for practical RT scales which, if confirmed as such, could definitely impede the NEMS' route toward realistic biosensing applications. To overcome this challenge, one possible trade-off strategy would be to implement NEMS devices as individual constituents of a functional array of similar devices [147]. While preserving the benefits of high MR and low MDM of a single device, this paradigm results in a considerably higher capture area of the NEMS array because the RT reaches a level of practical relevance. However, in this case the non-reactive areas of the chip containing multiple sensors functionalized with a single type of probe molecule must be adequately coated with an anti-fouling film in order to lower the probability of adsorption of target molecules at locations other than the sensitive areas and hence to permit ultra-low concentration detection.

To address the production of massively parallel arrays of NEMS for bio-recognition applications, one must be able to perform uniform, reliable bio-functionalization of nanoscale devices at a large scale (the array level), while being able to obtain an anti-fouling surface everywhere else. For this to become a reality, a major challenge is the functionalization of closely packed nanostructures in such a way that biological receptors are precisely located solely on the active biosensing areas, thus preventing the waste of biological matter and enabling the subsequent biological blocking of the passive parts of the chip. Thus far, the issue of freestanding nanostructures functionalization has been seldom addressed because of the absence of generic tools or techniques allowing large-scale molecular delivery at the nanoscale. One way to circumvent this difficulty would be to perform the functionalization step before completing the fabrication of the NEMS. This strategy can typically be used in a top–down NEMS fabrication process by protecting the biological layer during the subsequent NEMS fabrication steps. This may be achieved by placing the functionalized nanostructures at specific locations on a substrate and releasing them [148]. The main limitation of this strategy is the trade-off between the choice of the post-functionalization processing steps and the resilience of the chosen biological receptors to such technological constraints, which for most of them are biologically unfriendly.

9.3.1
NEMS-Based Energy Harvesting: an Emerging Field

The impact of NEMS technology in the energy-harvesting field, that is, a discipline aiming to convert ambient energy into useful electrical energy to power ultralow consumption ICT devices, is still incipient. A notable level of maturity has been achieved in this field by MEMS applications, where energy is extracted from ambient vibrations. Several mechanical-to-electrical transduction methods have been applied so far, but piezoelectricity has been demonstrated as the preferred solution because of the increased integrability of piezomaterials and especially the simplicity of the associated power management circuitry. Although most of the technologies and concepts that have been demonstrated to be feasible at the MEMS scale are not improved when dimensions are reduced to the NEMS scale, the combination of piezoelectric and nanowire technologies becomes relevant. A recent study [149] theoretically demonstrated the giant piezoelectricity of ZnO and GaN nanowires, which is due to the effect that charge redistribution on the free surfaces produces the local polarization. This effect, which has been theoretically reported previously [150], demonstrates that it is more efficient to fill a certain volume with a compact array of nanowires than to use a bulk thin film piezoelectric substrate, a clear enhancement produced by downscaling to the nanometer scale. An exhaustive study of the potential performance of piezoelectric nanostructures for mechanical energy harvesting is provided by Sun *et al.* [151]. In this paper, the authors compare rectangular and hexagonal nanowires and 2-D vertical thin films (nanofins), as well as different piezomaterials such as ZnO and $BaTiO_3$, and conclude that the power density ideally obtainable by filling the whole volume is in the range of $10^3 – 10^4$ W cm^{-3}.

This concept is in fact experimentally exploited in many different ways to implement energy nanoconverters or energy nanoharvesters based on arrays of piezoelectric nanowires. One specific technology, called nanopiezotronics, is based on combining the piezoelectric properties of ZnO nanowires or fine-wires (their micro-scale version) with the rectifying characteristics of the Schottky barrier, formed between the ZnO (semiconductor) and a metal. Most recent work demonstrates biomechanical-to-electrical conversion using a single wire generator, which is able to produce output voltages around 0.1 V from human finger tapping or from the body movement of a hamster [152], or even from the breathing and heartbeat of a rat [153], which demonstrate the potential applicability of NEMS to self-powering implanted nanodevices. High-output power nanogenerators have also been obtained by a rational assembly of ZnO nanowires in a 2-D array. The obtained nanogenerators are able to power real devices such as a LED [154] or an LCD [155]. Power densities of $11 \, \text{mW cm}^{-3}$ are experimentally demonstrated and by multilayer integration $1.1 \, \text{W cm}^{-3}$ are predicted. A complete review of nanopiezotronics technology for energy harvesting is made in [156].

Additional research in the field of biomechanical energy harvesting includes a recent study proposing a solution for the performance trade-off between the piezoelectric coefficient and stretchability [157]. Typically, organic piezoelectric materials like PVDF (polyvinylidene fluoride) are flexible, but show weak piezoelectricity and inorganic ceramic materials such as PZT, ZnO, or $BaTiO_3$ have piezoelectric coefficients one order of magnitude higher, but are brittle. PZT ribbons buckled by the attachment on a pre-stretched PDMS substrate display simultaneously high piezoelectricity and integrity under stretching and flexing operations [157]. Also the embedding of PZT nanofibers into a PDMS substrate is used to generate peaks of voltage and power around 1.6 V and 30 nW from external vibrations, respectively [158]. But organic does not necessarily mean low piezoelectricity. A method to directly write PVDF nanofibers with energy conversion efficiencies one order of magnitude higher than those of thin films was developed [159]. The method, based on near-field electrospinning, allows the mechanical stretching, polling, and positioning of the nanofibers. An exhaustive description of this technique can be found in the review paper of Chang *et al.* [160]. Finally, an interesting example of inorganic piezoelectric materials being embedded into an organic polymer is provided by the embedding of ZnO nanowires into a PVC substrate [161]. The collective stretching of the nanowires produced by the temperature-induced polymer shape-change allows power densities around $20 \, \text{nW cm}^{-3}$ to be achieved at $65 \, ^\circ\text{C}$ by means of a nonconventional thermoelectric effect.

Solutions to unsolved challenges, such as a real co-design of NEMS energy transducers and power management circuitry or the introduction of unexplored materials to span the sensitivity to new energy sources, will define the future research tendencies in this field. As an example, suspended graphene nanoribbons have been demonstrated to efficiently harvest the energy from thermal fluctuations due to the mechanical bistability induced by a controlled compressive stress [162]. In Figure 9.12, the simulation results of the out-of-plane

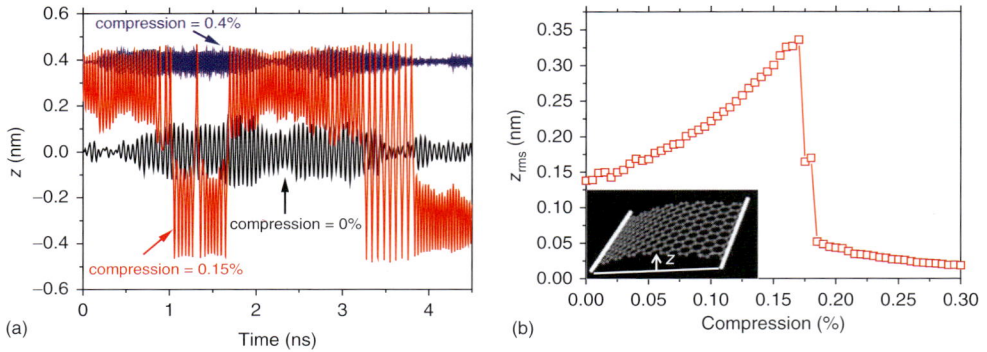

Figure 9.12 (a) Time evolution of the out-of-plane movement, z, of a 17 nm long and 1 nm wide graphene-suspended nanoribbon. (b) Root mean square value of the out-of-plane movement as a function of the compression level [163]. Inset: sketch of the suspended nanoribbon with the out-of-plane movement coordinate, z. (Reprinted with permission from Ref. [163]. Copyright 2011 American Physical Society.)

movement of an ideal 1 nm-by-17 nm clamped–clamped graphene nanoribbon (inset in Figure 9.12b) are shown. When the nanoribbon is compressed along the longitudinal direction, a mechanical buckling-like bistability is induced. For zero or excessively large (0.4%) compression conditions, the graphene response to the white Gaussian noise corresponding to the thermal fluctuations is confined to the bottom of the potential well (black and blue traces in Figure 9.12a); however, for a 0.15% compression, the noise is able to make the grapheme nanoribbon jump between the two wells of the bistable potential placed at $z_+ = 0.3$ nm and $z_- = -0.3$ nm, as the red trace of Figure 9.12a shows. A study of the out-of-plane movement dependence on the compression level reveals that 0.15% is the optimal value in terms of the rms value of the movement (Figure 9.12b).

An unsolved issue in the framework of the proposed harvesters, which are based on 2D materials like graphene, is the integration of a mechanism that could convert the harvested mechanical energy into the electrical domain. In this direction, the works of Chandratre and Sharma [162] and Ong and Reed [164] propose solutions based on geometrical local modifications or ion doping to provide graphene with piezoelectric properties. As an alternative, 2D monoatomic boron nitride layers could be used to define similar compression-induced bistable NEMS structures, but having in this case the natural piezoelectric properties of the material [163].

9.4
Summary and Outlook

The progress in the field of NEMS has continued at a rapid pace over the past decade. The area of NEMS is now entering a more mature stage, addressing real applications in the areas of sensing, telecommunications, information processing, and energy harvesting. A step forward is being taken for obtaining a better understanding of nonlinear behavior within the realm of sensing and logic applications. Moreover, the easy access to the nonlinear regime and defect-free material properties make NEMS an excellent platform for studying nonlinear dynamics in a more general context. Signal detection and sensitivity limits together with NEMS integration into complex systems at a larger scale are continuously being enhanced. Incorporation of new materials into NEMS will further improve device performance in terms of sensitivity, working range, and efficiency.

While device concepts and fundamental knowledge of the properties of NEMS structures are relatively advanced, a fabrication technology that would fulfill the requirements for high-dimensional precision, material compatibility, and high throughput is still a limiting factor for commercial applications. As a consequence, more effort in developing suitable fabrication methods adapted to industry is needed to ensure the future success of the field.

The specialization of nonlinear NEMS continues to emerge, with new phenomena being discovered, which will be applicable to ultrasensitive detectors, mechanical signal processors, and efficient energy harvesters. In nonlinear NEMS, weak signals can be amplified by making beneficial use of environmental noise, by employing processes like stochastic resonance.

Several challenges need to be addressed in order to utilize the potential of nonlinear NEMS. Modeling the nonlinear behavior requires numerical methods, which can be computationally intensive even for simple beam structures. Tight fabrication tolerances are demanded in order to predict and/or reproduce the dynamic behavior in the nonlinear regime within a workable tolerance window. To reduce the electrical power consumed by the strongly vibrating NEMS, devices may be optimized as to achieve nonlinear characteristics at weak driving. To this end, it is essential to understand the dissipation mechanisms in NEMS resonators. In order to employ noise-enhanced detection schemes, the barrier between the stable states of the bistable NEMS should be reduced. New ways to couple nonlinear NEMS in an efficient and adjustable way will further expand the NEMS toolbox, thereby allowing the construction of complicated dynamic systems that will lead to new concepts being discovered and a wide horizon to be explored. Carbon-based mechanics is a relatively new research field, but the push for refining detection schemes and integrating carbon-based materials into silicon technology will undoubtedly lead to the construction of better sensors, which may eventually be quantum-limited. A bright future thus seems to lie ahead for these miniature devices. From a fundamental physics point of view, challenges lie in improving detection schemes so that thermal motion at low temperatures and eventually zero-point motion can be detected. Carbon-based resonators also provide a unique system to study the nonlinear properties of mechanical resonators, especially in the quantum regime. Furthermore, it currently remains unknown what the limiting factor is regarding the intrinsic damping of carbon-based resonators. This is a more general issue in NEMS as damping in silicon resonators is also not understood in detail.

In the field of energy harvesting, the challenges to be faced are related not only to the efficient conversion, but also to the management and storage of the harvested energy at the nanoscale. Novel concepts and devices based on NEMS technology and oriented to the management/storage of the energy in a pure mechanical form would improve the energy efficiency of the overall harvesting process, since no conversions from the mechanical to the electrical domain would be needed.

Challenges making use of NEMS arrays in the biosensing realm can be foreseen both at the front-end for differential functionalization of closely packed sensors and at the back-end, for the integration of actuation and sensing capabilities at nanodevice arrays levels.

References

1. Cleland, A.N. and Roukes, M.L. (1998) A nanometre-scale mechanical electrometer. *Nature*, **392** (6672), 160.
2. Fon, W., Schwab, K.C., and Worlock, J.M. (2002) Phonon scattering mechanisms in suspended nanostructures from 4 to 40 K. *Phys. Rev.*, **B 66** (4), 045302.
3. Schwab, K., Arlett, J.L., and Worlock, J.M. (2001) Thermal conductance through discrete quantum channels. *Physica E*, **9** (1), 60.
4. Schwab, K., Henriksen, E.A., and Worlock, J.M. (2000) Measurement of the quantum of thermal conductance. *Nature*, **404** (6781), 974.
5. Schwab, K., Fon, W., and Henriksen, E. (2000) Quantized thermal conductance: measurements in nanostructures. *Physica B*, **280** (1-4), 458.
6. Cleland, A.N., O'Connell, A.D., and Hofheinz, M. (2010) Quantum ground state and single-phonon control of a mechanical resonator. *Nature*, **464** (7289), 697.
7. Teufel, J.D., Donner, T., and Li, D.L. (2011) Sideband cooling of micromechanical motion to the quantum ground state. *Nature*, **475** (7356), 359.
8. Roukes, M.L., LaHaye, M.D., and Suh, J. (2009) Nanomechanical measurements of a superconducting qubit. *Nature*, **459** (7249), 960.
9. Bachtold, A., Lassagne, B., and Tarakanov, Y. (2009) Coupling mechanics to charge transport in carbon nanotube mechanical resonators. *Science*, **325** (5944), 1107.
10. Steele, G.A., Huttel, A.K., and Witkamp, B. (2009) Strong coupling between single-electron tunneling and nanomechanical motion. *Science*, **325** (5944), 1103.
11. Roukes, M.L., Naik, A.K., and Hanay, M.S. (2009) Towards single-molecule nanomechanical mass spectrometry. *Nat. Nanotechnol.*, **4** (7), 445.
12. Roukes, M.L., Yang, Y.T., and Callegari, C. (2011) Surface adsorbate fluctuations and noise in nanoelectromechanical systems. *Nano Lett.*, **11** (4), 1753.
13. Dykman, M.I., Khasin, M., and Portman, J. (2010) Spectrum of an oscillator with jumping frequency and the interference of partial susceptibilities. *Phys. Rev. Lett.*, **105** (23), 230601.
14. Lifshitz, R. and Cross, M.C. (2008) *Nonlinear Dynamics of Nanomechanical and Micromechanical Resonators*, Wiley-VCH Verlag GmbH, Weinheim.
15. Weick, G., Pistolesi, F., and Mariani, E. (2010) Discontinuous Euler instability in nanoelectromechanical systems. *Phys. Rev. B*, **81** (12), 121409.
16. Atalaya, J., Isacsson, A., and Dykman, M.I. (2011) Diffusion-induced bistability of driven nanomechanical resonators. *Phys. Rev. Lett.*, **106** (22), 227202.
17. Karabalin, R.B., Cross, M.C., and Roukes, M.L. (2009) Nonlinear dynamics and chaos in two coupled nanomechanical resonators. *Phys. Rev. B*, **79** (16), 165309

18. Karabalin, R.B., Lifshitz, R., and Cross, M.C. (2011) Signal amplification by sensitive control of bifurcation topology. *Phys. Rev. Lett.*, **106** (9), 094102.

19. Westra, H.J.R., Poot, M., and van der Zant, H.S.J. (2010) Nonlinear modal interactions in clamped-clamped mechanical resonators. *Phys. Rev. Lett.*, **105** (11), 117205.

20. Bachtold, A., Eichler, A., and Moser, J. (2011) Nonlinear damping in mechanical resonators made from carbon nanotubes and graphene. *Nat. Nanotechnol.*, **6** (6), 339.

21. Ekinci, K.L., Sampathkumar, A., and Murray, T.W. (2011) Multiplexed optical operation of distributed nanoelectromechanical systems arrays. *Nano Lett.*, **11** (3), 1014.

22. Basarir, O., Bramhavar, S., and Ekinci, K.L. (2010) Near-field optical transducer for nanomechanical resonators. *Appl. Phys. Lett.*, **97** (25), 3530432.

23. Freeman, M.R., Liu, N., and Giesen, F. (2008) Time-domain control of ultrahigh-frequency nanomechanical systems. *Nat. Nanotechnol.*, **3** (12), 715.

24. Tang, H.X., Li, M., and Pernice, W.H.P. (2009) Broadband all-photonic transduction of nanocantilevers. *Nat. Nanotechnol.*, **4** (6), 377.

25. Tang, H.X., Li, M., and Pernice, W.H.P. (2008) Harnessing optical forces in integrated photonic circuits. *Nature*, **456** (7221), 480.

26. Okamoto, H., Ito, D., and Onomitsu, K. (2011) Vibration amplification, damping, and self-oscillations in micromechanical resonators induced by optomechanical coupling through carrier excitation. *Phys. Rev. Lett.*, **106** (3), 036801.

27. Li, M., Pernice, W.H.P., and Tang, H.X. (2009) Tunable bipolar optical interactions between guided lightwaves. *Nat. Photonics*, **3** (8), 464.

28. Cleland, A.N. and Roukes, M.L. (1999) External control of dissipation in a nanometer-scale radiofrequency mechanical resonator. *Sens. Actuators, A*, **72** (3), 256.

29. Bargatin, I., Kozinsky, I., and Roukes, M.L. (2007) Efficient electrothermal actuation of multiple modes of high-frequency nanoelectromechanical resonators. *Appl. Phys. Lett.*, **90** (9), 2709620.

30. Masmanidis, S.C., Karabalin, R.B., and De Vlaminck, I. (2007) Multifunctional nanomechanical systems via tunably coupled piezoelectric actuation. *Science*, **317** (5839), 780.

31. Karabalin, R.B., Matheny, M.H., and Feng, X.L. (2009) Piezoelectric nanoelectromechanical resonators based on aluminum nitride thin films. *Appl. Phys. Lett.*, **95** (10), 103111.

32. Truitt, P.A., Hertzberg, J.B., and Huang, C.C. (2007) Efficient and sensitive capacitive readout of nanomechanical resonator arrays. *Nano Lett.*, **7** (1), 120.

33. Tang, H.X., Bhaskaran, H., and Li, M. (2011) Active microcantilevers based on piezoresistive ferromagnetic thin films. *Appl. Phys. Lett.*, **98** (1), 013502.

34. Kotthaus, J.P., Unterreithmeier, Q.P., and Weig, E.M. (2009) Universal transduction scheme for nanomechanical systems based on dielectric forces. *Nature*, **458** (7241), 1001.

35. Bargatin, I., Myers, E.B., and Arlett, J. (2005) Sensitive detection of nanomechanical motion using piezoresistive signal downmixing. *Appl. Phys. Lett.*, **86** (13), 1896103.

36. Sazonova, V., Yaish, Y., and Ustunel, H. (2004) A tunable carbon nanotube electromechanical oscillator. *Nature*, **431** (7006), 284.

37. Chen, C.Y., Rosenblatt, S., and Bolotin, K.I. (2009) Performance of monolayer graphene nanomechanical resonators with electrical readout. *Nat. Nanotechnol.*, **4** (12), 861.

38. Ayari, A., Gouttenoire, V., and Barois, T. (2010) Digital and FM demodulation of a doubly clamped single-walled carbon-nanotube oscillator: towards a nanotube cell phone. *Small*, **6** (9), 1060.

39. Kacem, N., Hentz, S., and Pinto, D. (2009) Nonlinear dynamics of nanomechanical beam resonators: improving the performance of NEMS-based sensors. *Nanotechnology*, **20** (27), 275501.

40. Karabalin, R.B., Masmanidis, S.C., and Roukes, M.L. (2010) Efficient parametric amplification in high and very high

frequency piezoelectric nanoelectrome-chanical systems. *Appl. Phys. Lett.*, **97** (18), 3505500.

41. Ekinci, K.L., Yang, Y.T., and Huang, X.M.H. (2002) Balanced electronic detection of displacement in nano-electromechanical systems. *Appl. Phys. Lett.*, **81** (12), 2253.

42. Feng, X.L., White, C.J., and Hajimiri, A. (2008) A self-sustaining ultrahigh-frequency nanoelectromechanical oscillator. *Nat. Nanotechnol.*, **3** (6), 342.

43. Arcamone, J., van den Boogaart, M.A.F., and Serra-Graells, F. (2008) Full-wafer fabrication by nanostencil lithography of micro/nanomechanical mass sensors monolithically integrated with CMOS. *Nanotechnology*, **19** (30), 305302.

44. Faucher, M., Grimbert, B., and Cordier, Y. (2009) Amplified piezoelectric trans-duction of nanoscale motion in gallium nitride electromechanical resonators. *Appl. Phys. Lett.*, **94** (23), 233506.

45. Grogg, D. and Ionescu, A.M. (2011) The vibrating body transistor. *IEEE Trans. Electron Devices*, **58** (7), 2113.

46. Weinstein, D. and Bhave, S.A. (2010) The resonant body transistor. *Nano Lett.*, **10** (4), 1234.

47. Colinet, E., Durand, C., and Audebert, P. (2008) Measurement of nano-displacement based on in-plane suspended-Gate MOSFET detection compatible with a front-end CMOS process. IEEE International Solid-State Circuits Conference (ISSCC), p. 332.

48. Guerra, D.N., Bulsara, A.R., and Ditto, W.L. (2010) A noise-assisted repro-grammable nanomechanical logic gate. *Nano Lett.*, **10**, 1168.

49. Liu, N., Giesen, F., and Belov, M. (2008) Time-domain control of ultrahigh-frequency nanomechanical systems. *Nat. Nanotechnol.*, **3**, 715.

50. Unterreithmeier, Q.P., Weig, E.M., and Kotthaus, J.P. (2009) Universal trans-duction scheme for nanomechanical systems based on dielectric forces. *Nature*, **458**, 1001.

51. Suh, J., LaHaye, M.D., and Echternach, P.M. (2010) Parametric amplification and back-action noise squeezing by a qubit-coupled nanoresonator. *Nano Lett.*, **10**, 3990.

52. Karabalin, R.B., Feng, X.L., and Roukes, M.L. (2009) Parametric nanomechanical amplification at very high frequency. *Nano Lett.*, **9**, 3116.

53. Naik, A.K., Hanay, M.S., and Hiebert, W.K. (2009) Towards single-molecule nanomechanical mass spectrometry. *Nat. Nanotechnol.*, **4**, 445.

54. Li, M., Myers, E.B., Tang, H.X., Aldridge, S.J., McCaig, H.C., Whiting, J.J., Simonson, R.J., Lewis, N.S., and Roukes, M.L. (2010) Nanoelectrome-chanical resonator arrays for ultrafast, gas-phase chromatographic chemical analysis. *Nano Lett.*, **10**, 3899.

55. Lee, T.-H., Bhunia, S., and Mehregany, M. (2010) Electromechanical computing at 500°C with silicon carbide. *Science*, **329** (5997), 1316.

56. Guthy, C., Das, R.M., Drobot, B., and Evoy, S. (2010) Resonant characteristics of ultranarrow SiCN nanomechanical resonators. *J. Appl. Phys.*, **108**, 014306.

57. Gil-Santos, E., Ramos, D., and Martínez, J. (2010) Nanomechanical mass sensing and stiffness spectrometry based on two-dimensional vibrations of resonant nanowires. *Nat. Nanotechnol.*, **5**, 641.

58. Ivaldi, P., Abergel, J., and Matheny, M.H. (2011) 50 nm thick AlN film-based piezoelectric cantilevers for gravimetric detection. *J. Micromech. Microeng.*, **21**, 085023.

59. Sökmen, Ü., Stranz, A., and Waag, A. (2010) Evaluation of resonating Si cantilevers sputter-deposited with AlN piezoelectric thin films for mass sensing applications. *J. Micromech. Microeng.*, **20**, 064007.

60. Sulkko, J., Sillanp, M.A., and Hkkinen, P. (2010) Strong gate coupling of high-Q nanomechanical resonators. *Nano Lett.*, **10**, 4884.

61. Vick, D., Sauer, V., and Fraser, A.E. (2010) Bulk focused ion beam fabri-cation with three-dimensional shape control of nanoelectromechanical sys-tems. *J. Micromech. Microeng.*, **20**, 105005.

62. Sievilä, P., Chekurov, N., and Tittonen, I. (2010) The fabrication of silicon

nanostructures by focused-ion-beam implantation and TMAH wet etching. *Nanotechnology*, **21**, 145301.

63. Bischoff, L., Schmidt, B., and Langea, H. (2009) Nano-structures for sensors on SOI by writing FIB implantation and subsequent anisotropic wet chemical etching. *Nucl. Instrum. Methods Phys. Res., Sect. B*, **267**, 1372.

64. Rius, G., Llobet, J., and Borrisé, X. (2009) Fabrication of complementary metal-oxide-semiconductor integrated nanomechanical devices by ion beam patterning. *J. Vac. Sci. Technol. B*, **27**, 2691.

65. Caltech-KNI Leti-Minatec *www.nanovlsi.com* (accessed 2 September 2014).

66. Martin-Fernandez, I., Borrisé, X., and Lora-Tamayo, E. (2010) Batch wafer scale fabrication of passivated carbon nanotube transistors for electrochemical sensing applications. *J. Vac. Sci. Technol., B*, **28**, C6P1.

67. Lopez, J.L., Verd, J., and Teva, J. (2009) Integration of RF-MEMS resonators on submicrometric commercial CMOS technologies. *J. Micromech. Microeng.*, **19**, 015002.

68. Arcamone, J., Sansa, M., and Verd, J. (2009) Nanomechanical mass sensor for spatially resolved ultrasensitive monitoring of deposition rates in stencil lithography. *Small*, **5**, 176.

69. Lopez, J.L., Verd, J., and Uranga, A. (2009) A CMOS–MEMS RF-tunable bandpass filter based on two high-Q 22-MHz polysilicon clamped-clamped beam resonators. *IEEE Electron Device Lett.*, **30**, 718.

70. Verd, J., Sansa, M., and Uranga, A. (2011) Metal microelectromechanical oscillator exhibiting ultra-high water vapor resolution. *Lab Chip*, **11**, 2670.

71. Chen, W.-C., Fang, W., and Li, S.-S. (2011) A generalized CMOS-MEMS platform for micromechanical resonators monolithically integrated with circuits. *J. Micromech. Microeng.*, **21**, 065012.

72. Englander, O., Christensen, D., and Lin, L.W. (2003) Local synthesis of silicon nanowires and carbon nanotubes on microbridges. *Appl. Phys. Lett.*, **82** (26), 4797.

73. Engstrom, D.S., Rupesinghe, N.L., and Teo, K.B.K. (2011) Vertically aligned CNT growth on a microfabricated silicon heater with integrated temperature control-determination of the activation energy from a continuous thermal gradient. *J. Micromech. Microeng.*, **21** (1), 7.

74. Kallesoe, C., Wen, C.Y., and Molhave, K. (2010) Measurement of local Si-nanowire growth kinetics using in situ transmission electron microscopy of heated cantilevers. *Small*, **6** (18), 2058.

75. Molhave, K., Wacaser, B.A., and Petersen, D.H. (2008) Epitaxial integration of nanowires in microsystems by local micrometer-scale vapor-phase epitaxy. *Small*, **4** (10), 1741.

76. Sosnowchik, B.D., Lin, L.W., and Englander, O. (2010) Localized heating induced chemical vapor deposition for one-dimensional nanostructure synthesis. *J. Appl. Phys.*, **107** (5), 14.

77. Zhang, K.L., Yang, Y., and Pun, E.Y.B. (2010) Local and CMOS-compatible synthesis of CuO nanowires on a suspended microheater on a silicon substrate. *Nanotechnology*, **21** (23), 7.

78. Englander, O. (2010) *Sensing with Locally Self-Assembled One-Dimensional Nanostructures*, SPIE-International Society for Optical Engineering, Bellingham.

79. Luo, L. and Lin, L. (2007) Self-assembled ZNO nanowires via local vapor transport synthesis as UV sensor. TRANSDUCERS 2007, International Solid-State Sensors, Actuators and Microsystems Conference, 2007, p. 403.

80. Engstrom, D.S., Savu, V., and Zhu, X.N. (2011) High throughput nanofabrication of silicon nanowire and carbon nanotube tips on afm probes by stencil-deposited catalysts. *Nano Lett.*, **11** (4), 1568.

81. Kallesoe, C., Molhave, K., and Larsen, K.F. (2010) Integration, gap formation, and sharpening of III-V heterostructure nanowires by selective etching. *J. Vac. Sci. Technol., B*, **28** (1), 21.

82. Burg, B.R., Helbling, T., and Hierold, C. (2011) Piezoresistive pressure sensors

with parallel integration of individual single-walled carbon nanotubes. *J. Appl. Phys.*, **109** (6), 064310.

83. Sillanpää, M.A., Sarkar, J., and Sulkko, J. (2009) Accessing nanomechanical resonators via a fast microwave circuit. *Appl. Phys. Lett.*, **95**, 011909.

84. Kacem, N., Arcamone, J., and Perez-Murano, F. (2010) Dynamic range enhancement of nonlinear nanomechanical resonant cantilevers for highly sensitive NEMS gas/mass sensor applications. *J. Micromech. Microeng.*, **20** (4), 045023.

85. Tosolini, G., Villanueva, G., and Perez-Murano, F. (2010) Silicon microcantilevers with MOSFET detection. *Microelectron. Eng.*, **87**, 1245.

86. Wenzler, J.-S., Dunn, T., and Erramilli, S. (2009) Nanoelectromechanical system-integrated detector with silicon nanomechanical resonator and silicon nanochannel field effect transistor. *Appl. Phys. Lett.*, **105**, 094308.

87. Unterreithmeier, P., Faust, T., and Manus, S. (2010) On-chip interferometric detection of nanomechanical motion. *Nano Lett.*, **10**, 887.

88. Fong, K.Y., Pernice, W.H.P., and Li, M. (2010) High Q optomechanical resonators in silicon nitride nanophotonic circuits. *Appl. Phys. Lett.*, **97**, 073112.

89. Li, M., Pernice, W.H.P., and Tang, H.X. (2010) Ultrahigh-frequency nano-optomechanical resonators in slot waveguide ring cavities. *Appl. Phys. Lett.*, **95**, 183110.

90. Lee, J., Shen, W., and Payer, K. (2010) Toward attogram mass measurements in solution with suspended nanochannel resonators. *Nano Lett.*, **10**, 2537.

91. Sadek, A.S., Karabalin, R.B., and Du, J. (2010) Wiring nanoscale biosensors with piezoelectric nanomechanical resonators. *Nano Lett.*, **10**, 1769.

92. Scheible, D.V. and Blick, R.H. (2010) A mode-locked nanomechanical electron shuttle for phase-coherent frequency conversion. *New J. Phys.*, **12**, 023019.

93. Feng, X.L., Matheny, M.H., and Zorman, C.A. (2010) Low voltage nanoelectromechanical switches based on silicon carbide nanowires. *Nano Lett.*, **10**, 2891.

94. Ayari, A., Vincent, P., and Perisanu, S. (2007) Self-oscillations in field emission nanowire mechanical resonators: a nanometric dc-ac conversion. *Nano Lett.*, **7**, 2252.

95. Perisanu, S., Barois, T., and Poncharal, P. (2011) The mechanical resonances of electrostatically coupled nanocantilevers. *Appl. Phys. Lett.*, **98**, 063110.

96. Colinet, E., Duraffourg, L., and Labarthe, S. (2009) Self-oscillation conditions of a resonant nanoelectromechanical mass sensor. *J. Appl. Phys.*, **105**, 124908.

97. Huang, W.-L., Ren, Z., and Lin, Y.-W. (2008) Technical Digest, 21st IEEE International Conference on Micro Electro Mechanical Systems (MEMS'08), Tucson, AZ, January 13–17, 2008, p. 10.

98. Verd, J., Uranga, A., and Abadal, G. (2008) Monolithic CMOS MEMS oscillator circuit for sensing in the attogram range. *IEEE Electron Device Lett.*, **29**, 146.

99. Zalalutdinov, M.K., Cross, J.D., and Baldwin, J.W. (2010) CMOS-integrated RF MEMS resonators. *J. Microelectromech. Syst.*, **19**, 807.

100. Roodenburg, D., Spronck, J.W., and van der Zant, H.S.J. (2009) Buckling beam micromechanical memory with on-chip readout. *Appl. Phys. Lett.*, **94** (18), 183501.

101. Liu, T.-J.K., Jeon, J., and Nathanael, R. (2010) Prospects for MEM logic switch technology. IEEE International Electron Devices Meeting (IEDM'10) Technical Digest, December 2010, pp. 424–427.

102. Chen, F., Spencer, M., and Nathanael, R. (2010) IEEE International Solid-State Circuits Conference, p. 150.

103. Akarvardar, K., Elata, D., and Parsa, R. (2009) Design considerations for complementary nanoelectromechanical logic gates. In Electron Devices Meeting IEDM, 2009, p. 299.

104. Akarvardar, K. and Wong, H.S.P. (2009) Analog nanoelectromechanical relay with tunable transconductance. *IEEE Electron Device Lett.*, **30**, 1143.

105. Postma, H.W.C., Kozinsky, I., and Husain, A. (2005) Dynamic range of

nanotube- and nanowire-based electromechanical systems. *Appl. Phys. Lett.*, **86** (22), 223105.

106. Kozinsky, I., Postma, H.W.C., and Bargatin, I. (2006) Tuning nonlinearity, dynamic range, and frequency of nanomechanical resonators. *Appl. Phys. Lett.*, **88** (25), 253101.

107. Nichol, J.M., Hemesath, E.R., and Lauhon, L.J. (2009) Controlling the nonlinearity of silicon nanowire resonators using active feedback. *Appl. Phys. Lett.*, **95** (12), 123116.

108. Karabalin, R.B., Cross, M.C., and Roukes, M.L. (2009) Nonlinear dynamics and chaos in two coupled nanomechanical resonators. *Phys. Rev. B*, **79** (16), 165309.

109. Chen, Q., Huang, L., and Lai, Y.-C. (2010) Extensively chaotic motion in electrostatically driven nanowires and applications. *Nano Lett.*, **10** (2), 406.

110. Hälg, B. (1990) On a microelectromechanical nonvolatile memory cell. *IEEE Trans. Electron Devices*, **37** (10), 2230.

111. Charlot, B., Sun, W., and Yamashita, K. (2008) Bistable nanowire for micromechanical memory. *J. Micromech. Microeng.*, **18** (4), 045005.

112. Weick, G., von Oppen, F., and Pistolesi, F. (2011) Euler buckling instability and enhanced current blockade in suspended single-electron transistors. *Phys. Rev. B*, **83** (3), 035420.

113. Badzey, R.L., Zolfagharkhani, G., and Gaidarzhy, A. (2004) A controllable nanomechanical memory element. *Appl. Phys. Lett.*, **85** (16), 3587.

114. Venstra, W.J., Westra, H.J.R., and van der Zant, H.S.J. (2010) Mechanical stiffening, bistability, and bit operations in a microcantilever. *Appl. Phys. Lett.*, **97** (19), 193107.

115. Perisanu, S., Barois, T., and Ayari, A. (2010) Beyond the linear and Duffing regimes in nanomechanics: circularly polarized mechanical resonances of nanocantilevers. *Phys. Rev. B*, **81** (16), 165440.

116. Mahboob, I. and Yamaguchi, H. (2008) Bit storage and bit flip operations in an electromechanical oscillator. *Nat. Nanotechnol.*, **3** (5), 275.

117. Mahboob, I., Froitier, C., and Yamaguchi, H. (2010) A symmetry-breaking electromechanical detector. *Appl. Phys. Lett.*, **96** (21), 213103.

118. Cross, M.C., Rogers, J.L., and Lifshitz, R. (2006) Synchronization by reactive coupling and nonlinear frequency pulling. *Phys. Rev. E*, **73** (3), 036205.

119. Mahboob, I., Flurin, E., and Nishiguchi, K. (2011) Interconnect-free parallel logic circuits in a single mechanical resonator. *Nat. Commun.*, **2**, 198.

120. Freeman, M. and Hiebert, W. (2008) NEMS - taking another swing at computing. *Nat. Nanotechnol.*, **3** (5), 251.

121. Gammaitoni, L., Hanggi, P., and Jung, P. (1998) Stochastic resonance. *Rev. Mod. Phys.*, **70** (1), 223.

122. Shim, S.-B., Imboden, M., and Mohanty, P. (2007) Synchronized oscillation in coupled nanomechanical oscillators. *Science*, **316** (5821), 95.

123. Badzey, R.L. and Mohanty, P. (2005) Coherent signal amplification in bistable nanomechanical oscillators by stochastic resonance. *Nature*, **437** (7061), 995.

124. Almog, R., Zaitsev, S., and Shtempluck, O. (2007) Signal amplification in a nanomechanical Duffing resonator via stochastic resonance. *Appl. Phys. Lett.*, **90** (1), 013508.

125. Guerra, D.N., Dunn, T., and Mohanty, P. (2009) Signal amplification by 1/f noise in silicon-based nanomechanical resonators. *Nano Lett.*, **9** (9), 3096.

126. Cottone, F., Vocca, H., and Gammaitoni, L. (2009) Nonlinear energy harvesting. *Phys. Rev. Lett.*, **102** (8), 080601.

127. van der Zande, M., Barton, R.A., and Alden, J.S. (2010) Large-scale arrays of single-layer graphene resonators. *Nano Lett.*, **10** (12), 4869.

128. Poot, M. and van der Zant, H. (2011) Mechanical systems in the quantum regime. *Phys. Rep.*, **511** (5), 273.

129. Hüttel, A.K., Steele, G.A., and Witkamp, B. (2009) Carbon nanotubes as ultra-high quality factor mechanical resonators. *Nano Lett.*, **9**, 2547.

130. Babić, B., Furer, J., and Sahoo, S. (2003) Intrinsic thermal vibrations of suspended doubly clamped singe-wall carbon nanotubes. *Nano Lett.*, **3**, 1577.

131. Poncharal, P., Wang, Z.L., and Ugarte, D. (2003) Electrostatic deflections and electromechanical resonances of carbon nanotubes. *Science*, **283** (199), 1513.

132. Purcell, S.T., Vincent, P., and Journet, C. (2002) Tuning of nanotube mechanical resonances by electric field pulling. *Phys. Rev. Lett.*, **89**, 276103.

133. Jensen, K., Weldon, J., and Garcia, H. (2007) Nanotube radio. *Nano Lett.*, **7**, 3508.

134. Lassagne, B., Garcia-Sanchez, D., and Aguasca, A. (2008) Ultrasensitive mass sensing with a nanotube electromechanical resonator. *Nano Lett.*, **8**, 3735.

135. Witkamp, B., Poot, M., and vanderZant, H.S.J. (2006) Bending-mode vibration of a suspended nanotube resonator. *Nano Lett.*, **6**, 2904.

136. Gouttenoire, V., Barois, T., and Perisanu, S. (2010) Digital and FM demodulation of a doubly clamped single-walled carbon-nanotube oscillator: towards a nanotube cell phone. *Small*, **6**, 1060.

137. Witkamp, B., Poot, M., and Pathangi, H. (2008) Self-detecting gate-tunable nanotube paddle resonators. *Appl. Phys. Lett.*, **93**, 111909.

138. Lassagne, B., Tarakanov, Y., and Kinaret, J. (2009) Coupling mechanics to charge transport in carbon nanotube mechanical resonators. *Science*, **325**, 1107.

139. Eichler, A., Moser, J., and Chaste, J. (2011) Nonlinear damping in mechanical resonators made from carbon nanotubes and graphene. *Nat. Nanotechnol.*, **6**, 339.

140. Jensen, K., Kim, K., and Zettl, A. (2008) An atomic-resolution nanomechanical mass sensor. *Nat. Nanotechnol.*, **3**, 533.

141. Ekinci, K.L., Yang, Y.T., and Roukes, M.L. (2004) Ultimate limits to inertial mass sensing based upon nanoelectromechanical systems. *J. Appl. Phys.*, **95**, 2682.

142. Squires, T.M., Messinger, R.J., and Manalis, S.R. (2008) Making it stick: convection, reaction and diffusion in surface-based biosensors. *Nat. Biotechnol.*, **26**, 417.

143. Ilic, B., Yang, Y., and Aubin, K. (2005) Enumeration of DNA molecules bound to a nanomechanical oscillator. *Nano Lett.*, **5**, 925.

144. Ilic, B., Yang, Y., and Craighead, H.G. (2004) Virus detection using nanoelectromechanical devices. *Appl. Phys. Lett.*, **85**, 2604.

145. Sheehan, P.E. and Whitman, L.J. (2005) Detection limits for nanoscale biosensors. *Nano Lett.*, **5**, 803.

146. Nair, P.R. and Alam, M.A. (2006) Performance limits of nanobiosensors. *Appl. Phys. Lett.*, **88**, 698.

147. Sampathkumar, A., Ekinci, K.L., and Murray, T.W. (2011) Multiplexed optical operation of distributed nanoelectromechanical systems arrays. *Nano Lett.*, **11**, 1014.

148. Li, M., Bhiladvala, R.B., and Morrow, T.J. (2008) Bottom-up assembly of large-area nanowire resonator arrays. *Nat. Nanotechnol.*, **3**, 88.

149. Agrawal, R. and Espinosa, H.D. (2011) Giant piezoelectric size effects in zinc oxide and gallium nitride nanowires. A first principles investigation. *Nano Lett.*, **11**, 786.

150. Xiang, H.J., Yang, J., and Hou, J.G. (2006) Piezoelectricity in ZnO nanowires: a first-principles study. *Appl. Phys. Lett.*, **89**, 223111.

151. Sun, C., Shi, J., and Wang, X. (2010) Fundamental study of mechanical energy harvesting using piezoelectric nanostructures. *J. Appl. Phys.*, **108**, 034309.

152. Yang, R., Qin, Y., and Li, C. (2009) Converting biomechanical energy into electricity by a muscle-movement-driven nanogenerator. *Nano Lett.*, **9**, 1201.

153. Li, Z., Zhu, G., and Yang, R. (2010) Muscle-driven in vivo nanogenerator. *Adv. Mater.*, **22**, 2534.

154. Zhu, G., Yang, R., and Wang, S. (2010) Flexible high-output nanogenerator based on lateral ZnO nanowire array. *Nano Lett.*, **10**, 3155.

155. Hu, Y., Zhang, Y., and Xu, C. (2010) High-output nanogenerator by rational

unipolar assembly of conical nanowires and its application for driving a small liquid crystal display. *Nano Lett.*, **10**, 5025.

156. Kumar, B. and Kim, S.-W. (2012) Energy harvesting based on semiconducting piezoelectric ZnO nanostructures. *Nano Energy*, **1** (3), 342.

157. Qi, Y., Kim, J., and Nguyen, T.D. (2011) Enhanced piezoelectricity and stretchability in energy harvesting devices fabricated from buckled PZT ribbons. *Nano Lett.*, **11**, 1331.

158. Chen, X., Xu, S., and Yao, N. (2010) 1.6 V nanogenerator for mechanical energy harvesting using PZT nanofibers. *Nano Lett.*, **10**, 2133.

159. Chang, C., Tran, V.H., and Wang, J. (2010) Direct-write piezoelectric polymeric nanogenerator with high energy conversion efficiency. *Nano Lett.*, **10**, 726.

160. Chang, J., Dommer, M., and Chang, C. (2012) Piezoelectric nanofibers for energy scavenging applications. *Nano Energy*, **1** (3), 356.

161. Wang, X., Kim, K., and Wang, Y. (2010) Matrix-assisted energy conversion in nanostructured piezoelectric arrays. *Nano Lett.*, **10**, 4901.

162. Chandratre, S. and Sharma, P. (2012) Coaxing graphene to be piezoelectric. *Appl. Phys. Lett.*, **100** (2), 023114.

163. Qi, J., Qian, X., and Qi, L. (2012) Strain-engineering of band gaps in piezoelectric boron nitride nanoribbons. *Nano Lett.*, **12** (3), 1224.

164. Ong, M.T. and Reed, E.J. (2011) Engineered piezoelectricity in graphene. *ACS Nano*, **6** (2), 1387.

10
Organic Resonant MEMS Devices

Silvan Schmid

10.1
Introduction

With the introduction of SU-8 [1], an epoxy-type highly crosslinked photoresist, polymer materials started to be used as the actual building material of micromechanical sensors. Before, polymers had, besides acting as photoresist in photolithography, been mainly used as sensitive receptor layers in silicon-based micro gas sensors [2–5]. As a structural material, polymers benefit from their compliance compared to typical MEMS materials such as silicon. The relatively low stiffness enables micromechanical sensors with large deflections for small applied forces. Polymers have been used as the primary structural material, for example, in micromanipulators [6–9] or in large displacement microvalves in microfluidic systems [10–12]. Hybrid Si-polymer actuators, allowing for large strokes of a stiff silicon actuator element suspended with a soft metallized polymer hinge [13, 14], have successfully been used as a millimeter-wave phase shifter [15, 16] or as a power divider [17]. The metallized SU-8 hinges have been shown to withstand 800 million cycles without fatigue failure. Another application in which a flexible structure capable of large displacement amplitudes is desired is in micromechanical resonant energy harvesting. This specific application is discussed in Section 10.4.2.

Due to their low flexural rigidity, quasi-static polymeric micromechanical sensors can be more sensitive than Si-based static-mode sensors [18]. This has led to the development of SU-8 microcantilever sensors for biochemical sensing [19, 20] or stress sensing [21], and for soft probes in scanning force microscopy [22].

The present chapter is focused on organic micromechanical resonators. The figure of merit of a mechanical resonator is typically the quality factor (Q). In a high-Q resonator, the vibrational energy is efficiently stored, such that only a relatively minute amount of energy is lost during a cycle of oscillation. At high Q, the resonance peak is sharp, that is, the resonance frequency is defined with high fidelity. Hence, a high Q is desired for most resonator applications, especially for micro and nanomechanical resonant sensors for which the sensor's resolution Δx (limit of detection) is limited by the smallest change in resonance frequency

Resonant MEMS – Fundamentals, Implementation and Application, First Edition.
Edited by Oliver Brand, Isabelle Dufour, Stephen M. Heinrich and Fabien Josse.
© 2015 Wiley-VCH Verlag GmbH & Co. KGaA. Published 2015 by Wiley-VCH Verlag GmbH & Co. KGaA.

Δf that can be detected. Therefore, materials with low intrinsic damping, such as crystalline silicon, are typically used as the primary structural material for micro and nanomechanical resonators.

Polymers are known to exhibit large intrinsic damping and are therefore seldom the first choice for the design of mechanical resonators. But the superiority in Q of Si-based microresonators applies only for applications in which the intrinsic damping mechanisms are dominant, that is in vacuum or near-vacuum conditions. For many applications, for example, chemical sensing, the micromechanical resonators are operated in gas or liquid in which case, external viscous damping is responsible for the majority of the energy loss. In such a viscous environment, the quality factor is mainly a function of the resonator geometry and viscosity of the medium rather than the intrinsic energy loss. Hence, in viscous media, polymeric micromechanical resonators can show quality factor values of similar magnitude as silicon based resonators, with the exception of, for example, bulk-mode resonators such as quartz crystal microbalances (QCM) or thin-film bulk acoustic resonators (FBAR), which are designed to minimize viscous damping. Furthermore, in the case of resonant sensors, the resolution Δx of a measurement unit is not only given by the frequency resolution Δf but also the sensitivity $S = \partial f / \partial x$ (change of frequency per unit change of the measurement unit x).

$$\Delta x = S^{-1} \Delta f \qquad (10.1)$$

A micromechanical resonant sensor that is entirely built from a sensitive polymer can exhibit a superior sensitivity compared to a traditional Si sensor with a polymer coating and, thus, by Eq. (10.1), can yield a better limit of detection for a given frequency resolution. For example, it was shown that fully polymeric resonant micromechanical humidity sensors can have an unprecedented resolution [23]. In addition, the application of SU-8 microresonators as biomolecular sensors working in an aqueous buffer has been demonstrated [24], as have humidity sensing applications, which will be discussed in Section 10.4.1.

In special cases, micromechanical resonators with moderate-to-low-quality factors are required, for example, for mimicking the human cochlea for the design of auditory prostheses. The cochlea is the spiral-shaped cavity inside the inner ear in which the hair cells are located, which are responsible for detecting frequency-specific audio waves. Low-Q polymer micromechanical resonators have been shown to be well suited to mimic the behavior of these hair cells [25, 26]. This special application of polymeric microresonators is discussed in Section 10.4.3.

The stiffness (Young's modulus) of polymers is not only frequency-dependent (see Figure 10.1a) but also strongly temperature-sensitive, as shown in Figure 10.1b. The Young's modulus of SU-8 was found to soften by $-0.31\%°C^{-1}$, which results in a frequency detuning of $-0.17\%°C^{-1}$. This is a substantial temperature sensitivity, which makes the design of reliable polymeric microsystems challenging. Despite some possible sensitivity advantages, polymeric micromechanical resonators have remained a niche with only a handful of devices introduced to date. Section 10.2 gives an overview of these few existing

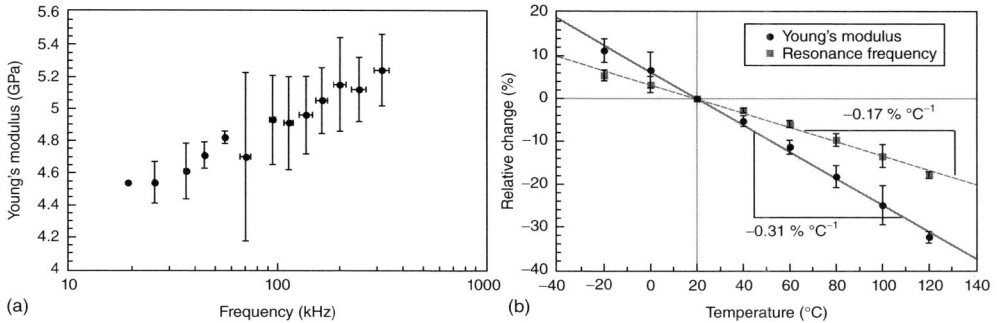

Figure 10.1 (a) The mean Young's modulus evaluated by iterative FEM from the resonant frequency of 38 SU-8 microcantilevers with lengths from 30 to 140 µm, measured in high vacuum at 25 °C. (b) The average relative Young's modulus and resonant frequency f change of 35 SU-8 microcantilevers with lengths from 13 to 180 µm with respect to the reference temperature $T_0 = 20 °C$. (Reproduced with permission from Ref. [27]).

polymeric micromechanical resonator designs. In Section 10.3, the quality factor in these highly damped resonators is discussed, followed by Section 10.4, in which three applications of organic micromechanical resonators are presented.

10.2
Device Designs

This section introduces four design schemes for polymeric micromechanical resonators. The schemes showcase different actuation methods and the specific fabrication techniques involved.

10.2.1
Conductive Polymer with Electrostatic Actuation

The pioneering work on polymeric micromechanical resonators was published by Conde's group in 2005 [28]. The use of a conductive polymer material enabled the application of the common electrostatic actuation scheme. An SEM image of a resonator and a schematic drawing explaining the transduction principle are shown in Figure 10.2a and b. The electrostatic actuation force was generated by applying a driving signal between the conductive polymer resonator and the metal gate on the glass wafer. The vibrational response was read out optically by the laser deflection method.

The conductive micromechanical resonators were fabricated from a polymer sandwich structure consisting of poly(3,4-ethylenedioxythiophene) / polystyrene sulfonate / polymethyl metacryalte (PEDOT/PSS/PMMA) [28 – 30]. The fabrication process was based on aluminum as a sacrificial layer material. The corresponding fabrication process is depicted in Figure 10.2c.

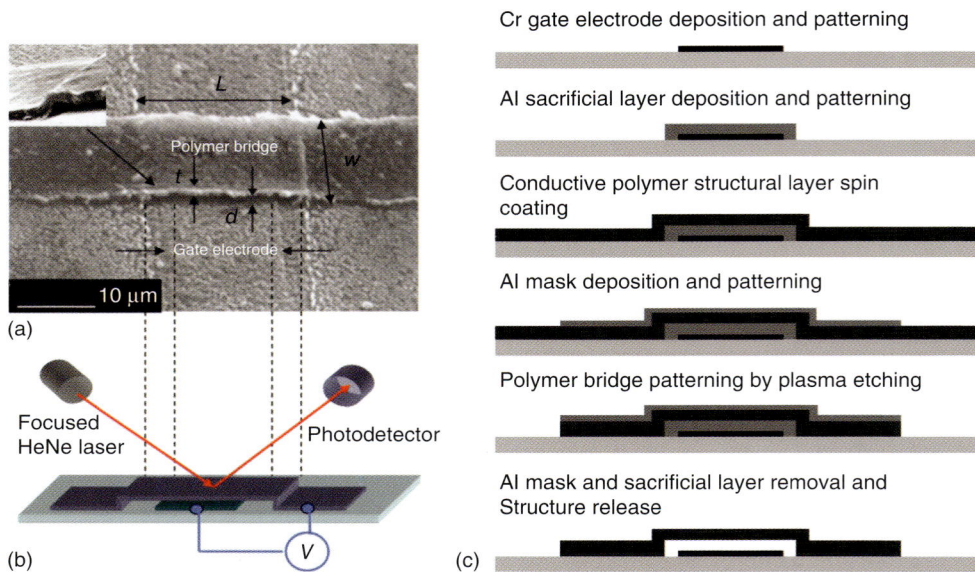

Figure 10.2 (a) SEM micrograph of a conductive polymer (PEDOT/PSS/PMMA) microresonator. (b) Schematic drawing of electromechanical polymer microresonator with electrostatic actuation and optical read-out. (c) Fabrication outline for conductive polymer micromechanical resonators. (Reproduced with permission from Ref. [29].)

The properties of polymeric resonators can be tuned by varying the amount of cross-linker [30]. Larger amounts of cross-linker have been shown to improve the polymer conductivity and its long-term stability. But the addition of cross-linker unfortunately results in a reduction of the quality factor. In order to enhance the mechanical properties, polyimide (PI)-reinforced PEDOT/PSS/PMMA resonators were studied [31]. The additional PI layer slightly reduced the temperature sensitivity.

A more significant improvement of the mechanical properties of the PEDOT/PSS/PMMA micromechanical resonators was achieved by adding a layer of amorphous carbon nanotubes (CNTs) [32]. The addition of CNTs increased the tensile stress in the doubly clamped resonators and thus increased the resonance frequency. The increased frequency also resulted in higher quality factors. This effect is discussed in more detail in Section 10.3.3. The CNT-related improvement of the resonance frequency and quality factor is shown in Figure 10.3.

10.2.2
Dielectric Polymer with Polarization Force Actuation

SU-8 has excellent mechanical properties and is therefore often the microme-chanical polymer of choice [1, 33, 34]. But in order to use SU-8 as the structural

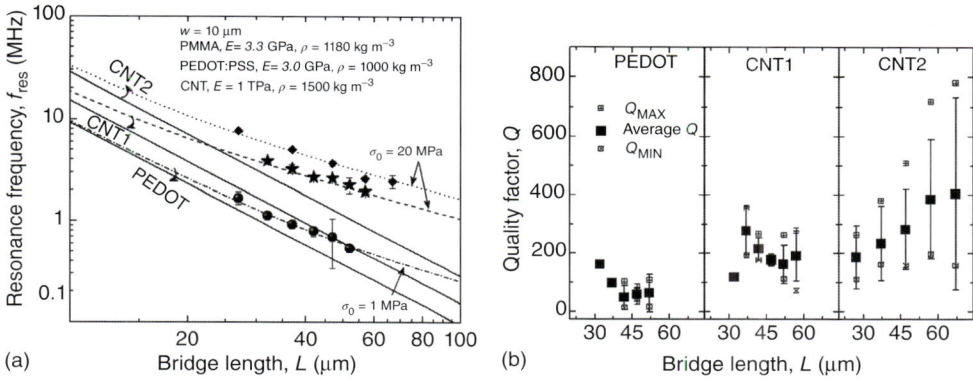

(a)

(b)

Figure 10.3 (a) Resonance frequency of conductive PEDOT micromechanical resonators with different CNT loading (blank: PEDOT, single layer of CNTs: CNT1, double layer CNTs: CNT2) as a function of length, measured in high vacuum. The solid lines represent the theoretical response of an Euler-Bernoulli beam ($\propto L^{-2}$). The dashed lines represent frequency values for resonators with a tensile stress σ. (b) Quality factor for the different polymer resonators. (Reproduced with permission from Ref. [32].)

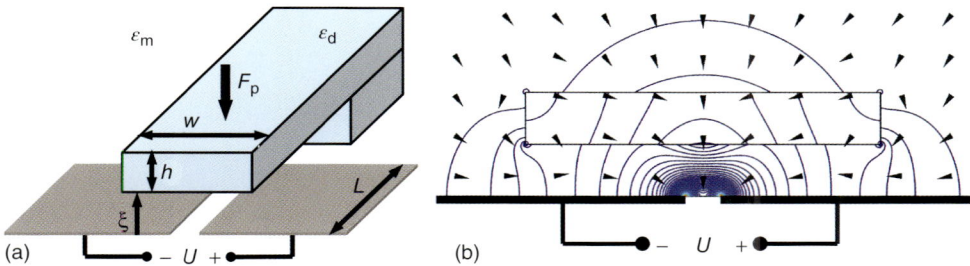

(a)

(b)

Figure 10.4 Schema for actuation based on dielectric polarization force. (a) 3D schematic and (b) 2D front view with electric field lines and field gradient direction (arrows). (Reproduced with permission from Ref. [37].)

material, an adequate actuation scheme is needed. The addition of a layer of conductive polymer would deteriorate the excellent mechanical properties of SU-8, as would a metallization [35]. Furthermore, adhesion of the metal layer to polymers can degrade over time at high actuation cycles [36].

A novel actuation scheme based on dielectric polarization forces was introduced for SU-8 micromechanical resonators in order to overcome some of these challenges [37]. This scheme does not require any modifications of the structural resonator material. The actuation scheme is shown in Figure 10.4. The dielectric micromechanical polymer resonator is placed over a pair of coplanar electrodes with a potential difference U. The resonator structure is attracted toward the highest electric field intensity, which exists over the gap of the electrodes. Assuming the beam width to be large compared to its thickness h and its distance to the electrodes ξ, the force can be simplified to [38]

$$F_\mathrm{p}(\xi) = -\frac{1}{2\pi}\varepsilon_0(\varepsilon_\mathrm{d} - \varepsilon_\mathrm{m})\alpha^2 L \frac{h}{\xi(\xi + h)}U^2 \tag{10.2}$$

Equation (10.2) indicates that in vacuum or air, F_p is negative, meaning that it acts toward the electrodes. If the dielectric is surrounded by water, which has a very high permittivity ($\varepsilon_\mathrm{m} \approx 80$), the force is positive and the beam is pushed away from the electrodes. To use this expression to determine the polarization force, the weighting factor α must be known. α depends on the dielectric constant and the thickness of the dielectric resonator material. Numerically determined values for α are published in [38].

The fabrication of such suspended SU-8 micromechanical resonators is based on a polymeric sacrificial layer made from a so-called lift-off resist (LOR). LOR does not dissolve in organic solvents and therefore does not intermix with the solvent-based SU-8 resin. There are two fabrication schemes based on LOR. The first requires the pre-patterning of the LOR, which results in micromechanical resonators with perfect clamping. The corresponding fabrication process is outlined in Figure 10.5a and a scanning electron micrograph is shown in Figure 10.6a. The second process requires fewer steps, but produces structures with suspended anchors due to an associated timed etching of the LOR layer. The corresponding fabrication scheme and image are shown in Figures 10.5b and 10.6b, c, respectively.

10.2.3
Superparamagnetic Nanoparticle Composite with Magnetic Actuation

Both the electrostatic actuation of conductive and dielectric polymer microresonators, as introduced in Sections 10.2.1 and 10.2.2, require electrodes placed in close proximity to the resonator. The small gap d between the resonator and the transduction electrodes is needed in order to achieve a sufficiently high electromechanical coupling, which is proportional to $1/d^2$ [38]. Because magnetic forces tend to have a longer range than electrostatic forces, the design of a magnetic polymeric actuator would facilitate the external actuation of the micromechanical resonator without the need of integrated transduction elements.

Magnetic micromechanical polymer resonators have been successfully built and actuated with an external coil by Suter *et al.* [39-41]. Their resonators were fabricated from a photo-patternable SU-8 composite containing superparamagnetic magnetite nanoparticles with a concentration of up to 5 vol.%. It was shown that a homeogenous dispersion of the nanoparticles in the SU-8 matrix is crucial for obtaining reproducible microresonators with few defects [42]. The developed magnetic polymer composite can be processed like a regular resist. The magnetic composite is applied by spin-coating and patterned by UV-photolithography. Because there is no need for electrodes in close proximity, the paramagnetic microcantilevers are designed as freestanding structures, being fully released from the supporting wafer. The corresponding fabrication process is schematically depicted in Figure 10.7, while the magnetic micromechanical resonator is shown in Figure 10.8. These types of polymer-based devices have

Process with perfect clamping   **Process with underetched clamping**

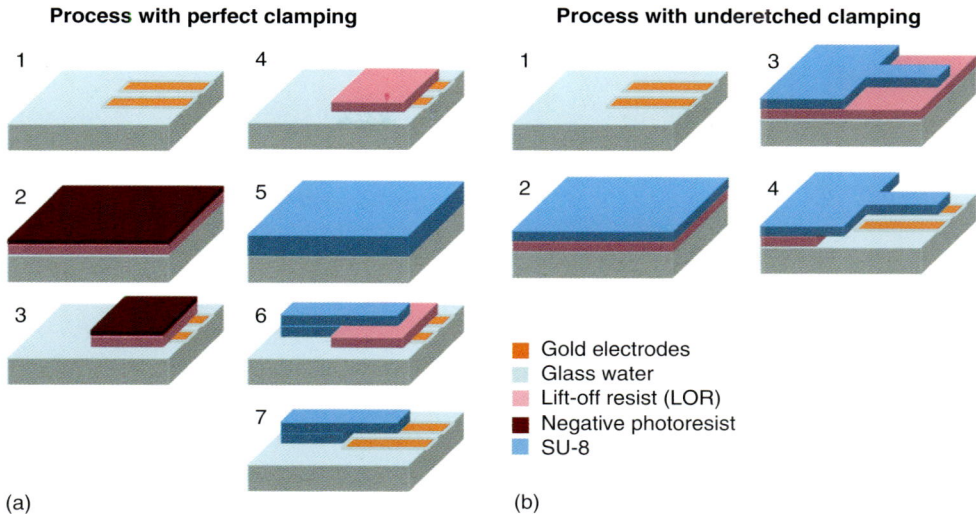

Gold electrodes
Glass water
Lift-off resist (LOR)
Negative photoresist
SU-8

(a)   (b)

Figure 10.5 (a) Fabrication outline of SU-8 micromechanical resonators for dielectric polarization actuation scheme, producing perfectly anchored structures [27]. 1, Gold electrodes patterned(lift-off) on glass wafer; 2, Spin-coating of LOR and negative resist; 3, Patterning (photo-lithographically) of negative resist and development of both, LOR and resist layer; 4, Selective removal of negative resist; 5, Spin-coating of SU-8; 6, Patterning of SU-8 (photo-lithographically); 7, Total dissolving of LOR layer. (b) Fabrication outline of SU-8 micromechanical resonators for dielectric polarization actuation scheme, producing a freestanding clamping [27]. 1, Gold electrodes patterned (lift-off) on glass wafer; 2, Spin-coating of LOR and SU-8; 3, Patterning (photo-lithographically) of SU-8 layer; 4, Partial dissolving of LOR layer..

been successfully actuated in vacuum, air, and water [40, 42]. It is also possible to replace the SU-8 with PMMA [43].

10.2.4
Metallized Polymer with Lorentz Force Actuation

In the actuation scheme discussed in the previous section, a paramagnetic micromechanical resonator was driven by an external modulated magnetic field. The generation of a high-frequency magnetic field is challenging and usually requires superconducting coils. Dubourg *et al.* have implemented a common magnetic actuation scheme to actuate polymeric microcantilevers where only a static magnetic field is required instead of a dynamic field [44]. Such a static magnetic field can be easily generated by rare-earth magnets. In this scheme, an AC current was passed through a wire at the tip of a polymeric cantilever. When placed in a perpendicularly aligned static magnetic field, a Lorentz force is generated in the direction of the desired cantilever vibrations. A schematic of the actuation scheme and the corresponding polymeric microcantilevers, featuring a metallic wire (conducting path), are shown in Figure 10.9. The fabrication of

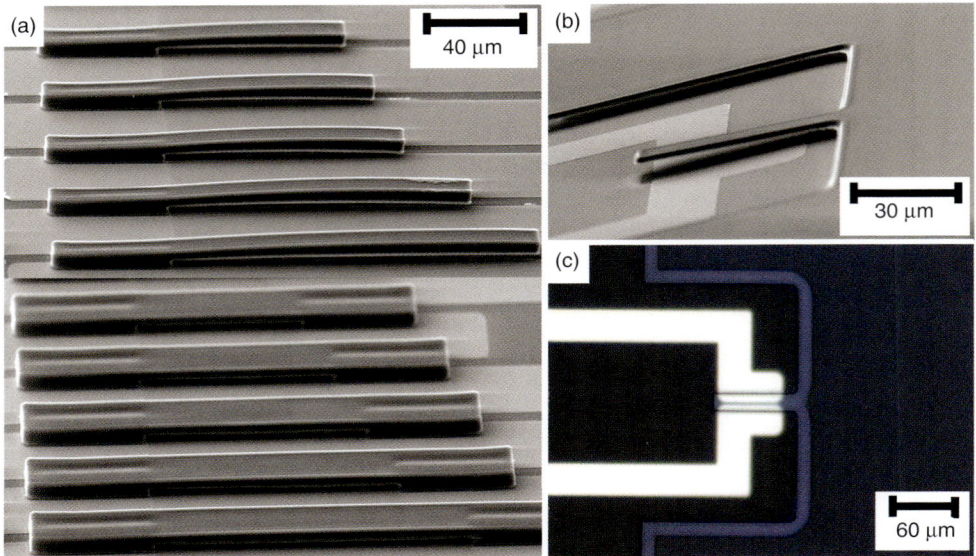

Figure 10.6 (a) SEM images of SU-8 micro-cantilevers and strings fabricated following the process outlined in Figure 10.5a. (b) SEM image of SU-8 microcantilever fabricated according to the process outlined in Figure 10.5b. (c) Microscope image of similar structure as shown in (b).

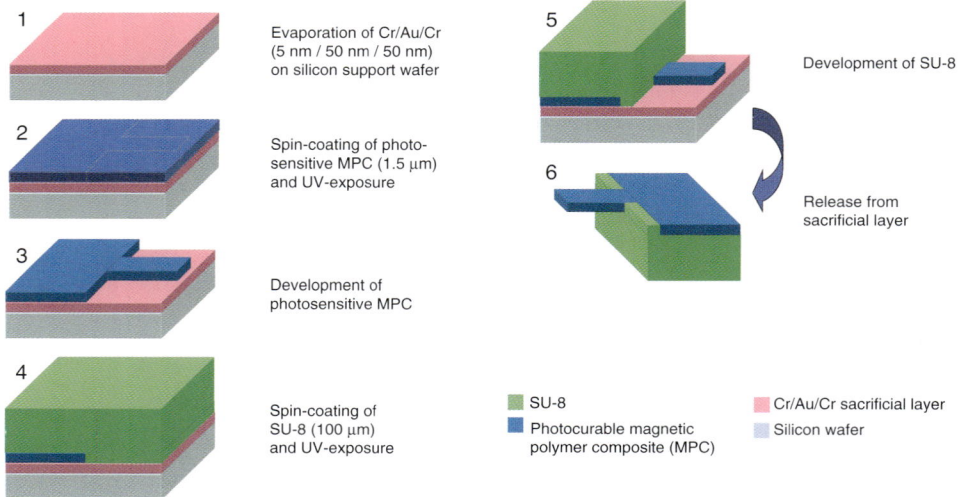

Figure 10.7 Schematic drawing of the fabrication process of micromechanical resonators made from a superparamagnetic SU-8 composite. (Reprinted with permission from Ref. [40].)

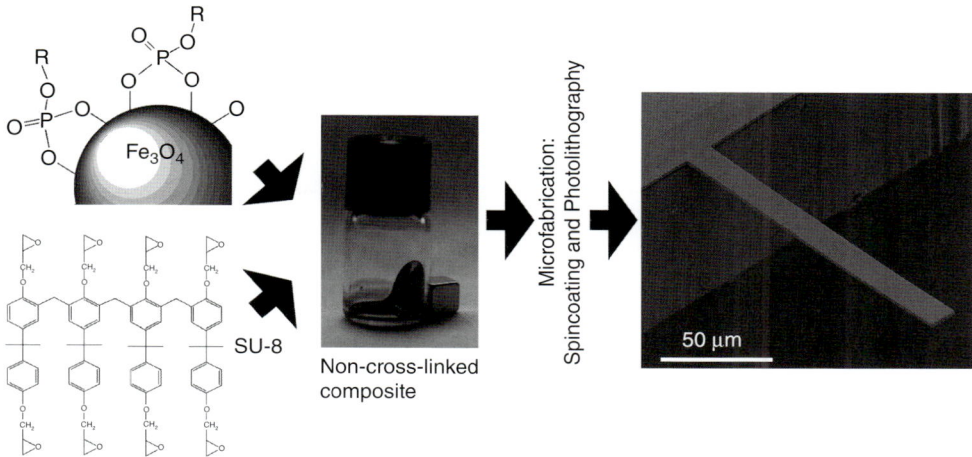

Figure 10.8 Superparamagnetic 13 nm magnetite nanoparticles are mixed with the UV-sensitive epoxy SU-8 to obtain a magnetic polymer composite (MPC). Suspended micromechanical resonators have been fabricated with the MPC using conventional microfabrication processes such as spin-coating and UV-photolithography. (Reprinted with permission from Ref. [42].)

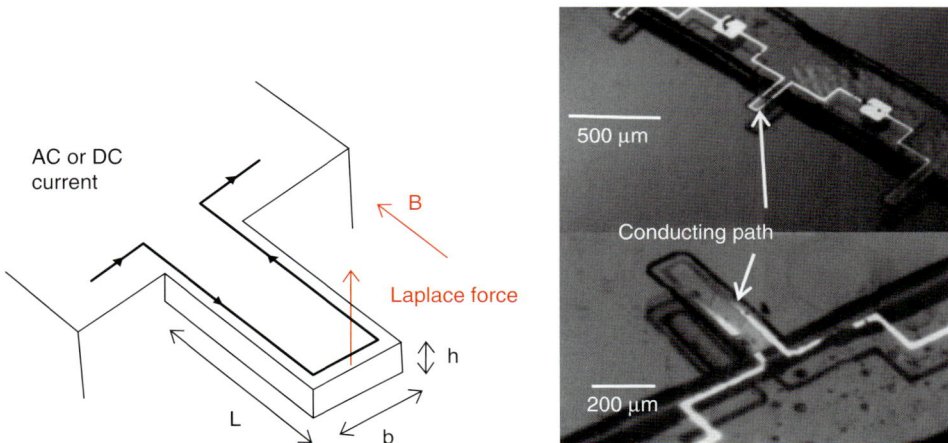

Figure 10.9 Schematic drawing and SEM images of SU-8 micromechanical resonators actuated by the Lorentz force (also called *Laplace force*). (Reproduced with permission from Ref. [44].)

these structures is based on a wafer-transfer process as it is described in [45]. The conductive path was further used to resistively heat the microresonator in order to tailor its mechanical properties. The effect of local heating on the material damping of these polymeric resonators is specifically discussed in Section 10.3.2.

10.3
Quality Factor of Polymeric Micromechanical Resonators

In this section, the quality factor of polymeric micromechanical resonators is discussed for three scenarios. In the first scenario, the resonator is immersed in a viscous environment, such as gas or a liquid. The second and third subsections discuss the scenario of relaxed (no residual stress) and unrelaxed (with residual stress) polymeric micromechanical resonators in vacuum. Relaxed structures are, for example, beams and plates, whose mechanical behavior is dominated by their flexural rigidity. Unrelaxed structures are, for example, strings and membranes, and their mechanics is dominated by tensile stress.

10.3.1
Quality Factor in Viscous Environment

When driven in a viscous environment, polymeric micromechanical resonators have been shown to be dominantly damped by viscous forces when operated in the conventional transverse flexural mode of vibration [46]. The resulting quality factors can be calculated by specific viscous damping models in gas [47] and liquid [48, 49] (see Chapter 2 for more details).

10.3.2
Quality Factor of Relaxed Resonators in Vacuum

In this section, the polymer-specific, high-intrinsic material damping and its relation to the quality factor of relaxed micromechanical resonators are discussed. When operating an organic resonator in vacuum, the quality factor has been shown to be mainly limited by intrinsic material damping [46].

Polymers are viscoelastic such that their response is not immediate, that is, the deformation induced by an external force takes time to occur due to the partly viscous material behavior. In a viscoelastic mechanical resonator, the displacement-induced stress is sinusoidal. For such an oscillatory stress $\sigma(t)$, the accompanying strain $\varepsilon(t)$ lags behind the stress by the phase δ [50].

$$\varepsilon = \varepsilon_0 \sin(\omega t)$$
$$\sigma = \sigma_0 \sin(\omega t + \delta) \tag{10.3}$$

The stress can be expanded to

$$\sigma = \sigma_0 \cos\delta \ \sin\omega t + \sigma_0 \sin\delta \ \cos\omega t \tag{10.4}$$

By defining two quantities

$$E' = \frac{\sigma_0}{\varepsilon_0} \cos\delta \ , \quad E'' = \frac{\sigma_0}{\varepsilon_0} \sin\delta \tag{10.5}$$

Equation (10.4) becomes

$$\sigma = \varepsilon_0 E' \sin\omega t + \varepsilon_0 E'' \cos\omega t \tag{10.6}$$

It can be seen that the E' term is in phase with the strain and the E'' term is $\pi/2$ out of phase with the strain. E' is called the *storage Young's modulus* and E'' is called the *loss Young's modulus*. The ratio of the two is

$$\frac{E''}{E'} = \frac{\sin \delta}{\cos \delta} = \tan \delta \qquad (10.7)$$

which is called the *loss tangent*.

The quality factor is defined in terms of the energy stored versus the energy lost during one cycle of oscillation

$$Q = 2\pi \frac{W}{\Delta W} \qquad (10.8)$$

Thus, the quality factor can be derived by calculating both the energy stored and energy lost during one cycle of oscillation in a volume unit of the resonator material. When doing so, the quality factor (Eq. (10.8)) due to intrinsic material loss, Q_{mat}, becomes [50]

$$Q_{mat} = \frac{E'}{E''} = \frac{1}{\tan \delta} \qquad (10.9)$$

Material damping in viscoelastic materials is frequency- and temperature-dependent. The energy loss is connected to submolecular vibrations, which are activated (absorb energy) at specific frequencies and temperatures [51]. Such activations, also called *transitions*, cause a loss tangent peak. The main transition, or α transition, in a polymer is the so-called glass transition at which the polymer changes from a frozen or crystalline state to an amorphous state. Figure 10.10a shows the loss tangent of SU-8 measured at different frequencies and temperatures. The data were obtained from the corresponding quality factors of SU-8 microcantilevers. SU-8 has a glass transition temperature of $T_g = 210\,^\circ C$. It can be seen that SU-8 has an inter-transition zone with minimum material damping at around $90\,^\circ C$. Dubourg *et al.* have used this temperature zone to minimize material damping of a micromechancial SU-8 cantilever [44]. The SU-8 microcantilevers of that study were heated locally by passing a current through an integrated metal wire in order to reach the inter-transition temperature (Figure 10.9), thereby doubling Q of the second flexural mode, as shown in Figure 10.10b.

10.3.3
Quality Factor of Unrelaxed Resonators in Vacuum

In two separate studies, it was demonstrated that both doubly clamped SiN [52] and SU-8 [46] resonators with a tensile pre-stress exhibited exceptionally high quality factors. When the tensile stress of these resonators is dominating the mechanical behavior over the flexural stress such structural elements are called *strings*. Q values of up to 200 000 have been measured for doubly clamped SiN nanostrings, while Q was lower than 6000 for relaxed singly clamped SiN cantilevers of similar dimensions. Similarly, Q values of up to 800 were measured for SU-8 microstrings, as can be seen in Figure 10.11a (values of SU-8 microstrings).

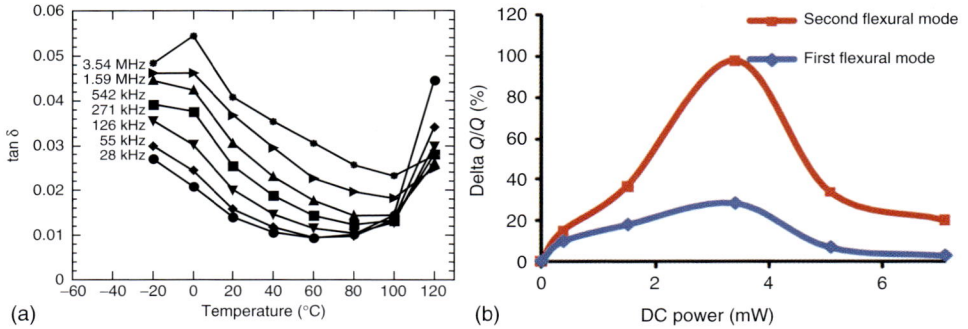

(a)

(b)

Figure 10.10 (a) Average loss tangents of 3 μm thick SU-8 cantilevers with lengths from 15 to 185 μm as a function of temperature. The values were measured at vacuum below 0.05 Pa. (Reproduced with permission from Ref. [46]. (b) Relative quality factor change of an SU-8 micromechanical cantilever resonator as a function of DC heating power. The initial quality factors of the microcantilevers are 16 and 18 for the first flexural mode and the second flexural mode, respectively. The first mode resonance frequency is approximately 17 kHz. (Reproduced with permission from Ref. [44].)

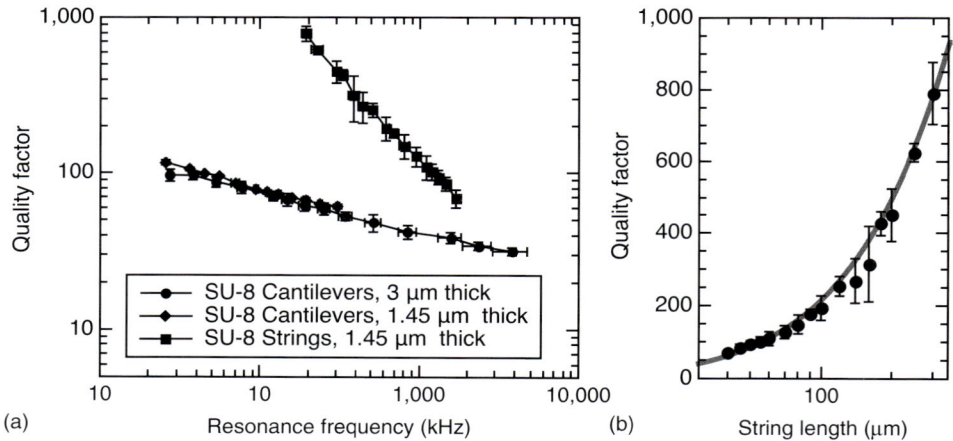

(a)

(b)

Figure 10.11 (a) Quality factor comparison of SU-8 microcantilevers and SU-8 microstrings. (b) Quality factors of the SU-8 microstrings as a function of the string length (same data as shown in (a)). The gray line represents the string quality factor model (Eq. (10.15)), for a tensile stress of 19 MPa, a Young's modulus of 5.2 GPa (see Figure 10.1) and a loss tangent of 0.025 (which is equal to a quality factor of 40). The measurements were done in high vacuum at room temperature. (The dataset is taken from Ref. [46].)

This was especially peculiar since the intrinsic material damping of the polymeric micromechanial resonators allowed for maximum Qs of approximately 100 (as shown in Figure 10.11a). In this section, the effect of a tensile stress on Q for a mechanical resonator is discussed in detail.

As discussed in Section 10.3.2, in a relaxed polymeric resonator such as a cantilever, energy is stored in the bending-related deformation. But in an unrelaxed mechanical resonator involving a tensile-stress, not only is the energy stored in the bending-related deformation, but also a large amount of energy is associated with the tensile stress. This additional stored energy must be added in the calculation of Q (10.8). Neglecting the contribution of energy loss due to the elongational stretching of the resonator, the quality factor of a polymeric microstring, Q_{string}, can be described by Schmid and Hierold [46]

$$Q_{string} = 2\pi \frac{W_{tension} + W_{bending}}{\Delta W_{bending}}$$

$$\approx \frac{W_{tension}}{W_{bending}} \cdot Q_{bending} \qquad (10.10)$$

where $W_{tension}$ is the stored elastic energy associated with string tension and $\Delta W_{bending}$ and $W_{bending}$ are the lost and stored energy, respectively, due to bending. According to this equation, it is obvious that Q_{string} is enhanced by the stored tensile energy. The approximation in the second line is based on the assumption that the magnitude of the tensile pre-stress is such that it dominates the bending contribution to the stored energy. $Q_{bending}$ is the quality factor due to bending related damping mechanisms in the relaxed state. Since the material damping is the dominating energy loss in polymeric micromechanical resonators $Q_{bending} \approx Q_{mat}$ [46]. The tensile-stress and bending contributions to the stored energy can be calculated, knowing the shape of deflection $u(x)$, and Eq. (10) can be written as [53]

$$Q_{string} \approx \frac{\frac{1}{2}\sigma A \int_0^L \left[\frac{\partial}{\partial x} u(x)\right]^2 dx}{\frac{1}{2}EI_z \int_0^L \left[\frac{\partial^2}{\partial x^2} u(x)\right]^2 dx} \cdot Q_{mat} \qquad (10.11)$$

where σ is the tensile stress, A the cross-section area, L is the length, E the storage Young's modulus, and I_z the geometrical moment of inertia. From (10.11) it can be seen that Q_{string} is a function of the tensile stress and the curvature of $u(x)$.

The effect of the additional energy stored in the tension described in Eq. (10) can also be examined from a different perspective. The quality factor may be calculated as the resonance frequency f_r divided by the peak width Δf [47]

$$Q = \frac{f_r}{\Delta f} = 2\pi \frac{W}{\Delta W} \qquad (10.12)$$

The axial tension increases the resonance frequency (which is related to the stored energy) without significantly influencing the peak width (which is related to the damping or energy lost).

For an ideal string, the flexural rigidity is zero and the corresponding mode shape is sinusoidal [54]. In reality, however a micro or nanomechanical string consists of a beam with a specific (non-zero) flexural rigidity. If the tensile stress in such a doubly clamped beam is sufficiently large, that is, if $\sigma \gg (n\pi)^2 Eh^2/(12L^2)$ [55], where n is the mode number and h is the beam thickness, the axial stress is dominating the mechanical behavior and the flexural rigidity can be neglected. This simplification results in accurate modeling of, for example, the eigenfrequency of a string. But the damping due to the flexural bending near the clamping is the dominant source of dissipation in string resonators [56] and therefore, must to be taken into account in a model for the quality factor of string resonators.

The string mode shape can be modeled as a sinusoidal function with an exponential correction for the clamping boundary conditions at the edges [57]

$$u(x) = u_0 \left[\sin\left(\frac{n\pi x}{L}\right) + n\pi\, \lambda_{\text{edge}} \left(\exp\left(-\frac{x}{L\lambda_{\text{edge}}}\right) - \cos\left(\frac{n\pi x}{L}\right) \right) \right] \quad (10.13)$$

with the normalized edge correction decay length defined by

$$\lambda_{\text{edge}} = \frac{h}{L} \sqrt{\frac{E}{12\sigma}} \quad (10.14)$$

u is defined between $0 \leq x \leq L/2$ as symmetry or *anti*-symmetry may be employed to obtain the shape in the remaining half of the string domain.

With the mode shape function (10.13), the quality factor of a polymeric string (10.11) can now be calculated and results in

$$Q_{\text{string}} = \left(\frac{(n\pi)^2}{12} \frac{E}{\sigma} \left(\frac{h}{L}\right)^2 + \frac{1}{\sqrt{3}} \sqrt{\frac{E}{\sigma}} \frac{h}{L} \right)^{-1} \cdot Q_{\text{mat}} \quad (10.15)$$

The final equation comprises two terms in the brackets. The left term is coming from the *anti*-nodal bending in the string center, while the right term is related to the bending near the clamped supports. A schematic explanation of the equation is shown in Figure 10.12. The right term outweighs the left term (since typically $h \ll L$). The bending near the clamping has a maximal curvature and hence it is the dominant contributor to the total stored bending energy. The left term becomes important for short lengths and higher mode numbers. Eq. 10.15 has also been expanded to the 2D case of a membrane [57].

In Figure 10.11b the quality factor of SU-8 microstring resonators with different lengths is shown and compared to the model (10.15). The comparison shows that the model does an excellent job of predicting the qualitative dependence of Q on length L and fits the measured values perfectly if one specifies a value of $Q_{\text{mat}} = 40$. This quality factor value for the intrinsic damping is consistent with the measured Qs of SU-8 microcantilevers, as shown in Figure 10.11a. A similar stress-related Q enhancement was also found with doubly-clamped microbeam resonators made from a CNT polymer composite, as can be seen in Figure 10.3b.

In summary, it is possible to overcome the limitations on Q given by the high-intrinsic material damping of polymer materials by designing polymeric micromechanical string resonators. It has been found that the quality factor of polymeric string resonators can exceed the quality factors of similar relaxed resonators (e.g., cantilevers) by one order of magnitude. The quality factor of mechanical string resonators can be calculated with the model (10.15) presented here.

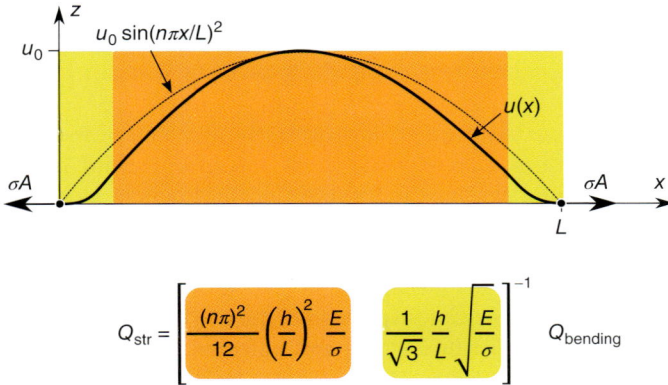

$$Q_{str} = \left[\frac{(n\pi)^2}{12} \left(\frac{h}{L}\right)^2 \frac{E}{\sigma} \quad \frac{1}{\sqrt{3}} \frac{h}{L} \sqrt{\frac{E}{\sigma}} \right]^{-1} Q_{bending}$$

Figure 10.12 Schematic drawing explaining Eq. (10.15) for generic bending-related damping mechanisms $Q_{bending}$. The first term comes from the damping due to the sinusoidal string bending. The second term comes from the local bending near the clamping.

10.4
Applications

This section showcases three successful applications that make use of the particular properties of polymeric micromechanical resonators. In Section 10.4.1, an all-polymer micromechanical resonant hygrometric humidity sensor is discussed. Instead of adding a moisture-sensitive polymer film onto a silicon-based sensor, this resonant sensor uses a moisture-sensitive polymer material (SU-8) directly as the structural material. In Section 10.4.2, the use of piezoelectric nanocomposite micromechanical resonators as energy harvesters is discussed. Here, soft polymeric resonators allow for the realization of small low-frequency resonators vibrating with large displacement amplitudes. These structures are ideal to efficiently absorb low-frequency ambient vibrations. In Section 10.4.3, the application of polymeric micromechanical resonators as an artificial cochlea for the next generation implantable bionic ears is discussed. Here, low-Q soft polymeric resonators were chosen to mimic the highly damped human cochlea.

10.4.1
Humidity Sensor

Polymers often are the key element in humidity sensors because of their ability to absorb water molecules. The absorption of water vapor changes the physical properties of the polymer. Humidity sensors measure a specific property change as a function of humidity [58]. When water gets absorbed, a polymer becomes heavier. This mass increase can be detected with gravimetric methods, that is, by detecting the resonance frequency change as a function of humidity, for example, with resonant microcantilevers [59, 60] or QCMs [61] coated with a thin polymer film. The absorbed water also changes the dielectric constant of the polymer. This change can be measured capacitively [60, 62, 63]. The absorption of water molecules also changes the resistance of a conductive polymer, which can be used as a metric for a humidity sensor [63].

One of the oldest techniques to measured humidity is the hygrometric method. Water absorption can result in a volume change or so-called swelling of the polymer. This humidity-related swelling can be measured, for example, by means of the deflection of bi-material structures such as silicon membranes [64, 65] or thin single-clamped beams [59] coated with a thin polymer film. With increasing humidity, the polymer swells and the mechanical bi-layer structure deflects due to the different coefficient of hygroscopic expansion of the two materials. But the combination of a stiff support structure with a soft polymer film restrains the expansion due to swelling of the polymer. This limits the maximal strain and thus limits the sensitivity of such a humidity sensor.

String resonators are highly sensitive stress sensors. In a pure polymeric string resonator, the hygrometric volume change directly translates into a change in tensile stress and thus into a shift in eigenfrequency. The humidity sensitivity of different SU-8 micromechanical resonators showed that SU-8 microstrings have a humidity sensitivity one order of magnitude higher than SU-8 microcantilevers [66]. Relative frequency shifts of strings from −0.34% per percent relative humidity up to −0.78% per percent relative humidity have been measured. The relative frequency change as a function of relative humidity of both micromechanical resonators is shown in Figure 10.13a.

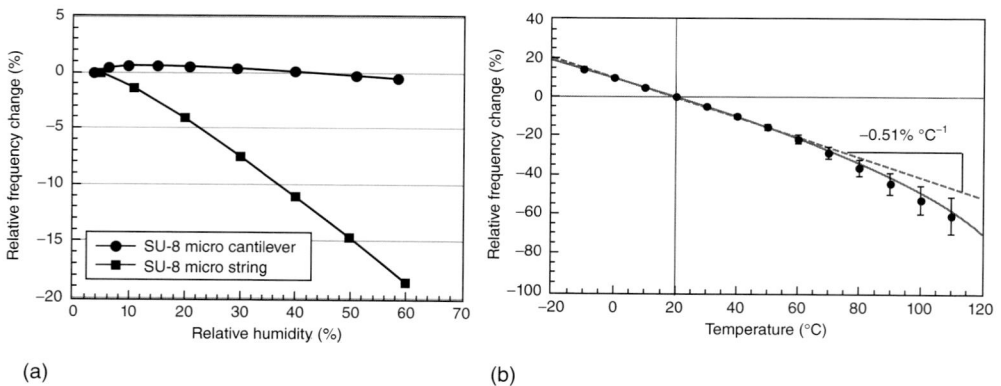

(a)   (b)

Figure 10.13 (a) Comparison of frequency response of an SU-8 microcantilever to an SU-8 microstring as a function of relative humidity at room temperature. (b) The relative change of the resonant frequency δf of 16 thin d-c microbeams with respect to temperature change. The relative change is calculated for a reference temperature $T_0 = 20\,°C$. The measurements were performed at a pressure below 0.05 Pa. The error bars represent the standard deviation from the 16 different microbeams. (Reproduced with permission from Ref. [27].)

The eigenfrequency f_0 of a string is a function of the tensile stress σ

$$f_0 = \frac{n}{2L}\sqrt{\frac{\sigma}{\rho}} \tag{10.16}$$

where L is the string length, n the mode number, and ρ is the mass density. The volume expansion with respect to the moisture content has been found to be linear to a good approximation for polyimide [67], different epoxies [68, 69] and also for SU-8 [70]. Therefore, the hygroscopic strain ε_{hyg} can be related to a linear coefficient α_{hyg} of the humidity-induced volume expansion according to [67]

$$\varepsilon_{\text{hyg}} = \alpha_{\text{hyg}}\text{RH} \tag{10.17}$$

with the relative humidity RH. The total strain in a polymer string is the sum of the intrinsic pre-strain ε_0 at $RH = 0\%$ and the hygroscopic strain

$$\varepsilon = \varepsilon_0 - \varepsilon_{\text{hyg}} = \varepsilon_0 - \alpha_{\text{hyg}}\text{RH} \tag{10.18}$$

It has been shown that the humidity influence on the Young's modulus E and mass density ρ of microstrings is negligible compared to the influence of the hygroscopic strain change [66]. Therefore, using the linear stress-strain relation, the stress in the polymer string as a function of the relative humidity can be written as

$$\sigma(\text{RH}) = E(\varepsilon_0 - \alpha_{\text{hyg}}\text{RH}) = \sigma_0 - E\alpha_{\text{hyg}}\text{RH} \tag{10.19}$$

Inserting Eq. (10.19) into Eq. (10.16) gives the eigenfrequency of a string as a function of the relative humidity [66]

$$f_0 = \frac{n}{2L}\sqrt{\frac{\sigma_0 - E\alpha_{\text{hyg}}\text{RH}}{\rho}} \tag{10.20}$$

It has been shown that the absorption of moisture and the corresponding volume change in SU-8 is completely reversible [70]. This is an important prerequisite for the design of a reproducible SU-8 humidity sensor. Such a humidity sensor based on an SU-8 micromechanical string resonator was presented by Schmid *et al.* [23]. Without a stiff support structure of bi-material hygrometric sensors, the expansion of the polymer is not restrained.

A common way to operate a micro or nanomechanical resonator is to let it oscillate at its resonance frequency by means of a phase-locked loop [59, 71] or a self-exciting positive feedback [72–74]. The self-excitation technique is especially suited to drive highly damped resonators, and it was successfully applied to drive SU-8 microresonators in air and water [75].

In Figure 10.14 the response of the SU-8 microstring sensor is compared to a reference humidity sensor. A sensitivity of 0.354% relative change of the resonant frequency per 1%RH was determined. With a frequency resolution of 6.0 Hz a relative humidity resolution of 0.006% was calculated. The sensitivity is one order of magnitude higher than the sensitivity of commercial sensors (Sensirion AG) with typical humidity resolutions of 0.05% , or of a resonant glass cantilever humidity sensor covered with a polymer layer with a sensitivity of 0.024% per %RH [76]. It

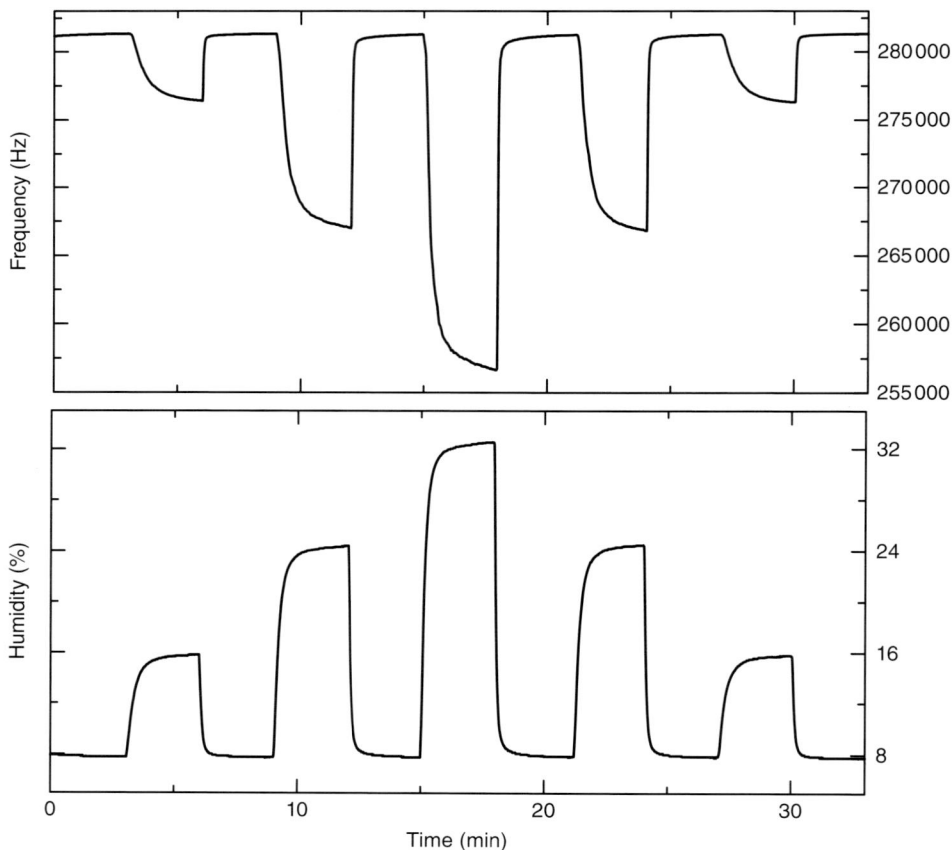

Figure 10.14 Self-sustained oscillation frequency response of an SU-8 microstring (200 μmm long, 1.45 μm thick) humidity sensor for varying relative humidity measured at room temperature. The SU-8 sensor response is compared to the response of a reference humidity sensor (from Sensirion GmbH). (Reproduced with permission from Ref. [23].)

was also found that the absorption of moisture is completely reversible, confirming the finding of [70]. Also, the time response was found to be comparable to the commercial reference sensor.

In Figure 10.15 a measurement over 14 h is shown to illustrate how thermal drift may be accounted for. The SU-8 microstring is highly temperature-sensitive and its oscillation frequency follows the temperature pattern, as seen in the upper graph of Figure 10.15. After temperature correction, the oscillation frequency follows the measured relative humidity without showing a drift due to material relaxation over the entire measurement period. The long-term

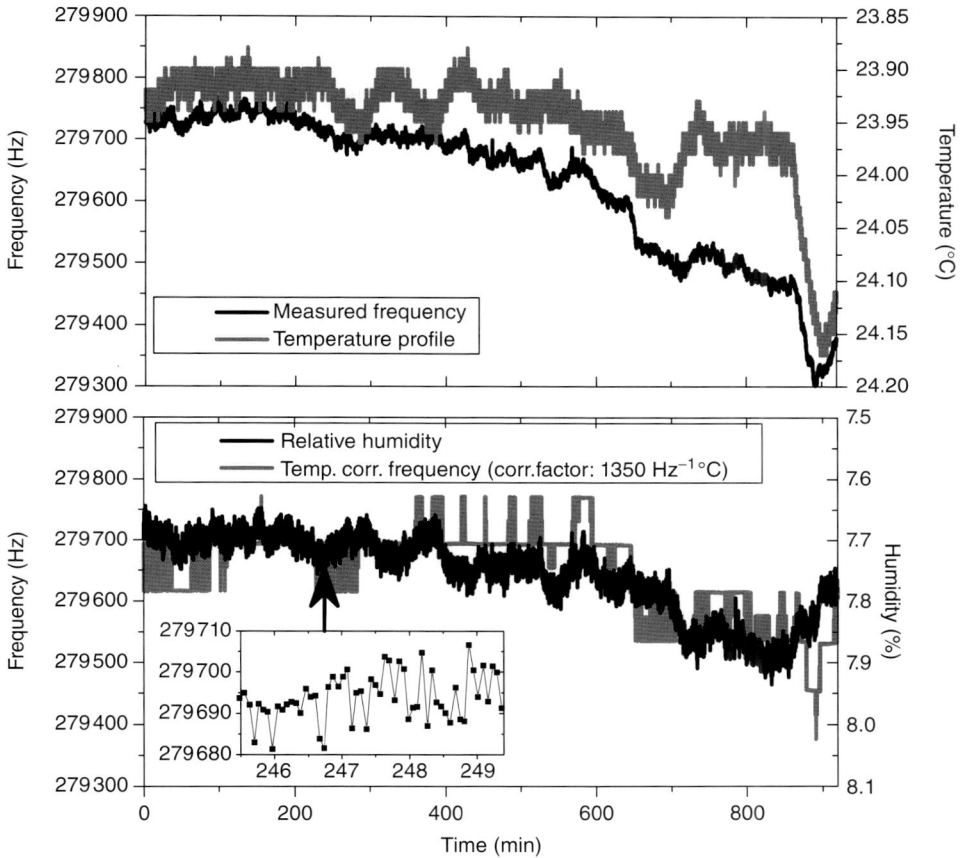

Figure 10.15 Oscillation frequency of an SU-8 micromechanical string resonator (200 μm long, 1.45 μm thick). In the top graph, the actual measured frequency response and the corresponding temperature profile are shown. In the lower graph, the temperature corrected frequency sequence and the measured relative humidity is shown. The inset shows a close-up view of the frequency. (Reproduced with permission from Ref. [23].)

stress relaxation behavior of the crosslinked SU-8 strings was evaluated and a typical exponential stress relaxation was observed with a relaxation time of 53 days.

Accuracy is the key characteristic of a humidity sensor. Micromechanical SU-8 string resonators have an unprecedented humidity resolution. But the high temperature sensitivity and material relaxation require an exact temperature correction and regular calibrations in order to be accurate. Such polymeric humidity sensors are especially interesting for applications that require the detection of minute humidity changes in a given short time period.

10.4.2
Vibrational Energy Harvesting

Piezoelectric cantilever resonators are promising structures to be used for direct energy harvesting from ambient vibrations for the production of electrical energy, for example, for autonomous wireless sensors [77]. In a micromechanical cantilever, the ambient vibrational energy is harvested from the inertially driven resonant motion of the cantilever mass. The typical vibration frequency in household appliances and everyday objects is typically lower than 200 Hz [77]. Hence, the resonance frequency of a piezoelectric microcantilever is favorably designed to be lower than this frequency. 200 Hz is rather low for a micro-scale resonator and demands a low flexural rigidity, which in the case of a cantilever can be achieved either with a small thickness or with a low modulus. Soft piezoelectric polymer microcantilevers typically have a low resonance frequency and allow large strains for an intertial load. Furthermore, polymeric resonators can be fabricated inexpensively in different dimensions, which is highly desirable for the large-scale fabrication of low-cost vibrational energy harvester modules.

Prashanthi *et al.* have published the application of piezoelectric polymeric micromechanical resonators as energy harvesters [78]. They developed an SU-8/zinc oxide (ZnO) nanoparticle composite that is directly photopatternable, which allows for a seamless process integration [79], similar to the magnetic composite resonators discussed in Section 10.2.3. It has been found that the piezoelectric performance of a ZnO nanocomposite improves with the weight fraction of ZnO nanoparticles [80]. Prashanti *et al.* could increase the weight percentage of the piezoelectric ZnO nanoparticles up to 20%. At higher particle concentrations, the composite lost the transparency in the optical UV region required for photolithography. The 20 wt% SU-8/ZnO nanocomposite remained photopatternable. It showed a piezoelectric coefficient of 6 to 8 pmV^{-1}. This value is of the same order of magnitude as the piezoelectric coefficient of bulk ZnO of 9.93 pmV^{-1} [81].

For the design of the microcantilever energy harvester, the piezoelectric composite was sandwiched with metal layers to pick up the bending induced

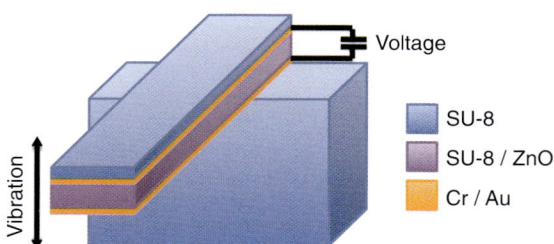

Figure 10.16 Schematic drawing of a piezoelectric SU-8 / ZnO nanocomposite microcantilever for energy harvesting. (According to Ref. [78].)

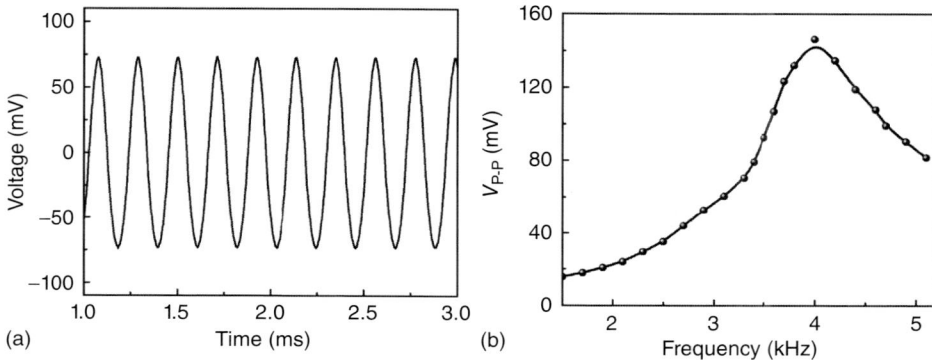

Figure 10.17 (a) Output voltage of an SU-8/ZnO nanocomposite cantilever during periodic bending at resonance. (b) The generated output voltage as a function of external vibration frequency. (Reproduced with permission from Ref. [78].)

voltage over the composite layer. A schematic drawing of the SU-8/ZnO cantilever is shown in Figure 10.16. In Figure 10.17a, the measured output voltage from such a piezoelectric microresonator is shown when vibrated at resonance with an external piezo-shaker. As can be seen in Figure 10.17b, the resonance frequency of the measured cantilever was at around 4 kHz. The measured voltage relates to a maximum output power of 0.025 μW for a resistive load of 100 kΩ. The resonance frequency of this SU-8/ZnO piezoelectric generator is still above the upper limit of 200 Hz, but it readily can be lowered by changing the geometry and dimensions.

10.4.3
Artificial Cochlea

The cochlea is the part of the inner ear of mammals that transduces acoustic waves to an electrical signal that stimulates auditory nerves. It is a spiral-shaped, hollow bony structure (similar to a snail shell) filled with lymph fluid. Fixed inside the cochlea and surrounded by lymph fluid is the basilar membrane (BM). This membrane is a frequency-selective acoustic sensor. It has local eigenfrequencies determined by varying mechanical system parameters, such as thickness and diameter. A specific acoustic frequency traveling inside the lymph fluid produces a local resonance on the membrane, which stimulates pressure-sensitive hair cells at this location. There are hair cells for the entire audible range that frequency-specifically stimulate the auditory nerves.

Hearing loss is often caused by the damage of the hair cells inside the cochlea. One surgical treatment for hearing loss is the cochlear implant [82]. These implants directly electrically activate frequency-specific nerves inside the cochlea. The electrical signal for electric nerve stimulation is generated from an external device including a microphone and sound processor. Cochlear implants

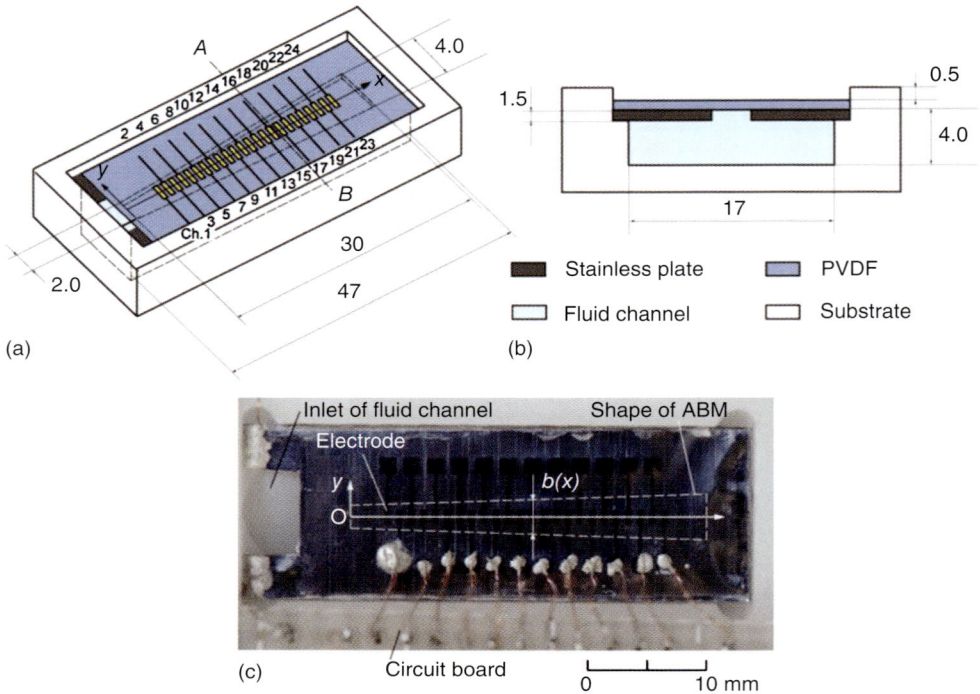

Figure 10.18 Schematic drawing and photograph of polymeric piezoelectric artificial basilar membrane (ABM); (a) 3D view, (b) cross sectional view at AB, and (c) photograph. (Reproduced with permission from Ref. [26].)

have been a great success and have partially restored hearing to more than 200 000 deaf people to date [83]. The implants, having up to 22 stimulating electrodes, re-establish speech recognition of 70–80% in a quiet environment. An approach for next-generation auditory prostheses with an even better hearing restoration prospects is the implantation of an entirely artificial cochlea based on MEMS technology.

The center piece of an artificial cochlea is the design of an optimized artificial basilar membrane (ABM). The BM is a relatively soft organic resonator immersed in a fluid. It is highly damped and has a quality factor between 1 and 10 [84]. This need for a highly damped system therefore suggests basing the ABM on polymer micromechanical resonators.

In 2004, Xu *et al.* fabricated an SU-8 microcantilever array with integrated optical readout, mimicking the function of the BM [25]. The array consisted of 4–5 cantilevers with resonance frequencies from 200 Hz to 7 kHz with quality factors around 10 in air. The polymeric cantilevers picked up the specific audio frequencies with little cross-talk. In order to better mimic the real cochlea, which is immersed in lymph fluid, discrete micromechanical resonators have been fabricated on tapered polymeric membranes. Wittbrodt *et al.* fabricated a cochlea model with discrete aluminum microbeams with varying length placed on a fully immersed polyimide membrane as ABM [85]. The ABM response was tested with a magneto-acoustic actuation and optical readout of the ABM vibration. Chen *et al.* used discrete Ni/Cu resonators on a piezoelectric PVDF (polyvinylidine difluoride) membrane [86]. They could directly readout the metal beam vibrations via the vibrational strain-induced piezoelectric voltage. They successfully detected an audio signal with their ABM that was in contact with silicone oil on one side.

In 2010, Shintaku *et al.* built an artificial cochlea based on a continuous (non-discrete) tapered piezoelectric polymer membrane [26]. They used thin metal contacts (with a negligible mechanical stiffness) for the piezoelectric readout of the local membrane vibrations. In Figure 10.18, their tapered PVDF membrane-based artificial cochlea is shown. They measured the acoustic frequency response of the membrane in air and in silicone oil. Figure 10.19 shows the vibration amplitude map of the membrane when it is in contact with silicone oil on the backside. They noted that the viscosity of the oil helps to decouple the local eigenmodes. They tested a prototype of their artificial cochlea in deafened guinea pigs [87]. The piezoelectrically generated electric potentials induced a response in their auditory

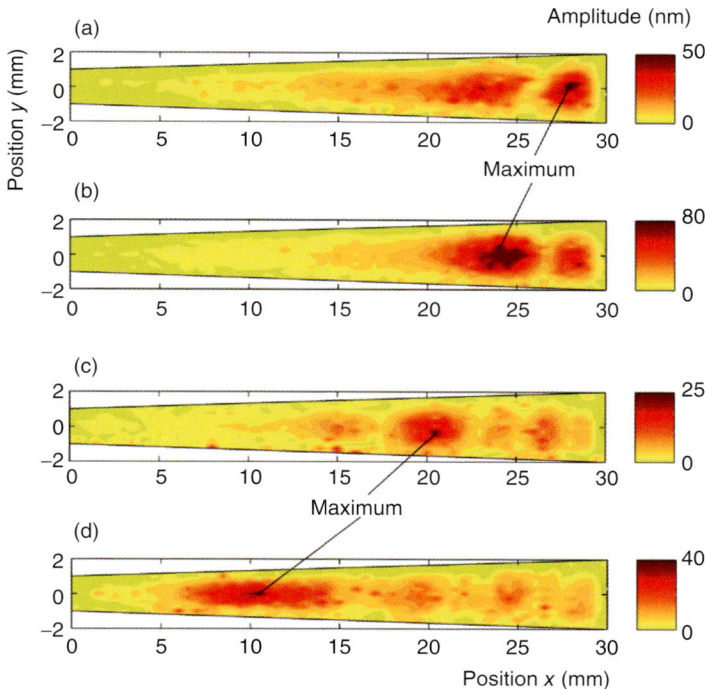

Figure 10.19 Contour maps of vibrational amplitude of a tapered piezoelectric polymer film for an acoustic actuation at (a) 1.5 kHz, (b) 2.0 kHz, (c) 3.0 kHz, and (d) 4.0 kHz in silicone oil with a viscosity of 1.75×10^{-2} Pa s. (Reproduced with permission from Ref. [26].)

brainstem, but the piezoelectric voltage was too low to directly stimulate the auditory nerves and had to be amplified 1000-fold. In the future, passive piezoelectric-polymer artificial cochlea implants may be capable of direct electric activation of the auditory nerves.

MEMS-based artificial cochleas featuring a polymeric ABM are not ready for human trials yet, but the results from different research groups show the feasibility of replacing a damaged cochlea with an artificial MEMS device in the future.

References

1. Lorenz, H., Despont, M., Fahrni, N., LaBianca, N., Renaud, P., and Vettiger, P. (1997) SU-8: a low-cost negative resist for MEMS. *J. Micromech. Microeng.*, **7** (3), 121.

2. Judy, J.W. (2001) Microelectromechanical systems (MEMS): fabrication, design and applications. *Smart Mater. Struct.*, **10** (6), 1115–1134.

3. Baltes, H., Brand, O., Hierlemann, A., Lange, D., and Hagleitner, C. (2002) CMOS MEMS - present and future. The 15th IEEE International Conference on Micro Electro Mechanical Systems, 2002, pp. 459–466.

4. Arshak, K., Moore, E., Lyons, G.M., Harris, J., and Clifford, S. (2004) A review of gas sensors employed in electronic nose applications. *Sens. Rev.*, **24** (2), 181–198.

5. Länge, K., Rapp, B.E., and Rapp, M. (2008) Surface acoustic wave biosensors: a review. *Anal. Bioanal. Chem.*, **391** (5), 1509–1519.

6. Zhou, J.W.L., Chan, H.Y.Y., To, T.K.H., Lai, K.W.C., and Li, W.J. (2004) Polymer MEMS actuators for underwater micromanipulation. *IEEE/ASME Trans. Mechatron.*, **9** (2), 334–342.

7. Chronis, N. and Lee, L.P. (2005) Electrothermally activated SU-8 microgripper for single cell manipulation in solution. *J. Microelectromech. Syst.*, **14** (4), 857–863.

8. Jager, E.W.H., Smela, E., and Inganas, O. (2000) Microfabricating conjugated polymer actuators. *Science*, **290** (5496), 1540–1545.

9. Nguyen, N.T., Ho, S.S., and Low, C.L.N. (2004) A polymeric microgripper with integrated thermal actuators. *J. Micromech. Microeng.*, **14** (7), 969–974.

10. Goll, C., Bacher, W., Bustgens, B., Maas, D., Ruprecht, R., and Schomburg, W.K. (1997) An electrostatically actuated polymer microvalve equipped with a movable membrane electrode. *J. Micromech. Microeng.*, **7** (3), 224–226.

11. Quake, S.R. and Scherer, A. (2000) From micro- to nanofabrication with soft materials. *Science*, **290**, 1536–1540.

12. Seidemann, V., Rabe, J., Feldmann, M., and Buettgenbach, S. (2002) SU8-micromechanical structures with in situ fabricated movable parts. *Microsyst. Technol.*, **8**, 348–350.

13. Bachmann, D., Schöberle, B., Kühne, S., Leiner, Y., and Hierold, C. (2006) Fabrication and characterization of folded SU-8 suspensions for MEMS applications. *Sens. Actuators, A*, **130-131**, 379–386.

14. Bachmann, D. and Hierold, C. (2008) Determination of the pull-off forces and pull-off dynamics of an electrostatically actuated silicon disk. *J. Microelectromech. Syst.*, **17** (3), 643–652.

15. Psychogiou, D., Li, Y., Hesselbarth, J., Kühne, S., Peroulis, D., Hierold, C., and Hafner, C. (2013) Millimeter-wave phase shifter based on waveguide-mounted RF-MEMS. *Microwave Opt. Technol. Lett.*, **55** (3), 465–468.

16. Li, Y., Psychogiou, D., Kühne, S., Hesselbarth, J., Hafner, C., and Hierold, C. (2013) Large stroke staggered vertical comb-drive actuator for the application of a millimeter-wave tunable phase shifter. *J. Microelectromech. Syst.*, Early access online.

17. Li, Y., Kühne, S., Psychogiou, D., Hesselbarth, J., and Hierold, C. (2011)

A microdevice with large deflection for variable-ratio RF MEMS power divider applications. *J. Micromech. Microeng.*, **21** (7), 074 013.

18. Tamayo, J., Ramos, D., Mertens, J., and Calleja, M. (2006) Effect of the adsorbate stiffness on the resonance response of microcantilever sensors. *Appl. Phys. Lett.*, **89**, 224 104.

19. Zhang, X.R. and Xu, X.F. (2004) Development of a biosensor based on laser-fabricated polymer microcantilevers. *Appl. Phys. Lett.*, **85** (12), 2423–2425.

20. Calleja, M., Nordstrom, M., Alvarez, M., Tamayo, J., Lechuga, L.M., and Boisen, A. (2005) Highly sensitive polymer-based cantilever-sensors for DNA detection. *Ultramicroscopy*, **105** (1-4), 215–222.

21. Thaysen, J., Yalcinkaya, A.D., Vettiger, P., and Menon, A. (2002) Polymer-based stress sensor with integrated readout. *J. Phys. D: Appl. Phys.*, **35** (21), 2698–2703.

22. Genolet, G., Brugger, J., Despont, M., Drechsler, U., and Vettiger, P. (1999) Soft, entirely photoplastic probes for scanning force microscopy. *Rev. Sci. Instrum.*, **70** (5), 2398–2401.

23. Schmid, S., Wägli, P., and Hierold, C. (2008) All-polymer microstring resonant humidity sensor with enhanced sensitivity due to change of intrinsic stress. Proceedings of the EUROSENSORS XXII Conference, Dresden, pp. 697–700.

24. Schmid, S., Wägli, P., and Hierold, C. (2009) Biosensors based on all-polymer resonant microbeams. Proceedings of the MEMS Conference, pp. 300–303.

25. Xu, T., Bachman, M., Zeng, F.G., and Li, G.P. (2004) Polymeric micro-cantilever array for auditory front-end processing. *Sens. Actuators, A*, **114** (2-3), 176–182.

26. Shintaku, H., Nakagawa, T., Kitagawa, D., Tanujaya, H., Kawano, S., and Ito, J. (2010) Development of piezoelectric acoustic sensor with frequency selectivity for artificial cochlea. *Sens. Actuators, A*, **158** (2), 183–192.

27. Schmid, S. (2009) *Electrostatically Actuated All-Polymer Microbeam Resonators - Characterization and Application*, Scientific Reports on Micro and Nanosystems, vol. **6**, Der Andere Verlag.

28. Zhang, G., Gaspar, J., Chu, V., and Conde, J.P. (2005) Electrostatically actuated polymer microresonators. *Appl. Phys. Lett.*, **87** (10), 104 104.

29. Zhang, G., Chu, V., and Conde, J.P. (2007) Electrostatically actuated conducting polymer microbridges. *J. Appl. Phys.*, **101**, 64 507.

30. Zhang, G., Chu, V., and Conde, J.P. (2007) Conductive blended polymer MEMS microresonators. *J. Microelectromech. Syst.*, **16** (2), 329–335.

31. Zhang, G., Chu, V., and Conde, J.P. (2007) Electrostatically actuated bilayer polyimide-based microresonators. *J. Micromech. Microeng.*, **17**, 797–803.

32. Sousa, P.M., Gutiérrez, M., Mendoza, E., Llobera, A., Chu, V., and Conde, J.P. (2011) Microelectromechanical resonators based on an all polymer/carbon nanotube composite structural material. *Appl. Phys. Lett.*, **99** (4), 044 104.

33. Conradie, E.H. and Moore, D.F. (2002) SU-8 thick photoresist processing as a functional material for MEMS applications. *J. Micromech. Microeng.*, **12** (4), 368.

34. Zhang, J., Tan, K.L., and Gong, H.Q. (2001) Characterization of the polymerization of SU-8 photoresist and its applications in micro-electro-mechanical systems (MEMS). *Polym. Test.*, **20** (6), 693–701.

35. Sandberg, R., Mølhave, K., Boisen, A., and Svendsen, W. (2005) Effect of gold coating on the Q-factor of a resonant cantilever. *J. Micromech. Microeng.*, **15** (12), 2249–2253.

36. Kim, Y.J. and Allen, M.G. (1999) In situ measurement of mechanical properties of polyimide films using micromachined resonant string structures. *IEEE Trans. Compon. Packag. Technol.*, **22** (2), 282–290.

37. Schmid, S., Wendlandt, M., Junker, D., and Hierold, C. (2006) Nonconductive polymer microresonators actuated by the Kelvin polarization force. *Appl. Phys. Lett.*, **89** (16), 163 506.

38. Schmid, S., Hierold, C., and Boisen, A. (2010) Modeling the Kelvin polarization force actuation of micro-and nanomechanical systems. *J. Appl. Phys.*, **107** (5), 054 510.

39. Suter, M., Graf, S., Ergeneman, O., Schmid, S., Camenzind, A., Nelson, B.J., and Hierold, C. (2009) Superparamagnetic photosensitive polymer nanocomposite for microactuators. Transducers 2009: The 15th International Conference on Solid-State Sensors, Actuators and Microsystems, pp. 869–872.

40. Suter, M., Ergeneman, O., Zürcher, J., Schmid, S., Camenzind, A., Nelson, B.J., and Hierold, C. (2011) Superparamagnetic photocurable nanocomposite for the fabrication of microcantilevers. *J. Micromech. Microeng.*, **21** (2), 025 023.

41. Suter, M. (2011) Photopatternable superparamagnetic nanocomposite for the fabrication of microstructures. PhD thesis, ETH Zurich.

42. Suter, M., Ergeneman, O., Zürcher, J., Moitzi, C., Pané, S., Rudin, T., Pratsinis, S., Nelson, B., and Hierold, C. (2011) A photopatternable superparamagnetic nanocomposite: material characterization and fabrication of microstructures. *Sens. Actuators, B*, **156** (1), 433–443.

43. Suter, M., Li, Y., Sotiriou, G.A., Teleki, A., Pratsinis, S.E., and Hierold, C. (2011) Low-cost fabrication of PMMA and PMMA based magnetic composite cantilevers. 2011 16th International Solid-State Sensors, Actuators and Microsystems Conference (TRANSDUCERS), pp. 398–401.

44. Dubourg, G., Dufour, I., Pellet, C., and Ayela, C. (2012) Optimization of the performances of SU-8 organic microcantilever resonators by tuning the viscoelastic properties of the polymer. *Sens. Actuators, B*, **169**, 320–326.

45. Dubourg, G., Fadel-Taris, L., Dufour, I., Pellet, C., and Ayela, C. (2011) Collective fabrication of all-organic microcantilever chips based on a hierarchical combination of shadow-masking and wafer-bonding processing methods. *J. Micromech. Microeng.*, **21** (9), 095 021.

46. Schmid, S. and Hierold, C. (2008) Damping mechanisms of single-clamped and prestressed double-clamped resonant polymer microbeams. *J. Appl. Phys.*, **104** (9), 093 516.

47. Bao, M. (2005) *Analysis and Design Principles of MEMS Devices*, Elsevier.

48. Green, C.P. and Sader, J.E. (2005) Frequency resonse of cantilever beams immersed in viscous fluids near a solid surface with applications to the atomic force microscope. *J. Appl. Phys.*, **98**, 114 913.

49. Sader, J.E. (1998) Frequency response of cantilever beams immersed in viscous fluids with applictions to the atomic force microscope. *J. Appl. Phys.*, **84** (1), 64–76.

50. Ward, I.M. and Sweeney, J. (2004) *An Introduction to the Mechanical Properties of Solid Polymers*, 2nd edn, John Wiley & Sons, Inc.

51. Weaver, W., Timoshenko, S.P., and Young, D.H. (1990) *Vibration Problems in Engineering*, Wiley Interscience.

52. Verbridge, S.S., Parpia, J.M., Reichenbach, R.B., Bellan, L.M., and Craighead, H.G. (2006) High quality factor resonance at room temperature with nanostrings under high tensile stress. *J. Appl. Phys.*, **99**, 124 304.

53. Schmid, S., Jensen, K.D., Nielsen, K.H., and Boisen, A. (2011) Damping mechanisms in high-Q micro and nanomechanical string resonators. *Phys. Rev. B*, **84** (16), 1–6.

54. Magnus, K. and Popp, K. (2005) *Schwingungen*, 7th edn, Teubner.

55. Boisen, A., Dohn, S., Keller, S.S., Schmid, S., and Tenje, M. (2011) Cantilever-like micromechanical sensors. *Rep. Prog. Phys.*, **74** (3), 036 101.

56. Unterreithmeier, Q.P., Faust, T., and Kotthaus, J.P. (2010) Damping of nanomechanical resonators. *Phys. Rev. Lett.*, **105** (2), 27 205.

57. Yu, P.L., Purdy, T., and Regal, C. (2012) Control of material damping in high-Q membrane microresonators. *Phys. Rev. Lett.*, **108** (8), 1–5.

58. Rittersma, Z.M. (2002) Recent achievements in miniaturised humidity sensors—a review of transduction techniques. *Sens. Actuators, A*, **96** (2-3), 196–210.

59. Battiston, F., Ramseyer, J.P., Lang, H., Baller, M., Gerber, C., Gimzewski, J., Meyer, E., and Güntherodt, H.J. (2001) A chemical sensor based on a microfabricated cantilever array with simultaneous resonance-frequency and bending readout. *Sens. Actuators, B*, **77** (1-2), 122–131.

60. Kurzawski, P., Hagleitner, C., and Hierlemann, A. (2006) Detection and discrimination capabilities of a multi-transducer single-chip gas sensor system. *Anal. Chem.*, **78** (19), 6910–6920.

61. Pascal-Delannoy, T., Sorli, B., and Boyer, A. (2000) Quartz crystal microbalance (QCM) used as humidity sensor. *Sens. Actuators, A*, **84** (3), 285–291.

62. Sakai, Y., Sadaoka, Y., and Matsuguchi, M. (1996) Humidity sensors based on polymer thin films. *Sens. Actuators, B*, **35** (1-3), 85–90.

63. Li, Y., Yang, M., Camaioni, N., and Casalbore-Miceli, G. (2001) Humidity sensors based on polymer solid electrolytes: investigation on the capacitive and resistive devices construction. *Sens. Actuators, B*, **77** (3), 625–631.

64. Sager, K., Gerlach, G., and Schroth, A. (1994) A humidity sensor of a new type. *Sens. Actuators, B*, **18** (1-3), 85–88.

65. Buchhold, R., Nakladala, A., Gerlach, G., and Neumann, P. (1998) Design studies on piezoresistive humidity sensors. *Sens. Actuators, B*, **53** (1-2), 1–7.

66. Schmid, S., Kühne, S., and Hierold, C. (2009) Influence of air humidity on polymeric microresonators. *J. Micromech. Microeng.*, **19**, 065 018.

67. Buchhold, R., Nakladal, A., Gerlach, G., Sahre, K., Eichhorn, K.J., and Müller, M. (1998) Reduction of mechanical stress in micromachined components caused by humidity-induced volume expansion of polymer layers. *Microsyst. Technol.*, **5**, 3–12.

68. Ardebili, H., Wong, E.H., and Pecht, M. (2003) Hygroscopic swelling and sorption characteristics of epoxy molding compounds used in electronic packaging. *IEEE Trans. Compon. Packag. Technol.*, **26** (1), 206–214.

69. Vanlandingham, M.R., Eduljee, R.F., and Gillespie, J.W. (1999) Moisture diffusion in epoxy systems. *J. Appl. Polym. Sci.*, **71**, 787–798.

70. Feng, R. and Farris, R.J. (2003) Influence of processing conditions on the thermal and mechanical properties of SU8 negative photoresist coatings. *J. Micromech. Microeng.*, **13** (1), 80–88.

71. Schmid, S., Kurek, M., Adolphsen, J.Q., and Boisen, A. (2013) Real-time single airborne nanoparticle detection with nanomechanical resonant filter-fiber. *Sci. Rep.*, **3**, 1288. DOI: 10.1038/srep01288.

72. Tamayo, J., Calleja, M., Ramos, D., and Mertens, J. (2007) Underlying mechanisms of the self-sustained oscillation of a nanomechanical stochastic resonator in a liquid. *Phys. Rev. B*, **76**, 180 201.

73. Feng, X.L., White, C.J., Hajimiri, A., and Roukes, M.L. (2008) A self-sustaining ultrahigh-frequency nanoelectromechanical oscillator. *Nat. Nanotechnol.*, **3** (6), 342–346.

74. Tamayo, J. (2005) Study of the noise of micromechanical oscillators under quality factor enhancement via driving force control. *J. Appl. Phys.*, **97**, 44 903.

75. Schmid, S., Senn, P., and Hierold, C. (2008) Electrostatically actuated non-conductive polymer microresonators in gaseous and aqueous environment. *Sens. Actuators, A*, **145-146**, 442–448.

76. Glück, A., Halder, W., Lindner, G., Müller, H., and Weindler, P. (1994) PVDF-excited resonance sensors for gas flow and humidity measurements. *Sens. Actuators, B*, **18/19**, 554–557.

77. Roundy, S., Wright, P.K., and Rabaey, J. (2003) A study of low level vibrations as a power source for wireless sensor nodes. *Comput. Commun.*, **26** (11), 1131–1144.

78. Prashanthi, K., Miriyala, N., Gaikwad, R., Moussa, W., Rao, V., and Thundat, T. (2013) *Vibrational Energy Harvesting Using Photo-Patternable Piezoelectric Nanocomposite Cantilevers*, Nano Energy.

79. Prashanthi, K., Naresh, M., Seena, V., Thundat, T., and Rao, R.V. (2012) A novel photoplastic piezoelectric nanocomposite for MEMS applications. *JMEMS Lett.*, **21** (2), 259–261.

80. Loh, K.J. and Chang, D. (2011) Zinc oxide nanoparticle-polymeric thin films for dynamic strain sensing. *J. Mater. Sci.*, **46** (1), 228–237.

81. Zhao, M.H., Wang, Z.L., and Mao, S.X. (2004) Piezoelectric characterization of individual zinc oxide nanobelt probed by piezoresponse force microscope. *Nano Lett.*, **4** (4), 587–590.

82. Zeng, F.G. (2004) Trends in cochlear implants. *Trends Amplif.*, **8** (1), 1–34.

83. Zeng, F.G. (2011) New horizons in auditory prostheses. *Hear. J.*, **64** (11), 24–27.

84. Robles, L. and Ruggero, M.A. (2001) Mechanics of the mammalian cochlea. *Physiol. Rev.*, **81** (3), 1305–1352.

85. Wittbrodt, M., Steele, C., and Puria, S. (2006) Developing a physical model of the human cochlea using microfabrication methods. *Audiol. Neurotol.*, **11** (2), 104–112.

86. Chen, F., Cohen, H.I., Bifano, T.G., Castle, J., Fortin, J., Kapusta, C., Mountain, D.C., Zosuls, A., and Hubbard, A.E. (2006) A hydromechanical biomimetic cochlea: experiments and models. *J. Acoust. Soc. Am.*, **119** (1), 394–405.

87. Inaoka, T., Shintaku, H., Nakagawa, T., Kawano, S., Ogita, H., Sakamoto, T., Hamanishi, S., Wada, H., and Ito, J. (2011) Piezoelectric materials mimic the function of the cochlear sensory epithelium. *Proc. Natl. Acad. Sci. U.S.A.*, **108** (45), 18 390–18 395.

11
Devices with Embedded Channels

Thomas P. Burg

11.1
Introduction

Time and frequency are among the most precisely measurable quantities in experimental physics. Micro- and nanomechanical resonators with embedded channels uniquely exploit this precision for measurements in liquid media, enabling many fascinating applications in the life sciences, chemistry, biotechnology, and physics. The sensing principle of these devices is based on the transduction of minute changes in mass into changes in mechanical resonance frequency. Owing to the small size and low stiffness of the resonators, their natural frequency is also highly sensitive to other parametric perturbations, such as temperature, external field gradients, or applied strain. By optimization of the device geometry, careful system design, and adequate referencing it is possible to single out a specific quantity of interest in the measurement.

Due to the intimate coupling between nanomechanical resonators and their surrounding medium, the operating environment is a dominating factor in the overall performance. In high vacuum or air, nanomechanical resonators allow measurements with extraordinary precision due to the combination of high sensitivity, imparted by their small size, and low external damping. For example, carbon nanotubes configured as mechanical resonators have been used to detect the adsorption of mass down to the atomic level [1, 2]. The work of Naik *et al.* [3] provides a fascinating glimpse at potential applications of such measurements by illustrating the ability of these devices to function as direct mass sensitive detectors in biological mass spectrometry.

In the pursuit of applications in chemical and biological sensing, it is attractive to consider the operation of nanomechanical resonators directly in the liquid phase. By exposing the resonator to molecules in solution, it becomes possible to detect and characterize binding reactions, as the attachment of molecules leads to mass loading of the device with a concomitant reduction in resonance frequency. Properties of the liquid, such as density and viscosity, also affect the resonance frequency and can therefore be detected.

Resonant MEMS – Fundamentals, Implementation and Application, First Edition.
Edited by Oliver Brand, Isabelle Dufour, Stephen M. Heinrich and Fabien Josse.
© 2015 Wiley-VCH Verlag GmbH & Co. KGaA. Published 2015 by Wiley-VCH Verlag GmbH & Co. KGaA.

Unfortunately, the viscosity of liquids presents a strong impediment to precision frequency-based measurements using conventional micro- and nanomechanical resonators. Because of their small mass, the ratio of stored energy to dissipation per cycle scales unfavorably with the size of the resonator. For example, very strong damping has been experimentally confirmed for carbon nanotube resonators in water by Sawano *et al.* [4], and Q-values less than 1 are typical for conventional atomic force microscope cantilevers in liquid [5]. Quality factors of $\sim$100 in water have been attained by using in-plane rather than out-of-plane flexural modes [6], but despite such progress in the engineering of liquid-optimized micro- and nanoresonators, their Q-values are still far from the range of 10 to 100 000 routinely achieved under vacuum. As a result, the frequency response of devices in liquid is significantly broadened and the vibration amplitude is strongly attenuated.

For measurement purposes, it is possible to counteract the dissipation by incorporating the resonator into an electronic feedback circuit, so that the frequency response of the combined system is sharply peaked. This method is known as Q-control in the context of atomic force microscopy [7]. The required driving force may be provided by, for example, electrostatics, magnetism, piezoelectricity, or thermoacoustics. Although the fundamental noise floor of the resonant sensor cannot be improved by external feedback or parametric amplification, such techniques can provide practical advantages for precision measurements of resonance frequency [8, 9].

Micro- and nanoresonators with embedded channels are attractive because they unite high mechanical quality factors normally associated only with vacuum environments with the ability to analyze samples in the liquid phase. This is enabled by confining the sample to the inside of the resonator structure while the outside can be surrounded by air or vacuum. Since the liquid sample constitutes a significant fraction of the mass of the resonator body, the resonance frequency depends strongly on fluid density. At the same time, bulk flow may be accurately measured via the Coriolis force. Although measurements of bulk fluid properties and flow do not strictly require miniaturization to attain high sensitivity, scaling is often advantageous for reasons of low sample consumption, speed, and high volume manufacturing. In contrast, miniaturization is paramount for measurements of surface bound molecules and suspended micro/nanoparticles or cells. In surface adsorption measurements, it is favorable to have a large ratio of inner channel surface area to resonator mass. In measurements of particle mass, the total weight of the resonator should be as small as possible compared to the particles under investigation. At the same time, practical applications typically dictate certain minimum requirements on fluidic throughput, ultimately limiting the minimum embedded channel size that can still be considered to be practically useful [10].

This chapter will first give a general introduction to the theory of microfluidic resonators and the different types of measurements that can be conducted with these versatile devices. The effects of fluid density and flow are considered first, followed by a brief account of the influence of fluid viscosity. This latter aspect is a topic of great complexity, and readers are referred to the references for a more comprehensive discussion of this problem. Next, the sensitivity to localized mass changes is described. Such localized loading commonly arises in both surface based measurements and in the detection of micro- and nanoparticles by mass. In the second part, the technology for fabricating and packaging embedded channel resonators is described. Finally, a short overview of the applications is presented and illustrated with examples from the current literature.

11.2
Theory

11.2.1
Effects of Fluid Density and Flow

To describe the effects of mass loading on the natural frequency of microfluidic resonators, we consider the generic geometry depicted in Figure 11.1. A suspended microfluidic channel is embedded inside an otherwise solid beam, which is thin compared to its length and width. Cantilevers, double-clamped strings, and many other shapes of embedded channel resonators vibrating in a transverse (flexural) mode can be mapped to this geometry. The mode of vibration is described by the displacement field $A \cdot u_z(x, y) \cdot \sin(\omega t)$, where the

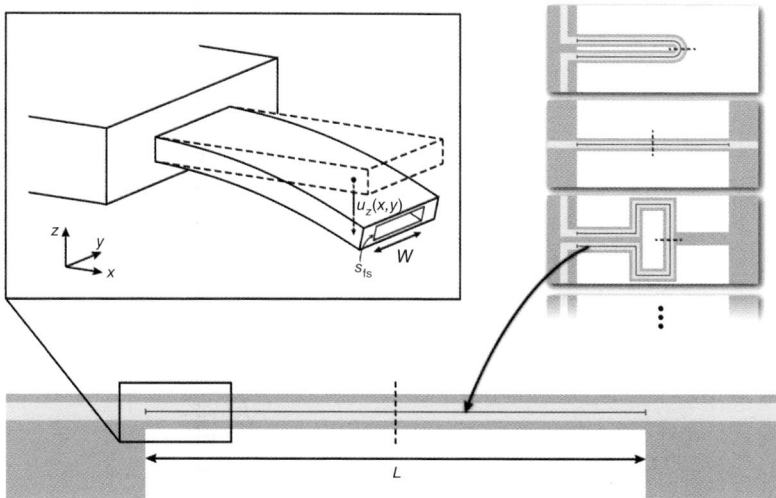

Figure 11.1 Model view of micromechanical resonators with embedded fluidic channels. A wide range of different geometries such as cantilevers, double-clamped strings, and torsional resonators can be mapped to a one-dimensional beam (bottom). The deflection function of the beam is u_z. This simplified model assumes a constant cross section with channel width W and a solid-fluid interface line of length s_{fs}.

amplitude A is assumed to be much smaller than the thickness of the resonator and u_z is the shape function normalized to $\max|u_z(x, y)| = 1$. Given u_z, the effect of small alterations in mass loading at different locations can be calculated by the Rayleigh-Ritz method. The method assumes that the mode shape is not significantly altered by the mass increment, which is typically well justified. Equating the maximum kinetic energy with the maximum potential energy (V_{max}) in one cycle of the resonator yields

$$\omega^2 = \frac{V_{max}}{\frac{1}{2}A^2 \iiint \rho(x, y, z) u_z(x, y)^2 \, dz \, dy \, dx} \tag{11.1}$$

where ρ is the local mass density. Changes in fluid density, adsorption of molecules, or the presence of particles suspended in the fluid are described by the mass density distribution $\rho(x, y, z) = \rho_0(x, y, z) + \Delta\rho(x, y, z)$. To first order, the increment $\Delta\rho$ yields a change in resonance frequency of

$$\frac{\Delta\omega}{\omega} \approx -\frac{1}{2} \frac{\iiint \Delta\rho(x, y, z) u_z(x, y)^2 \, dz \, dy \, dx}{m^*} \tag{11.2}$$

where the effective mass m^* is given by $m^* = \iiint \rho_0(x, y, z) u_z(x, y)^2 \, dz \, dy \, dx$.

Given the generic geometry of Figure 11.1, one can divide the integral over the resonator volume into a region occupied by the channel interior and a region occupied by the structural solid. With ρ_s denoting the density of the solid, ρ_f the initial density of the fluid, and $A_{s,f}$ the respective cross sectional areas, Eq. (11.2) is reduced to:

$$\frac{\Delta\omega}{\omega} \approx -\frac{1}{2} \frac{\Delta\rho_f}{\rho_f} \frac{1}{(A_s/A_f)(\rho_s/\rho_f) + 1} \tag{11.3}$$

Equation 11.3 shows that it is an advantage for the sensitivity of vibrating tube densitometers to have a large inner cross section A_f compared to the cross-sectional area of the channel walls (A_s). From a sensitivity point of view, miniaturization of these devices is thus not fundamentally advantageous. Nonetheless, a number of system level merits, such as readout precision, immunity to environmental disturbances, and the reduction in sample volume may outweigh aspects of sensitivity and advocate miniaturization, even for bulk measurements.

Internal hydrostatic pressure and fluid flow can strongly influence the dynamics of embedded channel resonators. Pressure couples to the resonance frequency primarily by altering the effective spring constant and the total mass of the device through bulging of the channel walls. The spring constant increases with pressure due to the enlarged channel cross section, which leads to an increase in resonance frequency. At the same time, the total mass increases as the channel volume expands. This results in a decrease in resonance frequency, thereby opposing the increase due to stiffening of the spring constant. The relative significance of these effects is strongly dependent on the device geometry and generally must be determined by numerical modeling.

Flow rates through embedded channel resonators can span several orders of magnitude due to the strong dependence of flow resistance on the size of the channel. For many applications, it is important to understand not just the integral volumetric flow, but also the velocity distribution inside the channel. Assuming laminar flow, the exact velocity profile in a rectangular channel of width W, height h, and cross-sectional aspect ratio $a = h/W$ is given by [11]

$$v(y, z) = \frac{\Delta P}{L} \frac{4}{\eta \pi^3} W^2 \sum_{k=1,3,5,\ldots} \frac{(-1)^{\frac{k-1}{2}}}{k^3} \left[1 - \frac{\cosh\left(k\pi a \left(\frac{z}{h}\right)\right)}{\cosh\left(\frac{k\pi}{2} a\right)} \right] \cos\left(k\pi \frac{y}{W}\right)$$

$$\text{for } |y| < \frac{W}{2} \text{ and } |z| < \frac{h}{2} \tag{11.4}$$

and the volume flow is

$$q = \frac{W \cdot h^3}{12\eta} \frac{\Delta P}{L} \frac{1}{a^2} \left[1 - \frac{192}{\pi^5 a} \sum_{k=1,3,5,\ldots} \frac{1}{k^5} \tanh\left(\frac{k\pi}{2} a\right) \right] \tag{11.5}$$

Here η is the viscosity of the fluid, ΔP is the applied pressure difference between inlet and outlet, and L is the length of the channel. When $a \ll 1$, it is often sufficiently accurate to use the parallel plate approximation

$$q_{pp} = \frac{W \cdot h^3}{12\eta} \frac{\Delta P}{L} \tag{11.6}$$

This approximation always overestimates the flow, but for aspect ratios $a < 0.1$ the error is below 7%, and for $a < 0.5$ it is below 50%. Figure 11.2 shows the flow rate for rectangular channels of different aspect ratios at an applied pressure differential of 1 bar between inlet and outlet. One bar represents an order of magnitude that is common in microfluidic experiments. However, this can in principle be extended to several hundred bars through the use of high pressure liquid chromatography (HPLC) pumps and suitable fittings.

An important effect resulting from the fluid-structure interaction under flow is the Coriolis force-induced twisting of embedded channel resonators. Following Enoksson *et al.* [12], this effect is illustrated in Figure 11.3. Vibration of the loop perpendicular to the plane causes a twisting force proportional to the instantaneous angular velocity ω, which is greatest when the loop passes through the zero-point, $\theta_y = 0$. At this point, the twisting force F is

$$F = 4\pi f \theta_{y0} \cdot L\varphi \tag{11.7}$$

where φ is the mass flow rate in kilogram per second, f the resonance frequency, and θ_{y0} is the maximum deflection angle. Use of a lump angle instead of the actual local deflections associated with the vibration mode works well for simple cantilever-like geometries vibrating in the first mode. If the transverse vibration frequency is lower than the resonance frequency of the torsional mode, the loop twist angle $\theta_x = 2 \cdot Fd/K_s$ is in phase with the force and 90° out-of-phase with

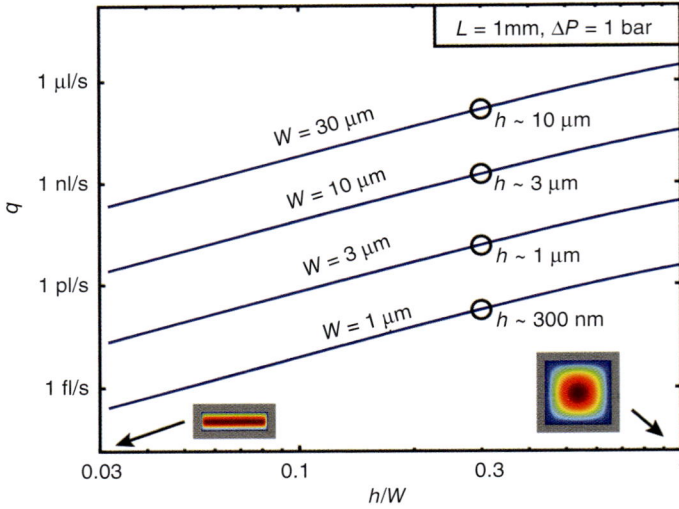

Figure 11.2 The flow rate through channels embedded into micro- and nanomechanical resonators depends strongly on channel height. Due to the size constraints imposed by the resonator dimensions, rates on the scale of pl/s to nl/s are typical at common operating pressures of ~1 bar. The icons at the bottom axes illustrate the laminar flow velocity distribution in channels of different aspect ratio.

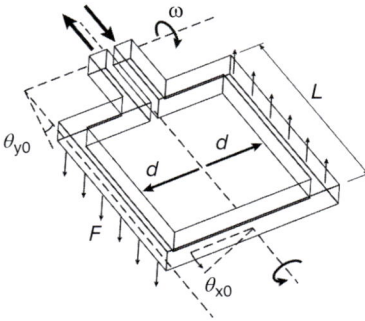

Figure 11.3 The Coriolis force induces twisting of a fluid conducting loop resonating at angular frequency ω. Measuring the twist angle θ_{x0} allows calculation of the total mass flow.

the transverse vibration. K_s here denotes the torsional stiffness of the loop. In principle, detection of the twist can be facilitated if the resonance frequencies of the transverse and twisting modes are matched. In practice, however, this is difficult to ensure in general, as both frequencies depend on the effective mass of the device, which scales differently for the two modes when the fluid density changes.

11.2.2
Effects of Viscosity on the Quality Factor

Damping of micro- and nanomechanical resonators is of great significance for precision measurements based on resonance frequency. The quality factor

$$Q = 2\pi \cdot E_{\text{stored}}/E_{\text{diss}} \tag{11.8}$$

is defined as the ratio between the energy stored in the resonator (E_{stored}) and the energy that is dissipated in one cycle of vibration (E_{diss}). While energy dissipation due to the viscosity of the medium is by far the most significant contribution to damping in resonators which are submersed in fluid, microfluidic resonators are not strongly affected by viscosity. Interestingly, filling of the embedded channel can even enhance the quality factor if the increase in E_{stored} due to the added mass exceeds the increase in E_{diss} due to viscosity [13, 14].

Particles suspended in the fluid can also lower the quality factor, which has been strikingly observed in experiments by Sparks *et al.* [15]. However, a detailed treatment of this effect is beyond the scope of this chapter.

In the following, we outline the effect of fluid viscosity on the quality factor of embedded channel cantilever resonators vibrating in the first flexural mode. The dependence of damping on viscosity and device size reveals a rich interplay of different effects, which together yield a non-monotonic dependence of damping on viscosity. To better understand this behavior, which is highly unusual in the world of nanomechanics, it is instructive to first compare the Stokes length $\delta = \sqrt{\eta/(\rho_f \omega)}$ to the channel height h. Here η is the fluid viscosity, ρ_f is the fluid density, and ω is the resonance frequency. The dimensionless frequency $\beta = (h/\delta)^2$ indicates the significance of inertia-induced shear in the fluid. The quality factor limitation of a microfluidic cantilever resonator has been calculated to first order by Sader *et al.* [16] and is given by

$$Q_{\text{fluid}} = F(\beta)\frac{\rho_{\text{cant}}}{\rho_f}\left(\frac{h_{\text{cant}}}{h_{\text{fluid}}}\right)\left(\frac{W_{\text{cant}}}{W_{\text{fluid}}}\right)\left(\frac{L_{\text{cant}}}{h_{\text{fluid}}}\right)^2 \tag{11.9}$$

Q_{fluid} enters the total quality factor through $Q^{-1} = Q_{\text{fluid}}^{-1} + Q_{\text{int}}^{-1}$, where Q_{int} represents the quality factor limitation due to mechanisms intrinsic to the solid resonator itself, such as residual air damping or clamping losses. The normalized quality factor $F(\beta)$ depends on the mode number and is influenced by the position of the channel relative to the neutral axis of the beam. If the channel axis coincides precisely with the neutral plane, $F(\beta)$ for the first mode is well approximated by $F(\beta) \approx 0.152\sqrt{\beta} + 38.7/\beta$, as shown in Figure 11.4. The minimum at $\beta_{\text{min}} \approx 46$ corresponds to the transition between a regime of low inertia ($\beta \ll \beta_{\text{min}}$) and a regime where fluid inertia dominates ($\beta \gg \beta_{\text{min}}$). In the first (high inertia) regime, the quality factor decreases with increasing viscosity. Interestingly, this trend reverses in the second (low inertia) regime. Experiments have confirmed this surprising prediction [14].

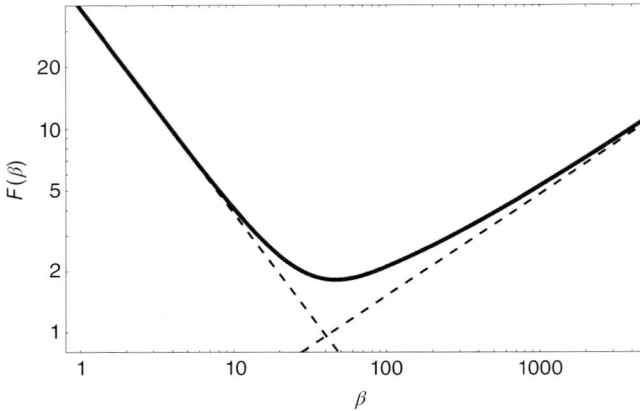

Figure 11.4 The normalized quality factor $F(\beta)$ is a non-monotonic function of the dimensionless frequency $\beta = (h/\delta)^2$, where δ is the Stokes length. Operation in the low-β regime ($\beta < 46$) is advantageous due to the increasing quality factor with increasing viscosity and/or decreasing channel height.

For higher order modes, Sader *et al.* [17] have shown that the quality factor drops significantly as a function of mode number, and that the decrease is more pronounced for devices operating in the regime $\beta \ll \beta_{min}$.

So far, all considerations regarding the quality factor have been limited to the case in which the fluidic channel is centered precisely on the neutral plane of a cantilever beam. Unfortunately, this condition is very difficult to satisfy in practice due to finite fabrication tolerances. In the case $\beta \ll \beta_{min}$, damping tends to increase strongly with off-axis placement, also showing an interesting dependence on fluid compressibility. In contrast, devices operating at $\beta \gg \beta_{min}$ are only minimally affected by off-axis placement. This behavior may be explained by the fact that the off-axis channel effectively expands and contracts during bending, forcing fluid in and out of the device. Energy is required for this pumping action, which leads to a decrease in quality factor with increasing viscosity. Another interesting result is that the detrimental effect of off-axis placement can partly be compensated by using a beam material with a high Poisson's ratio. Off-axis effects are not included in the simple model above, but an in-depth discussion of their many fascinating facets can be found in the specialized literature [14, 16, 18].

From an understanding of the principal mechanisms of energy dissipation in the liquid layer, some general guidelines for the design of microfluidic resonators emerge. Importantly, the operating regime around β_{min} should be avoided. If the application does not require a large channel height, for example, in order to pass eukaryotic cells, then one can take advantage of the strong confinement of the liquid in the low inertia regime to maximize Q. Care must be taken, however, to avoid dissipation due to the pumping effect if the channel cannot be centered precisely about the neutral plane. The opposite extreme of high inertia can be implemented by designing devices with a high-resonance frequency. However, due to the $F(\beta) \sim \sqrt{\beta}$ dependence in this regime, this can only provide an incremental improvement in Q.

11.2.3
Effect of Surface Reactions

A unique characteristic of microfluidic resonators is their very high surface-to-volume ratio, with channel heights in the micrometer and sub-micrometer regime and channel lengths on the order of several hundred micrometers. Here we look at the impact of this special sensor geometry on surface-based molecular interaction measurements. After considering sensitivity, we will turn to the connections between fluid flow, reaction kinetics, and solute transport, which are important in making accurate quantitative measurements of the rate constants of chemical and biomolecular interactions.

Surface adsorbed mass can be treated analogous to a change in fluid density provided that the adsorption proceeds uniformly along the full length of the channel. The main difference to bulk measurements here is that the mass increment is confined to the solid–liquid interface rather than being distributed throughout the channel volume. This difference plays only a minor role in devices that are thin compared to their length and width. In a device of uniform cross section, one can write the adsorbed mass per unit length of the channel as $\Delta\mu_A = s_{fs}\Delta\sigma$, where s_{fs} describes the perimeter of the channel cross-section (Figure 11.1) and $\Delta\sigma$ is the adsorbed mass per area. With these definitions one can write

$$\frac{\Delta\omega}{\omega} \approx -\frac{1}{2}\frac{s_{fs}\,\Delta\sigma}{A_s\rho_s + A_f\rho_f} \tag{11.10}$$

with $A_{s,f}, \rho_{s,f}$ as in Eq. (11.3). Sensitivity to surface adsorbed mass effectively improves with the ratio of internal surface area to total mass of the resonator. Miniaturization therefore provides an attractive route for improving sensitivity. For example, decreasing the height of the fluidic channel and the wall thickness makes the denominator of Eq. (11.10) smaller, yet the length of the solid–liquid interface remains large.

By functionalizing the inner surfaces of microfluidic resonators, selective real-time measurements of molecular interactions can be made. Such measurements are important for determining the characteristic rate constants of chemical and biomolecular binding reactions with high accuracy. Micromechanical resonators with embedded channels have a unique advantage for this purpose due to their extremely low volume consumption compared to state-of-the-art quartz crystal microbalance (QCM) or surface plasmon resonance (SPR) instruments.

A key question is under what conditions the accumulation of mass measured by the sensor accurately reflects the intrinsic rate of the reaction. To answer this question, we recapitulate below some of the criteria developed in the excellent review by Squires *et al.* [19]. Given a reaction of the form $A + B \underset{k_{off}}{\overset{k_{on}}{\rightleftharpoons}} AB$, where species A is the mobile species and species B has been immobilized to the channel surface, equilibrium is reached according to $\frac{\partial b}{\partial t} = k_{on}c_s \cdot (b_m - b) - k_{off}b$ if the concentration c_s of species A near the solid–liquid interface is constant. Here, $b = [AB]$ is the area density of molecules A bound to the surface, and b_m denotes the immobilized area density of species B. k_{on} and k_{off} denote the on-rate and the off-rate of the binding interaction between species A and B.

The first step is to recognize that the rate of binding to the surface is fundamentally limited in two ways: First, the chemical reaction itself imposes an upper limit on binding. Second, the replenishment of molecules by convection and diffusion must be sufficiently fast to match the adsorption to the surface. If the rate of binding exceeds the rate at which A can be replenished, significant depletion results and the observed kinetics is not accurately described by a first-order model. Otherwise, the solution concentration will be relatively constant along the channel, causing mass to adsorb homogeneously inside the device.

To arrive at a precise criterion for convection and diffusion to be "sufficiently fast," Squires *et al.* [19] first considered the maximum flux of molecules to the surface in a purely transport limited system where $k_{on} \to \infty$ and $k_{off} \to 0$. This means that all molecules that collide with the wall are adsorbed, and all other molecules pass through the channel unaffected. Molecules at the channel center require an average time $t_{diff} \sim h^2/D$ to diffuse to one of the walls (D denotes the diffusion coefficient). It would take the same molecules a time $t_{conv} \sim h/v_{center}$ to be carried a distance h along the axial direction (v_{center} denotes the flow velocity at the center of the channel). The ratio between these timescales is the Peclet number $Pe_H = h \cdot v_{center}/D$. For a channel of aspect ratio $\lambda = L/h$, one can see that if $\lambda/Pe_H \gg 1$, wall collisions are frequent, whereas if $\lambda/Pe_H \ll 1$, a significant fraction of molecules pass through the channel without interacting with the surface. In the latter case there is a depletion zone just above the surface. Although the thickness of this zone grows with distance from the channel entrance, it does not extend to the center of the channel. The total flux of molecules diffusing through the depletion zone is well approximated by $J_D \approx 0.81 \cdot Dc_0 W(6l^2 Pe_H)^{\frac{1}{3}}$, where c_0 is the initial concentration [19].

Comparing the transport limited flux J_D with the flux $J_R \approx L \cdot W \cdot k_{on} \cdot c_0 \cdot b_m$ in a hypothetical reaction limited system provides a quantitative measure of the relative significance of mass transport in a given experiment. This ratio between J_R and J_D is known as the Damkohler number, and for the specific case considered here we find

$$Da = \frac{Lk_{on}b_m}{0.81 \cdot D(6\lambda^2 Pe_H)^{1/3}} \tag{11.11}$$

With this on hand, it can be judged quickly whether a system operates in the reaction-limited regime, where $Da \ll 1$ or in the mass transport-–limited regime $Da \gg 1$.

Based on the requirement $Da \ll 1$ for a reaction-limited system, Arlett *et al.* [20] have suggested $L^* = 1.2D^2q/(b_m{}^3h^2k_{on}{}^3W)$ as a guideline for the maximum channel length for which mass accumulation on the surface can be considered

reaction limited. The expression for L^* highlights the sensitivity of sample depletion on the density of binding sites and reaction rate. Due to the cubic dependence
on the on-rate (k_{on}) and the surface density of available binding sites (b_m), L^* can
span an enormous range from less than 1 μm to several meters, even in the same
geometry. Fortunately, all that is important in most practical applications is that L^*
is longer than the embedded channel, and a lower bound estimate is usually sufficient. One should also note that in real implementations, the maximum pressure
that can be applied is typically limited, and with the volumetric flow rate $q \sim h^3$,
one finds $L^* \sim h$.

11.2.4
Single Particle Measurements

The direct weighing of micro- and nanoparticles suspended in a carrier liquid
represents a particularly interesting and useful type of measurement that
can be conducted with embedded channel microresonators. Here $\Delta \rho$
in Eq. (11.2) is approximately described by a delta function $\Delta \rho(x, y, z) =
m_p \cdot \delta(x - x_p)\delta(y - y_p)\delta(z - z_p)$. Again, it should be pointed out that the influence
of $\Delta \rho$ on u_z can usually be neglected if the particle mass is much smaller than the
mass of the resonator.

Using Eq. (11.2), this leads to the following formula for the effect of the particle
on the resonance frequency:

$$\frac{\Delta \omega}{\omega} = -\frac{1}{2}\frac{m_p}{m^*} u_z(x_p, y_p)^2 \tag{11.12}$$

As expected, there is a pronounced dependence on particle position through the
local amplitude of oscillation [21]. Another noteworthy point is that, to first order,
the frequency shift resulting from multiple particles is the linear superposition of
the frequency shifts for the individual particles.

Choosing a long and thin cantilever resonator as example, the displacement field
is obtained by solving the Euler-Bernoulli equation [22]

$$EI\frac{\partial^4 u_z}{\partial x^4} = \mu_A \omega^2 u_z \tag{11.13}$$

which we write here in the frequency domain. In Eq. (11.13), E is Young's
modulus, I the area moment of inertia, and μ_A denotes the cross-section mass
$\mu_A = \iint \rho(y, z)\,dy\,dz$. Here we only consider beams of constant cross-section,
and the dependence of ρ on x has therefore been dropped. The solution to Eq.
(11.13) for a beam with one clamped and one free end is given by

$$u_z(x) = A \cdot \frac{\cos\left(\lambda_n {}^x\!/_L\right) - \cosh\left(\lambda_n {}^x\!/_L\right) + a_n \cdot \sin\left(\lambda_n {}^x\!/_L\right) - a_n \cdot \sinh\left(\lambda_n {}^x\!/_L\right)}{|\cos(\lambda_n) - \cosh(\lambda_n) + a_n \cdot \sin(\lambda_n) - a_n \cdot \sinh(\lambda_n)|}$$

$$\tag{11.14}$$

where the constants λ_n and a_n depend on the mode of vibration as given in
Table 11.1 [22].

Table 11.1 Parameters λ_n and a_n for different modes of a vibrating lever. The parameter λ_n is related to the resonance frequency by $\omega = \lambda_n^2 \left(\frac{1}{L}\right)^2 \sqrt{\frac{EI}{\mu_A}}$.

Mode number n	λ_n	a_n	ω/ω_1
1	1.875	−0.7341	1
2	4.694	−1.0185	6.267
3	7.855	−0.9992	17.551
4	10.996	−1	34.393
5	14.137	−1	56.848
>5	$\sim \pi(2n-1)/2$	−1	$(\lambda_n/\lambda_1)^2$

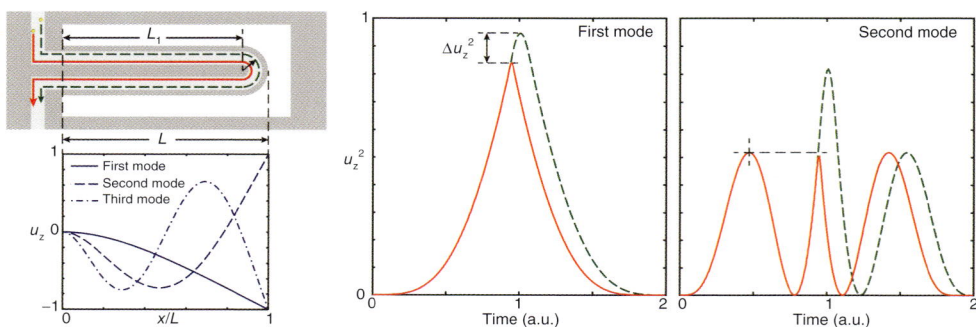

Figure 11.5 The sensitivity profile of resonance frequency to particles in the channel is given by the square of the normalized deflection function u_z. The first three modes for a cantilever resonator of length L are plotted on the left, and the corresponding single-particle signatures for modes 1 and 2 are shown on the right. Particles traveling along the inner path (red solid line) give rise to a smaller center peak than particles following the outer streamline (green dashed line). The magnitude of peaks at other *anti*-node locations is independent of the transversal particle position for modes $n \geq 2$.

The position dependence of the sensitivity is one of the most significant limitations in particle mass measurements with embedded channel cantilever resonators. As shown in Figure 11.5, particles can approach the apex of the cantilever most closely on the outer (green dashed line) path. The inner (red solid line) path turns at a point that is closer to the cantilever base by the channel width W. This ambiguity in position inevitably translates into a difference in the measured frequency shift. For example, Figure 11.5 illustrates a device in which the straight portion of the channel has the length $L_1 = 0.93\,L$ and the channel width was chosen as $W = 0.04\,L$. The difference Δu_z^2 is approximately 12% of the mean. An accurate fit of the data with a model in which the exact transverse position of the particle in the channel is left as a parameter could, in principle, recover the precise amplitude. In practice, however, this may be confounded by noise or by systematic errors, such as an unsteady flow velocity. An elegant solution to this problem has been demonstrated by Lee *et al.* [23] using higher vibrational modes. Their method exploits the existence of local extrema in $u_z(x)^2$ in modes $n \geq 2$ to decouple particle mass and position with high fidelity in the presence of noise.

11.3
Device Technology

11.3.1
Fabrication

The integration of microfluidic channels inside micromechanical resonators poses a unique combination of technological challenges. Most importantly, robust operation requires the liquid to be reliably separated from the exterior of the resonator. In addition, as the embedded channel size is obviously limited by the size of the resonator, the flow resistance can be exceedingly large. On-chip integration of a hierarchical microfluidic network with large and small channels is thus required in order to enable rapid exchange of samples [24]. The challenge in implementation consequently comprises the fabrication of the resonator and embedded channel itself, and the connection of this device to the macroscopic world. Practical applications often depend critically on proper packaging and fluidic integration, and this important aspect is easily overlooked during device engineering.

The two principal strategies for fabricating micro- and nanomechanical resonators with embedded fluidic channels can be divided into sacrificial layer etching and wafer bonding. Table 11.2 summarizes the general structures, common material systems, and key design constraints of the two approaches. Sacrificial layer methods are attractive when the mass of the resonator is to be minimized. To form the channel, a material which is easily etched is deposited and patterned on a substrate and subsequently encapsulated with the structural material of the resonator. After patterning openings by photolithography, the sacrificial layer is removed by chemical etching, leaving an empty shell. A portion of the channel is then undercut to create the mechanical resonator. Due to the substantial length of the channel, the material system must ensure high etch rate selectivity between the sacrificial layer and the wall material. For example, Westberg *et al.* [25] fabricated resonators with embedded channels of several hundred micrometers length using the Aluminum metallization of a CMOS process as sacrificial material and the passivation dielectrics as structural material (Figure 11.6a). Several other groups have employed the combination of low-stress, low pressure chemical vapor deposited silicon nitride and polysilicon, using silicon nitride as the structural material and polysilicon as sacrificial layer (Figure 11.6b) [24, 26–30]. While patterning of the sacrificial layer is commonly done by photolithography, the Craighead laboratory has demonstrated an elegant alternative using sacrificial electrospun nanofibers for the templated physical vapor deposition of glass (Figure 11.6c) [31].

Table 11.2 Resonators with embedded channels can be fabricated by sacrificial layer etching or via wafer bonding. Sacrificial layer processes typically result in topography (left), but can also incorporate a planarization step (middle) to facilitate subsequent packaging. Wafer bonding technology inherently results in a planar topography and places very few restrictions on the embedded channel geometry (right).

	Sacrificial layer processes			Wafer bonding processes
	Structural (walls)	**Sacrificial**	**Etch selectivity**	
Materials	PVD or PECVD SiO_2, Si_xN_y	Aluminum	Very high	Single-crystal silicon
		Polycarbonate	Very high	
	LPCVD Si_xN_y	Polysilicon	$\sim 1000 : 1$(KOH)	
		SiO_2	$\sim 100 : 1$ (HF)	
		PSG	$\sim 100 : 1$ (HF)	
	Parylene	Positive photoresist	Very high	
Channel height	~ 50 nm to 2 µm			~ 500 nm to 1000 µm
Channel length	~ 100 µm – several millimeters (process dependent)			Not limited by process
Sidewalls thickness	~ 100 nm to 1 µm			> 1 µm
Top/bottom thickness	~ 100 nm to 1 µm			> 300 nm
Actuation	Electrostatic, thermomechanical (other possible)			Electrostatic (other possible)
Detection	Optical (external), thin film piezoresistors			Optical (external), capacitive, silicon piezoresistors

There are two important limitations to the sacrificial layer technology. First, the channel height cannot exceed the thickness range of common thin film deposition technology. This may in principle be overcome by electroplating, but so far no devices based on this process have been reported. The second limitation concerns the formation of a hermetic seal for vacuum packaging. This is challenging due to the non-planar surface inherent to sacrificial layer technologies. Surface planarization can alleviate this problem, but strong restrictions in suitable packaging materials and processes remain [26].

Wafer bonding processes provide an attractive alternative to sacrificial layer technology for the fabrication of microfluidic resonators from single crystal silicon. In this approach, microchannels are first defined by dry or wet etching in a silicon substrate and subsequently sealed by fusion bonding to another silicon wafer. This sequence yields engineered substrates with embedded channels, which are then processed to define micromechanical resonators by releasing sections using bulk or surface micromachining. Among the first applications of this technology were the fluid density and mass flow meters developed by Enoksson *et al.* and Corman *et al.* (Figure 11.6e) [12, 36, 37]. Sparks *et al.* [38] later used the wafer bonding approach in combination with deep reactive ion etching to create fluid density sensors with sub-microliter swept volume (Figure 11.6f). By packaging these resonators under vacuum and using on-chip getters to counteract outgassing, high-quality factors in the range of 5000–20 000 could be maintained.

Ease of packaging constitutes an important strength of the wafer bonding approach. Moreover, the use of single crystal silicon as structural material and the flexibility in buried channel layout greatly contribute to precision and stability of the overall measurement system. On the other hand, designs based on wafer bonding technology do not scale to small size as well as devices made by sacrificial layer methods. The main reason for this limitation is that the sidewalls of bonded channels are difficult to shrink below a few micrometers. Nonetheless, very precise measurements have been obtained with bonded devices such as the ones shown in Figure 11.6g, h [33, 34, 39].

An interesting combination of some of the advantages of both wafer bonding and sacrificial layer methods is the technique proposed by Dijkstra *et al.* [40]. The process starts with a sequence similar to the conventional SCREAM (single crystal reactive etching and metallization) technique to form deep channels buried under the surface of a silicon substrate [41]. A layer of low pressure chemical vapor deposition (LPCVD) silicon nitride is then conformally deposited through the openings inside the buried channels. Portions of these channels can then be suspended by bulk etching of the silicon substrate. Due to the nature of the process, this method is best suited for the fabrication of channels with thin (<1 μm) sidewalls and a cross section of several 10 μm.

11.3.2
Packaging Considerations

Adequate packaging of embedded channel microresonators is extremely important to render these devices practical for real-world applications. The fluidic interface requires particular attention. This interface needs to bridge the difference in volume between macroscopic reservoirs and on-chip fluidic channels of micrometer or sub-micrometer size. Conventional liquid handling equipment can manipulate volumes down to approximately 1 μl. A multiple of this is generally required to completely fill the connections between the chip and external pumps and valves. Any exchange of sample thus requires several microliters to be displaced through the microfluidic chip. As long as the size of the embedded resonator channel is large enough, this is easily possible by a direct in-line connection scheme (Figure 11.7a). A useful rule-of-thumb is that the channel height and width should be greater than approximately 15–30 μm if standard low-pressure pumps and fittings are used.

To circumvent the need for high pressure in embedded channel resonators with micrometer-scale cross-sections, an on-chip microfluidic bypass network should be employed (Figure 11.7b). The resonator, which has a high-flow resistance, is connected between two larger microfluidic channels acting as on-chip sample loops. This scheme allows samples to be rapidly exchanged in the bypass channels without the need to flow large volumes of liquid through the picoliter-scale resonator.

The packaging technology of choice for a specific application depends on the types of samples to be used, the actuation and detection principle, and on whether or not vacuum packaging is required. Anodic bonding of Pyrex to Silicon excels both in chemical compatibility and hermeticity [33, 36]. Devices packaged in this way can almost always be regenerated by cleaning with aggressive chemicals. This is significant, since reusability and a long lifetime of the MEMS part is essential for efficient and economical application of the technology. In addition, a chemically inert flow path is highly desirable when the internal surfaces of the device need to be modified, either for the purpose of chemical or biomolecular recognition, or for passivation against non-specific binding. Silicon and glass surfaces are also advantageous in this regard, as they can be derivatized efficiently by a range of covalent and non-covalent coupling agents.

A planar surface topography is required for packaging by wafer bonding without intermediate layers, especially for forming a vacuum cavity around the microresonator. Electrical feedthroughs for actuation and/or detection increase the complexity of the process [42]. It is possible to circumvent the problem of creating feedthroughs by moving the electrical connections away from the bond interface, either to the outer surface of the glass [36], or to underneath the resonator [33], where soft intermediate layer bonding can be used without interfering with the fluidics.

An elegant method of packaging microfluidic resonators by intermediate layer bonding has been described by Westberg *et al.* [25]. Thick gold frames were electroplated around the inlet and outlet of a fluidic resonator to allow in-line connections to reservoirs by solder or thermocompression bonding (Figure 11.6a). Gold is an excellent material for such interfaces from a chemical

---

Figure 11.6 Micro- and nanomechanical resonators with embedded fluidic channels fabricated by sacrificial layer etching (a–d) and wafer bonding methods (e–h). (a) CMOS passivation layers as structural material/aluminum sacrificial [32]. (b) LPCVD low-stress Si_xN_y structural material/LPCVD polysilicon sacrificial [26]. (c) Sputtered glass as structural material/electrospun polycarbonate fibers as sacrificial [31]. (d) LPCVD Si_3N_4 structural material/LPCVD polysilicon sacrificial [27]. (e–h) Silicon–silicon direct bonding with enclosed cavities [12, 33–35]. (Images reproduced with permission.)

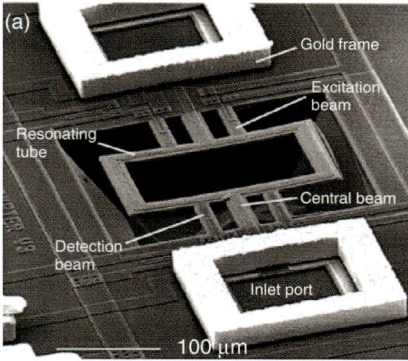

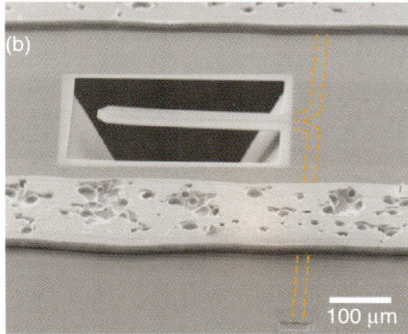

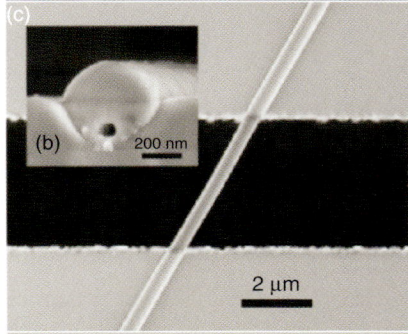

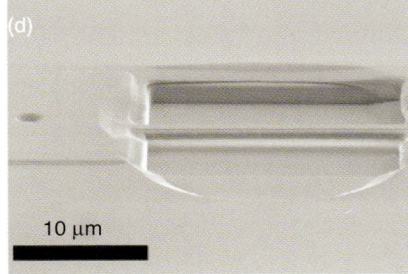

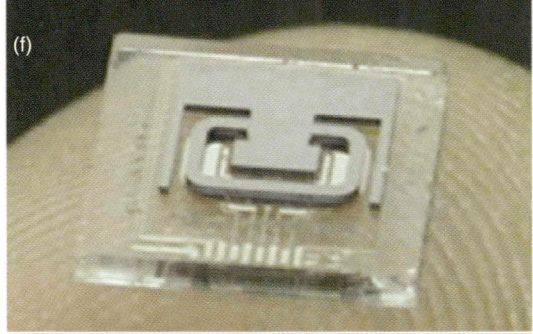

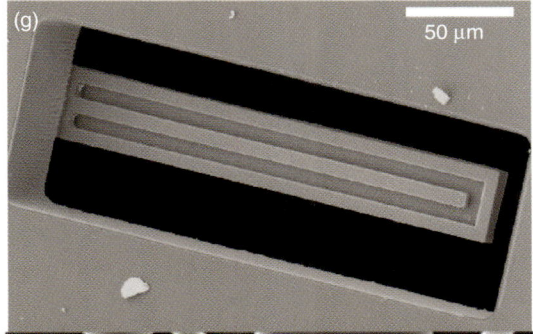

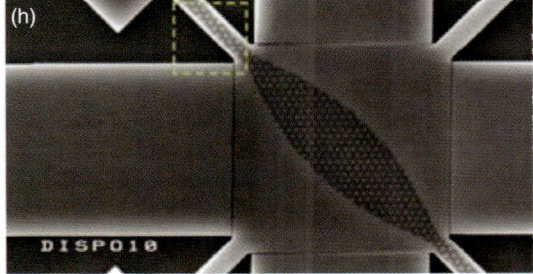

(a)

(b)

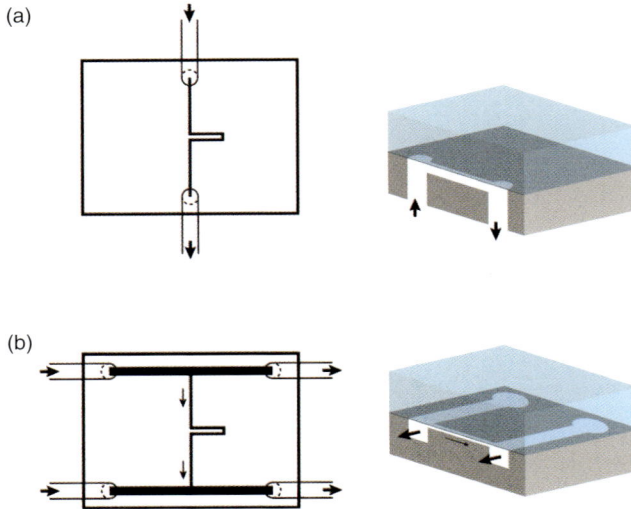

Figure 11.7 Fluidic channels embedded inside micromechanical resonators often have extremely high-flow resistance. Up to channel heights of ~30 µm, samples can be efficiently delivered through direct in-line connection with external tubing (a). Smaller channels should employ an on-chip microfluidic bypass scheme (b).

resistance point of view. However, for corrosive applications one must carefully ensure that adhesion and/or seed layers are adequately protected from contact with the fluid to avoid delamination.

Alternatively, glass frit bonding has been used by different groups for the packaging of embedded channel microresonators [26, 43]. Bonding is carried out at a temperature between 400 and 450 °C, at which the frit material is soft and can form a hermetic seal even over significant surface topography [44]. The work by Sparks *et al.* employed an in-line connection scheme which allowed the flow path to be separated from the frit material. In contrast, the silicon nitride devices by Burg *et al.* [26] incorporated microfluidic bypass channels, which were defined by lines of glass frit, as seen in Figure 11.6b.

If vacuum encapsulation and high chemical resistance are not required, room temperature packaging using polymers can be an attractive alternative for packaging microfluidic resonator. For example, Agache *et al.* [34] have created microfluidic delivery channels with in-line connected hollow plate resonators using a dry film photoresist in combination with a glued Pyrex capping wafer. Conventional microfluidics based on polydimethylsiloxane (PDMS) can also be adequate for some applications, although the simplicity of fabrication must be carefully weighed against the drawbacks of lower durability and porosity of this polymer [45].

11.4
Applications

11.4.1
Measurements of Fluid Density and Mass Flow

Measurements of fluid properties were among the first applications of microme-chanical resonators with embedded fluidic channels [37]. After the pioneering work by Enokssen *et al.* in the mid-1990s, academic laboratories and commercial ventures continued to advance these devices by further miniaturization and sys-tems integration [25, 43]. Integrated fluid density meters offer a distinct advantage for measurements with rare and precious samples, as they require significantly smaller volume than conventional vibrating tube fluid densitometers. Although physical sensitivity does not fundamentally improve with scaling, other impor-tant device parameters do benefit from miniaturization. In particular, the quality factor, temperature control, and sensitivity to the environment scale favorably with size. Applications of fluid density measurements are numerous and encom-pass the range from monitoring purity of products in the petrochemical and fuel industry to basic research in biology, chemistry, and physics [46]. An interest-ing application of density measurements for biological and chemical analysis has been described by Son *et al.* [47], who incorporated Suspended Microchannel Res-onators of 10 pl volume as a universal detector in line with a gel filtration column in high-performance liquid chromatography (HPLC). This method exhibited a sim-ilar limit of detection (LOD) to UV-visible optical absorbance detectors and an over 10-fold better LOD than refractive index detection using glycine as a stan-dard.

Related to integrated fluid density sensors are Coriolis mass flow meters, which also sense a volumetric property of the medium. As flow meters are generally used in an in-line configuration, a small internal volume translates directly into the abil-ity to measure low flow rates. The range of applications again is extremely broad, ranging from analytical instrumentation to process monitoring and precision fluid dispensing.

11.4.2
Single Particle and Single Cell Measurements

In recent years, the ability of microfluidic resonators to measure the mass and density of single micrometer and sub-micrometer particles has emerged as an enabling technology. Depending on the size of particles, one can delineate two major threads of applications. Nanoparticle characterization constitutes the first category. One common problem in this field is the detection of protein aggregates or other impurities in biotechnological products and pharmaceutical formulations [48–50].

The second category comprises measurements on sub-visible particles spanning the range from 100 nm to ~30 μm diameter. These include nearly all types of biological cells from bacteria to fungi and most mammalian cells. There are many fascinating questions in cell biology, physiology, and pharmacology in which precision measurements of mass at the single-cell level may provide valuable information. Some examples of such questions, which have been investigated through the pioneering work of Scott Manalis and his team at MIT, are described in the following.

Knudsen *et al.* [51] have described the use of mass measurements of bacteria to rapidly detect antibiotic resistance. This approach, which was based on monitoring osmotic shock response, was successful for some but not all strains of bacteria tested. In another study, Bryan *et al.* [52] measured for the first time the evolution of cell mass, volume, and density during the cell cycle of yeast with high precision. Both of these studies were based on population averages. Even richer information can be gained from single-cell measurements of cell growth rate, but such measurements have long been challenging to achieve. The difficulty is that microfluidic resonators inherently operate in a flow-through mode in which particles pass the detector only once. To overcome this limitation, the Manalis laboratory developed two complementary approaches.

In the first approach, mechanical traps were incorporated inside the embedded channel to stably maintain cells at the point of highest device sensitivity. This made it possible to weigh cells in different fluids to determine their mass density with high precision [53].

In the second approach, cells were repeatedly passed through the resonator by periodically reversing the flow direction using a sophisticated closed-loop control system [54, 55]. Using this method, Son *et al.* [56] have for the first time been able to conduct high-resolution mass measurements showing evidence that the G1-S transition in L1210 mouse lymphocytes is triggered when the cells achieve a critical growth rate rather than a critical size threshold. Several key technical advancements allowed this observation. First, the use of hydrodynamic focusing allowed a drastic reduction in position-dependent variability in mass measurements. Second, an advanced fluid control system minimizing shear stress enabled the observation of cells over several cell cycles, and third, integration of the mass measurement setup with an optical microscope allowed the simultaneous measurement of mass and fluorescent cell-cycle markers.

11.4.3
Surface-Based Measurements

Surface-based measurements with Suspended Microchannel Resonators allow the characterization of molecular interactions by monitoring the rate of adsorption and desorption of mass at the channel walls in real time. By maintaining a high-flow velocity, the influence of convective and diffusive mass transport limitations can be minimized. A significant advantage of embedded channel resonator devices for surface-based biomolecular interaction analysis is their low

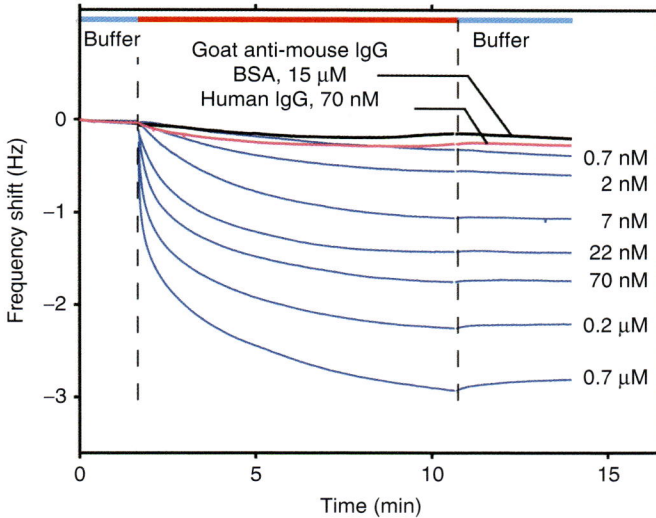

Figure 11.8 Real-time measurement of binding of IgG proteins in solution to immobilized *anti*-IgG at the inner surfaces of suspended microchannel resonators. (Reproduced with permission from Ref. [33].)

sample consumption. Figure 11.8 shows an example of the binding between IgG proteins and immobilized *anti*-IgG antibodies. Measurements were conducted at a flow rate on the order of $100 \, \text{nl} \, \text{min}^{-1}$, which is two to three orders of magnitude smaller than rates commonly used with commercial SPR instruments. A critical requirement for such measurements is that one of the binding partners is immobilized in a way that minimally interferes with the interaction, and that nonspecific binding is small. Biochemical functionalization of embedded channel resonators is especially challenging due to the inaccessibility of the interior device surfaces. In addition, it is difficult to quantitatively validate the success of functionalization *in situ*. Electrostatic adsorption of poly-l-lysine (PLL) grafted with PEG-biotin has been used in the measurements shown in Figure 11.8. An alternative to PLL is 3,4-Dihydroxy-l-phenylalanine (DOPA), which can strongly bond to a wide range of surfaces. DOPA has been combined with a highly efficient *anti*-fouling matrix, poly(carboxybetaine methacrylate) (pCBMA) by the groups of Scott Manalis and Shaoyi Jiang for the specific SMR-based detection of cancer biomarkers in undiluted serum at the ng/ml level [57].

Micromechanical resonators with embedded channels have shown significant potential in proof-of-concept studies to work as sensitive detectors for specific biological molecules. However, any label-free detection technology is limited in selectivity by the surface functionalization. This must be kept in mind when considering the viability of any physical detection technology for diagnostic applications in which complex samples, such as patient serum, are used. The particular strength of embedded channel devices is their extraordinarily small volume and inherent compatibility with microfluidic systems. It is likely that this aspect will prove to be of great use for the development of new detectors for the biochemical analysis of small organism and perhaps single cells.

11.5
Conclusion

In summary, there are several aspects that make fluidic micro- and nanomechanical resonators an exciting class of microelectromechanical systems with unique physical characteristics and a wide body of applications. Of particular significance is that these devices maintain a high quality factor even when they are filled with a viscous fluid. While the fluid-induced damping can be undetectably small in some device geometries, its magnitude is dependent on the device design and becomes more prominent in resonators with large channel cross sections and high resonance frequencies. In general, the physics of damping in these systems is complex and exhibits an intriguing, non-monotonic dependence on fluid viscosity.

Through their unique design and high quality factors, embedded channel resonators have opened the door to measurements of fluid density and flow, characterization of nanoparticles and cells, and to ultra-low volume mass-based molecular interaction studies. They have thus become versatile instruments for measurements in many fields including physical chemistry, materials science, and biology.

Future developments will likely continue to advance the boundaries of miniaturization and frequency readout precision for improved mass sensitivity. Challenges that need to be addressed in this direction lie in the fabrication of mechanically robust suspended nanofluidic channels, more compact packaging, and in the design of improved displacement sensing technologies. The displacement sensing system is of particular importance, since a reduction in internal volume will have to be accompanied either by large scale parallelization or by an increase in the linear flow velocity per device in order to maintain a practically useful sample throughput. The readout of small devices therefore must be able to operate at a comparatively high bandwidth while providing a noise floor near the thermomechanical noise of the resonator. Ultimately, it is conceivable that the technology could approach the levels of resolution that can today be attained by solid nanomechanical resonator sensors in vacuum. This would provide fascinating opportunities for the mass spectrometric characterization of biological macromolecules and large biomolecular machines in their native solution environment.

Devices with channel cross sections in the micrometer range have already shown great potential as universal detectors for the characterization of physical, chemical, and biological processes by way of fluid density or single-particle mass measurements. This technology has become mature and reliable over the past 10 years, and it is beginning to be available commercially today. The widespread availability should lead to new uses in laboratory assays, for example, in the context of studies on cell growth regulation (microbiology and cancer research), and in many fields where micro- and nanoparticle characterization is of interest.

The latter domain may include research on new vectors for drug delivery or the use of dispersed particles as microcarriers for surface-based molecular interaction measurements.

Research in microfluidic resonators will continue along several axes from fundamental theory to fabrication technology and instrumentation development. The technology is thus at the interface of several disciplines, and its advancement will likely continue to enable a diverse range of applications that benefit from precision frequency-based measurements in liquid.

References

1. Jensen, K., Kim, K., and Zettl, A. (2008) An atomic-resolution nanomechanical mass sensor. *Nat. Nanotechnol.*, **3** (9), 533–537.
2. Lassagne, B. *et al.* (2008) Ultrasensitive mass sensing with a nanotube electromechanical resonator. *Nano Lett.*, **8** (11), 3735–3738.
3. Naik, A.K. *et al.* (2009) Towards single-molecule nanomechanical mass spectrometry. *Nat. Nanotechnol.*, **4** (7), 445–450.
4. Sawano, S., Arie, T., and Akita, S. (2010) Carbon nanotube resonator in liquid. *Nano Lett.*, **10** (9), 3395–3398.
5. Sader, J.E. (1998) Frequency response of cantilever beams immersed in viscous fluids with applications to the atomic force microscope. *J. Appl. Phys.*, **84** (1), 64–76.
6. Beardslee, L.A. *et al.* (2010) Liquid-phase chemical sensing using lateral mode resonant cantilevers. *Anal. Chem.*, **82** (18), 7542–7549.
7. Humphris, A.D.L., Tamayo, J., and Miles, M.J. (2000) Active quality factor control in liquids for force spectroscopy. *Langmuir*, **16** (21), 7891–7894.
8. Tamayo, J. (2005) Study of the noise of micromechanical oscillators under quality factor enhancement via driving force control. *J. Appl. Phys.*, **97** (4), 044903.
9. Cleland, A.N. (2005) Thermomechanical noise limits on parametric sensing with nanomechanical resonators. *New J. Phys.*, 7, 235.
10. Arlett, J.L. and Roukes, M.L. (2010) Ultimate and practical limits of fluid-based mass detection with suspended microchannel resonators. *J. Appl. Phys.*, **108** (8), 084701.
11. White, F.M. (1991) *Viscous Fluid Flow*, McGraw-Hill Professional Publishing.
12. Enoksson, P., Stemme, G., and Stemme, E. (1997) A silicon resonant sensor structure for Coriolis mass-flow measurements. *J. Microelectromech. Syst.*, **6** (2), 119–125.
13. Blanco-Gomez, G. and Agache, V. (2012) Experimental study of energy dissipation in high quality factor hollow square plate MEMS resonators for liquid mass sensing. *J. Microelectromech. Syst.*, **21** (1), 224–234.
14. Burg, T.P., Sader, J.E., and Manalis, S.R. (2009) Nonmonotonic energy dissipation in microfluidic resonators. *Phys. Rev. Lett.*, **102** (22), 228103.
15. Sparks, D., Cruz, V., and Najafi, N. (2007) The resonant behavior of silicon tubes under two-phase microfluidic conditions with both microbeads and gas bubbles. *Sens. Actuators, A-Phys.*, **135** (2), 827–832.
16. Sader, J.E., Burg, T.P., and Manalis, S.R. (2010) Energy dissipation in microfluidic beam resonators. *J. Fluid Mech.*, **650**, 215–250.
17. Sader, J.E., Lee, J., and Manalis, S.R. (2010) Energy dissipation in microfluidic beam resonators: dependence on mode number. *J. Appl. Phys.*, **108** (11), 114507.
18. Sader, J.E. *et al.*, (2011) Energy dissipation in microfluidic beam resonators: effect of Poisson's ratio. *Phys. Rev. E*, **84** (2), 026304.
19. Squires, T.M., Messinger, R.J., and Manalis, S.R. (2008) Making it stick: convection, reaction and diffusion in

surface-based biosensors. *Nat. Biotechnol.*, **26** (4), 417–426.

20. Arlett, J.L., Myers, E.B., and Roukes, M.L. (2011) Comparative advantages of mechanical biosensors. *Nat. Nanotechnol.*, **6** (4), 203–215.

21. Dohn, S. *et al.* (2007) Mass and position determination of attached particles on cantilever based mass sensors. *Rev. Sci. Instrum.*, **78** (10), 103303.

22. Sarid, D. (1994) *Scanning Force Microscopy*, Oxford Series in Optical and Imaging Sciences, Oxford University Press, New York.

23. Lee, J., Bryan, A.K., and Manalis, S.R. (2011) High precision particle mass sensing using microchannel resonators in the second vibration mode. *Rev. Sci. Instrum.*, **82** (2), 023704.

24. Burg, T.P. and Manalis, S.R. (2003) Suspended microchannel resonators for biomolecular detection. *Appl. Phys. Lett.*, **83** (13), 2698–2700.

25. Westberg, D. *et al.* (1997) A CMOS-compatible fluid density sensor. *J. Micromech. Microeng.*, 7 (3), 253–255.

26. Burg, T.P. *et al.* (2006) Vacuum-packaged suspended microchannel resonant mass sensor for biomolecular detection. *J. Microelectromech. Syst.*, **15** (6), 1466–1476.

27. Barton, R.A. *et al.* (2010) Fabrication of a nanomechanical mass sensor containing a nanofluidic channel. *Nano Lett.*, **10** (6), 2058–2063.

28. Khan, M.F. *et al.* (2011) Fabrication of resonant micro cantilevers with integrated transparent fluidic channel. *Microelectron. Eng.*, **88** (8), 2300–2303.

29. Berenschot, J.W. *et al.* (2002) Advanced sacrificial poly-Si technology for fluidic systems. *J. Micromech. Microeng.*, **12** (5), 621–624.

30. Stern, M.B., Geis, M.W., and Curtin, J.E. (1997) Nanochannel fabrication for chemical sensors. *J. Vac. Sci. Technol., B*, **15** (6), 2887–2891.

31. Verbridge, S.S. *et al.* (2005) Suspended glass nanochannels coupled with microstructures for single molecule detection. *J. Appl. Phys.*, **97** (12), 124317-1–12437-4.

32. Westberg, D. *et al.* (1999) A CMOS-compatible device for fluid density measurements fabricated by sacrificial aluminium etching. *Sens. Actuators, A-Phys.*, **73** (3), 243–251.

33. Burg, T.P. *et al.* (2007) Weighing of biomolecules, single cells and single nanoparticles in fluid. *Nature*, **446** (7139), 1066–1069.

34. Agache, V. *et al.* (2011) An embedded microchannel in a MEMS plate resonator for ultrasensitive mass sensing in liquid. *Lab Chip*, **11** (15), 2598–2603.

35. Smith, R. *et al.* (2009) A MEMS-based Coriolis mass flow sensor for industrial applications. *IEEE Trans. Ind. Electron.*, **56** (4), 1066–1071.

36. Corman, T. *et al.* (2000) A low-pressure encapsulated resonant fluid density sensor with feedback control electronics. *Meas. Sci. Technol.*, **11** (3), 205–211.

37. Enoksson, P., Stemme, G., and Stemme, E. (1995) Fluid density sensor-based on resonance vibration. *Sens. Actuators, A: Phys.*, **47** (1-3), 327–331.

38. Sparks, D. *et al.* (2003) Measurement of density and chemical concentration using a microfluidic chip. *Lab Chip*, **3** (1), 19–21.

39. Lee, J. *et al.* (2010) Toward attogram mass measurements in solution with suspended nanochannel resonators. *Nano Lett.*, **10** (7), 2537–2542.

40. Dijkstra, M. *et al.* (2007) A versatile surface channel concept for microfluidic applications. *J. Micromech. Microeng.*, **17** (10), 1971–1977.

41. Shaw, K.A., Zhang, Z.L., and Macdonald, N.C. (1994) Scream-I – a single mask, single-crystal silicon, reactive ion etching process for microelectromechanical structures. *Sens. Actuators, A: Phys.*, **40** (1), 63–70.

42. Lee, J. *et al.* (2011) Suspended microchannel resonators with piezoresistive sensors. *Lab Chip*, **11** (4), 645–651.

43. Sparks, D. *et al.* (2003) A variable temperature, resonant density sensor made using an improved chip-level vacuum package. *Sens. Actuators, A: Phys.*, **107** (2), 119–124.

44. Knechtel, R., Wiemer, M., and Fromel, J. (2006) Wafer level encapsulation of microsystems using glass frit bonding. *Microsyst. Technol. Micro Nanosyst. Inf. Storage Process. Syst.*, **12** (5), 468–472.

45. Burg, T.P. and Manalis, S.R. (2005) Microfluidic packaging of suspended microchannel resonators for biomolecular detection. 2005 3rd IEEE/EMBS Special Topic Conference on Microtechnology in Medicine and Biology, pp. 264–267.

46. Sparks, D. *et al.* (2008) Embedded MEMS-based concentration sensor for fuel cell and biofuel applications. *Sens. Actuators, A: Phys.*, **145**, 9–13.

47. Son, S. *et al.* (2008) Suspended microchannel resonators for ultralow volume universal detection. *Anal. Chem.*, **80** (12), 4757–4760.

48. Mach, H. *et al.* (2011) The use of flow cytometry for the detection of subvisible particles in therapeutic protein formulations. *J. Pharm. Sci.*, **100** (5), 1671–1678.

49. Patel, A.R., Lau, D., and Liu, J. (2012) Quantification and characterization of micrometer and submicrometer subvisible particles in protein therapeutics by use of a suspended microchannel resonator. *Anal. Chem.*, **84** (15), 6833–6840.

50. Zolls, S. *et al.* (2012) Particles in therapeutic protein formulations, part 1: overview of analytical methods. *J. Pharm. Sci.*, **101** (3), 914–935.

51. Knudsen, S.M. *et al.* (2009) Determination of bacterial antibiotic resistance based on osmotic shock response. *Anal. Chem.*, **81** (16), 7087–7090.

52. Bryan, A.K. *et al.* (2010) Measurement of mass, density, and volume during the cell cycle of yeast. *Proc. Natl. Acad. Sci. U.S.A.*, **107** (3), 999–1004.

53. Weng, Y.C. *et al.* (2011) Mass sensors with mechanical traps for weighing single cells in different fluids. *Lab Chip*, **11** (24), 4174–4180.

54. Godin, M. *et al.* (2010) Using buoyant mass to measure the growth of single cells. *Nat. Methods*, **7** (5), 387–390.

55. Grover, W.H. *et al.* (2011) Measuring single-cell density. *Proc. Natl. Acad. Sci. U.S.A.*, **108** (27), 10992–10996.

56. Son, S. *et al.* (2012) Direct observation of mammalian cell growth and size regulation. *Nat. Methods*, **9** (9), 910–912.

57. von Muhlen, M.G. *et al.* (2010) Label-free biomarker sensing in undiluted serum with suspended microchannel resonators. *Anal. Chem.*, **82** (5), 1905–1910.

12
Hermetic Packaging for Resonant MEMS

Matthew William Messana, Andrew Bradley Graham, and Thomas William Kenny

12.1
Introduction

One of the very first MEMS devices built was the resonant gate transistor [1, 2]; this device featured a released micromechanical resonator. Even back in 1965, the authors were aware of many factors that could influence the accuracy and stability of the resonant frequency of the MEMS device, including process parameter control, temperature, materials characteristics, fatigue and aging, and the environment that the resonator operates in. In the 50 years since the demonstration of early MEMS resonators, there has been significant progress on all of these issues, with much of that progress described in other chapters of this book. In this chapter, we focus on Hermetic Packaging of resonant micro electro mechanical systems (MEMS) devices, such as gyroscopes, resonators for time references, and other devices that rely on the relationship between the natural frequency of a hermetically sealed resonator and some signal of interest, such as temperature, stress, acceleration, or pressure.

One of the most important considerations in the design of any MEMS resonator is the selection and construction of the package. The characteristics of the package can determine many of the important characteristics of the resonator. Additionally, the package can represent a significant portion of the cost of the overall product. In an example from the book by Senturia, *Microsystem Design*, packaging for MEMS products can represent almost half of the overall cost of a representative pressure sensor product, while the silicon (MEMS and CMOS (complementary metal-oxide-semiconductor) circuit) may cost 1/3, and calibration and testing the balance [3]. The main reason for the high cost of packaging for MEMS is that, unlike electronics packaging, there is a much broader set of requirements for a MEMS package, making standardization difficult. This particular example of a packaging cost breakdown is for a pressure sensor, which must include a mounting site for the sensor element, while also providing an access port for external pressure, as well as all the usual electrical connections, electrical components, and similar quality control tests. Many of the same issues and cost

Resonant MEMS – Fundamentals, Implementation and Application, First Edition.
Edited by Oliver Brand, Isabelle Dufour, Stephen M. Heinrich and Fabien Josse.
© 2015 Wiley-VCH Verlag GmbH & Co. KGaA. Published 2015 by Wiley-VCH Verlag GmbH & Co. KGaA.

components can be expected for resonators, which have similar complexity and similar requirements.

Packaging for hermetically sealed resonating devices must include some sort of electrical interface to allow power connections and oscillator signal outputs. In the specific cases of packaging that are the focus of this chapter, the resonator package must provide a hermetic vacuum environment with little or no trapped gas that would adsorb and desorb from the micromechanical elements. The resonator package must provide a mounting structure upon which the resonating device is attached; this mounting structure should isolate the resonator from stresses applied to the outside of the package. The resonator package must protect the resonator from chemical attack, infrared and optical radiation, particles, fluids, and all other external phenomena that could lead to significant changes in the resonant frequency of the MEMS resonator or the failure of the MEMS resonator.

Even after those basic interface requirements are met, the package will play an important role in the long-term stability and reliability of the MEMS resonator. The stability of the MEMS resonator is important so that devices operate with the same resonant frequency and quality factor many years from now as they do today. For example, we would not want the frequency of a resonator-based timing product to drift over the lifetime of a product, such as a cellphone, a watch, or other portable digital accessory. Similarly, we would not want the dynamical characteristics of a gyroscope used for automotive vehicle stability control to change so much that the device failed to detect vehicle rotation at some extreme temperature.

The operation of a MEMS resonator can be significantly affected by the pressure and temperature inside the package as well as the gas species. The temperature of the device is usually determined by the temperature of the environment, because it is probably too expensive to consider an active temperature control system with a miniature oven. Temperature is usually measured and then compensated for in the circuitry that operates the resonator. Pressure and gas species can be controlled by making the package hermetic and controlling the gas species during final sealing.

For MEMS resonators, the pressure within the package is especially important, as it can contribute to the damping processes that limit the quality factor of the resonator. An earlier chapter in this book on Damping describes the concepts and issues in a general way. For a MEMS resonator, it is usually important that the dynamics be stable over long time periods. Therefore, it is usually unwise to have the air pressure-induced damping play a significant role in determining the quality factor of the resonator, because the pressure can be hard to specify and control. Therefore, designers usually select a design where Q is limited by a stable dissipation mechanism, such as thermoelastic dissipation, and select a package and other parameters to make sure that the pressure is so low that small changes in the pressure do not cause changes in Q. Normally, this requires pressures near 1 Pa or lower within the package.

Additionally, it is important to minimize the possibility that gas molecules can adsorb onto or desorb from the MEMS resonators discussed in this chapter. Of course, there are other kinds of resonant sensors where molecular adsorption is desired, such as resonant chemical sensors, but devices that require exposed resonators are not the focus of this chapter.

For hermetically sealed resonators, molecular species can be a significant source of unwanted frequency drift. Even though the mass of these molecules is incredibly small, these small changes can cause small, but noticeable shifts in the resonant frequency during temperature cycles. Temperature-induced hysteresis in MEMS resonators can be minimized by building packages with exceptionally low pressure ($\ll 0.01$ Pa), or by making sure that the residual gas in the package is nonadsorbing. Nobel gases (He, Ne, Ar, Kr) are good examples of molecules that cannot adhere to any surface at room temperatures. Also, relatively inert molecules, such as H_2 and N_2 can serve in this role, since they cannot form chemical bonds on MEMS surfaces at modest temperatures. In general, it is hard to remove all adsorbing molecules during a low-temperature packaging process, as these molecules can adhere to all the surfaces inside the package during the process, and are available to migrate to and from the resonator during temperature cycles. In some cases, a chemically active film, or a "getter" is included in the package to capture unwanted molecules over the life of a product. Otherwise, packaging for ultra-stable MEMS resonators probably must be carried out at very high temperatures and in very clean environments.

Lastly, the package is responsible for protecting the MEMS resonator. Because of the very small size of the mechanical element, MEMS resonators can be destroyed by exposure to direct contact with structures or contact with particles, or liquids. In this regard, the package plays a large role in protecting the MEMS and minimizing failures.

12.2
Overview of Packaging Types

This chapter will primarily discuss packaging techniques for MEMS resonators, which require vacuum encapsulation and hermetic sealing. Many of these methods are also in use for packaging of gyroscopes, which have similar, though less stringent requirements. The two broad classes of MEMS packaging techniques that will be described in this chapter are die-level packaging and wafer-level packaging. For additional background on MEMS packaging techniques, see Refs [4, 5]. Electrical interconnect techniques, including isolation and metallization, vary widely and will not be discussed in detail as a part of the examples described in this chapter.

In all forms of MEMS packaging, the fundamental challenge is that the MEMS devices produced by the MEMS fabrication process generally consist of extremely fragile structures on the top surface of a wafer, as shown in Figure 12.1. The MEMS resonator is likely to consist of structures that are $1-10\,\mu m$ thick, separated from the underlying substrate in some locations by a gap of less than $1\,\mu m$.

MEMS resonator

Substrate

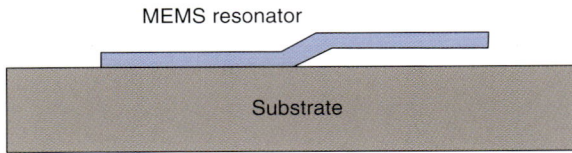

Figure 12.1 Illustration of released MEMS resonator on the surface of the wafer at the end of the MEMS fabrication process.

In fact, the fabrication process usually produces many thousands of these devices arrayed across the surface of the wafer. Separation of the die into individual resonators is a critical step, and it is critical to protect the MEMS device from the environment during die separation. In any die-level packaging approach, the separation of the die happens prior to packaging, so the packaging process also includes the use of some temporary protection structures so that the devices are not destroyed during die separation. Wafer-level packaging approaches encapsulate the resonator prior to die separation, and eliminate the need for temporary protection during die separation.

In die-level packaging approaches, resonator die are individually placed into their final package. This final package provides all of the necessary elements of a MEMS resonator package: environmental interface, hermeticity, protection, and electrical interface. The MEMS structure itself is directly exposed to the environment inside this package. Vacuum cans, similar to those used to package quartz crystals, are the most common die-level package and will be described further in Section 12.3. Die-level packaging can be very costly because it requires the careful handling of individually released devices.

Wafer-level packaging implies that the MEMS resonators are packaged as a part of their fabrication or while they are still part of a whole wafer. Even with this approach, individual resonators will still need to be handled after die separation when the final product is assembled with the circuit. The important distinction here is that the very fragile MEMS resonator is fully encapsulated and protected as a part of its fabrication. After fabrication and die separation, the only additional step is molding into a standard electrical package for integration onto a printed circuit board, flex circuit, or some mounting scheme in a real product. In wafer-level packaging, the MEMS wafer is singulated and the individual devices are packaged in roughly the same manner as occurs in a conventional CMOS electronics packaging process. This approach allows the use of all existing conventional tools and processes for handling die, mounting, bonding, and overmolding. Therefore, these approaches can be completed at lower cost and in higher volume.

Within the category of wafer-level packaging, we will discuss two distinct approaches: wafer bonding and thin-film encapsulation. In wafer-bonding, the package is formed by bonding a "cap wafer" on top of the MEMS wafer prior to die separation. In thin-film encapsulation, the MEMS devices are buried under a series of thin films and released within this film structure, prior to die separation. A unique process developed in our group at Stanford University, called "epi-seal" falls into the class of wafer-scale thin-film encapsulation. Resonators produced using this fabrication process are extremely stable and low cost without the need for a getter [6].

All of these packages must provide a stable environment for the MEMS resonator. Getters are commonly added to packages to absorb contaminants such as trace gases, moisture, and particles. While effective, they add cost and complexity to the product. Getters will be discussed in Section 12.6

In addition to the processing described in this chapter, most complete MEMS products require some interface electronics to sense and/or actuate the MEMS device. The electronics, typically implemented in a CMOS process flow, are almost always fabricated on a completely separate chip from the MEMS. The two chips are both bonded into their final package and wirebonded together for electrical contact. There are many processes in which MEMS can be built either before or after CMOS on the same wafer [7–10]. These processes typically have very strict thermal budgets and many limitations on the materials and processes for the MEMS devices. Most of these integrated processes result in MEMS + CMOS devices on the same wafer, with the MEMS devices released and in need of protection prior to die separation.

12.3
Die-Level Vacuum-Can Packaging

One of the most common die-level packages is the vacuum can. A similar package has been used for many years to hermetically seal quartz crystals. In the early days of MEMS, this was the preferred package for any device that needed a low-pressure hermetic environment. Even today, some very high-performance (and expensive) gyroscopes and accelerometers are still packaged in this manner [11]. For this technique, individual MEMS devices are placed into a metal container that is sealed shut using either solder or a welding process. An example cross section of a final package is shown in Figure 12.2.

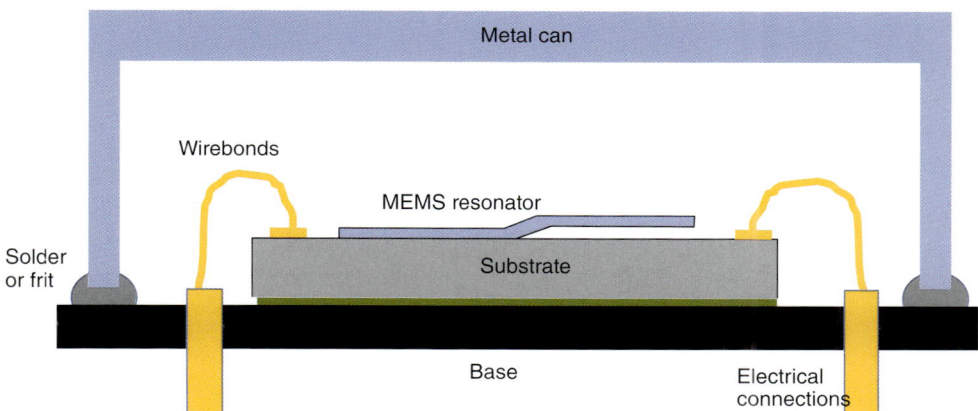

Figure 12.2 Cross section of a MEMS device packaged in a vacuum canister.

The typical fabrication and assembly process for MEMS in die-level package approach begins with either a silicon-on-insulator (SOI) wafer or the deposition of a sacrificial material and then the device material onto a conventional silicon substrate wafer. In the MEMS process, structures are patterned and etched into the device layer of the wafer. The sacrificial material is removed beneath the devices, thus releasing the MEMS resonators and allowing them to move freely. These released devices are extremely fragile and susceptible to contamination.

These devices must now be separated from each other. For CMOS-only wafers, the die separation step is most commonly carried out using a wafer saw. This involves cutting the devices apart with an abrasive diamond saw while spraying coolant (usually water) on the wafer. This process cannot be directly applied without modification to most released MEMS devices. Not only is the cooling water destructive to the MEMS, but the debris from the abrasive blade would also contaminate the released devices. In an effort to protect the fragile MEMS devices during wafer sawing, some groups have proposed dicing the wafer while it is adhered face-down on the dicing tape. This requires special adhesive tape with relief holes cut in it where the MEMS devices are situated [12]. Also there needs to be some alignment between the wafer and the adhesive tape with its release holes. Alignment features on the backside of the wafer are also needed for alignment to the wafer saw blade. There are many complex approaches to this process, and many companies have developed their own proprietary processes. Unfortunately, these processes all impose additional development time on the product, require development and operation of nonstandard tools, and regularly reduce the overall yield of the process. Nevertheless, this approach does preserve maximum diversity of the MEMS process, and has enabled the development of a great variety of MEMS products.

Laser dicing has become a viable alternative to wafer sawing for separation released MEMS resonator die [13]. This approach requires a high-power laser with a wavelength that is strongly-absorbed in the silicon substrate. Molten silicon ejected from the dicing lane can re-deposit on the released resonators if not adequately protected. This potential for deposition of material onto the resonators is especially undesirable, as this can cause the resonant frequency of the released devices to change, and is probably likely to cause significant part-to-part variations in the frequency. A recently introduced "stealth dicing" process avoids this problem by focusing the laser energy at an internal point in wafer and introducing localized damage that allows the die to be separated at the next step in the process [14]. Some groups and companies have demonstrated die separation methods that are incorporated into the fabrication process. In one case, this can involve plasma etching most of the way though the wafer and snapping the devices apart afterward [15]. Alternatively, one could use a diamond scribe to score the wafer and then break it [16]. All of these customized processes for die separation cause the overall packaged device to be more expensive, because they involve exposing the completed wafer to custom tools and nonstandard processes at the most fragile and important moment in the MEMS process.

The individual dies must now be bonded onto the base of the vacuum can. This movement of these devices from the dicing frame to the base must be executed very delicately. The individual dies can only be handled from the sides, bottom, or a place on the top of the die where there are no released structures. Adhesives for bonding to the base must be carefully chosen to be conductive if necessary and not to outgas. Outgassed materials may change the pressure inside the final package and could stick to the surface of the MEMS device, altering its operation. Once bonded, electrode pads on the MEMS die can be wire bonded to the electrical feedthroughs on the vacuum can.

This assembly and a lid for the canister are then placed into a vacuum chamber. The chamber is pumped down and likely purged several times with an inert gas to minimize trace reactive gasses in the final packaged device chamber. While pumped down to the desired pressure, the lid must either be welded or soldered to the base, forming the hermetic seal around the device and completing the vacuum can-packaging process.

Examples of devices packaged in a vacuum can be found in Refs [17, 18]. CERDIPs (ceramic dual in-line packages) are also similar to the vacuum can packages and are described in [19]. Traditionally, quartz crystal resonators for precision time references were packaged in this way.

There are many variations to get around some of these challenges described above. For example, the die can be separated prior to the release of the MEMS elements. Prior to the release step, the MEMS elements are reasonably protected, and can withstand immersion in fluids, high-pressure airflow, and can probably survive a dicing saw environment as well, as long as there can be thorough cleaning after the dicing step. The individual dies will be released just prior to placement into the vacuum can. Having to individually release each device could be quite costly and possibly result in significant process variation among devices. Also, any other materials used in the devices, such as the metals for electrical contact, must be resistant to the etchant used for the release.

All of this individual die handling and processing steps combine to reduce device yield. Thus, more testing is required after packaging, further increasing costs. While vacuum cans have been a viable option for packaging MEMS, all trends in packaging for MEMS devices, such as accelerometers and gyros and resonators are pointing toward wafer-level approaches. For resonators, all of the disadvantages of die-level packaging contribute to cost, and since cost is one of the main drivers for innovation in consumer electronics, all commercially-produced MEMS resonators for timing products are packaged using the wafer-level approaches.

12.4
Wafer Bonding for Device Packaging

Many consumer products are turning to wafer bonding as a means of packaging MEMS devices at the wafer level. This allows for the hermetic sealing of a whole wafer's worth of devices at one time. Devices produced in this manner are

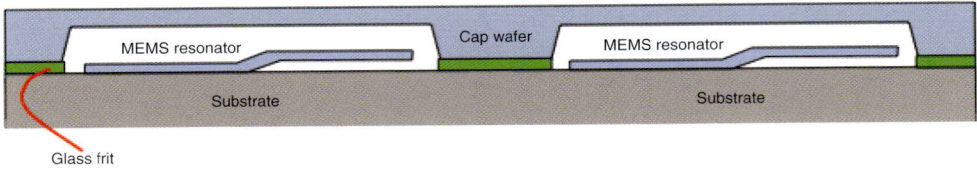

Figure 12.3 Cross section of a MEMS device encapsulated using wafer bonding.

much more robust to standard CMOS handling techniques. Figure 12.3 shows a cross section of a completely packaged MEMS device using wafer bonding. In this illustration, the "Cap Wafer" is another silicon wafer with pre-etched cavities positioned to align with the released devices. Wet etching methods are available for the inexpensive formation of the cap wafer, using some of the same techniques and tools used for the production of the MEMS resonators, so this is a very convenient method.

A wafer bond–packaging process, by definition, starts out with at least two wafers. Similar to the vacuum can packaging, MEMS structures are patterned and etched on the first wafer. At this point, some processing must be performed on this wafer to define either horizontal or vertical electrical feedthroughs. The devices on this wafer can now be released.

In parallel with this processing, a bond wafer is prepared for encapsulating the MEMS structures. Relief cavities are etched into this wafer so as not to interfere with the movement of the MEMS beneath. The thermal expansion coefficient of this wafer should be roughly the same as the handle wafer to minimize stress at various operating and fabrication temperatures, and to eliminate temperature-induced errors in the resonator characteristics during operation. Typical wafer pairs include silicon-Pyrex*, silicon-quartz, and silicon–silicon. For packaging of MEMS resonators, silicon–silicon pairs are certainly the most common in order to reduce strains from thermal mismatch that can cause frequency errors [20].

At the time of the wafer–wafer bonding, the surface of both the device wafer and the bond wafer must be planar and smooth in the bond area. The two wafers are aligned to each other and bonded using one of many different bonding methods. Most common among these methods are eutectic, anodic, and fusion bonding, which are described below.

Eutectic wafer–wafer bonding usually takes advantage of the gold-silicon eutectic system. Similar to ordinary tin-lead solder, the gold-silicon system melts at a much lower temperature (363 °C) than either pure gold or pure silicon [21]. To perform this type of bond, gold is deposited on the areas of the wafer where the bond is to be made (typically on the bond wafer). These opposing, facing areas must have freshly exposed silicon, without any coatings. Wafers pairs are placed in a vacuum aligner-bonder, and the pressure is pumped down to below the vacuum requirement for the resonators. Often modest heating is used to remove adsorbed films, such as water, from the all the surfaces. The wafers are brought into contact, and then heated to a temperature above the eutectic temperature, to form the gold-to-silicon bond. As an example, an accelerometer, packaged using a eutectic bond on both sides of the device wafer, is described in Ref. [22]. Similar methods can be used for packaging of resonators. One challenge is that this approach might still lead to modest rates of outgassing within the bonded chamber, so it is common to deposit a getter in the top surface of the cavities in the cap wafer. In this case, it is necessary to be sure that the getters activation temperature and the eutectic temperature are compatible – usually, this requires that the getter activation temperature is below the eutectic temperature.

Anodic bonding uses increased temperature and high voltage to bond a glass wafer to the silicon wafer. Typically a Pyrex wafer is bonded to a silicon wafer. A large negative bias (on the order of -300 V) is applied to the glass wafer relative to the silicon wafer. This attracts positively charged sodium ions away from the bond interface, leaving extra oxygen molecules for the formation of SiO_2, resulting in a strong covalent bond to the silicon substrate. Pyrex glass is transparent to visible light and therefore, this packaging technique can be adapted to microscale optical devices. Some characterization of anodic bonded packages is described in Ref. [23]. An accelerometer packaged using multiple anodic bonds is described in Ref. [24].

Materials required for anodic and eutectic bonding are typically prohibited from foundries running any type of CMOS process. Specifically gold and sodium are detrimental to CMOS devices. Without dedicated contaminated equipment to continue processing, a eutectic or anodic bond would need to be the last step of a fabrication process. Additionally, these bonding techniques typically require large areas for the bond (on the order of 40% or more of the die, as estimated from [25]). This reduces the number of devices that can be fabricated on one wafer, further increasing product costs.

Fusion bonding could also be used to encapsulate MEMS devices like the more common anodic and eutectic bonds. Fusion bonding relies on the surfaces being very smooth and the spontaneous locking of these two surfaces together when brought into contact, followed by a high-temperature annealing step (typically at 1000° C). This process is very sensitive, and wafers that are too rough or too curved will not bond. Bond strength can be improved by a higher-temperature anneal, sometimes up to 1150° C. Usually other materials involved with the fabrication process limit this anneal temperature to much lower values. In the MEMS literature, most devices fabricated using fusion bonding are pressure sensors, and this step takes place early in the fabrication process. The fusion bonded wafer simultaneously forms a vacuum cavity and a movable membrane that can be used to measure pressure [26]. There are few, if any, examples of fusion bonding being used to for packaging of released devices, mainly because of the extreme temperatures required in this process.

Now that the released devices are fully encapsulated, wafers can be handled in much the same way as a CMOS wafer, except that the total thickness is higher. Assuming the MEMS structures can withstand the vibrations, standard wafer saw techniques can be used to separate the die. Also, standard pick-and-place tools can be used to put the MEMS dies into their final package, which can be as simple and as inexpensive as a plastic injection molded package commonly used for CMOS devices. For injection molding, dies can be subject to pressures of 1000 psi. The cavity dimension and bond wafer thickness must be carefully considered to ensure that the microcavity can withstand these pressures, but these considerations can easily be managed through simulation and testing.

Processing costs for these processes are not optimal because of the significant amount of real estate on the wafer surface required for the wafer bonding area – often as much as 80% of the total area of the wafer can be occupied by these bond-related features. Additionally, a specialized vacuum bonding tool is needed to perform the following functions: control pressure, control gas species, perform alignment, bring wafers together, apply heat, and apply bias voltage. Also, there may be a modest reduction in total device yield because of the bonding must be performed while the resonators are released, and there is risk of particulates being trapped and released within the cavities.

Despite this, wafer bonding typically results in substantially lower product cost than the vacuum can packaging techniques. The main cost advantage comes from the use of standardized tools for dicing, handling, mounting, bonding, and overmolding of MEMS resonators fabricated in such a process. Many consumer products are fabricated using similar techniques to those described above. Recent examples of the use of such a process include an accelerometer from Motorola [27], a gyroscope from Bosch [28], and a gyroscope from Invensense [29]. Discera (a division of Micrel) has large volume production of MEMS-based timing products, in which their resonators are packaged in a process similar to that described above [21]. At the time of writing, Discera has a diverse product portfolio of MEMS-based timing products, and has shipped more than 20 million oscillators based on this process. In many cases, the average selling price of the Discera products made using this approach are well below $1.00, and this includes the MEMS, the CMOS, the packaging materials and all costs associated with the packaging process, as well as testing, sales and distribution costs, other fixed costs, and (hopefully) some profit.

12.5
Thin Film Encapsulation-Based Packaging

Unlike the other encapsulation techniques, thin film methods use only the deposition of thin films to fully encapsulate the MEMS devices. Thin films can be deposited using chemical vapor deposition (CVD), evaporation, sputtering, or atomic layer deposition (ALD). An example cross section of a MEMS device encapsulated with thin films is shown in Figure 12.4.

Like the other fabrication techniques, the film-encapsulation fabrication process begins with the definition of the MEMS structures. After the devices are patterned and etched into the device layer, a sacrificial layer is deposited over all of the devices. It is important to choose the sacrificial material, such that an etchant exists that selectively etches it over the MEMS device material. SiO_2 is a very common material for the sacrificial material for a few reasons. First, it can

Film encapsulation layer   MEMS resonator

Substrate

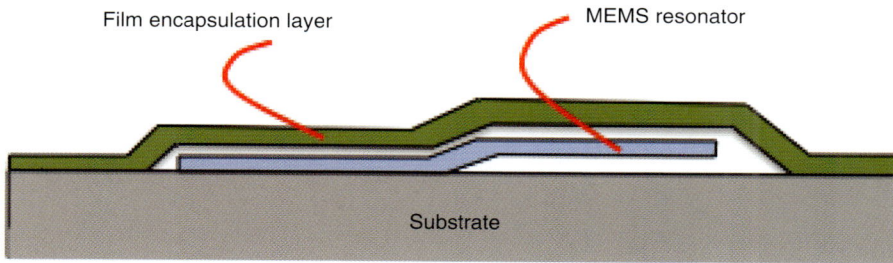

Figure 12.4 Cross section of a MEMS device fabricated using thin film encapsulation.

easily be deposited at relatively low temperatures. Also, it can easily be etched with hydrofluoric acid (HF), which is highly selective to etching silicon dioxide over silicon. Also it is CMOS compatible and readily available in most foundries. If the devices are to move in the plane of the wafer, this material must bridge over or fill in all of the trenches defining lateral movement. For devices intended to operate out-of-plane, this requirement can be relaxed.

An encapsulation material can now be deposited on top of the sacrificial layer and the MEMS beneath. This material should also be resistant to the etchant that will be used later to etch the sacrificial material. Vent holes are then etched into the encapsulation layer for access to the sacrificial material below. The sacrificial material is removed around the MEMS devices with an isotropic etchant. To control the area that is released, this could be a timed etch or use an etch stop.

Once the devices are released, the vent holes need to be sealed with some thin film. Ideally the seal material would not be deposited on the MEMS below. Also, the material should result in a hermetic seal. Many different materials have been used in for this application. For example, the encapsulation film may be composed of aluminum [30], polysilicon [31], silicon dioxide [32], and silicon nitride [8, 33]. The sacrificial material may be silicon, silicon dioxide, or organics. Ayazi and Kohl describe a process in which a sacrificial polymer is sublimated to release a MEMS device, with the sublimation products diffusing through a different polymer overcoat encapsulation layer [34]. Many of these deposition processes have been adapted to be used on many wafers at the same time. With such a process, not only is a whole wafer sealed at once, but several *wafers* may be sealed at the same time.

For resonators, it is important that the final chamber be hermetically sealed in as clean a process as is possible. Deposition of metal films as a seal layer, such as in [30] can be advantageous as those metals can act as a getter for residual contaminants, and the Ultra High Vacuum (UHV) environment used for the deposition can produce a very low pressure environment after the seal. In contrast LPCVD (low pressure chemical vapor deposition) sealing methods are lower temperature and will leave behind some quantity of LPCVD process gas molecules, which may adsorb and desorb from the encapsulated resonator, leading to temperature-induced frequency hysteresis.

After encapsulation, the wafer can be singulated into individual dies using standard wafer sawing techniques. The dies can be integrated into their final packages using standard processes. For injection molded packages, special consideration for the cap dimensions is needed to ensure that the membrane can withstand the high pressures involved. Most deposited materials (such as SiN and Metals) are limited to very thin layers because of stress issues and slow deposition rates. Depending on the span and thickness of the encapsulation layer, it may be difficult to injection mold thin film encapsulated MEMS into their final package. Fortunately there are other standard low-cost packages that can be used for this purpose.

Thin film encapsulation processes hold the promise of being very inexpensive. As mentioned before, many of the deposition and sealing processes can be performed on multiple wafers at the same time. Also, unlike wafer bonding methods, encapsulation processes use much less of the wafer area for bond rings or other features associated with routing of electrodes from one side of the bond ring to the other, enabling smaller die for the same MEMS function. By reducing the footprint required for a die, the per-wafer product yield can be greatly increased, which can allow reductions in the product cost. The main problem with these types of processes is that the design rules are often very limiting. High-performance MEMS resonators may require thick layers, large features, complex geometries or specific materials that prevent thin film encapsulation from being used as a part of the fabrication process.

12.6
Getters

Protecting MEMS resonators from contamination is essential to their stability and longevity. Getters are materials that act as "magnets" for various contaminants, such as trace gasses, moisture, and particles, and can permanently trap these contaminants within the package. Getters are commonly added to some MEMS packages for accelerometers, gyros, and resonators with the intent that the getter material will attract the contaminants more than the MEMS structure itself. Table 12.1 summarizes the various types of getters and the materials they target.

Exotic metals or their alloys are typically used to absorb trace gasses inside the package [35]. Gasses can enter the package through a leak or outgas from one of the materials inside the package. The addition of a getter helps maintain an equilibrium pressure and gas concentration inside the package. Some of these types of getters need a high temperature activation to speed up the absorption process when the devices are first packaged.

Desiccants can be included in a package to absorb water vapor. Moisture is a particular problem for MEMS devices. It can condense on the surface, causing a change in mass and thus a change in resonant frequency of a device. Also, moisture could oxidize the surface, changing the stiffness of the structure or allowing electrical charges to get trapped on the surface. Surface tension from condensed water could easily break a delicate MEMS structure. The amount of moisture a desiccant can absorb reduces as temperature increases. This could limit the operating

Table 12.1 Summary of getters available for MEMS packaging.

Class	Materials	Purpose
Metals	Titanium, alloys of zirconium, aluminum, and iron	Absorb trace gasses such as hydrogen, ammonia, oxygen, and more
Desiccants	Zeolite, CaO	Absorb moisture and maintain a dry environment
Particle getters	Tacky polymers	Attract particles and keep them from the MEMS device

temperature of the device because water vapor may be released from the desiccant if the temperature is elevated.

Particles can change the mass of a structure or get stuck in a small feature, rendering a device useless. A tacky material can be included in the package to attract these particles before they get to the MEMS. The material for this type of getter must be chosen carefully to not outgas any moisture or trace gasses that could affect device operation.

While getters are a viable alternative to making stable MEMS devices, they add cost and complexity to the product. Many different types of getters are available from Cookson Electronics Semiconductor Packaging Materials [36] and SAES Getters [37].

12.7
The "Stanford epi-Seal Process" for Packaging of MEMS Resonators

The epi-seal process falls into the category of a thin film encapsulation packaging technique. The basic concept for this process was conceived by Markus Lutz, and was developed jointly through a collaboration between the Micro Structures and Sensors Lab at Stanford University and Bosch Research and Technology Center (RTC) in Palo Alto, CA [38, 39]. This collaboration initially involved many people, including Gary Yama, Aaron Partridge, Rob Candler, Bongsang Kim, Matt Hopcroft, WooTae Park, and Thomas Kenny.

The process begins with a SOI substrate. A Bosch deep reactive ion etch (DRIE) defines the devices in the device layer of the SOI. Next, a sacrificial layer of SiO_2 is deposited on the wafer using LPCVD, as shown in Figure 12.5. This oxide sacrificial material must either fill in all of the trenches on the device layer or bridge over them to form a planar surface on top of the wafer, as shown in Figure 12.5. This requirement limits the size of the gaps that can be formed between the fixed and moving elements in this process. For high-frequency resonators to be used as time references, this restriction is not an important limitation. Typically, the trenches are all drawn at approximately $1\,\mu m$ and the SiO_2 is deposited to be $2\,\mu m$ thick,

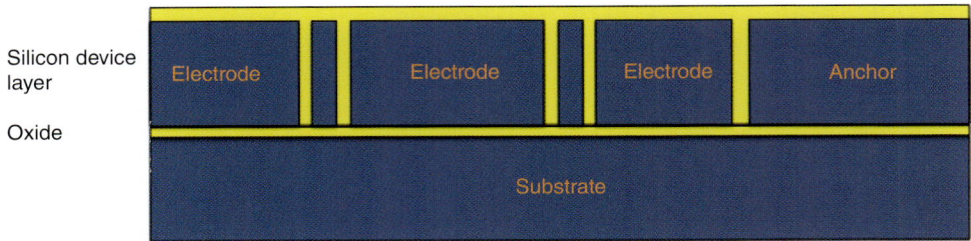

Figure 12.5 Illustration of the early stage of the "epi-seal" encapsulation process, after etch definition of the structures, and the deposition of the sacrificial oxide film.

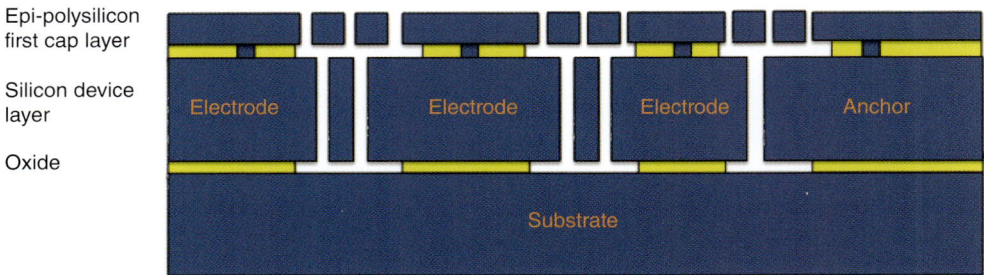

Figure 12.6 Illustration of the MEMS encapsulation process after HF vapor release etch.

which is more than enough to bridge or fill in all of the narrow trenches. After the deposition of the SiO_2, contact vias are plasma etched into this SiO_2 layer to allow subsequent formation of electrical contacts from the surface down to the device layer.

A thin film ($\sim 3\,\mu m$) of epitaxial polysilicon is then deposited on the wafer to serve as the beginning of the encapsulation structure. After deposition, the wafers are polished using chemical mechanical polishing (CMP) to clean up this surface for subsequent lithography. A series of narrow "vents" are then etched into the encapsulation layer, over the locations where underlying devices are to be released. The width of these vents is typically $<0.7\,\mu m$, with this dimension selected to guarantee that they will be sealed in the subsequent film deposition without allowing deposits to grow on the underlying structures. These vents allow access to the buried devices to remove the sacrificial material and release the devices. The SiO_2 is etched from the top and bottom of the devices using a time HF vapor etch. The HF vapor etch is a "dry etch," which does not cause formation of liquid residue within the encapsulation. Figure 12.6 shows the situation at the completion of the HF vapor release etch.

After the HF vapor release etch, the wafers are put directly into the epitaxial silicon reactor. At this point, all of the SiO_2 *should* be removed from the device surfaces. However, during transport from the vapor etcher to the epi reactor it is likely that a thin native oxide layer will have formed on the surface of the silicon devices. Fortunately, this oxide layer is easily dissociated from the silicon surface

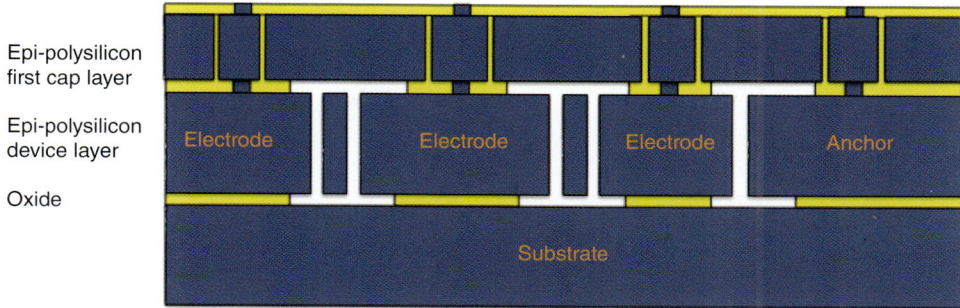

Figure 12.7   Encapsulated resonators at the completion of the epi-seal process.

in the epitaxial chamber by using a reduced pressure, high temperature hydrogen bake immediately prior to silicon deposition. This *in situ* clean results in a pristine single-crystal silicon surface and has been used for many years in the epitaxial growth community to achieve good epitaxial alignment of the deposited films to the substrate [40]. These clean surfaces are likely responsible for the long-term stability of the epi-sealed devices. This process also has the beneficial effect that the etched silicon sidewalls are smoothed; we believe that this smoothing of the silicon sidewalls has contributed to the extraordinary long-term stability observed for resonators packaged in this process.

After a brief H_2 exposure, processes gasses are introduced to initiate the polysilicon growth. At least 1 µm of silicon is deposited to seal the vents and an additional 10–20 µm can be deposited to strengthen the cap layer. During the deposition, at least 90% of the gas flowing is hydrogen. The other gasses in the chamber are dichlorosilane ($SiCl_2H_2$), phosphine (PH_3), and hydrochloric acid (HCl).

Once the microcavities are sealed, several additional steps are needed to make isolated electrical contact to the buried electrodes. Isolation trenches are etched around the contacts and filled with LPCVD SiO_2. Contact vias are etched into this layer and aluminum is deposited and patterned. These integrated vertical interconnects allow the dies to be very small, enabling the production of more devices on a single wafer. The resulting structures look as shown in Figure 12.7.

The limited exposure and handling while the devices are released result in a very high-yield process. Dies separation can be easily accomplished using standard wafer sawing with little or no reduction in yield. MEMS devices fabricated in this manner can then be wire bonded and molded into their final standard electrical package, just like a CMOS device, using handling and packaging tools that are fully standard. The epi-seal packaging process has also proven to be a good solution for the fabrication of MEMS resonators. The long-term frequency stability is far superior to other MEMS resonators [5] and the Kenny group has used this process as a "baseline process" to support many generations of improved resonators [39, 41–45].

All of the processing steps in the "epi-seal" process use standard tools and there are no boutique steps or tools needed for the fabrication. The materials are all CMOS compatible and circuits could be fabricated on the same wafer as the MEMS. Vertical interconnects allow the die size to be very small. Thin film encapsulation further reduces die size over bonding methods and allows for a whole wafer of devices to be sealed at one time. Once sealed, devices can be separated, pick-and-placed, wire bonded, and injection mold packaged using nothing but standard electronics packaging procedures, tools, and vendors. All of this results in a very low-cost product.

12.8
Conclusion

In the 50 years since the demonstration of early MEMS resonators, there has been significant progress on hermetic packaging of resonators. The characteristics of the package can determine many of the important characteristics of the resonator, including the cost, reliability, and performance. It is particularly important to control the pressure inside the package, and this chapter describes many examples of packaging approaches and processes that are focused on this aspect. Cleanliness is also critical, so selection of materials, temperatures, and processes is often constrained by the impact on cleanliness. The recent successes at commercialization of Resonant MEMS has resulted from the development of packaging methods, including some described in this chapter. In the future, the evolution of product requirements will continue to provoke research into improved packaging methods. This will be especially true for products that combine multiple kinds of resonant MEMS with different packaging requirements, such as chemical sensors and gyroscopes.

References

1. Nathanson, H.C., Newell, W.E., and Wickstrom, R.A. (1965) Tuning forks sound a hopeful note. *Electronics*, **38**, 84–87.
2. Nathanson, H.C., Newell, W.E., Wickstrom, R.A., and David, J.R. (1967) The resonant gate transistor. *IEEE Trans. Electron Devices*, **ED-14**, 117–133.
3. Senturia, S.D. (2001) *Microsystem Design*, Springer, New York.
4. Najafi, K. (2003) Micropackaging technologies for integrated microsystems: applications to MEMS and MOEMS. *Proc. SPIE*, **4979**, 1–19.
5. Gilleo, K. (2005) *MEMS/MOEMS Packaging*, McGraw-Hill, New York.
6. Kim, B. *et al.* (2007) Frequency stability of wafer-scale film encapsulated silicon based MEMS resonators. *Sens. Actuators, A: Phys.*, **136**, 125–131.
7. Fedder, G.K., Howe, R.T., Liu, T.-J.K., and Quevy, E.P. (2008) Technologies for cofabricating MEMS and electronics. *Proc. IEEE*, **96**, 306.
8. Lin, L., Howe, R.T., and Pisano, A.P. (1998) Microelectromechanical filters for signal processing. *J. Microelectromech. Syst.*, **7**, 286.
9. Provine, J. *et al.* (2010) Time evolution of released hole arrays into membranes via vacuum silicon migration. Solid-State Sensors, Actuators, and Microsystems Workshop, Hilton Head Island, SC, 2010, pp. 344–347.
10. Sato, T. *et al.* (2000) Micro-structure transformation of silicon: a newly developed transformation technology for

patterning silicon surfaces using the surface migration of silicon atoms by hydrogen annealing. *Jpn. J. Appl. Phys.*, **39**, 5033.

11. (2010) Systron Donner MEMS Micromachined Angular Rate Sensor – LCG50 Datasheet, ed. Concord, CA.

12. Roberts, C.M. *et al.* (1994) Method for separating circuit dies from a wafer. US Patent 5,362,681.

13. Krishnan, V. and Bo, T. (2007) Thin silicon wafer dicing with a dual-focused laser beam. *J. Micromech. Microeng.*, **17**, 2505.

14. Hamamatsu Photonics (2010) Stealth Dicing Technology, August 26, 2010, *http://jp.hamamatsu.com/products/ semicon-fpd/pd393/L9570-01/index_ en.html* (accessed 1 September 2014).

15. Douglas, S. and Timothy, H. (2004) Micromachined needles and lancets with design adjustable bevel angles. *J. Micromech. Microeng.*, **14**, 1230.

16. Acker, M.S. (2001) The backend process: step11-scribe and break. *Adv. Packag.*, **11** (1), 45–50.

17. Chau, K.H.L. and Sulouff, R.E. (1998) Technology for the high-volume manufacturing of integrated surface-micromachined accelerometer products. *Microelectron. J.*, **29**, 579–586.

18. Sherman, S.J. *et al.* (1992) A low cost monolithic accelerometer; product/technology update. Technical Digest of the International Electron Devices Meeting, 1992, pp. 501-504.

19. Gooch, R. *et al.* (1999) Wafer-level vacuum packaging for MEMS. *J. Vac. Sci. Tech.*, **A17**, 2295–2299.

20. Discera (now a division of Micrel, *www.discera.com* (accessed 6 September 2014)) and Silicon Clocks (now a division of SiLabs, *www.silabs.com* (accessed 6 September 2014)) are both representative examples of Si-Si bonding for hermetic resonator packaging.

21. Wolffenbuttel, R.F. and Wise, K.D. (1994) Low-temperature silicon wafer-to-wafer bonding using gold at eutectic temperature. *Sens. Actuators, A: Phys.*, **43**, 223–229.

22. Zavracky, P.M. *et al.* (1996) Design and process considerations for a tunneling tip accelerometer. *J. Micromech. Microeng.*, **6**, 352.

23. Henmi, H. *et al.* (1994) Vacuum packaging for microsensors by glass-silicon anodic bonding. *Sens. Actuators, A: Phys.*, **43**, 243–248.

24. Rudolf, F. *et al.* (1990) Precision accelerometers with [mu]g resolution. *Sens. Actuators, A: Phys.*, **21**, 297–302.

25. Mitchell, J. *et al.* (2005) Encapsulation of vacuum sensors in a wafer level package using a gold-silicon eutectic. The 13th International Conference on Solid-State Sensors, Actuators and Microsystems, 2005. TRANSDUCERS '05, Digest of Technical Papers, Vol. 1, pp. 928–931.

26. Petersen, K. *et al.* (1988) Silicon fusion bonding for pressure sensors. Technical Digest of the Solid-State Sensor and Actuator Workshop, 1988, IEEE, pp. 144-147.

27. Delpoux, A. (1997) Motorola and MEMMS: the way up to a surface micromachined accelerometer. *Microelectron. J.*, **28**, 381–387.

28. Lutz, M. *et al.* (1997) A precision yaw rate sensor in silicon micromachining. International Conference on Solid State Sensors and Actuators, 1997. TRANSDUCERS '97, Chicago, IL, 1997, Vol. 2, pp. 847–850.

29. Seeger, J. *et al.* (2010) Development of high-performance, high-volume consumer MEMS. Solid-State Sensors, Actuators, and Microsystems Workshop, Hilton Head Island, SC, 2010, pp. 61-64.

30. Bartek, M. *et al.* (1997) Vacuum sealing of microcavities using metal evaporation. *Sens. Actuators, A: Phys.*, **61**, 364–368.

31. Rihui, H. and Chang-Jin, K. (2007) On-wafer monolithic encapsulation by surface micromachining with porous polysilicon shells. *J. Microelectromech. Syst.*, **16**, 462–472.

32. Park, W.T., Partridge, A., Candler, R.N., Ayanoor-Vitikkate, V., Yama, G., Lutz, M., and Kenny, T.W. (2006) Encapsulated Sub-millimeter piezoresistive accelerometers. *J. Microelectromech. Syst.*, **15**, 507.

33. Liu, C. and Tai, Y.-C. (1999) Sealing of micromachined cavities using

chemical vapor deposition methods: characterization and optimization. *J. Microelectromech. Syst.*, **8**, 135–145.

34. Monajemi, P., Ayazi, F., Joseph, P.J., and Kohl, P.A. (2005) A low cost wafer-level MEMS packaging technology. 18th IEEE International Conference on Micro Electro Mechanical Systems, 2005. MEMS 2005, January 30–February 3, 2005, pp. 634–637.

35. Ramesham, R. and Kullberg, R.C. (2009) Review of vacuum packaging and maintenance of MEMS and the use of getters therein. *J. Micro/Nanolithogr. MEMS MOEMS*, **8**, 031307–031309.

36. Alpha Advanced Materials (2010) Cookson Semi Electronic Polymers: Cookson Electronics Assembly Materials, August 19, 2010, *http://www.cooksonsemi.com/products/polymer/hicap-staydry.asp* (accessed 1 September 2014).

37. SAES (2010) SAES Getters – Getters for MEMS, August 22, 2010, *http://www.saesgetters.comdefault.aspx?idPage=1476* (accessed 1 September 2014).

38. Kim, B. *et al.* (2004) Investigation of MEMS resonator characteristics for long-term and wide temperature variation operation. *ASME Conf. Proc.*, **2004**, 413–416.

39. Candler, R.N. *et al.* (2003) Single wafer encapsulation of MEMS devices. *IEEE Trans. Adv. Packag.*, **26**, 227–232.

40. Goulding, M.R. (1993) The selective epitaxial growth of silicon. *Mater. Sci. Eng., B*, **17**, 47–67.

41. Candler, R.N. *et al.* (2005) Hydrogen diffusion and pressure control of encapsulated MEMS resonators. The 13th International Conference on Solid-State Sensors, Actuators and Microsystems, 2005. TRANSDUCERS '05. Digest of Technical Papers, Vol. 1, pp. 920–923.

42. Kim, B. *et al.* (2007) Si-SiO2 composite MEMS resonators in CMOS compatible wafer-scale thin-film encapsulation. IEEE International Frequency Control Symposium, 2007 Joint with the 21st European Frequency and Time Forum, pp. 1214–1219.

43. Graham, A.B. *et al.* (2010) A method for wafer-scale encapsulation of large lateral deflection MEMS devices. *J. Microelectromech. Syst.*, **19**, 28–37.

44. Messana, M.W. *et al.* (2010) Packaging of large lateral deflection MEMS using a combination of fusion bonding and epitaxial reactor sealing. Solid-State Sensors, Actuators, and Microsystems Workshop, Hilton Head Island, SC, 2010, pp. 336-339.

45. Melamud, R. *et al.* (2007) Temperature-compensated high-stability silicon resonators. *Appl. Phys. Lett.*, **90**, 244107-3.

13
Compensation, Tuning, and Trimming of MEMS Resonators

Roozbeh Tabrizian and Farrokh Ayazi

As micro electro mechanical systems (MEMS) resonators transition from research laboratories to commercial products, stability and reproducibility across many performance parameters become increasingly important. The main performance metrics of a MEMS resonator include its center frequency (f_0), temperature coefficient of frequency (TCF), quality factor (Q), and motional resistance (R_m). Additional parameters related to power handling, linearity, and tunability may become important in some applications.

MEMS resonators are used either as stable frequency references (e.g., in timing and spectral processing applications) or as sensitive mechanical sensors (e.g., in environmental and inertial sensor applications). As a frequency reference, resonator parameters need to be insensitive to environmental conditions such as temperature, humidity, and vibration. Over the past decade, microfabrication techniques have advanced to realize micromechanical resonators with efficient electromechanical transducers. Because of their small mass and small damping, MEMS resonators are inherently sensitive to physical phenomena that impose fluctuations in their mass, stiffness, or damping. While this makes MEMS resonators potentially interesting for physical and chemical sensing, it also raises concerns in their functionality and reliability as accurate frequency references, especially for highly stable timing applications. Furthermore, process limitations and variations may impose technological barriers and excessive cost burdens on realization of high-precision resonators. Finally, being limited by the fundamental physical and chemical properties of current materials used for implementation of MEMS resonators, residual variations, and environmental instabilities remain inevitable. As a result, a recent wave of research and development in the field of resonant MEMS has concentrated on development of compensation, tuning, and trimming techniques through novel designs and material engineering.

Resonant MEMS – Fundamentals, Implementation and Application, First Edition.
Edited by Oliver Brand, Isabelle Dufour, Stephen M. Heinrich and Fabien Josse.
© 2015 Wiley-VCH Verlag GmbH & Co. KGaA. Published 2015 by Wiley-VCH Verlag GmbH & Co. KGaA.

13.1
Introduction

Depending on the specific application of the MEMS resonator, one or more of its metrics may require accurate control [1]. The large amount of research invested in precise control of MEMS resonator characteristics can be categorized into three groups: compensation, tuning, and trimming. Techniques developed under each of these categories may address resonator performance control at the device or system level, and are applied in the design stage and/or post-fabrication. The most preferred techniques can be applied in batch across an entire wafer at once or in one step and are referred to as wafer-level techniques.

13.2
Compensation Techniques in MEMS Resonators

Compensation is attributed to all design-level techniques that reduce or eliminate the sensitivity of one or several resonator performance metrics to environmental variations (i.e. temperature, pressure, etc.) or processing uncertainties (i.e. material properties, impurity concentration, lithography errors, etc.).

13.2.1
Compensation for Thermal Effects

Both f_0 and Q of the MEMS resonators typically show large temperature sensitivities. While thermal behavior of the Q varies depending on the geometrical design, resonance mode [2] and frequency of the resonator and is still under further investigation, the f_0 temperature dependency $-TCF-$ of the majority of mechanical resonators has been well characterized, and can be precisely formulated using high-order polynomials. Furthermore, while resonators implemented in different crystal cuts of quartz are showing relatively small temperature-induced frequency drifts, which can be well-characterized by third-order polynomials, uncompensated MEMS resonators are usually showing a large linear temperature dependency. The first-order (i.e., linear) temperature sensitivity of f_0 can be defined and related to the temperature sensitivity of the resonator's material properties by its *TCF*:

$$\text{TCF} = \frac{1}{f_0} \cdot \frac{\partial f_0}{\partial T} \approx \frac{1}{2}(\text{TCE} + \text{CTE}) \tag{13.1}$$

where *TCE*, the temperature coefficient of the Young's modulus and *CTE*, the coefficient of thermal expansion, are intrinsic material properties. Although Eq. (13.1) can be easily derived for MEMS resonators operating in their extensional bulk acoustic or flexural modes, a similarly simple relationship does not exist for all the resonance modes excited in micromechanical structures with finite dimensions. This is due to the fact that the resonance frequency of these modes cannot be related to the material properties and resonator dimensions through a simple closed-form expression; and unlike bulk modes, all different independent elastic constants may contribute effectively in defining the resonance frequency.

The large *TCF* of the MEMS resonators is mainly a result of high *TCE* of the materials commonly used to implement these devices. For native-silicon resonators operating in extensional and flexural modes, the large *TCF* of -30 ppm $°C^{-1}$ results in a frequency drift as large as 3750 ppm across the industrial temperature range of -40 to $85\,°C$. Major applications of MEMS resonators, such as temperature-compensated crystal oscillators (TCXOs) and thermally stable resonant sensors often require sub-parts per million overall instability levels over the entire operating range, which is very challenging, if not impossible, to be realized solely by active frequency tuning methods. Therefore, passive *TCF* compensation techniques that provide radical decrease in the *TCF* of MEMS resonators are of great interest. Some examples of these techniques are device geometry and acoustic engineering to operate in resonance modes with lower *TCF*s, compensation of semiconductor material *TCE* through engineering of doping profiles, and addition of a compensating material with a *TCE* of opposite sign to form a composite structure with reduced *TCF*.

13.2.1.1 Engineering the Geometry

Geometry of a resonator can be engineered for creation of resonance modes with lower *TCF*. Figure 13.1 shows the SEM image of a *TCF*-compensated concave silicon bulk acoustic resonator (CBAR) in comparison with a rectangular geometry silicon bulk acoustic resonator (SiBAR) [4, 5].

The concave resonator geometry has resulted in the excitation of a resonance mode with large energy concentration in shear acoustic fields and thus considerable reduction in the *TCF*. Furthermore, such geometry has resulted in significant improvement in device Q due to reduction of energy leakage from the resonator toward surrounding substrate [3].

13.2.1.2 Doping

Another group of *TCF* compensation techniques, which can be used for resonators implemented in semiconductors, is based on the introduction of dopants into the substrate to reduce the large *TCE* of different elastic constants by modifying the electronic energy level of material. These techniques can be broadly termed as doping-profile engineering. Different works falling under this category include the utilization of heavily *P*-doped [6, 7] and *N*-doped silicon substrates [8, 9], thermomigration of aluminum atoms into the silicon substrate [10], and carrier depletion of the device using multiple *PN* junctions [6]. It has been recently demonstrated that the efficiency of doping-profile engineering technique is considerably depending on the resonance mode as well as device relative crystallographic orientation [11]. Several designs are proposed to employ this dependency to implement devices with maximum compensation of the linear temperature dependency; thus resulting in a second-order parabolic

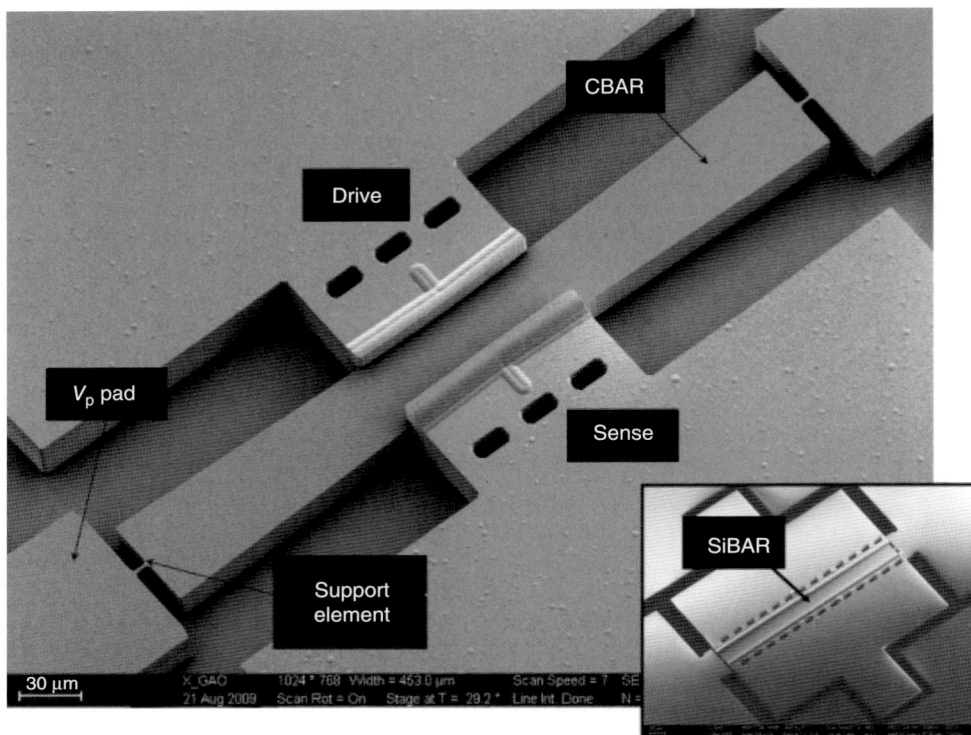

Figure 13.1 Geometry-engineered *TCF*-compensated concave SiBAR (CBAR). The device shows a *TCF* of −6 ppm °C^{-1} at 105 MHz with a *Q* of 100 000 [3].

temperature characteristic. This parabolic characteristic, besides showing considerable improvement in overall frequency drift in the desired temperature range, possesses a turn-over point with zero local linear *TCF*; thus facilitating the implementation of highly stable oven-controlled crystal oscillators (OCXO) operating at the turn-over point where temperature sensitivity to ambient thermal fluctuations is negligible. Figure 13.2 shows the temperature characteristic of a SiBAR with thin film aluminum nitride (AlN) transduction [11, 12], oriented in <100> crystal direction of a heavily doped silicon substrate, to achieve the maximum compensation of linear *TCF*.

Although doping-profile engineering has demonstrated considerable advancement in temperature desensitization of the resonance frequency, the repeatability of the temperature characteristic in devices compensated using this technique calls for accurate control of impurity concentration across the wafer. This not only imposes extra costs, but also introduces additional substrate thickness-control problems in the manufacturing process.

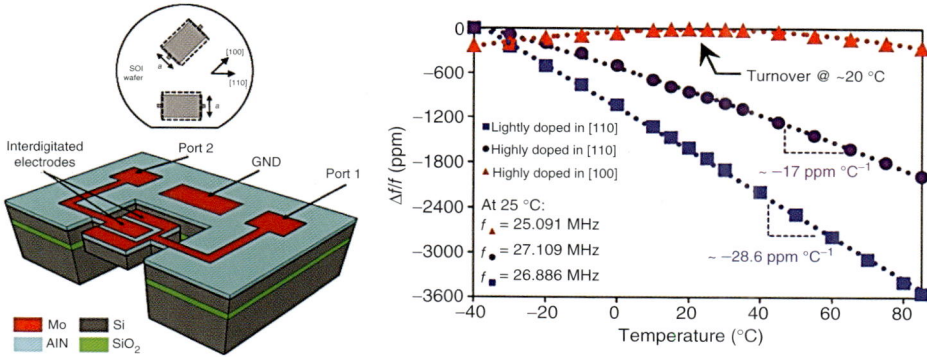

Figure 13.2 Temperature characteristics of extensional mode resonators aligned in different crystallographic direction of silicon and for different doping concentration [11].

13.2.1.3 Composite Resonators

More traditional *TCF* compensation techniques rely on the addition of a compensating material. In these approaches, the large negative *TCE* of native silicon is compensated by the addition of a material with positive *TCE* (usually silicon dioxide). Adding the correct amount of silicon dioxide (SiO_2) to the silicon resonator forms a composite Si/SiO_2 structure with temperature-compensated *TCE*. Figure 13.3 illustrates this concept schematically.

The compensating silicon dioxide can be added in layers to the peripheral surfaces of the resonator body (surface SiO_2 compensation) [13, 14]. In this technique, the compensating SiO_2 provides a parallel stiffness loading on the surface of silicon, which is inherently inefficient for compensation of in-plane resonance modes considering the smaller stiffness of SiO_2 compared to Si in the temperature range of interest. Although attempts in increasing the interface area between Si and SiO_2 has resulted in higher efficiencies [15], thick layers of SiO_2, comparable to that of silicon, are required to fully compensate the linear *TCF*. Therefore, surface SiO_2 compensation technique is mostly applicable to thin

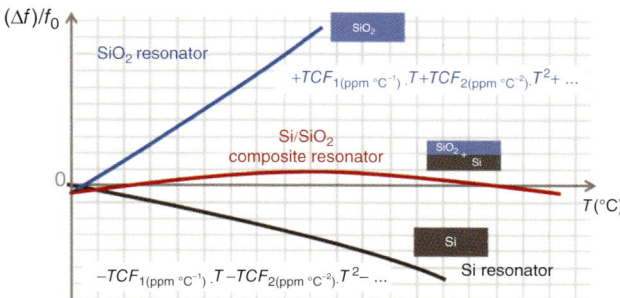

Figure 13.3 Schematic representation of compensating a silicon resonator with silicon dioxide. When first-order compensation has been achieved, a residual quadratic characteristic remains due to higher order temperature coefficients of the elastic constants.

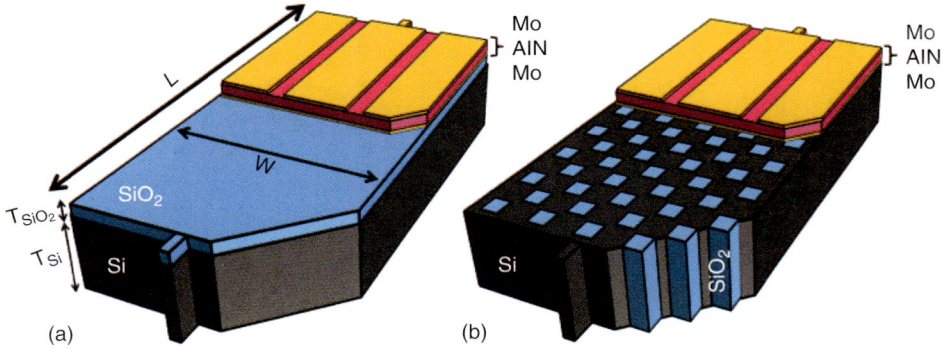

Figure 13.4 Silicon resonators with piezoelectric transduction: (a) with surface oxide compensation and (b) with bulk oxide compensation.

substrates and low-frequency flexural or extensional mode resonators, but more difficult to use for full *TCF* compensation of bulk acoustic wave (BAW) [4, 5, 16, 17] or high-frequency flexural devices with thick substrates, which are usually desirable to achieve high Q and superior power handling and linearity [7].

Alternatively, efficient *TCF* compensation of thick and high-frequency silicon resonators has been achieved by embedding compensating SiO_2 in the bulk of the resonant structure (bulk SiO_2 compensation). Figure 13.4 compares surface and bulk SiO_2 compensation schemes for a SiBAR with aluminum nitride (AlN) piezoelectric transduction.

In bulk compensation techniques, less amount of SiO_2 is required since it can be placed in series with particle polarization of desired resonance modes, hence providing considerably higher stiffness loading on the silicon microstructure. In these techniques, bulk SiO_2 is embedded inside the resonator body by carefully filling trenches with thermal or LPCVD (low pressure chemical vapor deposition) SiO_2 to provide a solid void-less platform.

Bulk SiO_2 compensation has been used for realization of temperature-stable acoustic platforms for implementation of temperature-compensated resonators operating in arbitrary resonance modes [18]. In this approach, a 2D array of silicon dioxide pillars has been uniformly distributed in the silicon substrate to provide a Si/SiO_2 homogenous composite platform (called hereafter SilOx) with temperature-compensated acoustic properties (Figure 13.5). The planar aspect ratio as well as the distribution density of pillars can be designed to simultaneously compensate several elastic moduli; thus providing an acoustic platform with temperature-insensitive properties to implement resonators with any desired resonance mode and particle polarization [18].

As verified experimentally, the addition of silicon dioxide does not degrade the main resonator characteristics such as Q, insertion loss (*IL*), and power handling. Full compensation of linear *TCF* for resonators implemented in such platforms has been demonstrated for different quasi-shear and extensional acoustic modes while Q remained high over the temperature range. Therefore, these devices are

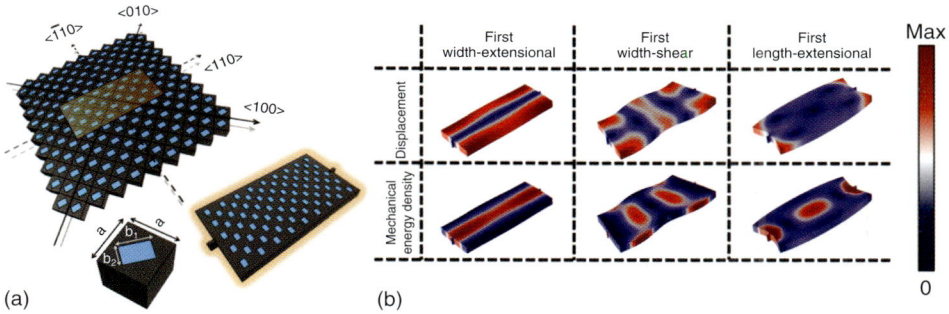

Figure 13.5 (a) Rectangular SilOx resonator. a and $b_{1,2}$ are SilOx unit and SiO_2 pillar lateral dimensions respectively. (b) Displacement and mechanical energy density of a rectangular SilOx resonator for its first width-extensional (WE_1), width-shear (WS_1), and length-extensional (LE_1) modes [18].

suitable for implementation of TCXOs with consistent phase-noise across the entire temperature range of operation [19]. Figure 13.6 shows the SEM picture of a rectangular SilOx resonator with thin-film AlN transduction and three of its temperature-compensated resonance modes. The cancelation of the linear *TCF* has resulted in a parabolic temperature characteristic showing a 40-fold improvement in the overall frequency drift across the temperature range of −20 to 100 °C

Furthermore, the relative volumetric ratio of silicon dioxide in SilOx cells can be used to tailor the temperature characteristic of f_0 to have a turn-over point at a desired temperature (Figure 13.7), facilitating implementation of OCXOs using these resonators.

In another approach, compensating bulk SiO_2 can be selectively embedded in specific regions of the microresonator with the highest acoustic energy density. Although this technique can only be applied to one specific resonance mode at a time, full compensation of linear *TCF* can be achieved with the smallest amount of SiO_2, which can simplify the fabrication process and design complexity of the device. Figure 13.8 shows the selective bulk SiO_2 compensation, where regions with high acoustic energy density of an in-plane, high-frequency flexural resonator have been used to amplify the stiffness loading of embedded silicon dioxide on the resonator and hence realizing full linear *TCF* cancelation with least amount of SiO_2 [20].

Although bulk SiO_2 compensation facilitates full *TCF* compensation of thick structures with the least amount of compensating material, considering the trench-refilling approaches used for embedding SiO_2 in the resonator body, the repeatability of temperature characteristic of these devices is limited by the fabrication lithography resolution. In fact, any uncertainty in definition of trenches directly translates into the variation of the relative volumetric ratio of SiO_2 in resonator body and changes the temperature characteristic of the resonance frequency.

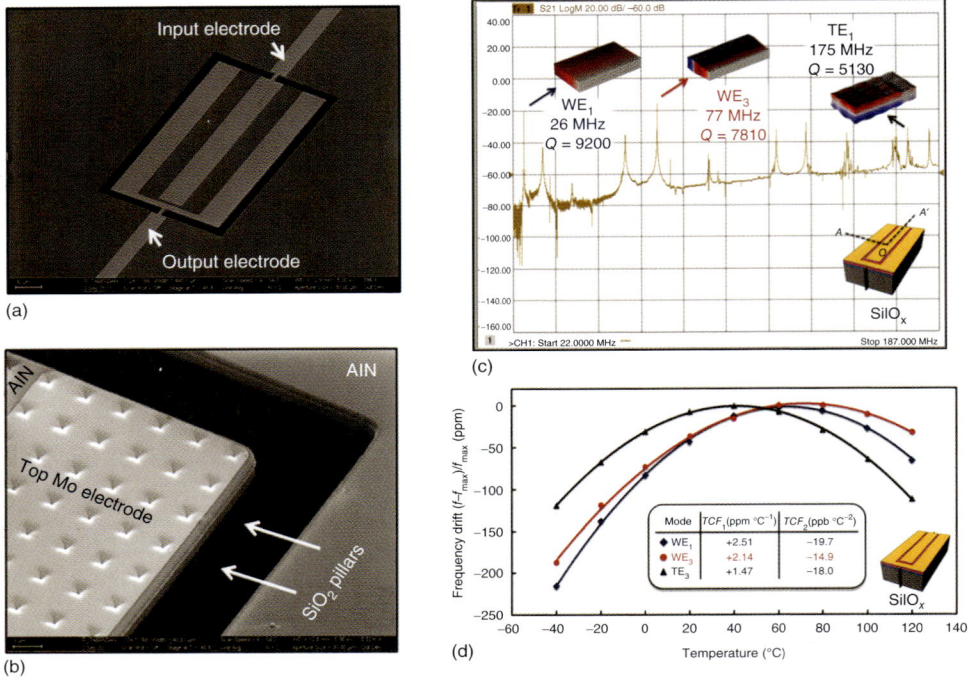

(a)

(b)

(c)

(d)

Mode	TCF_1 (ppm °C^{-1})	TCF_2 (ppb °C^{-2})
WE$_1$	+2.51	−19.7
WE$_3$	+2.14	−14.9
TE$_3$	+1.47	−18.0

Figure 13.6 (a, b) SEM image of temperature-compensated SilOx bulk acoustic resonator with thin-film AlN transduction and (c, d) large-span frequency response of the device and temperature characteristic of several modes with different particle polarization [18].

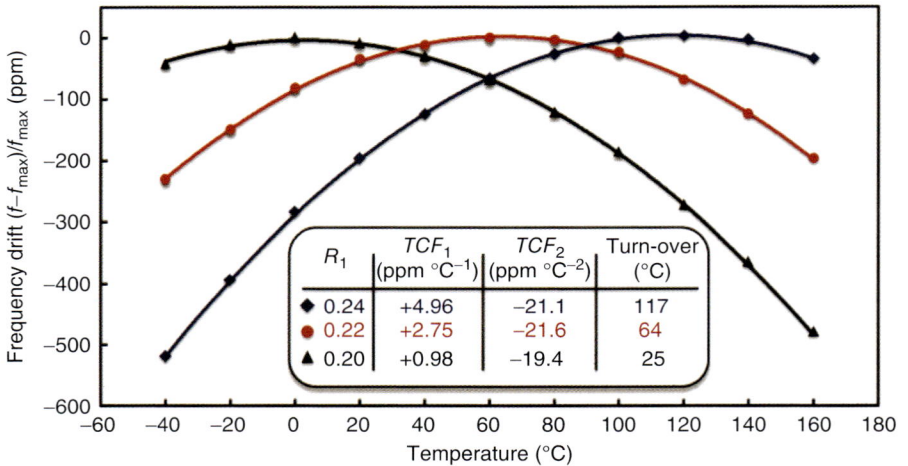

R_1	TCF_1 (ppm °C^{-1})	TCF_2 (ppm °C^{-2})	Turn-over (°C)
0.24	+4.96	−21.1	117
0.22	+2.75	−21.6	64
0.20	+0.98	−19.4	25

Figure 13.7 Turn-over temperature tailoring by changing the volumetric ratio of oxide in SilOx unit (R_1 in inset table), for a SilOx resonator operating in the WE$_1$ mode [18].

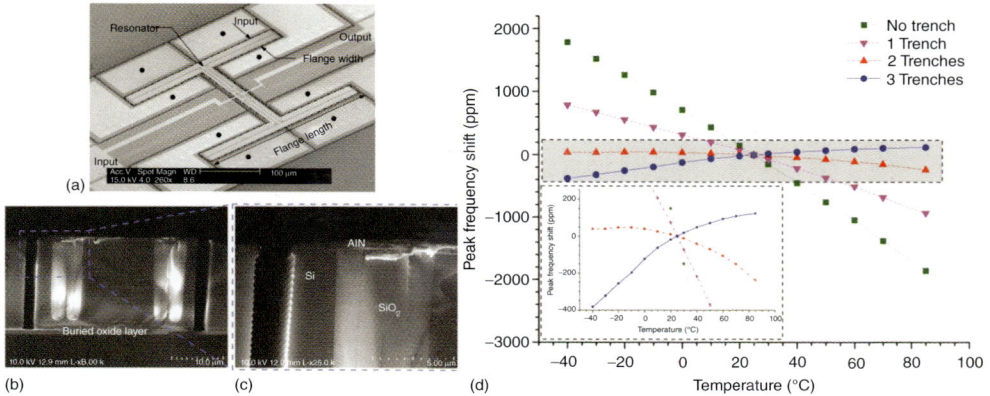

Figure 13.8 Flexural resonators with selective bulk SiO$_2$ temperature compensation (a–c) and their temperature characteristic of resonance frequency (d) [20].

13.2.2
Compensation for Manufacturing Uncertainties

In several applications of MEMS resonators such as real-time clocks (RTCs), the accuracy of the resonance frequency and its repeatability are of crucial importance. Therefore, desensitization of the resonator design and performance to manufacturing uncertainties become very important. This becomes even more critical considering the further potential frequency drifts induced by packaging stress and the limited amount of post-process frequency pulling/trimming budget available by tuning and trimming techniques.

The resonance frequency f_0 of resonators in arbitrary modes can be described by:

$$f_0 = \frac{1}{L} V_a = \frac{1}{2\pi} \sqrt{\frac{k_{\text{eff}}}{m_{\text{eff}}}} \tag{13.2}$$

where L is the characteristic length parameter, dependent on the geometry of the device, V_a is the effective acoustic velocity corresponding to L, and k_{eff} and m_{eff} are the effective stiffness and mass, respectively. Therefore, variations in the resonance frequency can arise due to process-induced variations in the device geometry and those material properties that affect the acoustic velocity. Several intrinsic and process-induced material properties of the substrates and films used for the implementation of MEMS resonators may affect the acoustic properties of the device and their repeatability; among them, the impurity concentration and its uniformity across the wafer, substrate misorientation, and thickness variation can be considered uncertainties intrinsic to the initial substrates, while film stress and in-plane misalignments can be considered process-induced uncertainties. For instance, the acoustic velocities experience a change as large as 1% when switching from intrinsic to heavily phosphorous-doped silicon substrates [21].

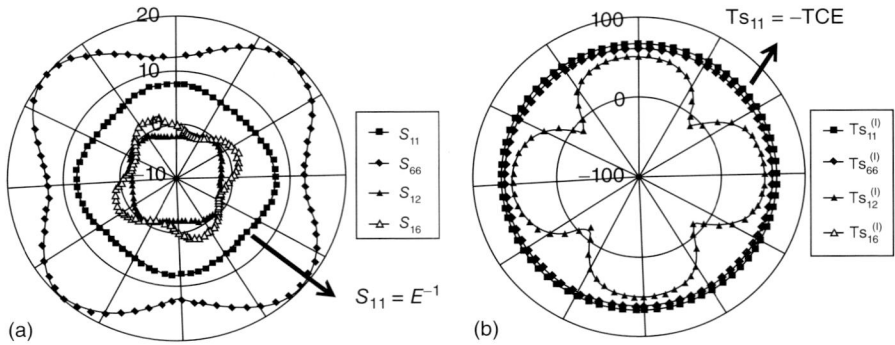

Figure 13.9 Crystallographic orientation dependency of different compliance constants S_{ij} (a) and their temperature coefficients (b), in a (100) silicon substrate. While the Young's modulus (*E*), which is the multiplicative inverse of S_{11}, shows an anisotropic profile, it's temperature coefficient is orientation-independent [22].

Also, an angular misalignment of $\pm 0.1°$ from the [110] axis of silicon can theoretically affect the acoustic velocity (and frequency) of longitudinal acoustic waves by 1.8 ppm. Doubling the misalignment to $\pm 0.2°$ increases the frequency shift to 7.3 ppm. This is due to the strong orientation dependency of different elastic/compliance constants in silicon (Figure 13.9).

Similarly, the sensitivity of *Q* to angular misalignment in SiBARs has been reported [23]. Therefore, excellent control over the dopant distribution and angular misalignments is crucial.

Besides frequency inaccuracies resulting from material uncertainties, geometrical variations may significantly contribute to the deviation of the resonance frequency from the targeted value. This can be due to lithographical errors as well as nonidealities in etching processes used to define the device on the substrate. In silicon-based MEMS resonators, the Bosch DRIE (deep reactive ion etching) process is typically used to create high-aspect ratio trenches, which define the in-plane dimensions of the resonator. Since the resonance frequency of in-plane modes is mainly defined by the trench-defined lateral dimensions of the resonator, systematic changes in these dimensions must be well understood so that they can be compensated for in the resonator design. Process compensation is achieved when the effective stiffness to mass ratio in Eq. (13.2) is maintained, even under geometric changes to the device. Several methods have been proposed to provide this condition in silicon resonators. For micromechanical I-shaped bulk acoustic resonators (IBARs), properly designed tapered flanks can be used to balance the contribution of mass and stiffness changes [21]. In contrast, for high-frequency SiBARs, perforations have been introduced, which accommodate these dimensional variations [24]. Figure 13.10 shows a process-compensated thin-piezoelectric-on-substrate (TPoS) resonator design in which 4 µm by 4 µm perforations have been introduced to balance the stiffness-to-mass changes. This is evidenced by the middle plot of Figure 13.10,

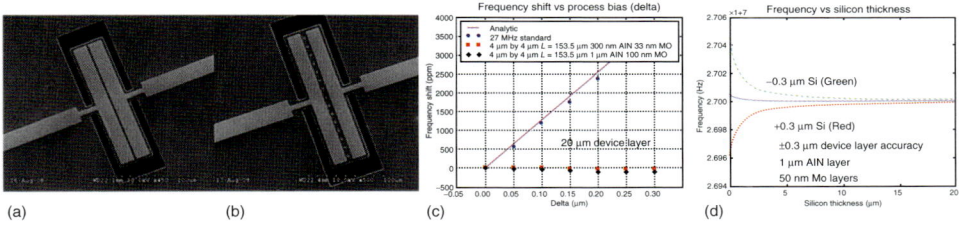

Figure 13.10 (a) SEM image of uncompensated TPoS resonator and (b) process-compensated 27 MHz TPoS resonator with a line of specially designed 4 μm by 4 μm perforations [24]. (c) Frequency trends for 27 MHz TPoS resonators versus DRIE trench critical dimension bias (delta). (d) Frequency trends of TPoS resonator versus SOI device thickness for ±0.3 μm layer tolerance.

in which the relationship between frequency variation and lithographic process bias offset has been calculated for the design using finite element (FE) simulations. It can be seen that the 4 μm by 4 μm square perforations slightly overcompensate the design (indicating a negative trend with process bias) versus the uncompensated design, even for different combinations of the piezoelectric transduction stack. Interestingly, the difference between the first-order analytical model from Eq. (13.2), solid-line in Figure 13.10c, and the triangles from FE simulation of the structure is accurately explained by the addition of support tethers, indicating that the tethers themselves have a slight compensating effect.

In addition to lithographic process compensation, attention must be given to the thickness dependencies as well. Although the thickness has a negligible effect for plain SiBARs in certain thickness regimes [25], this assumption does not hold for TPoS resonators. Since these resonators are formed from a composite stack of materials, including the silicon layer as the bulk of the device, the AlN piezoelectric layer, and molybdenum (Mo) top and bottom electrode layers, the effective acoustic velocity is a composite parameter of these layers, depending on the relative stack thicknesses. Since the AlN and Mo layers are formed by thin film deposition, their thicknesses can be controlled to high accuracy. Therefore, the dominant variations will come from the SOI (silicon-on-insulator) wafer, which are often toleranced to ±0.3 μm by the vendor. Substituting these numbers into a first-order analytical model, which takes into account the relative thickness of each layer, it is found that increasing the nominal silicon thickness can significantly reduce the resonators' sensitivity to the silicon layer thickness (Figure 13.10d).

13.2.3
Compensation and Control of Quality Factor

In certain applications of MEMS resonators, control and compensation of Q are paramount to obtaining reliable performance. For the case of MEMS vibratory gyroscopes, it can be desirable to operate at high frequencies so that high-Q devices can be used to get large sensitivities, while at the same time maintaining a large bandwidth for large dynamic range. One example of this approach is the Q-controlled spoke gyroscope, in which the Q was purposefully limited by the thermoelastic damping (TED) in the central spoke region [26]. Since the coupling between the thermal and elastic domains is directly proportional to the absolute temperature, this mechanism has a strong temperature dependency. It also depends strongly on the mode shape of the resonator, with high-frequency bulk acoustic modes generally being less subjected to TED than their low-frequency flexural beam counterparts [27]. However, selective addition of release holes can introduce additional TED by modifying the overall effective thermal paths [2].

Another important source of energy dissipation in MEMS resonators is energy leakage through their supporting tethers [2, 28–30]. These tethers connect the released resonant structure to its surrounding substrate. Since point supports are desirable, the finite size of these tethers limited by photolithography precision increases the energy leakage from the resonator. However, proper design of these supports can provide passive compensation for support loss. A group of these methods focuses on the reflection of leaked waves back in to the resonator by creating acoustic mismatch at the support/anchor region [31, 32]. Although these techniques provide considerable improvement in $Q_{support}$, substantial elimination of support leakage at the resonance frequency requires the engineering of the phononic band structure of the material. This has been realized by the implementation of acoustic bandgap structures [33, 34]. These structures provide a frequency band in which phonons cannot propagate. By selectively engineering the bandgap to include the resonance frequency of the resonators, full compensation of support loss can be achieved. Figure 13.11 shows an AlN-on-Si bulk acoustic resonator with linear acoustic bandgap supports as well as the corresponding phononic band diagram, where the bandgaps are shaded.

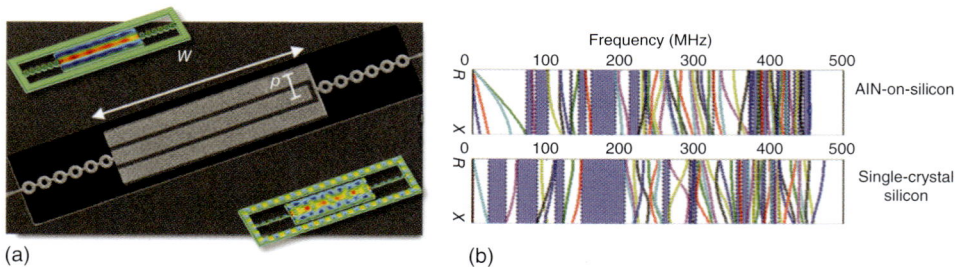

(a)   (b)

Figure 13.11 (a) AlN-on-silicon TPoS bulk acoustic resonator with linear acoustic bandgap (LAB) structures as supports providing support loss compensation [34] and (b) the dispersion diagrams of typical LAB structure showing the frequency of propagating phonons as a function of their wavelength for single crystal silicon and AlN-on-silicon substrates.

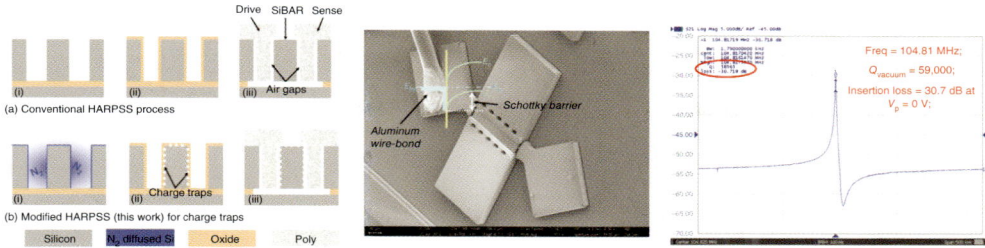

Figure 13.12 Charge trapping process and concept as well as frequency response of a V_p-less capacitive SiBAR [36].

13.2.4
Compensation for Polarization Voltage

Capacitive resonators require a DC polarization voltage (V_p) for operation. This translates into the requirement of additional circuitry resulting in incompatibility of these resonators with low-voltage CMOS processes. Moreover, since several characteristics of these devices are sensitive to V_p, stringent control is required to provide a constant DC voltage on-chip over all resonator operating conditions, which adds further complexity to the CMOS interface circuitry. Several techniques have been introduced for V_p compensation based on charge trapping inside the electrically isolated resonant silicon microstructures [35, 36]. Figure 13.12 shows an example of these techniques using charge injection through nitrogen (N_2) annealing besides charge trapping provided by formation of a Schottky barrier.

13.3
Tuning Methods in MEMS Resonators

Although *TCF* and process compensation techniques considerably reduce the amount of temperature and process-induced frequency drifts and uncertainties, full compensation of residual f_0 fluctuations as well as drifts caused by aging effects requires dynamic adjustment, which is provided by tuning mechanisms. Frequency tuning can be addressed on both device and system levels.

13.3.1
Device Level Tuning

These tuning mechanisms provide electronic f_0 pulling for individual resonators through electromechanical transducers and by changing the effective equivalent stiffness of the micromechanical resonator.

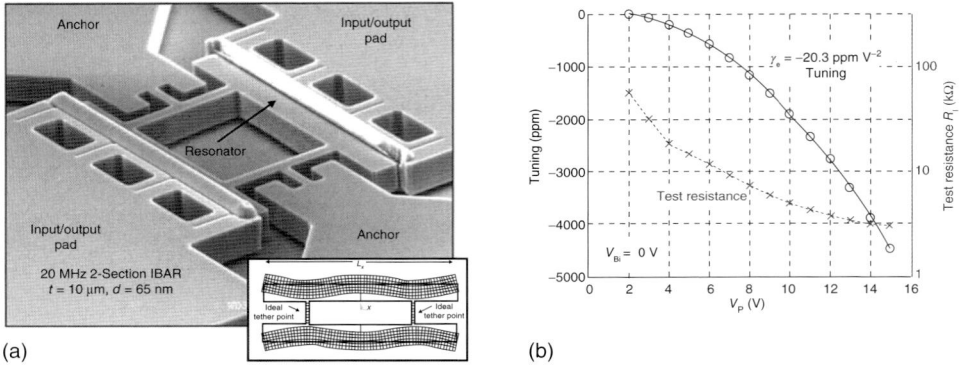

Figure 13.13 (a) A 20 MHz IBAR with large frequency tuning range and (b) the device tuning range and port-to-port resistance as a function of V_P [38].

13.3.1.1 Electrostatic Tuning

Electrostatic tuning has traditionally been the most commonly employed device-level tuning mechanism. In this technique, an electrical spring whose stiffness is tunable via a DC voltage is added in series with the effective mechanical stiffness of the resonator and provides continuous f_0 tuning with a quadratic characteristic. While the small stiffness of the electrical spring, resulting from low-efficiency electrostatic transduction, makes this tuning technique more appropriate for low-frequency flexural mode resonators [37], large tuning ranges have also been reported for high-frequency bulk acoustic devices with efficient sub-micrometer capacitive gaps [21]. Figure 13.13 shows a high-frequency (20 MHz) IBAR designed for maximum tunability, as well as the measured tuning range and device port-to-port test resistance as a function of V_P. Temperature-compensated oscillators have been built using this tuning mechanism in IBARs [38].

Although the application of electrostatic tuning is convenient in capacitive resonators, it results in fluctuations of resonator's *IL*, which is usually not desirable. This problem has been solved by separating the capacitive tuning electrodes from the AC electromechanical transducer [39]. Figure 13.14 shows a 32 kHz flexural resonator with AlN piezoelectric transduction and electrostatic tuning. Having a large tuning pad, which jointly serves as a frequency-loading mass, enables large f_0 tuning, capable of covering the temperature-induced frequency drift across the entire industrial temperature range.

13.3.1.2 Thermal Tuning

Another electronic device-level tuning mechanism, which is capable of providing large tuning ranges, is based on ovenization of the microresonator via Joule heating. In this technique, a tunable DC current passing through a heater element results in increased resonator temperature and hence tunable frequency changes due to the large *TCF* of uncompensated resonators [40–42]. The highest efficiency

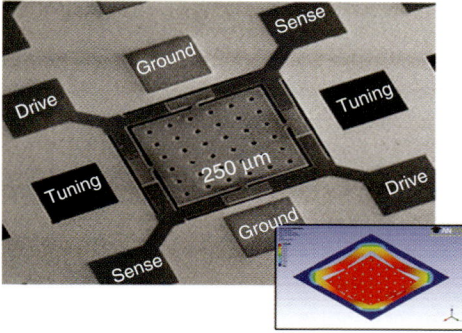

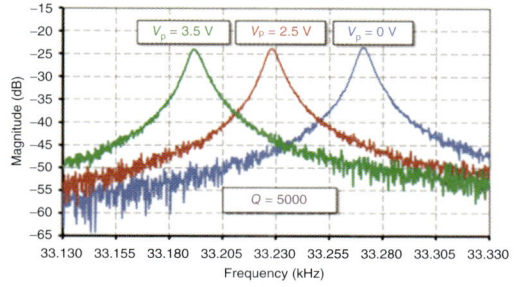

Figure 13.14 A 32 kHz flexural-mode resonator with AlN piezoelectric transduction and electrostatic tuning [39].

and agility of thermal tuning has been achieved by self-ovenization of silicon resonators. In this technique, the heating current passes through the device and the resonator serves as the micro-oven simultaneously [41, 42].

13.3.1.3 Piezoelectric Tuning

In addition to electrostatic and thermal tuning techniques, a tuning mechanism that is applicable to piezoelectrically-transduced MEMS resonators is based on the piezoelectric stiffening effect. In this technique, the stiffness of the piezoelectric film is controlled by tuning its electric termination at the tuning port [43, 44]. Figure 13.15 shows the concept and also an example of this technique, where a

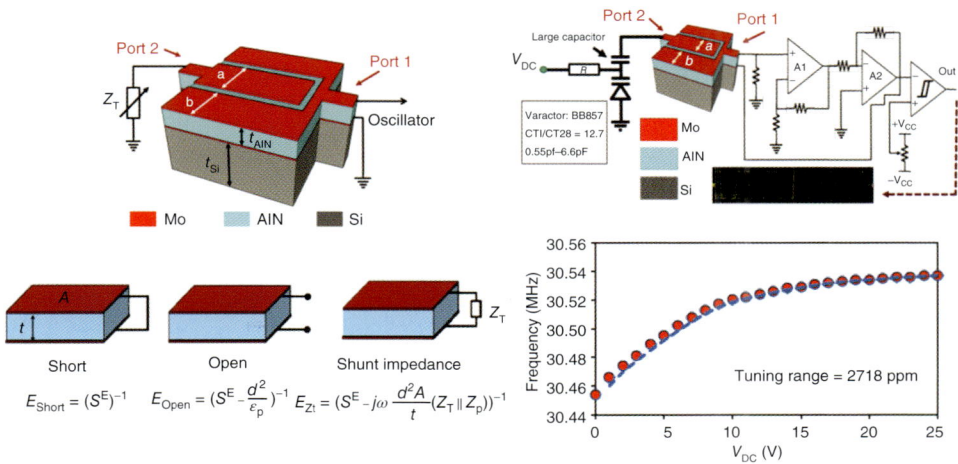

$$E_{Short} = (S^E)^{-1} \qquad E_{Open} = (S^E - \frac{d^2}{\varepsilon_p})^{-1} \qquad E_{Zt} = (S^E - j\omega \frac{d^2A}{t}(Z_T \| Z_p))^{-1}$$

Figure 13.15 Piezoelectric tuning concept and tunable oscillator implemented based on TPOS tunable device [43].

separate port is dedicated for tuning purposes, while another port is used for the implementation of the oscillator.

13.3.2
System-Level Tuning

System-level tuning and frequency-control mechanisms are applied to the systems where MEMS resonators serve as the frequency reference. These techniques include the use of phase-locked loops [45] or the addition of tunable electrical impedances and phase-shifters in series and/or in parallel with the resonator [19, 46, 47]. Systems employing these techniques can tune out the residual frequency drifts and uncertainties of TCF and process-compensated resonators, to form highly stable frequency references. Figure 13.16 shows the schematic design of a 27 MHz oscillator implemented from using a TCF-compensated SilOx resonator in addition to tunable capacitors and phase-shifters embedded in series with the resonator in the oscillator loop.

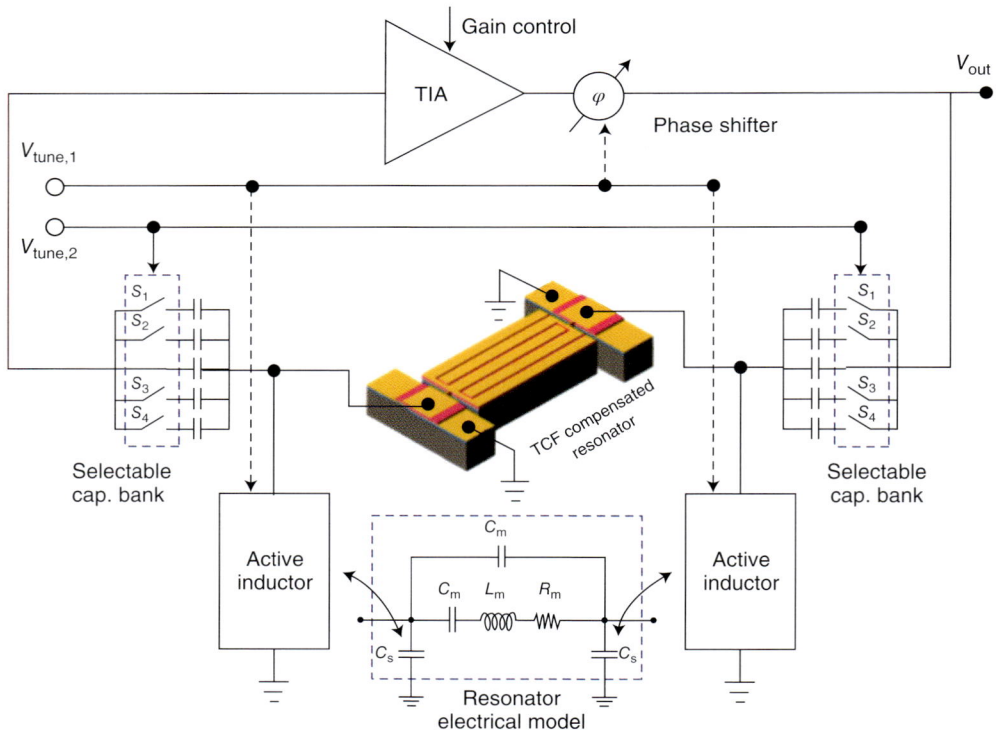

Figure 13.16 System diagram of a 27 MHz temperature-stable MEMS oscillator with subparts per million instability using SilOx resonator and system-level tuning [19].

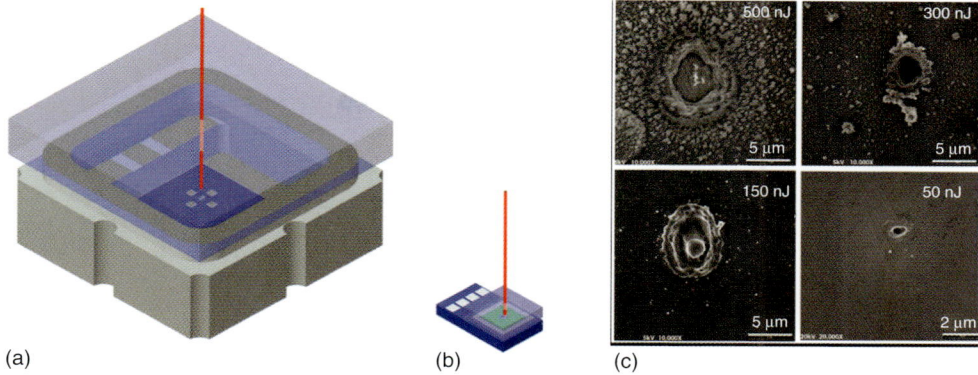

Figure 13.17 Post-packaging laser trimming through a transparent lid for (a) a ceramic package and (b) a wafer-level package. (c) Laser spot for different pulse energies. (The figures are taken from Ref. [49].)

13.4
Trimming Methods

Trimming refers to the permanent shift of a device parameter, usually performed after fabrication. Commonly, trimming is achieved using selective addition (e.g., metal deposition [48]) or removal (e.g., with laser ablation [49]) of material. Besides various processing errors and uncertainties, the packaging itself usually induces large drifts in the resonance frequency. Therefore, trimming techniques, which can be applied after device packaging, are desirable. An example of these techniques is selective laser trimming through a transparent lid of the resonator package (Figure 13.17).

Emerging methods include localized growth of oxide on thermally actuated silicon resonators [50] and metal diffusion via formation of metal-silicon eutectic alloys at elevated temperatures [51]. The latter is achieved by passing a controlled DC current through the body of the silicon resonator, causing its temperature to increase via Joule heating. This is an especially attractive technique because it allows post-vacuum-packaging trimming at the wafer or individual device levels. No material is vaporized within the vacuum package, allowing the package to maintain a high vacuum. Use of different metals allows either up or down trimming. For gold(Au), the resonance frequency increases after running the thermal current, since the Au–Si bond is higher in strength than the Si–Si bond (Figure 13.18). For aluminum (Al), the resonance frequency decreases after trimming because the Al–Si bond is lower in strength than both Au–Si and Si–Si bonds. Aluminum was also found to be preferred from a trimming time standpoint, since aluminum thermomigrates against thermal gradients, whereas Au–Si bonding requires reaching the eutectic temperature to activate the trimming process.

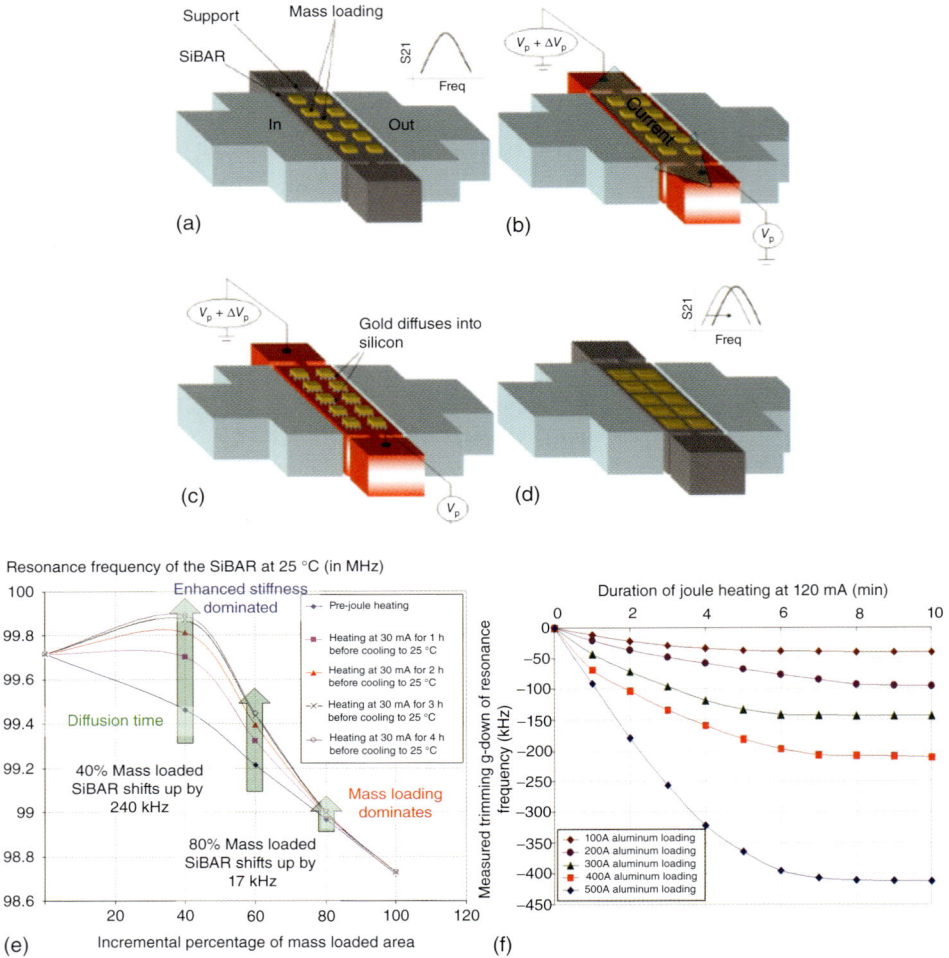

Figure 13.18 (a–d) Schematic representation of gold-diffusion-based trimming in gold-coated SiBARs. When a current is passed through the SiBAR body, the temperature is raised above the Au–Si eutectic temperature, allowing the gold layer to diffuse into silicon. (e) Frequency up-trimming is achieved with Au–Si eutectic, since the Au–Si bond is stronger than the Si–Si bond. (f) Frequency down-trimming is achieved with Al–Si thermomigration, since Al–Si bonds are weaker than Si–Si bonds [51].

References

1. Ayazi, F. (2009) MEMS for integrated timing and spectral processing. Custom Integrated Circuits Conference, pp. 65–72.

2. Ayazi, F., Sorenson, L., and Tabrizian, R. (2011) Energy dissipation in micromechanical resonators. SPIE Defense, Security and Sensing, pp. 1–13.

3. Samarao, A.K., Casinovi, G., and Ayazi, F. (2010) Passive TCF compensation in high Q silicon micromechanical resonators. IEEE International Conference on Micro

Electro Mechanical Systems (MEMS), pp. 116–119.

4. Pourkamali, S., Ho, G.K., and Ayazi, F. (2007) Low-impedance VHF and UHF capacitive silicon bulk acoustic wave resonators – part I: concept and fabrication. *IEEE Trans. Electron Devices*, **54**, 2017–2023.

5. Pourkamali, S., Ho, G.K., and Ayazi, F. (2007) Low-impedance VHF and UHF capacitive silicon bulk acoustic wave resonators – Part II: measurement and characterization. *IEEE Trans. Electron Devices*, **54**, 2024–2030.

6. Samarao, A.K. and Ayazi, F. (2010) Intrinsic temperature compensation of highly resistive high-Q silicon microresonatros via charge depletion. International Frequency Control Symposium, pp. 334–339.

7. Pan, W. and Ayazi, F. (2010) Thin-film piezoelectric-on-substrate resonators with Q enhancement and TCF reduction. Proceedings of the IEEE MEMS, pp. 727–730.

8. Hajjam, A., Rahafrooz, A., and Pourkamali, S. (2010) Sub-100 ppm/C temperature stability in thermally actuated high frequency silicon resonators via degenerate phosphorous doping and bias current optimization. IEDM, 2010, pp. 170–173.

9. Shahmohammadi, M., Harrington, B.P., Gonzales, J., and Abdolvand, R. (2012) Temperature-compensated extensional-mode MEMS resonators on highly N-type doped silicon substrates. Hilton Head, 2012, pp. 371–374.

10. Samarao, A.K. and Ayazi, F. (2009) Temperature compensation of silicon micromechanical resonators via degenerate doping. IEEE International Electron Devices Meeting, 2009, pp. 789–792.

11. Shahmohammadi, M., Harington, B.P., and Abdolvand, R. (2013) Turnover temperature point in extensional-mode highly doped silicon micro-resonators. *IEEE Trans. Electron Devices*, **60**, 1213–1220.

12. Ho, G.K., Abdolvand, R., Sivapurapu, A., Humad, S., and Ayazi, F. (2008) Piezoelectric-on-silicon lateral bulk acoustic wave micromechanical resonators. *IEEE J. Microelectromech. Syst.*, **17**, 512–520.

13. Melamud, R., Chandorkar, S.A., Kim, B., Lee, H.K., Salvia, J., Bahl, G., Hopcroft, M., and Kenny, T. (2009) Temperature insensitive composite micromechanical resonators. *J. Microelectromech. Syst.*, **18**, 1409–1419.

14. Abdolvand, R., Ho, G.K., Butler, J., and Ayazi, F. (2007) ZnO-on-nanocrystalline-diamond lateral bulk acoustic resonators. IEEE MEMS, 2007, pp. 795–798.

15. Grogg, D., Tekin, H.C., Ciressan-Badila, N.D., Tsamados, D., Mazza, M., and Ionescu, A.M. (2009) Bulk lateral MEM resonator on thin SOI with high Q-factor. *IEEE J. Microelectromech. Syst.*, **18** (2), 466–479.

16. Hao, Z., Pourkamali, S., and Ayazi, F. (2004) VHF single crystal silicon elliptic bulk-mode capacitive disk resonators; part I: design and modeling. *IEEE J. Microelectromech. Syst.*, **13**, 1043–1053.

17. Pourkamali, S., Hao, Z., and Ayazi, F. (2004) VHF single crystal silicon elliptic bulk-mode capacitive disk resonators; part II: implementation and characterization. *IEEE J. Microelectromech. Syst.*, **13**, 1054–1062.

18. Tabrizian, R., Casinovi, G., and Ayazi, F. (2013) Temperature-stable Silicon Oxide (SilOx) micromechanical resonators. *IEEE Trans. Electron Devices*, **60**, 2656–2663.

19. Tabrizian, R., Pardo, M., and Ayazi, F. (2012) A 27 MHz temperature compensated MEMS oscillators with sub-ppm instability. IEEE MEMS, 2012, pp. 23–26.

20. Takhar, Z.W.V.A., Peczalski, A., and Rais-Zadeh, M. (2013) Piezoelectrically transduced temperature-compensated flexural-mode silicon resonators. *IEEE J. Microelectromech. Syst.*, **22**, 815–823.

21. Ho, G.K., Perng, J.K.C., and Ayazi, F. (2010) Micromechanical IBARs: modeling and process compensation. *IEEE J. Microelectromech. Syst.*, **19**, 516–525.

22. Bourgeois, C., Steinsland, E., Blanc, N., and deRooij, N.F. (1997) Design of resonators for the determination of the temperature coefficients of elastic constants of monocrystalline silicon.

presented at the IEEE International Frequency Control Symposium, 1997.

23. Samarao, A.K. and Ayazi, F. (2010) Quality factor sensitivity to crystallographic axis misalignment in Silicon Micromechanical Resonators. Technical Digest Solid-State Sensors, Actuators, and Microsystems Workshop, Hilton Head, SC, June 2010, pp. 479–482.

24. Ayazi, F. and Sorenson, L. (2011) Thin-film piezoelectric-on-insulator resonators having perforated resonator bodies therein. US Patent 7,888,84315.

25. Casinovi, G., Gao, X., and Ayazi, F. (2010) Lamb waves and resonant modes in rectangular bar silicon resonators. *IEEE J. Microelectromech. Syst.*, **19**, 827–839.

26. Sung, W.K., Dalal, M.J., and Ayazi, F. (2010) A 3 MHz spoke gyroscope with wide bandwidth and large dynamic range. IEEE International Conference on Micro Electro Mechanical Systems (MEMS 2010), Hong Kong, January 2010, pp. 104–107.

27. Abdolvand, R., Johari, H., Ho, G.K., and Ayazi, F. (2006) Quality factor in trench-refilled polysilicon beam resonators. *IEEE J. Microelectromech. Syst.*, **15**, 471–478.

28. Hao, Z., Abdolvand, R., and Ayazi, F. (2006) A high-Q length-extensional bulk-mode mass sensor with annexed sensing platforms. IEEE International Conference on Micro Electro Mechanical Systems (MEMS '06), pp. 598–601.

29. Hao, Z. and Ayazi, F. (2007) Support loss in the radial bulk-mode vibrations of center-supported micromechanical disk resonators. *Sens. Actuators, A: Phys.*, **134**, 582–585.

30. Tabrizian, R., Rais-Zadeh, M., and Ayazi, F. (2009) Effect of phonon interactions on limiting the fQ product of micromechanical resonators. IEEE Internation Conference on Solid-State Sensors, Actuators and Microsystems (Transducers), 2009, pp. 2131–2134.

31. Wang, J., Butler, J., Feygelson, T., and Nguyen, C.T.-C. (2004) 1.51-GHz nanocrystalline diamond micromechanical disk resonator with material-mismatched isolating support. presented at the IEEE International Conference on Micro Electro Mechanical Systems (MEMS), 2004.

32. Harrington, B.P. and Abdolvand, R. (2009) Q-enhancement through minimization of acoustic energy radiation in micromachined lateral-mode resonators. presented at the Solid-State Sensors, Actuators and Microsystems (TRANSDUCERS), 2009.

33. Mohammadi, S., Eftekhar, A., Khelif, A., and Adibi, A. (2010) Support loss suppression in micromechanical resonators by the use of phononic band gap structures. SPIE Photonics West, 2010, pp. 76090 W-7.

34. Sorenson, L., Fu, J.L., and Ayazi, F. (2011) One-dimensional linear acoustic bandgap structures for performance enhancement of AlN-on-Si micromechanical resonators. presented at the International Conference on Solid-State Sensors, Actuators and Microsystems (TRANSDUCERS), 2011.

35. Li, S.-S., Lin, Y.-W., Xie, Y., Ren, Z., and Nguyen, C.T.C. (2005) Chargebiased vibrating micromechanical resonators. Presented at the IEEE Ultrasonics Symposium (IUS), 2005.

36. Samarao, A.K. and Ayazi, F. (2010) Self-polarized capacitive silicon micromechanical resonators via charge trapping. Presented at the IEEE International Electron Devices Meeting (IEDM), 2010.

37. Hashimoto, K. (2009) *RF Bulk Acoustic Wave Filters for Communications*, Artech House Publishers.

38. Sundaresan, K., Ho, G.K., Pourkamali, S., and Ayazi, F. (2007) Electronically temperature-compensated silicon bulk acoustic resonator reference oscillators. *IEEE J. Solid-State Circuits*, **42**, 1425–1434.

39. Serrano, D.E., Tabrizian, R., and Ayazi, F. (2012) Electrostatically tunable piezoelectric-on-silicon micromechanical resonator for real time clock. *IEEE Trans. Ultrason. Ferroelectr. Freq. Control*, **59**, 358–365.

40. Tazzoli, A., Rinaldi, M., and Piazza, G. (2011) Ovenized high frequency oscillators based on aluminum nitride contour mode MEMS resonators. IEEE International Electron Devices Meeting (IEDM), 2011, pp. 481–484.

41. Sundaresan, K., Ho, G.K., Pourkamali, S., and Ayazi, F. (2006) A low noise 100 MHz silicon BAW reference oscillator. Custom Integrated Circuits Conference, 2006, pp. 841–844.

42. Tabrizian, R. and Ayazi, F. (2012) Reconfigurable SiBAR filters with sidewall AlN signal transduction. Solid-State Sensors, Actuators, and Microsystems Workshop (Hilton Head 2012) Hilton Head, SC, 2012, pp. 98–99.

43. Shahmohammadi, M., Dikbas, D., Harrington, B.P., and Abdolvand, R. (2011) Passive tuning in lateral-mode thin-film piezoelectric oscillators. 2011 Joint Conference of the IEEE International Frequency Control Symposium(IFCS) and European Frequency and Time Forum(EFTF), San Francisco, CA, May 2011, pp. 1–5.

44. Tabrizian, R. and Ayazi, F. (2011) Tunable silicon bulk acoustic resonators with multi-face AlN transduction. Proceedings of the 2011 Joint IEEE International Frequency Control Symposium and European Frequency and Time Forum (IFCS/EFTF 2011), San Francisco, CA, May 2011, pp. 749–752.

45. Salvia, J.C., Melamud, R., Chandorkar, S.A., Lord, S.F., and Kenny, T. (2010) Real-time temperature compensation of MEMS oscillators using an integrated micro-oven and a phase lock loop. *IEEE J. Microelectromech. Syst.*, **19**, 192–201.

46. Lavasani, H.M., Pan, W., Harrington, B.P., Abdolvand, R., and Ayazi, F. (2010) An electronically temperature-compensated 427 MHz low phase-noise AlN-on-Si micromechanical reference oscillator. Presented at the IEEE Radio Frequency Integrated Circuits Symposium (RFIC), 2010.

47. Lavasani, H.M., Pan, W., Harrington, B.P., Abdolvand, R., and Ayazi, F. (2012) Electronic temperature compensation of lateral bulk acoustic resonator reference oscillators using enhanced series tuning technique. *IEEE J. Solid-State Circuits*, **47**, 1381–1393.

48. Courcimault, C.G. and Allen, M.G. (2005) High-Q mechanical tuning of MEMS resonators using a metal deposition-annealing technique. Presented at the International Conference on Solid-State Sensors, Actuators and Microsystems (TRANSDUCERS), 2005.

49. Hsu, W.-T. and Brown, A.R. (2007) Frequency trimming of MEMS resonator oscillators. Presented at the IEEE International Frequency Control Symposium, 2007.

50. Hajjam, A., Dietrich, K., Rahafrooz, A., and Pourkamali, S. (2012) A self-controlled frequency trimming technique for micromechanical resonators. Presented at the Solid-State Sensors, Actuators, and Microsystems Workshop, Hilton Head, SC, 2012.

51. Samarao, A.K. and Ayazi, F. (2009) Post-fabrication electrical trimming of silicon bulk acoustic resonators using Joule heating. Presented at the IEEE International Conference on Micro Electro Mechanical Systems (MEMS), 2009.

Part III
Application

Resonant MEMS – Fundamentals, Implementation and Application, First Edition.
Edited by Oliver Brand, Isabelle Dufour, Stephen M. Heinrich and Fabien Josse.
© 2015 Wiley-VCH Verlag GmbH & Co. KGaA. Published 2015 by Wiley-VCH Verlag GmbH & Co. KGaA.

14
MEMS Inertial Sensors

Diego Emilio Serrano and Farrokh Ayazi

14.1
Introduction

Inertial sensors are devices that respond to physical motion, such as linear displacement or angular rotation, and transform this response into electrical signals that can be amplified and processed by electronic circuits. Accelerometers and gyroscopes are among the most common types of micro-electromechanical systems (MEMS) inertial sensors [1]. Accelerometers detect axial acceleration along particular coordinates of interest; whereas gyroscopes, measure the rate of rotation – angular velocity – around the x-, y-, or z-axis (i.e., roll, pitch, and yaw rotation, respectively). Taken together, tri-axial accelerometers and gyroscopes form what is better known as a 6-degree-of-freedom (6-DOF) inertial measurement unit (IMU) [2]. In recent years, MEMS IMUs have become of particular interest because of their possible integration into hand-held devices to implement inertial navigation by means of dead reckoning. This capability can potentially enable pedestrian guidance in enclosed spaces or remote regions where Global Positioning System (GPS) signals are not available.

In this chapter, the operation principles of accelerometers and gyroscopes will be described, including the benefits and challenges of implementing them as resonant MEMS sensors. High-frequency bulk-acoustic wave (BAW) devices will be used as a case study to serve as a contrast to traditional MEMS architectures. Lastly, efforts on the development, miniaturization, and performance enhancement of integrated IMUs will also be highlighted.

14.2
Accelerometers

Since the early 1990s, stand-alone MEMS accelerometers have been utilized in a wide range of commercial applications such as airbag deployment in cars, free-fall detection in laptops, and motion recognition for gaming, among others. Thanks

Resonant MEMS – Fundamentals, Implementation and Application, First Edition.
Edited by Oliver Brand, Isabelle Dufour, Stephen M. Heinrich and Fabien Josse.
© 2015 Wiley-VCH Verlag GmbH & Co. KGaA. Published 2015 by Wiley-VCH Verlag GmbH & Co. KGaA.

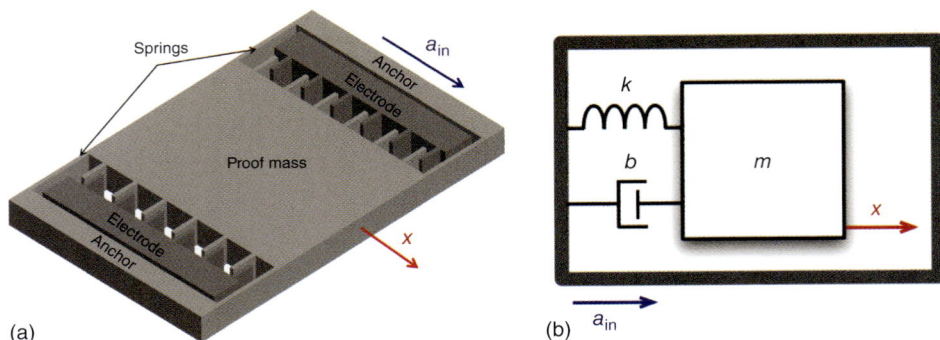

Figure 14.1 (a, b) Schematic representation and lumped-element model of a single-axis MEMS accelerometer. When exposed to an acceleration a_{in}, the mass translates a distance of x relative to its anchor frame.

to the rapid development and evolution of these devices, fundamental principles shared by many other MEMS structures have been thoroughly studied. The basic operation response of accelerometers is presented herein as an introduction to the understanding of resonant inertial sensors. The effects of squeeze-film damping on performance and stability will be briefly discussed, followed by a short review on accelerometer mechanical noise. Alternative techniques to operate accelerometers as resonant sensors are presented as a comparison to traditional low-frequency quasi-static devices.

14.2.1
Principles of Operation

Most MEMS accelerometers are implemented by the use of a moving proof-mass anchored to a reference frame through a spring mechanism. The structure is usually enclosed in a damped environment that determines the dynamic response of the device. Therefore, an accelerometer can be modeled as a second-order spring-mass-damper system that translates, relative to its frame of reference, when exposed to an external acceleration. Figure 14.1 illustrates the schematic diagram of a single-axis accelerometer and its equivalent lumped-element model.

In this system, the mass m is mainly determined by the volume and density of the proof-mass; the spring constant k by the dimensions, boundary conditions, and elastic properties of the tethers that support the mass, and the damping constant b by the gas or fluid surrounding the system (conventionally air). The equation of motion of the system is then given by:

$$m\frac{\partial^2 x}{\partial t^2} + b\frac{\partial x}{\partial t} + kx = F_{axl} \tag{14.1}$$

where $F_{axl} = ma_{in}$, a_{in} being the experienced input acceleration. When solving for x in the s-domain, the mechanical transfer function of the accelerometer can be

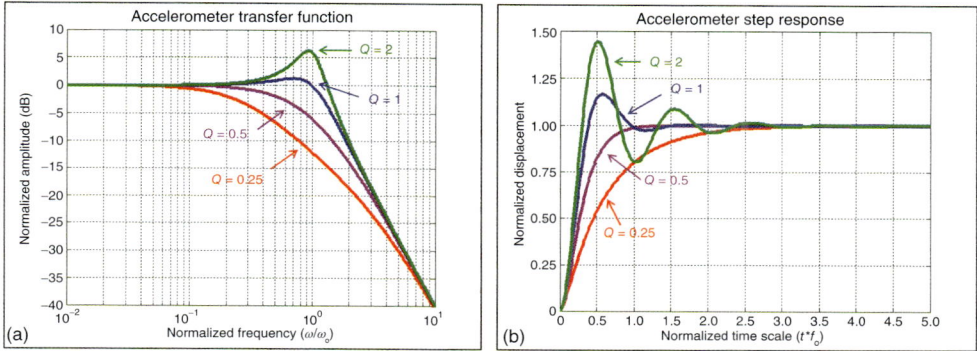

Figure 14.2 (a, b) Normalized magnitude transfer function (frequency response) and time-domain step response of a MEMS accelerometer for different values of Q.

expressed as:

$$\frac{X(s)}{A_{in}(s)} = \frac{1}{s^2 + 2\zeta\omega_0 s + \omega_0^2} \tag{14.2}$$

ζ and ω_0 being the system damping ratio and angular resonance frequency, respectively:

$$\zeta = \frac{1}{2Q} = \frac{b}{2\sqrt{km}} \quad \text{and} \quad \omega_0 = \sqrt{\frac{k}{m}} \tag{14.3}$$

Figure 14.2, shows the normalized frequency response ($s = j\omega$) and the equivalent time-domain step response of an accelerometer for different values of the quality factor Q. It is evident that the dynamics of the system are closely related to the frequency region of operation and the damping conditions to which the structure is exposed.

14.2.2
Quasi-Static Accelerometers

Most MEMS accelerometers are designed to operate as quasi-static devices in a low-frequency regime ($\omega \ll \omega_0$), where the mechanical transfer function can be considered an independent function of the frequency of operation ω:

$$\left.\left|\frac{X(s)}{A_{in}(s)}\right|\right|_{\omega \ll \omega_0} = \frac{x}{a_{in}} \approx \frac{1}{\omega_0^2} \tag{14.4}$$

From Eq. (14.4), it can be clearly inferred that there is always a trade-off between sensitivity and maximum achievable 3-dB bandwidth, which is directly proportional to the resonance frequency: $BW \propto \omega_0$. Thus, in order to extend their frequency range of operation, accelerometers should ideally work as critically damped systems, where $Q = 0.5$. For lower values of Q, their 3-dB cut-off frequency is dramatically reduced, affecting how fast the structure can

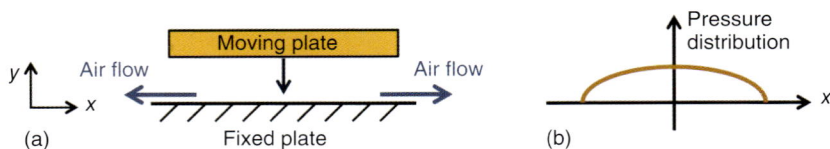

Figure 14.3 (a, b) Schematic representation of squeeze-film air damping effect in a closing-gap parallel plate structure and pressure distribution inside the channel.

respond. On the other hand, for $Q > 0.5$, the system becomes under-damped, which can result in a significant amount of ringing when exposed to a step or impulse acceleration – such as the one experienced from a shock or drop during operation. Ringing can be detrimental to an accelerometer not only because it increases its settling time (Figure 14.2), but also because if significant overshoot is experienced, the structure might exceed its maximum tolerable displacement, causing permanent damage. Thus, a good control of the system damping is necessary to guarantee a fast but stable response.

14.2.2.1 Squeeze-Film Damping

For almost every low-frequency micromechanical structure, the damping constant b, and thus Q, is mainly determined by the interaction of the structure with the gas or fluid surrounding it. Squeeze-film damping is a particular case in which adjacent plates moving toward or away from each other – like the ones between the proof-mass and the fixed anchors in Figure 14.1 – displace the gas molecules in the gap between them. This causes a pressure distribution inside the gap that generates a force opposing the movement of the plates (Figure 14.3).

This is a dissipative effect that takes energy away from the system, thus damping its response. Close-form expressions for b can be calculated from Reynold's equation by using the appropriate boundary conditions [3, 4]. For the case of a single pair of rectangular parallel plates, the damping constant is given by:

$$b = \frac{96\mu_{\text{eff}}LW^3}{\pi^4 g_0^3} \tag{14.5}$$

where μ_{eff} is the effective gas viscosity, g_0 is the initial gap size between the plates, and W and L are the plate width and length, respectively. This expression assumes that the displacement is much smaller than the initial gap, the gas is incompressible, the frequency of operation is low, and $W \ll L$. But more general theoretical and empirical models have been derived to account for nonidealities such as large displacements, border effects, and gas rarefaction [5–7].

From Eq. (14.5), it is clear that in microstructures, where W and L are small, a large value of μ_{eff} is desirable to provide the appropriate amount of damping. Since the effective viscosity is proportional to the pressure in the system, quasi-static accelerometers are generally implemented to work at atmospheric pressure. This poses a great challenge when trying to integrate these devices with resonant structures like gyroscopes, which require large values of Q to function. An alternative method to control squeeze-film damping is to reduce the gap size g_0.

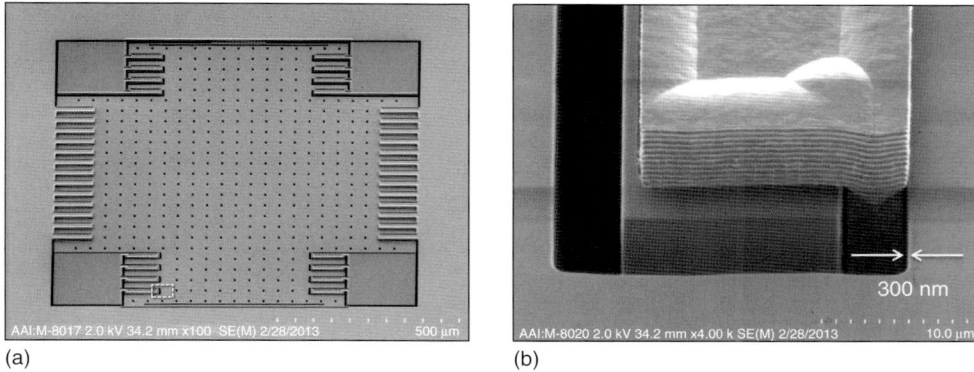

Figure 14.4 (a, b) SEM view of *x*-axis component of tri-axial accelerometer and close-up view of air nano-gaps for increased damping at low-pressure levels (1–10 Torr).

Unfortunately, in conventional microfabrication processes, the smallest achievable trench size is limited by the maximum allowable aspect ratio (trench depth to width) of the etching tools. On the other hand, alternative fabrication flows, like the high-aspect ratio poly and single-crystal silicon (HARPSS™) process [8], yield air nano-gaps that can supply the right amount of damping even at low pressure levels. This is attained by the use of thin sacrificial layers that can be precisely controlled via oxidation. Figure 14.4 shows an SEM view of the *x/y*-axis component of a tri-axial accelerometer designed to operate at pressure levels as low as 1–10 Torr [9]. Gaps of only 300 nm provide 300 times larger damping than a structure with the same dimensions and substrate thickness but with a typical gap size of 2 μm.

14.2.2.2 Electromechanical Transduction in Accelerometers

The translation experienced by an acceleration sensor has to be converted into an electrical signal that can then be processed by an interface circuit. To do so, electromechanical transduction mechanisms such as piezoelectric [10], piezoresistive [11], or electrostatic detection [12], are necessary. Due to its ease of implementation and good environmental stability, electrostatic detection (i.e., capacitive sensing) has been widely adopted in most commercial MEMS inertial sensors. The displacement generated in an accelerometer can be measured via electrostatic transduction by monitoring the change in gap that occurs between plates attached to the moving proof-mass and the stationary frame (Figure 14.1). The capacitance in a parallel plate capacitor is expressed as:

$$\frac{\partial C}{\partial x} = \frac{\varepsilon A}{(g_0 - x)^2} \approx \frac{\varepsilon A}{g_0}\left(\frac{1}{g_0} + \frac{2}{g_0^2}x + \frac{3}{g_0^3}x^2 + \cdots\right) \tag{14.6}$$

where C, ε, A, and g_0 are the capacitance, electric constant, area, and initial gap size of the capacitor, respectively. The rightmost term in Eq. (14.6) represents the Taylor series expansion of the expression. For small changes in gap ($x \ll g_0$), Eq. (14.6) can be linearized by neglecting its high order terms; and then by multiplying it with Eq. (14.4), the total transfer function of a capacitive accelerometer can be approximated to:

$$S = \frac{\partial C}{\partial a_{\text{in}}} \approx \frac{1}{\omega_0^2} \frac{\varepsilon A}{g_0^2} \tag{14.7}$$

Thus, by the use of air nano-gaps ($g_0 = 100 - 300\,\text{nm}$) a large electromechanical sensitivity can be achieved. An electronic circuit can then be used to detect the change in capacitance by applying a voltage between the capacitor and amplifying the varying charge [13].

14.2.2.3 Mechanical Noise in Accelerometers

Noise in micromechanical structures can be modeled as a force noise generator proportional to the loss mechanisms in the system. Similar to a resistor in an electrical circuit, the damping constant b has an associated thermal spot-noise (root-mean-square noise per unit of bandwidth) given by [14]:

$$\overline{f_n} = \sqrt{4 k_B T b} \tag{14.8}$$

where T is the temperature of operation and k_B the Boltzmann constant. The input referred mechanical noise equivalent acceleration (MNEA) can be calculated by finding the noise displacement generated by this force, and then dividing it by the accelerometer mechanical transfer function:

$$\text{MNEA} = \sqrt{\frac{4 k_B T \omega_0}{Qm}} \tag{14.9}$$

where MNEA has units of $(\text{m s}^{-2})_{\text{RMS}} (\sqrt{\text{Hz}})^{-1}$. Although increasing the value of Q is beneficial to reduce the spot-noise, larger Q values compromise both the overshoot and the settling-time of the system. Therefore, having a large proof-mass m is the most critical aspect necessary for the design of low-noise accelerometers. Lower resonance frequency ω_0 – achieved through lower stiffness – is also desirable, but making the structure too compliant can affect the shock survivability and vibration immunity of the device. Electronic filtering can always be utilized to reduce the effective bandwidth of integration, allowing for larger ω_0 without sacrificing integrated-noise performance.

14.2.3
Resonant Accelerometers

MEMS accelerometers can also be designed to operate as resonant sensors. Unlike quasi-static devices, which function under critically damped or over-damped conditions, resonant accelerometers require high quality factors that allow for the implementation of highly-selective frequency oscillators. Consequently, they require low-damping techniques, such as high-vacuum packaging. The value of their oscillation frequency can then be monitored to track for changes in the system caused by the experience of acceleration.

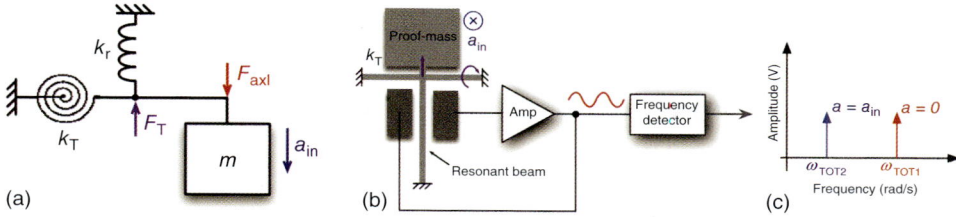

Figure 14.5 (a–c) Schematic diagram of a resonant accelerometer. Positive feedback is used to set the system into oscillation. The slow-changing acceleration a_{in} causes a quasi-static variation in the stiffness, which modulates the frequency of the resonant displacement x.

A conventional way of generating variations in the oscillation frequency of an electromechanical structure is by introducing an acceleration-dependent change in the resonator stiffness [15]. Consider the diagram shown in Figure 14.5. A high-Q capacitive beam resonator is brought into oscillation at its natural resonance frequency ω_{r0} by the use of positive feedback with electronics. In the presence of an out-of-plane acceleration a_{in}, a proof-mass attached to an end of the resonant beam will exert an inertial force F_{axl} on the torsion-spring k_T that will "push" or "pull" the beam, hence "softening" or "hardening" the resonator stiffness k_r. As shown in expression 14.3, the frequency of a resonator is a direct function of its stiffness, thus ω_{r0} will change when acceleration is applied. An analogy to this phenomenon is the change in frequency experienced by a guitar string as a result of the tensile stress applied when it is being tuned.

14.2.3.1 Electrostatic Spring-Softening

A more effective way of varying the stiffness of an electromechanical resonator is through an effect known as *electrostatic spring-softening*. A capacitive accelerometer structure – such as the proof-mass and anchored electrodes shown in Figure 14.1 – can be brought into resonance by applying an AC voltage v_{ac} at the resonance frequency ω_0 of the structure, and a biasing potential V_{DC} between the device and the electrodes. The generated electrostatic force will cause an apparent change in the stiffness of the structure. This can be explained by looking at the force expression:

$$F_{elec} = \frac{1}{2}(v_{ac} + V_{DC})^2 \frac{\partial C}{\partial x} \tag{14.10}$$

where $\partial C/\partial x$ is the change in capacitance given by Eq. (14.6). Neglecting the higher order terms, expression 14.10 can be simplified and reorganized as:

$$F_{elec} \approx \frac{1}{2}\frac{\varepsilon A}{g_0^2}\left(V_{DC}^2 + 2v_{ac}V_{DC} + 2\frac{V_{DC}^2}{g_0}x\right) \tag{14.11}$$

The first term in Eq. (14.11) corresponds to a constant electrostatic force that can be canceled when differential capacitive sensing is used: an opposite force of equal magnitude will be experienced by the structure. The second term is the alternating actuation force that keeps the device under resonance. And the third term, acts as an electrostatic stiffness given that it is proportional to the device displacement x.

14.2.3.2 Acceleration Sensitivity in Resonant Accelerometers

When expression 14.11 is added into Eq. (14.1) as an additional external force acting on the system, it can be clearly seen that the electrostatic stiffness force directly affects the overall stiffness of the resonator:

$$m\frac{d^2x}{dt^2} + b\frac{dx}{dt} + \left(k - \frac{\varepsilon A}{g_0^3}V_{DC}^2\right)x = ma_{in} + \frac{1}{2}\frac{\varepsilon A}{g_0^2}(V_{DC}^2 + 2v_{ac}V_{DC}) \quad (14.12)$$

Therefore, the total frequency of the system will be given by:

$$\omega_{TOT} = \sqrt{\frac{k - \frac{\varepsilon A}{g_0^3}V_{DC}^2}{m}} \approx \omega_0\left(1 - \frac{\varepsilon A}{2kg_0^3}V_{DC}^2\right) \quad (14.13)$$

Since the mechanical resonance frequency ω_0 is usually much higher than the frequency of the acceleration to be detected, the displacement generated by a_{in} can be considered a quasi-static change in the initial capacitive gap g_0, which will modulate the displacement x. The change in g_0 generated by a_{in} can be calculated using Eq. (14.4). Thus, the change in the total frequency, due to a slow-changing acceleration can be approximated to:

$$\frac{\partial\omega_{TOT}}{\partial a_{in}} = \frac{\partial g_0}{\partial a_{in}}\frac{\partial\omega_{TOT}}{\partial g_0} \approx -\frac{3}{2}\frac{\varepsilon A}{k\omega_0 g_0^4}V_{DC}^2 \quad (14.14)$$

This result gives the impression that having small gaps is extremely beneficial to increase the sensitivity of resonant accelerometers. However, it is important to highlight that in Eq. (14.14), g_0 is in fact the quasi-static varying gap due to an applied acceleration, which means that Eq. (14.14) is a highly non-linear function. Signal-processing techniques are then required to extend the linear range of this type of devices.

Another limitation of MEMS-resonant accelerometers is their high sensitivity to temperature fluctuations. For instance, if the structure is implemented in a single crystal silicon, it will have an intrinsic temperature coefficient of frequency (TCF) of approximately -30 ppm $°C^{-1}$ [16]. This large dependency demands temperature compensation schemes that increase the power consumption and complexity of the system. These are a few reasons why quasi-static accelerometers have been the choice of preference in the implementation of commercial products.

14.3
Gyroscopes

The unprecedented success of inertial sensors in portable electronics, and particularly in the smart phone business, has made micromechanical gyroscopes the fastest growing sector in the MEMS market. The demand for devices that are accurate enough to implement pedestrian navigation systems (PNSs), capable of operating in the absence of GPS, has accelerated the need for improvements in performance, power consumption, and miniaturization. Moreover, application spaces such as military and automotive, require enhancements in the robustness and reliability of existent designs. This section describes the operation principles of MEMS gyroscopes and how different device parameters and methods of operation affect their performance. BAW gyroscopes are used as a case study to highlight the differences between conventional vibratory mode-split gyros and high-frequency mode-matched resonant rotation-rate sensors.

14.3.1
Principles of Operation

Most MEMS gyroscopes are implemented by the use of vibratory structures that respond to rotation rate (i.e., angular velocity Ω) around a particular axis of interest. These systems experience an acceleration a_{cor} – orthogonal and proportional to both Ω and the velocity of the driving vibration v_{drv} – which is generated by what is known as the Coriolis effect:

$$\overrightarrow{a}_{cor} = -2\overrightarrow{\Omega} \times \overrightarrow{v}_{drv} \tag{14.15}$$

This phenomenon is best understood by analyzing a simple yet insightful example: the Foucault pendulum. A point-mass m attached with a string of length l to a fixed rod, constitutes a pendulum that can be set into a perpetual oscillatory swing (assuming there is no dissipation in the system) (Figure 14.6). If the pendulum is placed in the center of a rotating frame – like at the North Pole of the earth – an outside observer who does not rotate, will see the pendulum moving back and forth in the same direction independently of how fast the earth is revolving. On the other hand, if the observer is standing in the frame of rotation (i.e., somewhere in the earth), it would appear to him as if the pendulum swing is changing its oscillation trajectory at the same rate as the earth rotates. The perceived precession can be attributed to the fictitious Coriolis acceleration given by Eq. (14.15). Thus, if the earth spectator knows the velocity of the pendulum v_{drv}, he can indirectly calculate Ω by measuring a_{in}. This effect is dubbed as fictitious because it is only perceived in the rotation frame of reference rather than being generated by a physical interaction.

14.3.1.1 Vibratory Gyroscopes
Similar to the Foucault pendulum, vibratory gyroscopes detect rotation rate by the use of the Coriolis effect. Figure 14.7 shows a schematic representation and lumped element model of one of the most conventional types of rotation-rate sensors: the tuning-fork gyroscope (TFG) [17].

This device is composed of two suspended proof-masses (with total mass of m) anchored to a substrate through springs that allow them to displace in both x- and y-directions (k_x and k_y). The masses are excited into vibration along the

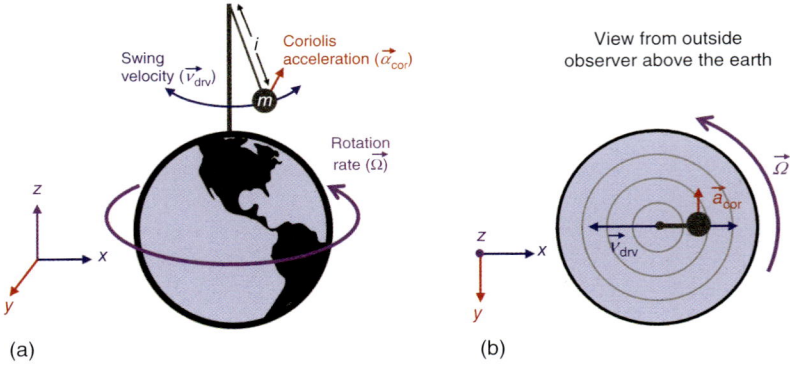

Figure 14.6 (a, b) The Foucault pendulum. For an observer outside the rotation frame (i.e., the earth) the pendulum swing looks fixed in space along a particular axis. For an observer inside the frame of rotation, the swing seems to precess due to the Coriolis effect.

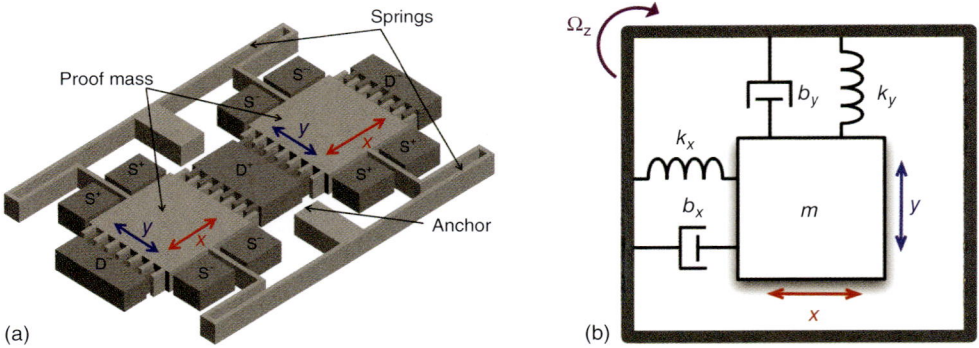

Figure 14.7 (a, b) Schematic representation and lumped-element model of a single-axis MEMS gyroscope. The device can be modeled as two independent second-order systems that couple through the Coriolis effect.

x-axis (or drive-axis) in *anti*-phase (when one displaces by x, the other displaces by $-x$) in order to reject common-mode signals, such as displacement due to linear acceleration. When an angular velocity Ω_z is applied around the z-coordinate, the structure will experience a Coriolis acceleration. The force induced by this acceleration will cause the proof-masses to displace along the y-axis (or sense-axis), also in contrary directions (y and $-y$). Electrostatic sensing can then be used to measure the differential change in capacitance between the structure and fixed electrodes (S^+ and S^-). Other electromechanical transducers such as piezoelectric or piezoresistive sensing can also be used to excite and/or sense the structure displacement.

Consequently, an ideal gyroscope can be modeled as two independent and orthogonal second-order systems that interact with each other only in the presence of the Coriolis effect:

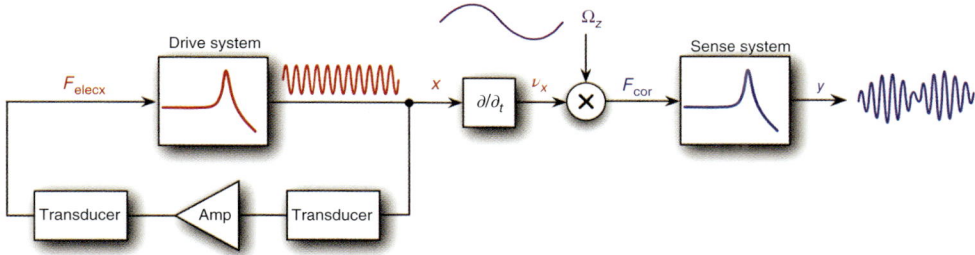

Figure 14.8 Block diagram representation of vibratory gyroscope. The drive-loop is excited into oscillation through positive feedback. The cross-product of the velocity of vibration v_x and the rotation-rate Ω_z, generates an orthogonal Coriolis force F_{cor} that excites the sense-system causing a displacement y. This can be viewed as an amplitude modulated signal with carrier frequency ω_{0x}.

$$m\frac{\partial^2 x}{\partial t^2} + b_x \frac{\partial x}{\partial t} + k_x x = F_{elecx} - 2\Omega_z m \frac{\partial y}{\partial t} \tag{14.16}$$

$$m\frac{\partial^2 y}{\partial t^2} + b_y \frac{\partial y}{\partial t} + k_y y = F_{elecy} + 2\Omega_z m \frac{\partial x}{\partial t} \tag{14.17}$$

The left-hand side of expressions 14.16 and 14.17 represent the dynamics of the two independent systems, whereas the right-hand side terms are external forces (F_{elecx}, F_{elecy}) such as electromechanical excitation or feedback, and the Coriolis forces that couple the two systems in the presence of the rotation rate.

In rotation-rate gyroscopes, one of the two systems is forced into resonance, conventionally by establishing self-oscillation with the help of electronic positive feedback. For instance, F_{elecx} in Eq. (14.16) can be utilized as a driving force to establish an alternating displacement x. In addition, the displacement amplitude and phase can be regulated to guarantee that the "drive-loop" oscillation is insensitive to any Coriolis excitation produced by Ω_z. If the oscillation is established at the resonance frequency ω_x, the force to displacement transfer function of the drive system can be expressed as:

$$\left|\frac{x(s)}{F_{elecx}(s)}\right| = \frac{Q_x}{m\omega_x^2} \quad \text{and} \quad \angle\frac{x(s)}{F_{elecx}(s)} = -90° \tag{14.18}$$

Once the drive displacement x – and thus the drive velocity $v_{drv} = \partial x/\partial t$ – has been established, the sense system given by Eq. (14.17), will respond to forces generated by the Coriolis effect. Figure 14.8 shows a simplified block-diagram representation of a vibratory gyroscope.

14.3.1.2 Mode-Split versus Mode-Matched Gyroscopes

The sensitivity of a gyroscope will be determined by the relative value of the resonance frequencies of the two coupled systems. When the frequencies of the drive (ω_x) and sense (ω_y) resonance modes are different from each other, the

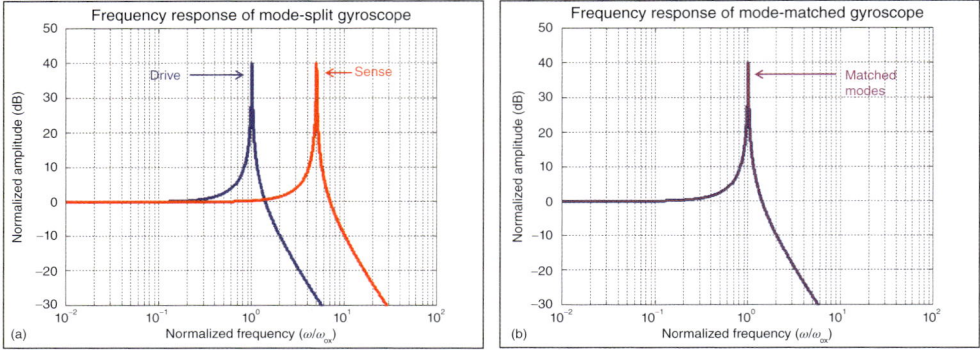

Figure 14.9 (a, b) Frequency response of a mode-split and a mode-matched gyroscope. In mode-split gyroscopes, the displacement generated by rotation-rate is given by the response of a quasi-static accelerometer. In mode-matched gyroscopes, this displacement is amplified by Q.

device is said to operate in a mode-split condition (Figure 14.9). In the particular case where $\omega_x \ll \omega_y$, the displacement response of the sense-axis system to a Coriolis force – given by Eq. (14.17) – will be that of a low-frequency quasi-static accelerometer. Assuming that there are no additional external forces affecting the sense mode ($F_{elecy} = 0$), the ratio of drive to sense displacement as a function of rotation-rate can be expressed as:

$$\left|\frac{y}{x}\right|_{split} = \frac{2\omega_x}{\omega_y^2}\Omega_Z \quad \text{and} \quad \angle\left(\frac{y}{x}\right)_{split} = -90° \tag{14.19}$$

On the other hand, devices in which the resonance frequencies of the drive and sense modes are equal ($\omega_x = \omega_y = \omega_0$) are known as mode-matched gyroscopes. These structures are advantageous given that the sense displacement y gets amplified by Q, allowing a maximum transfer of energy from the drive to the sense resonator. Thus for a mode-matched gyroscope with quality factor Q:

$$\left|\frac{y}{x}\right|_{mathced} = \frac{2Q}{\omega_0}\Omega_Z \quad \text{and} \quad \angle\left(\frac{y}{x}\right)_{mathced} = 0° \tag{14.20}$$

Most commercially available gyroscopes are low-frequency (tens of kilohertz) TFG-like devices that operate in a mode-split configuration [18, 19]. Moreover, structures that are theoretically designed to have resonance frequencies of equal values are exposed to imperfections during fabrication that cause mismatches in the parameters that define these two values ($k_x \neq k_y$). More symmetric devices, such as rings or other shell-like structures, can be used to distribute these process variations and reduce frequency splits [20]. Figure 14.10 shows a comparison between a TFG [21] and a shell-type device, the resonant star gyroscope (RSG) [22].

Unfortunately, even if mode-matching is achieved with these low-frequency flexural devices, the bandwidth of operation is significantly limited by the high Q required to resonate the structures (BW = ω_0/Q). Feedback electronics can be

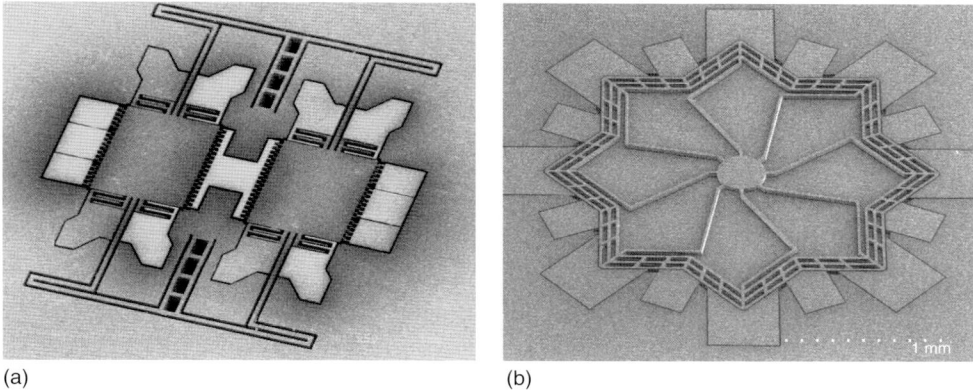

(a)   (b)

Figure 14.10   SEM view of low-frequency gyroscopes: (a) tuning-fork gyroscopes (TFG) and (b) resonating-star gyroscope (RSG).

used to further extend the band of operation, but this comes at the expense of higher complexity and higher system-level noise.

14.3.2
Bulk-Acoustic Wave (BAW) Gyroscopes

An alternative way of implementing gyroscopes is by the use of high-frequency, high-Q degenerate modes of a BAW resonator [23]. Unlike low-frequency flexural devices, high-frequency resonant structures are less prone to squeeze-film air damping, which relaxes the requirement of a high-vacuum environment to achieve larger Q values. Also, operating at higher frequencies (MHz range) allows for the design of gyroscopes with increased bandwidth even under mode-matched condition. Furthermore, the solid-state nature of BAW resonators gives them higher immunity to shock and random vibration, which is a key requirement for the implementation of high-performance inertial sensors.

The operation principles of rotation-rate BAW gyroscopes are similar to those described in Section 14.3.1. One of two degenerate modes of a disk is driven into resonance; in the presence of a Coriolis force, the second mode will be excited, generating a displacement proportional to the applied angular velocity [24]. A pair of modes are said to be degenerate when they have spatially orthogonal mode-shapes with equal resonance frequencies. Figure 14.11 shows as an example the first elliptical in-plane degenerate modes ($n = 2$) of a disk resonator. It can be clearly seen that the nodes (maximum points of displacement) of mode 1 are aligned with the antinodes (zero displacement points) of mode 2, and vice versa (i.e., they are orthogonal).

For the particular case of gyroscopes with cyclical symmetry, the dynamics of the system can be written in terms of the generalized radial displacements q_1 and q_2 as follows:

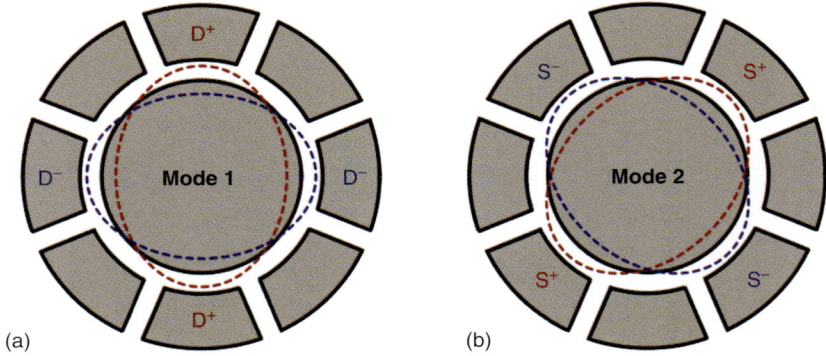

Figure 14.11 (a, b) First elliptical degenerate modes ($n = 2$) of a BAW resonator. Dotted lines represent the half-cycle deflection of the mode-shapes.

$$m\frac{\partial^2 q_1}{\partial t^2} + b_{11}\frac{\partial q_1}{\partial t} + k_{11}q_1 + k_{12}q_2 = F_1 - 2n\ A_g\Omega_Z m\frac{\partial q_2}{\partial t} \tag{14.21}$$

$$m\frac{\partial^2 q_2}{\partial t^2} + b_{22}\frac{\partial q_2}{\partial t} + k_{22}q_2 + k_{12}q_1 = F_2 + 2n\ A_g\Omega_Z m\frac{\partial q_1}{\partial t} \tag{14.22}$$

Two main differences are clear when comparing Eqs. (14.21) and (14.22) with the idealized equations given by Eqs. (14.16) and (14.17). First, the constants n and A_g now accompany the Coriolis force terms to account for the angular gain of the sensor. And second, an additional term, proportional to the constant k_{12}, was included to account for the intrinsic coupling between the two modes in the presence of imperfections. It is worth noting that mode coupling is also present in low-frequency gyros, but for convenience, its effects will be described in the context of BAW devices.

14.3.2.1 Angular Gain
In gyroscopes like the Foucault pendulum or the TFG, the direction of displacement of the resonance modes is typically well defined along perpendicular coordinates in the Cartesian plane (i.e., the x and y axes). On the other hand, the degenerate modes of ring or disk gyroscopes are orthogonal along the radial cylindrical coordinate r, but also have components of vibration on the circumferential or angular direction θ. In the presence of rotation – normal to the plane of vibration – these θ-oriented components do not contribute to the Coriolis coupling between the two modes because the cross product of their velocity and the rotation vector does not generate a force along the radial direction. The effective amount of coupling between two degenerate modes is known as the angular gain A_g, and is a function of the mode-shape order n. Hence – similarly to how Eq. (14.20) was derived – for mode-matched BAW disk gyroscopes, the ratio of the

drive and sense displacement as a function of rate can be expressed as:

$$\left| \frac{q_{2cor}}{q_1} \right|_{matched} = \frac{2nA_g Q}{\omega_0} \Omega_Z \quad \text{and} \quad \angle \left(\frac{q_{2cor}}{q_1} \right)_{matched} = 0° \qquad (14.23)$$

where q_{2cor} represents the sense-mode displacement generated only by the effect of the Coriolis force, ω_0 is the resonance frequency of the matched modes, and Q the sense-mode quality factor. The value of A_g can be directly calculated from the normal mode equations of the structure [25], and it is approximately 0.4 for the first elliptical mode ($n = 2$) and 0.2 for the second elliptical mode ($n = 3$).

14.3.2.2 Zero-Rate Output

In ideal vibratory gyroscopes, the drive and sense modes of vibration are always orthogonal to each other and perfectly aligned to the electromechanical transducer. For example, in a capacitive BAW gyroscope, the antinodes of the drive mode displacement will be in line with the electrodes utilized to excite the mode into resonance (D^+ and D^- in Figure 14.11). Similarly, the antinodes of the sense mode will be aligned with the readout electrodes (S^+ and S^-). This implies that in the absence of rotation rate, the displacement of the sense mode, and hence the signal coming out of its electrodes, will always be zero.

In reality, material and lithography imperfections encountered during fabrication, can cause both small frequency splits between the two modes ($k_{11} \neq k_{22}$), as well as a limited amount of misalignment between the resonant structure and its fixed electrodes. These nonidealities can produce an undesired excitation of the sense mode that will show up as a signal at the output even when no rotation-rate is applied; this effect is commonly known as zero-rate output (ZRO). Depending on the design and the fabrication process tolerances, this signal can be orders of magnitude larger than the Coriolis response, therefore cancelation techniques are necessary to suppress it. It is worth mentioning that ZRO is not particular to BAW gyroscopes and has been intensively studied in low-frequency flexural structures [26].

Direct Excitation of Sense-Mode In the presence of misalignment between the antinodes of the resonance mode-shapes and the fixed capacitive electrodes, the sense mode will experience a direct excitation from the drive electrodes (D^+ and/or D^-) that will generate an undesired sense displacement q_{2FT}. For instance, if the modes are misaligned by an angle θ_0 with respect to the electrode frame, both forces F_1 and F_2 from Eqs. (14.21) and (14.22) will have components of the drive electrostatic force f_d generated by the D^+ and/or D^- transducers:

$$F_1 = f_d \cos(\theta_0) \quad \text{and} \quad F_2 = f_d \sin(\theta_0) \qquad (14.24)$$

Thus, in the absence of rate, q_{2FT} can be represented as fraction of the drive displacement q_1:

$$\frac{q_2 FT}{q_1} = \kappa \qquad (14.25)$$

where value κ will be proportional to $\tan(\theta_0)$ and dependent on the ratio of damping and stiffness parameters of the two systems [27]. In the case of a mode-matched gyroscope, f_d will affect both modes proportionally, causing no phase difference between the drive displacement q_1 and the ZRO signal q_{2FT}. Hence, when comparing Eq. (14.25) with Eq. (14.23), it can be inferred that in a mode-matched device, q_{2FT} is in-phase with the Coriolis response q_{2cor}. In other words, this signal acts as feed-through (FT), giving the impression that a constant input rotation-rate is being applied (i.e., offset). In the case of mode-split gyroscopes, q_{2FT} is in fact 90° out-of-phase – or in quadrature – with q_1, explaining why sometimes ZRO is simply referred to as "quadrature." Ways to cancel or correct for these errors both in the sensor or electronically are described below.

Mode-to-Mode Coupling A second source of ZRO can emerge even when the resonator modes are perfectly aligned to their surrounding electrodes. In the presence of geometry or material imperfections, resonance mode-pairs can become non-degenerate. This not only results in a frequency split $\Delta\omega = |\omega_1 - \omega_2|$ between the modes due to differences in the direct stiffness terms ($k_{11} \neq k_{22}$), but also causes direct coupling from one to the other ($k_{12} \neq 0$). Therefore, if the drive-mode is perfectly aligned to the excitation source, but the modes are not decoupled, the residual internal stress generated by the drive displacement q_1 will stimulate the sense mode generating a ZRO signal q_{2Q}. Since the magnitude and phase of q_{2Q} depend on the same imperfections that generate frequency splits, their values can be expressed as a function of $\Delta\omega$ [28]. Figure 14.12 shows a simplified schematic representation of how the coupling between the two modes can be expressed.

When the frequency split is zero (no imperfections), the modes do not interact with each other, thus q_{2Q} will also be zero. As $\Delta\omega$ increases, q_{2Q} will raise at a rate proportional to the quality factors of the modes. The magnitude of q_{2Q} will drop when the drive and sense frequencies start getting far apart from their 3 dB bandwidth. Thus, in the case of mode-split gyroscopes ($\omega_1 \ll \omega_2$), these effects are negligible, thus q_{2FT} will be the main source of ZRO. The phase relation between

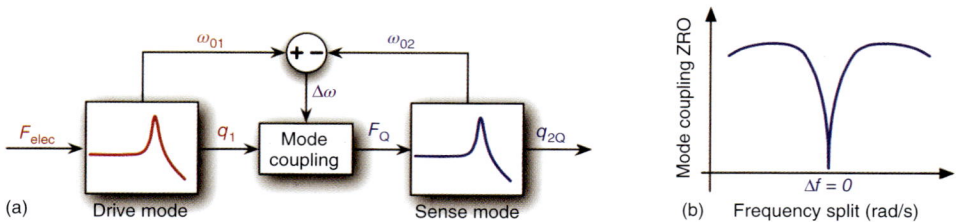

(a) Drive mode   Sense mode   (b) Frequency split (rad/s)

Figure 14.12 (a, b) Simplified representation of mode-to-mode coupling. For a frequency split of 0, the modes are degenerate so there will be no interaction between them. For finite values of Δf, the modes are no longer linearly independent causing the transfer of energy from one to the other.

q_{2Q} and q_1 can also be considered to be a function of the frequency split, approaching $-90°$ as $\Delta\omega$ gets closer to zero. Therefore, in mode-matched gyroscopes with small split values, q_{2Q} will show up in quadrature with respect to the Coriolis displacement. In literature, frequency split and mode coupling are often treated.

14.3.2.3 ZRO Cancelation

In mode-matched gyroscopes, ZRO due to misalignment and frequency split can show up as both in-phase and quadrature errors with respect to the Coriolis signal. Cancelation of these non-idealities can be performed via back-end nulling in the electronics, through electromechanical compensation in the sensor, or by a combination of both.

Since in vibratory gyroscopes the rotation-rate is modulated at the resonance frequency of the drive mode ω_1 (Figure 14.8), amplitude demodulation can be utilized to extract the rate information. If the drive signal is used as the reference local oscillator in an I/Q-demodulator, the ZRO components that are in quadrature with the Coriolis signal (i.e., q_{2Q}) will be rejected:

$$v_{\text{out}} = A_{\text{s}}((|q_{2\text{cor}}| + |q_{2\text{FT}}|)\sin(\omega_1 t) + q_{2Q}\ \cos(\omega_1 t)) \cdot A_{\text{d}}|q_1|\sin(\omega_1 t) \quad (14.26)$$

$$v_{out} \approx \frac{1}{2}\ A_{\text{s}}A_{\text{d}} \cdot (|q_{2\text{cor}}| + |q_{2\text{FT}}|) \cdot |q_1| \quad (14.27)$$

A_{d} and A_{s} are the displacement-to-voltage gain values for the drive and sense channel respectively, and v_{out} is the band-pass filtered output voltage amplitude of the demodulator. On the other hand, the in-phase ZRO signal $q_{2\text{FT}}$ can be rejected by the use of a feed-through cancelation circuit, where a scaled and inverted version of q_1 is directly added to the sense signal in order to null $q_{2\text{FT}}$.

To improve the signal-to-noise ratio of a gyroscope system, high gain amplifiers are usually required at the front-end of the interface circuit. If the amplitude of the ZRO signal is much larger than the detectable range of the Coriolis signal $q_{2\text{cor}}$, the circuitry will saturate, preventing the detection of any change in q_2 due to an applied rotation-rate. Thus, in order to reduce or completely suppress ZRO, an imperfect gyroscope must be brought into a mode-matched, mode-aligned condition.

14.3.2.4 Electromechanical Transduction in Gyroscopes

Large electromechanical transduction is essential to facilitate the establishment of the drive loop oscillation and to enhance the overall sensitivity of the gyroscope. Similar to MEMS accelerometers, piezoelectric, piezoresistive, and capacitive transducers have been utilized for the implementation of rotation-rate sensors. In the case of high-frequency BAW gyroscopes, electrostatic excitation and sensing by the use of capacitive nano-gaps has shown to provide high coupling coefficients [29]. As discussed in Section 14.2.3.1, the electrostatic force for a parallel plate capacitor is approximately given by Eq. (14.11). This same expression can be utilized for BAW resonators as long as the displacement at resonance – given by Eq. (14.18) – is much smaller than the initial gap size g_{o}.

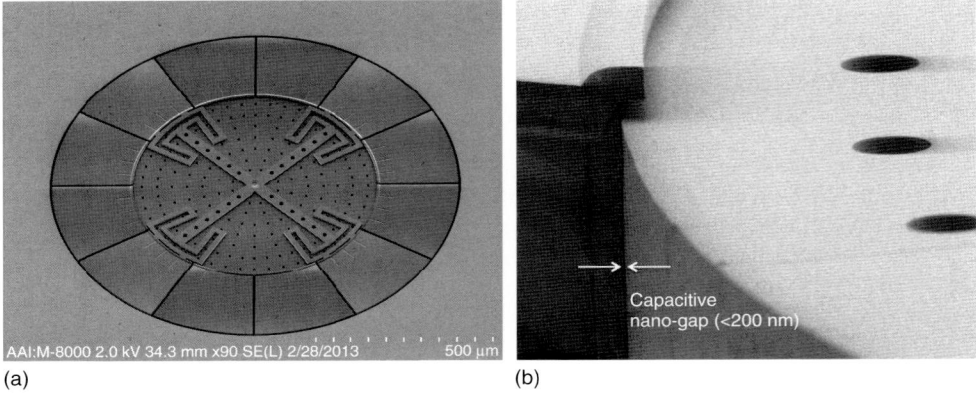

(a)   (b)

Figure 14.13   (a, b) SEM view of BAW disk gyroscope and close-up view of nano-gaps for reduced motional impedance and high rotation-rate sensitivity.

If the disk has a polarization voltage $V_{DC} = V_p$, and an AC drive voltage $v_{ac} = v_d$ is applied to one of the D^+ electrodes (Figure 14.11), a current i_d will be generated on each of the D^- electrodes. Considering only the time-varying AC terms in Eq. (14.11), the magnitude of i_d will be given by:

$$|i_d| = V_p \frac{\partial C_d}{\partial t} = V_p \frac{\partial C_d}{\partial q_1} \frac{\partial q_1}{\partial t} \approx \frac{Q}{m\omega_0} \left(\frac{\varepsilon A V_p}{g_0^2} \right)^2 |v_d| \tag{14.28}$$

where C_d is the capacitance of the drive electrode with initial gap g_0. The ratio of v_d to i_d is known as the motional resistance of the resonator, and corresponds to the amount of loss the drive electronics will need to compensate for in order to bring the drive-loop into oscillation.

Similarly, the Coriolis displacement q_{2cor} will cause a change in the sense capacitance C_s, generating a current i_s, proportional to Ω_z, on each of the sense electrodes S^+ and S^-:

$$|i_s| = V_p \frac{\partial C_s}{\partial t} = V_p \frac{\partial C_s}{\partial q_2} \frac{\partial q_2}{\partial t} = 2nA_g Q |q_1| \frac{\varepsilon A V_p}{g_0^2} \Omega_Z \tag{14.29}$$

Figure 14.13 shows an SEM view of a single-crystal silicon (SCS) BAW disk gyroscope, and a close-up of the capacitive nano-gaps (<200 nm) implement with the HARPSS process [8]. The polysilicon trace on top of resonator is implemented to provide electrical connection from the electrodes to the center of the disk.

14.3.2.5   Electrostatic Mode Matching and Mode Alignment

A frequency split $\Delta\omega$ can be represented as a mismatch in the stiffness of the resonance modes of interest ($k_{11} \neq k_{22}$). Electrostatic spring softening can be utilized to reduce the effective stiffness value of a particular mode, bringing the part into a mode-matched condition. For instance, if the frequency of the drive-mode ω_1 is larger than the sense-mode frequency ω_2, a voltage difference $V_{DC} = V_p - V_T$ can be applied between the disk and one of the drive electrodes (D^+ or D^-) to

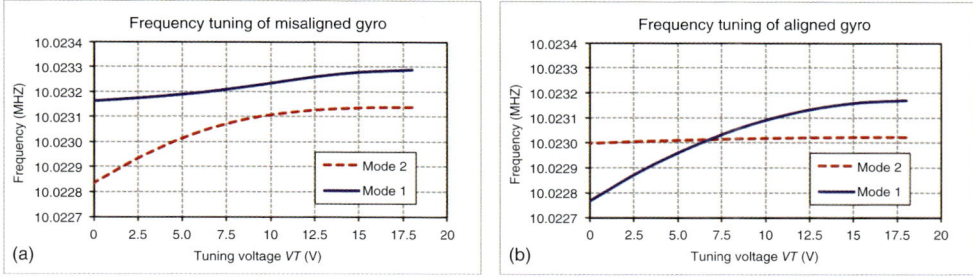

Figure 14.14 Frequency tuning for BAW gyroscopes before (a) and after (b) mode alignment was performed. For a misaligned part, both modes are affected by the drive tuning potential V_T, thus mode-matching cannot be achieved. Once aligned, only the drive-mode frequency is tuned by V_T and $\Delta\omega = 0$ is attained.

tune its value by controlling V_T. From Eq. (14.13) – and assuming $q_1 \ll g_0$ – the drive-mode mechanical resonance frequency ω_{01} can be adjusted to a value of:

$$\omega_1 = \omega_{01}\left(1 - \frac{\varepsilon A}{2k g_0^3}\left(V_P - V_T\right)^2\right) \tag{14.30}$$

Similarly, the modes of a disk gyroscope can be realigned with respect to the drive and sense electrodes, through the use of electrostatic spring softening. By introducing alignment electrodes in-between the nodal and *anti*-nodal locations of the modes, the effective stiffness along a particular direction can be increased, forcing the mode-shapes to rotate. Figure 14.14 shows the tuning response of an $n = 3$ BAW gyroscope before and after mode-alignment. When the part is misaligned, both modes get affected by the drive tuning potential V_T. After aligning the part, only the drive mode frequency is tuned allowing to mode-match the resonator.

After frequency-matching and mode-alignment are achieved, the gyroscope is perfectly balanced; thus, the ZRO will be completely canceled and the Coriolis coupling between the two modes will be maximized.

14.3.3
Mechanical Noise in Mode-Matched Gyroscopes

Similar to accelerometers, the mechanical noise in a mode-matched gyroscope can be modeled as a thermal noise force generator, given by Eq. (14.8), acting on the sense mode of the device. The input referred mechanical noise equivalent rotation-rate ($MNE\Omega$) can be calculated by taking the generated noise displacement q_{2n} and referring it to the input. This can be done by replacing q_{2n} in Eq. (14.23) and finding the corresponding Ω:

$$\text{MNE}\Omega = \frac{1}{2n A_g q_1}\sqrt{\frac{4k_B T}{m\omega_0 Q}} \tag{14.31}$$

From Eq. (14.31) it is clearly seen that by going with higher values of frequency ω_0 and Q, the spot noise of a gyroscope can be scaled down. In the case of mode-matched BAW gyroscopes – which operate in the megahertz range and yield Q values in the tens to hundreds of thousands at low-vacuum levels (1 – 10 Tor) – the noise can be considerably lower as compared to mode-split flexural structures that operate in the tens of kilohertz, and have extremely low effective values of Q due to the large frequency mismatch.

14.4
Multi-degree-of-Freedom Inertial Measurement Units

An IMU is a system composed of multi-DOF accelerometers and gyroscopes that, with the help of other sensors (e.g., magnetometers and barometers) and predictive algorithms, can be utilized to accurately determine the location and heading information of a moving object. In IMUs, tri-axial accelerometers are used to calculate displacement by integrating acceleration signals measured along the x-, y-, and z-axis. On the other hand, tri-axial gyroscopes provide the pitch, roll, and yaw direction information of the object. Lastly, tri-axial magnetic sensors can be added to measure orientation with respect to the earth magnetic poles, increasing the accuracy, and robustness of the dead-reckoning process. Therefore, unlike GPS receivers, IMUs can track an object without the need of communicating with an external source, such as a satellite or an antenna. This is indispensable for the implementation of navigation systems that need to be operated in enclosed spaces or remote areas.

14.4.1
System-in-Package IMUs

In recent years, the development and commercialization of MEMS-based IMUs has experienced an accelerated growth thanks to advancements in microfabrication, packaging, and system integration. Given that the progress of MEMS accelerometers and gyroscopes has happened almost independently, these first-generation off-the-shelf IMUs are being implemented by assembling individual sensors in a system-in-package (SiP) configuration [30, 31]. Tri-axial accelerometers, gyroscopes, and magnetometers – with their corresponding interface application-specific integrated circuits (ASICs) – are placed side-by-side or stacked inside a package to configure a 9-DOF inertial sensor [32].

The SiP approach can also be utilized to implement homogeneous IMUs, where identical individual single-axis components are placed in a package with their axis of detection oriented along each of the spatial coordinates. This enables a high-level of uniformity in terms of noise, sensitivity and dynamic range among all three axis of detection. Furthermore, by having discrete components, signal coupling, and cross-axis sensitivities are much reduced as compared to single-proof

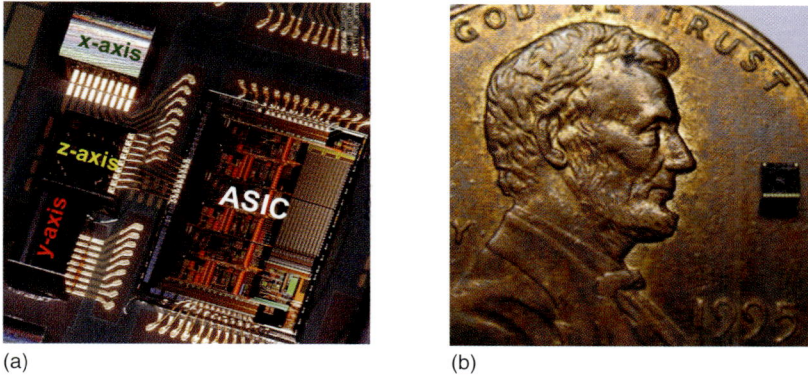

(a)   (b)

Figure 14.15 (a) Image of Qualtré's homogeneous tri-axial gyroscope composed of three single-axis devices (*x*/*y*-axis mounted on the side) and (b) view of individual wafer-level packaged BAW disk gyroscope.

mass or mechanically coupled designs. Figure 14.15 shows a homogeneous SiP tri-axial gyroscope (a) developed by Qualtré, Inc. [33]. Each individual axis utilizes a wafer-level packaged BAW disk gyroscope (b); the *x*- and *y*-axis components are mounted on the side to align the axis of detection with the coordinates of interest. The remarkably small size of BAW gyros facilitates the integration of the sensor into commercially competitive packages.

14.4.2
Single-Die IMUs

The demand for smaller and cheaper, high-performance MEMS IMUs is contin-uously on the rise. The need for ultra-thin, area-efficient sensors is becoming an imminent necessity as these devices are being incorporated into smaller portable electronics, such as watches and body-motion and health-monitoring gadgets. This trend incentivizes the development of single-die IMUs in which all tri-axial accelerometers, gyroscopes – and possibly magnetometers – are batch fabricated and wafer-level packaged on the same substrate. Likewise, single-die integration reduces the overall cost of the final product given that it minimizes assembly and packaging complexity, and potentially increases the number of sensors per unit area.

In order to detect rotation and acceleration along all three axes, multi-DOF single-die inertial sensors must incorporate structures capable of detection motion in the out-of-plane direction with respect to the substrate. In the particu-lar case of capacitive devices, top and/or bottom electrodes must be incorporated to maximize electromechanical coupling. A high-frequency annular gyroscope for pitch and roll detection is shown in Figure 14.16 [34]. The structure is process compatible with BAW disk gyroscopes for yaw detection, allowing the implementation of tri-axial rotation-rate sensors on a single-die.

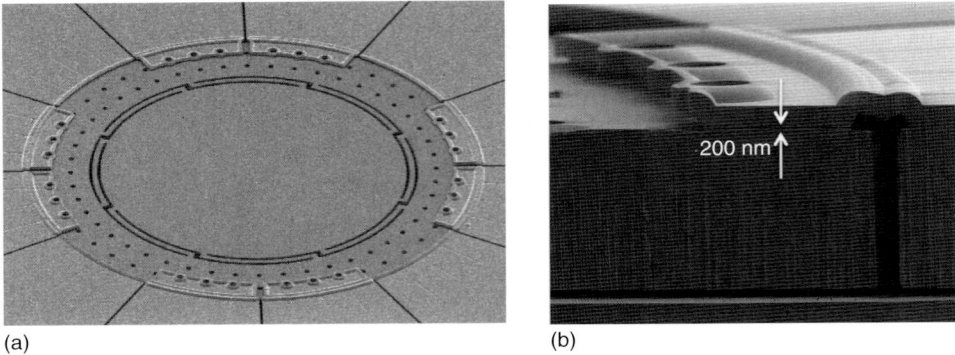

(a)   (b)

Figure 14.16 (a, b) Annulus high-frequency gyroscope for pitch and roll detection. Out-of-plane nano-gaps provide high electromechanical coupling for improved sensitivity.

Similarly, vertical acceleration detection can be achieved through the use of top capacitive electrodes, which in combination with in-plane devices, can be used to measure acceleration along all three axes [9]. Out-of-plane capacitive detection also facilitates the implementation of single proof-mass tri-axial accelerometers that have a much reduced die area as compared to utilizing separate masses for each axis. For instance, a pendulum-like structure, such as the one shown in Figure 14.17, can be used for this purpose. The device consists of proof-mass anchored to the substrate utilizing a cross-shaped spring [35]. Four pick-off electrodes placed on top of the moving structure are multiplexed to read out changes in capacitance generated by the x-, y-, and z-axis acceleration components. In the presence of acceleration along the x-axis, the tethers act as torsional springs, allowing the mass to tilt. This causes a differential change in capacitance $\Delta C_x = (C_1 + C_2) - (C_3 + C_4)$, where C_1 through C_4 correspond to the individual capacitances between the proof-mass and fixed electrodes. Similarly, acceleration along the y-axis causes a differential change of $\Delta C_y = (C_1 + C_4) - (C_2 + C_3)$. Lastly, z-axis acceleration produces out-of-plane translation of the proof-mass, causing an effective capacitance variation $\Delta C_z = (C_1 + C_2 + C_3 + C_4) - 4C_0$, where C_0 is the zero-input acceleration rest capacitance.

One of the main challenges for the implementation of process-compatible accelerometers and gyroscopes is that traditionally, the pressure levels required for each type of sensor are significantly different. In conventional quasi-static accelerometers, high-pressure levels (equal or close to atmospheric pressure: 760 Torr) are required to over-damp the system in order to avoid Q-peaking [9]. On the other hand, traditional low-frequency flexural-mode gyroscopes must operate at high vacuum (mTorr range) in order to attain high Q levels for enhanced sensitivity and noise. This suggests that accelerometers and gyroscopes must be packaged in different environments, complicating their integration into a single-die.

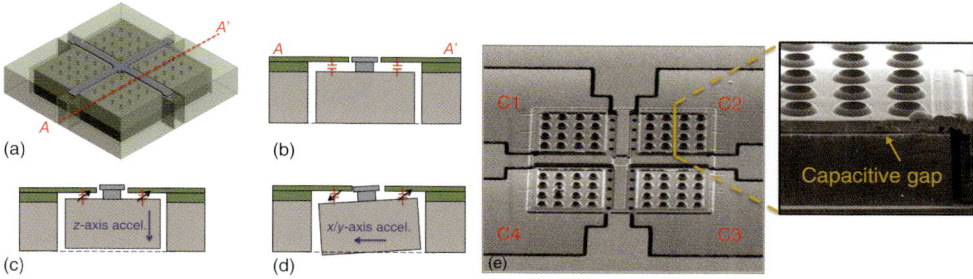

Figure 14.17 (a) Single-proof-mass tri-axial accelerometer composed of SCS mass supported by cross-shaped polysilicon tethers. (b) Fixed top electrodes form capacitive gaps with mass. (c) Mass translates under z-axis acceleration and (d) tilts with *x*/*y*-axis acceleration (e) SEM top view and cross-section of single-proof-mass tri-axial accelerometer.

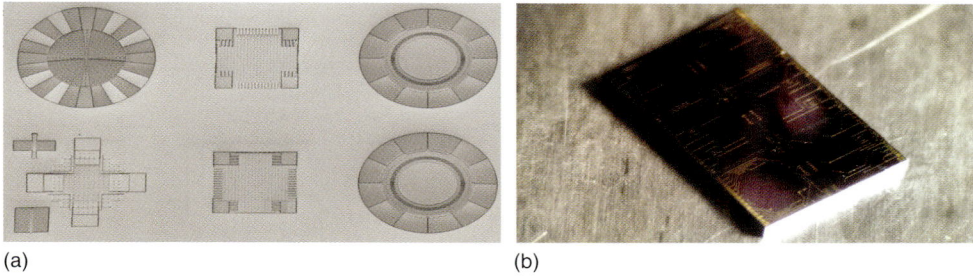

Figure 14.18 (a, b) SEM view and optical image of 6-DOF IMU ($4 \times 6 \times 0.4\,\text{mm}^3$). Single-axis accelerometers and gyroscopes were used to reduce cross-axis sensitivity and increase performance. The structures were co-fabricated and wafer-level packaged using the HARPSS™ process.

In the case of high-frequency gyroscopes, large Q values (tens to hundreds of thousands) can be achieved at moderate vacuum levels ($1-10\,\text{Torr}$). In combination with capacitive air nano-gaps that provide increased squeeze-film air damping at lower frequencies (see Section 14.2.2.1), these devices can be easily co-integrated with quasi-static accelerometers to implement single-die multi-DOF IMUs. Figure 14.18 shows an optical picture of a wafer-level packaged, 6-DOF IMU implemented using the HARPSS process. Single-axis components were utilized to reduce cross-axis sensitivity and improve the noise and scale-factor performance.

14.4.3
Future Trends in Sensor Integration

Even though miniaturization is a key parameter for integration, functionality, and performance will still be the main drivers in high-end applications such as inertial navigation. Unfortunately, the current specifications of off-the-shelf 6-DOF and 9-DOF IMUs do not meet the accuracy requirements for in-door pedestrian guidance, resulting in errors in the order of a few feet only after a few seconds. Therefore, further improvements in sensor design and multi-sensor fusion are necessary in order to demonstrate that MEMS technology can potentially be an enabler in this application space.

References

1. Yazdi, N., Ayazi, F., and Najafi, K. (1998) Micromachined inertial sensors. *Proc. IEEE*, **86**, 1640–1659.
2. Ayazi, F. (2011) Multi-DOF inertial MEMS: from gaming to dead-reckoning. Technical Digest of the 16th International Conference on Solid-State Sensors, Actuators and Microsystems (Transducers'11), Beijing, China, 2011, pp. 2805–2808.
3. Griffin, W., Richardson, H.H., and Yamanami, S. (1966) A study of fluid squeeze-film damping. *J. Basic Eng.*, **88**, 451–456.
4. Bao, M. and Yang, H. (2007) Squeeze film air damping in MEMS. *Sens. Actuators, A: Phys.*, **136**, 3–27.
5. Veijola, T., Pursula, A., and Råback, P. (2005) Extending the validity of squeezed-film damper models with elongations of surface dimensions. *J. Micromech. Microeng.*, **15**, 1624–1636.
6. Bao, M., Yang, H., Yin, H., and Sun, Y. (2002) Energy transfer model for squeeze-film air damping in low vacuum. *J. Micromech. Microeng.*, **12**, 341–346.
7. Sumali, H. (2007) Squeeze-film damping in the free molecular regime: model validation and measurement on a MEMS. *J. Micromech. Microeng.*, **17**, 2231–2240.
8. Ayazi, F. and Najafi, K. (1999) High aspect-ratio polysilicon micromachining technology. Proceedings of the 10th International Conference on Solid-State Sensors and Actuators (Transducers '99), 1999, pp. 320–323.
9. Jeong, Y., Serrano, D.E., Keesara, V., Sung, W.K., and Ayazi, F. (2013) Wafer-level vacuum-packaged triaxial accelerometer with nano airgaps. Proceedings of the 26th IEEE International Conference on Micro Electro Mechanical Systems (MEMS '13), 2013, pp. 33–36.
10. DeVoe, D.L. and Pisano, A.P. (2001) Surface micromachined piezoelectric accelerometers (PiXLs). *J. Microelectromech. Syst.*, **10**, 180–186.
11. Kwon, K. and Park, S. (1998) A bulk-micromachined three-axis accelerometer using silicon direct bonding technology and polysilicon layer. *Sens. Actuators, A: Phys.*, **66**, 250–255.
12. Chae, J., Kulah, H., and Najafi, K. (2005) A monolithic three-axis micro-g micromachined silicon capacitive accelerometer. *J. Microelectromech. Syst.*, **14**, 235–242.
13. Amini, B.V., Abdolvand, R., and Ayazi, F. (2006) A 4.5-mW closed-loop micro-gravity CMOS SOI accelerometer. *IEEE J. Solid-State Circuits*, **41**, 2983–2991.
14. Gabrielson, T.B. (1993) Mechanical-thermal noise in micromachined acoustic and vibration sensors. *IEEE Trans. Electron Devices*, **40**, 903–909.
15. Roessig, T.A., Howe, R.T., Pisano, A.P., and Smith, J.H. (1997) Surface-micromachined resonant accelerometer. International Conference on Solid State Sensors and Actuators, 1997. TRANSDUCERS'97, Chicago, IL, 1997, pp. 859–862.
16. Hopcroft, M.A. (2007) Silicon micromechanical resonators for frequency references. PhD dissertation. Stanford University.
17. Bernstein, J., Cho, S., King, A.T., Kourepenis, A., Maciel, P., and Weinberg, M. (1993) A micromachined comb-drive tuning fork rate gyroscope. Proceedings of the IEEE Micro Electro Mechanical Systems, 1993. (MEMS '93), pp. 143–148.
18. Seeger, J., Lim, M., and Nasiri, S. (2010) Development of high performance high volume consumer mems gyroscopes. Presented at the Technical Digest Solid-State Sensors, Actuators, and Microsystems Workshop, Hilton Head, SC, 2010.

19. Prandi, L., Caminada, C., Coronato, L., Cazzaniga, G., Biganzoli, F., Antonello, R. et al. (2011) A low-power 3-axis digital-output MEMS gyroscope with single drive and multiplexed angular rate readout. Solid-State Circuits Conference Digest of Technical Papers (ISSCC), 2011 IEEE International, pp. 104–106.

20. Silicon Sensing (2014) Single Axis Gyros, *http://www.siliconsensing.com/gyros* (accessed 2 September 2014).

21. Zaman, M.F., Sharma, A., and Ayazi, F. (2006) High performance matched-mode tuning fork gyroscope. Proceedings of the 19th IEEE International Conference on Micro Electro Mechanical Systems (MEMS '06), 2006, pp. 66–69.

22. Zaman, M.F., Sharma, A., and Ayazi, F. (2009) The resonating star gyroscope: a novel multiple-shell silicon gyroscope with sub-5 deg/hr Allan deviation bias instability. *IEEE Sens. J.*, **9**, 616–624.

23. Hao, Z., Pourkamali, S., and Ayazi, F. (2004) VHF single-crystal silicon elliptic bulk-mode capacitive disk resonators-part I: design and modeling. *J. Microelectromech. Syst.*, **13**, 1043–1053.

24. Johari, H. and Ayazi, F. (2007) High-frequency capacitive disk gyroscopes in (100) and (111) silicon. Proceedings of the 20th IEEE International Conference on Micro Electro Mechanical Systems (MEMS '07), 2007, pp. 47–50.

25. Gallacher, B.J. (2002) The design, fabrication and testing of a multi-axis vibrating ring gyroscope. PhD Thesis. University of Newcastle upon Tyne.

26. Weinberg, M.S. and Kourepenis, A. (2006) Error sources in in-plane silicon tuning-fork MEMS gyroscopes. *J. Microelectromech. Syst.*, **15**, 479–491.

27. Putty, M.W. (1995) Micromachined vibrating ring gyroscope. PhD dissertation. University of Michigan.

28. Lynch, D. (1995) Vibratory gyro analysis by the method of averaging. Proceedings of the 2nd St. Petersburg Conference on Gyroscopic Technology and Navigation, St. Petersburg, 1995, pp. 26–34.

29. Johari, H. and Ayazi, F. (2006) Capacitive bulk acoustic wave silicon disk gyroscopes. Technical Digest IEEE International Electron Devices Meeting (IEDM '06), 2006, pp. 1–4.

30. STMicroelectronics (2012) LSM333D – iNEMO Inertial Module: 9 Degrees of Freedom Sensing Solution, *http://www.st.com/web/en/catalog/sense_power/FM89/SC1448/PF253022* (accessed 12 September 2014).

31. Invensense (2014) MPU-9250 – Nine-Axis (Gyro + Accelerometer + Compass) MEMS MotionTracking™ Devices, *http://www.invensense.com/mems/gyro/documents/PS-MPU-9250A-01.pdf* (accessed 2 September 2014).

32. Vigna, B. (2012) It makes sense: how extreme analog and sensing will change the world. Presented at the Technical Digest Solid-State Sensors, Actuators, and Microsystems Workshop, Hilton Head, SC, 2012.

33. Qualtré, Inc. *http://www.qualtre.com* (accessed 2 September 2014).

34. Sung, W.K., Dalal, M., and Ayazi, F. (2011) A mode-matched 0.9 MHz single proof-mass dual-axis gyroscope. Technical Digest of the 16th International Conference on Solid-State Sensors, Actuators and Microsystems (Transducers '11), 2011.

35. Serrano, D.E., Jeong, Y., Keesara, V., Sung, W.K., and Ayazi, F. (2014) Single proof-mass tri-axial pendulum accelerometers operating in vacuum. Technical Digest of the IEEE International Micro Electro Mechanical Systems Conference (MEMS '14), San Francisco, CA, 2014, pp. 28–31.

15
Resonant MEMS Chemical Sensors

Luke A. Beardslee, Oliver Brand, and Fabien Josse

15.1
Introduction

Coated microcantilevers are a very promising sensor platform for the detection of trace amounts of bio-chemical analytes [1–3]. Applications of particular interest include environmental monitoring, biomedical applications, automotive sensors, national security, and many others. One specific example would be the detection and quantification of a potentially dangerous gas such as methane or carbon monoxide in a given environment. A second example would be the rapid detection of the level of a specific protein in a blood sample for diagnosis. What all of these potential applications have in common is that specific bio-chemical compounds must be identified, and in most cases, a concentration of that compound needs to be quantified (a simple positive/negative result is not sufficient).

For the applications listed above, extensive efforts have been focused on developing microsensor platforms that can be field-deployed and that can perform real-time measurements. This technology is not widely used currently, but with continued development could greatly aid a variety of applications. For example, current methods for monitoring chemical contaminants include gas/liquid chromatography and mass spectroscopy. Although these methods are very accurate, they often require samples to be taken to laboratories for analysis, are typically time-consuming, and are relatively expensive. Moreover, analytes can potentially degrade from the time a sample is collected until the time it is analyzed. Therefore, there is a need to develop *in situ* monitoring systems for rapid bio-chemical analysis and characterization of samples.

The advantages of microsensors include small size, low per-unit cost, low power consumption, and easy integration with wireless communication circuitry. Integration with wireless communications creates the possibility of sensor networks that could be monitored in real time by a central operator. Moreover, the small size of microsensors allows for many sensors, each sensitive to a specific analyte, to be placed in a single package. This opens the possibility for discrimination of the various components of the sample under study without the need for complex bench-top equipment.

Resonant MEMS – Fundamentals, Implementation and Application, First Edition.
Edited by Oliver Brand, Isabelle Dufour, Stephen M. Heinrich and Fabien Josse.
© 2015 Wiley-VCH Verlag GmbH & Co. KGaA. Published 2015 by Wiley-VCH Verlag GmbH & Co. KGaA.

Examining Figure 15.1, there are several key components to any bio-chemical microsensor system. The first element is the sensitive layer/film. This layer, for example, a polymer in the case of sensing chemicals, or an antibody (for example) in the case of sensing biomolecules, is used to sorb the analyte of interest from the surrounding medium. (Depending on the sensing film, analytes can be absorbed or adsorbed.) Ideally, this layer would have a high affinity for the analyte so that it will preferentially sorb the analyte and concentrate it on the surface of the transducer. In addition to a high affinity for the analyte, it is important that the sensitive layer shows a high selectivity for the chemical being measured. This means that the sensitive layer should have as high an affinity as possible for the analyte of interest and as low an affinity as possible for other substances. The selectivity of the film for the analyte becomes particularly important when trying to discriminate between analytes with similar physical or chemical properties. While the sensing film captures the analyte of interest, it also converts the information from the chemical domain into the physical domain, that is, the sorbed analyte changes a physical property (e.g., mass, refractive index, or dielectric constant) of the sensing film, which is then sensed by the physical transducer.

By definition, a transducer converts a signal from one energy domain into a different energy domain using one or several transducer effects. Of particular interest in this chapter are transducers that result in an electrical output signal, such as a change in resistance, capacitance, or voltage/current. In general, a wide variety of transducer effects and a wide variety of fabrication technologies can be explored for chemical sensors; depending on the sensing method, bio-chemical sensors are often classified as either optical, thermal, mass-sensitive, or electrochemical sensors [4]. In the case of resonant bio-chemical sensors, one is particularly interested in sensors that measure the mass uptake in the sensing film via a change in the resonance frequency of a particular microstructure. This resonant microstructure is often a microelectromechanical system (MEMS), fabricated using technologies adapted from the integrated circuit industry, thus resonant MEMS offer the opportunity of developing low-cost, portable sensor systems for the rapid detection and quantification of bio-chemical analytes. Moreover, resonant MEMS offer a stable sensor signal, that is, a resonant frequency that can be readily detected by a variety of circuit topologies. Finally, resonant MEMS are readily fabricated

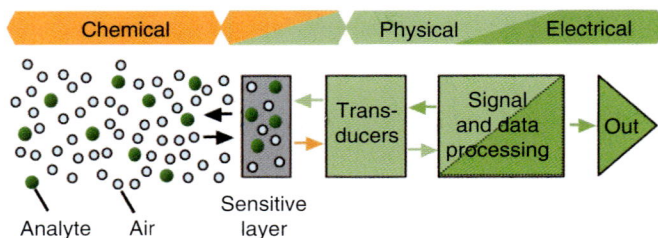

Figure 15.1 Diagram of the key components of an integrated bio-chemical sensor. The surrounding medium shown here is air, but could also be another gas or a liquid.

using silicon and other materials commonly used in the integrated circuit industry, which facilitates their integration with interface circuitry.

The output signal of the transducer can be further processed using analog and digital signal processing in order to generate the desired output signal. Signal processing can be as simple as an amplification stage, but can also include substantial digital signal processing to improve the selectivity of a chemical sensor system using feature extraction algorithms with signals from an array of sensors as the system input.

While there exist various types of resonant MEMS devices that could be used as sensors platforms, the focus of this chapter is the use of microcantilevers as resonant devices for the development of highly sensitive chemical sensing platforms. Microcantilevers can operate in either the dynamic mode or the static mode. When the device is operating in the dynamic mode, sorption of analyte changes the mass and possibly the material properties of the coating, which, in turn, causes the resonance frequency of the microcantilever to change. In the static mode, sorption of analyte leads to a change of surface stress, which yields a quasi-static bending of the cantilever. In this chapter, the characteristics of resonant MEMS devices, specifically microcantilevers, as bio-chemical sensors are analyzed.

15.2
Modeling of Resonant Microcantilever Chemical Sensors

Microcantilever-based chemical sensors can be considered to be made of two layers: a base layer of thickness h_1 and a chemically sensitive coating layer of thickness h_2 (Figure 15.2). The coating layer's properties (both mechanical

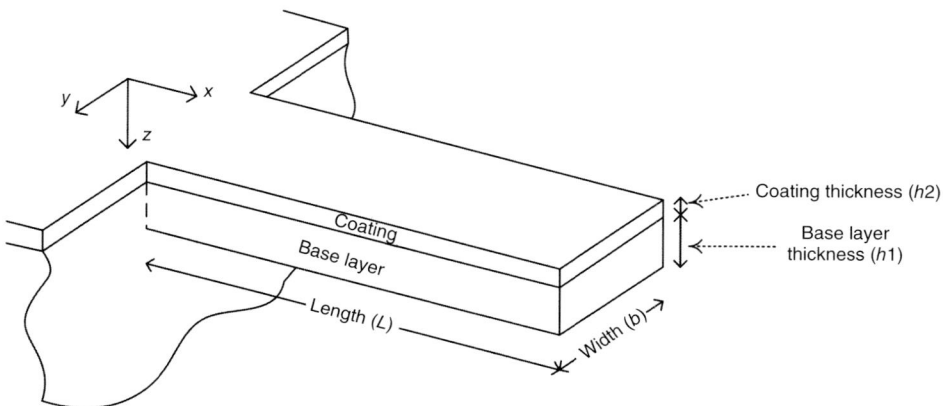

Figure 15.2 Schematic of a coated microcantilever chemical sensor platform. The clamped end of the cantilever is defined at $x = 0$ and the free end by $x = L$.

and chemical) ultimately determine the response of the sensor to an analyte. Therefore, in order to model and predict the response of a microcantilever sensor, it is necessary to understand the sorption process associated with an analyte/coating pair as well as the mechanical properties of the coating. In general, polymers are used to sense chemical analytes and are normally viscoelastic. When the device is operating in the dynamic mode, sorption of analyte changes the mass and viscoelastic properties of the coating, which, in turn, causes the resonant frequency of the microcantilever to change [5–9]. If the material in the second layer (i.e., the sensitive coating layer) were elastic, the model could be corrected to assume a thicker cantilever with a composite density and flexural rigidity. The addition of a viscoelastic coating adds damping to the system, which in some applications should be taken into account in the equation of motion [9]. The base layer is usually made out of an elastic material such as silicon (or a composite of elastic materials, i.e., Si, SiO_2, and SiN_x), whereas the polymer is chosen with regard to the analyte under study [10]. In the model, it will be assumed that the base material is purely elastic, meaning that no loss will be assumed from the base layer. The operating medium is assumed to be a viscous fluid, either a gas or a liquid. The difference between gas and liquid sensing with respect to coating selection can be demonstrated by considering partition coefficients for partitioning of an analyte from either the gas phase or from water into the polymer coating. It is recalled here that the partition coefficient (also known as the equilibrium constant), which provides insight into the extent of analyte partitioning into the coating, is a thermodynamic parameter that characterizes the distribution of organic analytes between the polymer coating and the gas or liquid solution in contact with the polymer-coated device [11, 12].

Figure 15.2 shows the length, L, width, b, and thickness, h_1, of the microcantilever as well as the polymer layer's thickness, h_2. The deflection function, $w(x, t)$, represents the vertical displacement (in z-direction) along the length of the beam as a function of time. Assuming that the microcantilever is a rectangular beam with $L \gg (h_1 + h_2)$ undergoing small transverse displacements, the equation of motion for a polymer-coated microcantilever operating in a viscous fluid medium is given by [11–14].

$$(EI)^* \frac{\partial^4 w(x, t)}{\partial x^4} + \frac{m}{L} \frac{\partial^2 w(x, t)}{\partial t^2} = F(x)e^{j\omega t} + F_{\text{fluid}}(x, t) \tag{15.1}$$

where $(EI)^*$ is the beam's complex flexural rigidity, m is the overall mass of the microcantilever/polymer layer system, $F(x)$ is the position-dependent forcing function per unit length applied at an angular frequency of ω, and $F_{\text{fluid}}(x, t)$ is the force (per unit length) exerted by the fluid medium on the microcantilever. It is noted that the above equation represents the equation of motion for a beam vibrating in the flexural, transverse, or lateral mode. In each case, the appropriate flexural rigidity, forcing function, and force exerted by the fluid medium on the microcantilever will apply [11, 15–20].

Since the polymer coating is usually viscoelastic, the flexural rigidity is complex and can be calculated by

$$(EI)^* = (\sqrt{(EI)'^2 + (EI)''^2})e^{j\theta} \tag{15.2}$$

with

$$(EI)' = E_1 I_1 + E_2' I_2 \tag{15.3}$$

and

$$(EI)'' = E_2'' I_2 \tag{15.4}$$

E_1 is the Young's modulus of the microcantilever base layer, which is assumed to be purely elastic, E_2' and E_2'' are the storage modulus and loss modulus of the polymer layer, respectively, and I_1 and I_2 are the moments of inertia of the micro-cantilever base and polymer layer, respectively. For a system with a polymer layer, the moment of inertia can be approximated assuming a time-invariant neutral axis, found by using the Cauchy principal value theorem on the position of the time-variant neutral axis location [9, 21]. The parameter θ in Eq. (15.2) is the loss angle of the composite beam and is defined as

$$\theta = \tan^{-1}\left(\frac{h_2 E_2''}{h_1 E_1 + h_2 E_2'}\right) \tag{15.5}$$

Although not shown explicitly in Eqs. (15.2)–(15.5), it is noted that the complex modulus of the coating may depend on the operating frequency. $F_{\text{fluid}}(x, t)$ is defined as [15]

$$F_{\text{fluid}}(x) = -g_1 \frac{\partial w(x, t)}{\partial t} - g_2 \frac{\partial^2 w(x, t)}{\partial t^2} \tag{15.6}$$

where g_1 and g_2 are time-independent coefficients associated with the fluidic damping force per unit length and the fluidic inertial force (effective fluidic mass) per unit length, respectively, and are defined as [12, 16, 20]

$$g_1 = \frac{\pi}{4} \rho_L b^2 \Gamma_I \left(Re, \frac{h}{b}\right) \omega \tag{15.7}$$

and

$$g_2 = \frac{\pi}{4} \rho_L b^2 \Gamma_R \left(Re, \frac{h}{b}\right) \tag{15.8}$$

respectively. Here, ρ_L is the density of the fluid and Γ_R (Re, h/b) and Γ_I (Re, h/b) are the real and imaginary parts of the hydrodynamic function, Γ (Re, h/b), of the microcantilever. The hydrodynamic function is dependent on the aspect ratio h/b and on the Reynolds number [12], which is a measure of the ratio of inertial forces to the viscous forces acting on the beam, and is defined as

$$Re = \frac{\rho_L \omega b^2}{4 \eta_L} \tag{15.9}$$

where η_L is the viscosity of the fluid. Assuming a sinusoidal time-dependence for the deflection function, the right-hand side of Eq. (15.6) can be written as

$\left(\frac{jg_1}{\omega} - g_2\right)\frac{\partial^2 w(x,t)}{\partial t^2}$. Substitution into Eq. (15.1) results in the following equation of motion,

$$(EI)^* \frac{\partial^4 w(x,t)}{\partial x^4} + m_B^* \frac{\partial^2 w(x,t)}{\partial t^2} = F(x)e^{j\omega t} \tag{15.10}$$

where m_B^* is the complex effective mass per unit length given by

$$m_B^* = \sqrt{\left(\frac{m}{L} + g_2\right)^2 + \left(\frac{g_1}{\omega}\right)^2} \; e^{j\phi} \tag{15.11}$$

where

$$\phi = -\tan^{-1}\left(\frac{g_1/\omega}{m/L + g_2}\right) \tag{15.12}$$

15.2.1
Generalized Resonant Frequency

The deflection of the coated microcantilever is composed of an infinite number of harmonic flexural modes. The resonant frequency for the ith mode of a coated microcantilever in a viscous fluid medium can be found using the generalized equation of motion for a vibrating beam, Eq. (15.10). It is commonly assumed that the Young's modulus of the base layer is frequency-independent. For the polymer layer, the complex Young's modulus could depend on the frequency. However, a first approximation is to use the polymer properties at the resonant frequency and neglect their variation in the vicinity of the resonant peak. Assuming a frequency-dependent hydrodynamic function, the resonant frequency obtained by solving Eq. (15.10) (subjected to the appropriate boundary conditions) is given by

$$f_{\mathrm{res},i} = \frac{\alpha_i^2}{2\pi}\sqrt{\frac{k}{M}} \tag{15.13}$$

where α_i is a number corresponding to a particular flexural mode and is 1.875 for the primary or fundamental flexural mode [9].

$$(EI)' - \frac{\frac{\left((g_1/\omega) + \left(\frac{\omega}{2}\right)\frac{\mathrm{d}}{\mathrm{d}\omega}(g_1/\omega)\right)}{\left(m/L + g_2 + \left(\frac{\omega}{2}\right)\frac{\mathrm{d}}{\mathrm{d}\omega}(g_2)\right)}(EI)''}{L^3} \tag{15.14}$$

$$M = (m + Lg_2) + \frac{\left((g_1/\omega) + \left(\frac{\omega}{2}\right)\frac{\mathrm{d}}{\mathrm{d}\omega}(g_1/\omega)\right)}{\left(m/L + g_2 + \left(\frac{\omega}{2}\right)\frac{\mathrm{d}}{\mathrm{d}\omega}(g_2)\right)}(g_1/\omega)L \tag{15.15}$$

It can be shown that Eq. (15.13) reduces to the well-known expression for the resonant frequency of a coated microcantilever in a vacuum ($g_1 = g_2 = 0$), given by

$$f_0 = \frac{\alpha_0^2}{2\pi}\sqrt{\frac{E_1 I_1 + E_2' I_2}{mL^3}} \tag{15.16}$$

with $\alpha_0 \cong 1.875$ corresponding to the fundamental flexural mode [9].

If there are only small dissipative effects from both the polymer and the viscous fluid medium, the resonant frequency simplifies to

$$f_{res} = f_0 \frac{1}{\sqrt{1 + \frac{Lg_2}{m}}} \tag{15.17}$$

This form is analogous to that obtained in [15] for an uncoated microcantilever in a viscous liquid medium. When the medium is assumed inviscid, Eq. (15.17) can be further simplified to a form analogous to the one presented in [13] as

$$f_{res} = f_0 \frac{1}{\sqrt{1 + L\pi\rho_L b^2/(4m)}} \tag{15.18}$$

If both the Young's modulus of the polymer layer and the Reynolds number are assumed frequency-independent near resonance (making the hydrodynamic function frequency-independent), the resonant frequency can be simplified to

$$f_{res,i} = \frac{\alpha_i^2}{2\pi L^2} \sqrt{\frac{\left(\frac{m}{L} + g_2\right)(EI)' - (g_1/\omega)(EI)''}{\left(\frac{m}{L} + g_2\right)^2 + (g_1/\omega)^2}} \tag{15.19}$$

In general, assuming a frequency-dependent hydrodynamic function and Young's modulus for the polymer layer, the resonant frequency is obtained from Eq. (15.13) through the use of an iterative process with an initial guess for ω (typically the resonant frequency in a vacuum).

15.3
Effects of Chemical Analyte Sorption into the Coating

15.3.1
Resonant Frequency

Introduction of a chemical analyte into the operational medium will not only change the medium's viscosity and density, but will also affect the characteristics of the coated microcantilever through chemical sorption into the polymer layer. For example, the mass and thickness of the polymer layer will increase; the moment of inertia for both layers will change due to a change in the neutral axis location; and the complex Young's modulus of the polymer layer will change. While the shifts in the resonant frequency due to some of these effects are negligible in comparison with the overall resonant frequency, they still can be significant with regard to the overall change in the resonant frequency, Δf_{res}, which is used to determine the chemical concentration of the analyte in the medium.

Assuming small variations in the properties of the polymer layer due to analyte sorption, the relative shift in the resonant frequency can be found as a function of the change in microcantilever mass, Δm, changes in the storage and loss moduli, $\Delta E_2'$ and $\Delta E_2''$, changes in the moments of inertia, ΔI_1 and ΔI_2, and changes in viscosity and density, which cause changes in the hydrodynamic function, thus Δg_1 and Δg_2 as [14]

$$\frac{\Delta f_{\text{res}}}{f_{\text{res}}} \cong (\lambda_m \Delta m) + (\lambda_{E'_2} \Delta E'_2 + \lambda_{E''_2} \Delta E''_2) + (\lambda_{I_1} \Delta I_1 + \lambda_{I_2} \Delta I_2) + (\lambda_{g_1} \Delta g_1 + \lambda_{g_2} \Delta g_2)$$

$$(15.20)$$

In Eq. (15.20), the λ coefficients describe the sensitivities to various changes in the system. It is noted that the swelling-induced change in the polymer thickness is included in the ΔI terms. All other terms (including higher-order terms) have been assumed negligible in deriving Eq. (15.20) for small frequency shifts. Also, the third term in parentheses (describing changes in the moments of inertia, ΔI_1 and ΔI_2) has been shown to make negligible contributions to the frequency shift because of the insensitivity of the neutral axis position to the analyte sorption [14]. (This is similar to what was shown in [9], in which swelling effects were neglected.) The effects due to ΔI_1 and ΔI_2 will therefore be disregarded. Variations in the viscosity and density can also affect the resonant frequency as indicated by the Δg_1 and Δg_2 terms in Eq. (15.20). However, in practical applications, these shifts are normally assumed to be accounted for using a differential measurement from a reference cantilever, a cantilever of identical geometry and polymer coating thickness, which is not sensitive to the analyte. Therefore, changes in the viscosity and density will not be included in the discussion dealing with Eq. (15.20). Thus, Eq. (15.20) reduces to

$$\frac{\Delta f_{\text{res}}}{f_{\text{res}}} \cong (\lambda_m \Delta m) + (\lambda_{E'_2} \Delta E'_2 + \lambda_{E''_2} \Delta E''_2)$$

$$(15.21)$$

where

$$\lambda_m = \left(\begin{array}{c} \dfrac{\left((g_1/\omega) + \left(\frac{\omega}{2}\right) \frac{d}{d\omega}(g_1/\omega) \right)(g_1/\omega)(EI)'}{2kLM \left(m/L + g_2 + \left(\frac{\omega}{2}\right) \frac{d}{d\omega}(g_2) \right)^2} \\ + \dfrac{\left((g_1/\omega) + \left(\frac{\omega}{2}\right) \frac{d}{d\omega}(g_1/\omega) \right)(m/L + g_2)(EI)''}{2kLM \left(m/L + g_2 + \left(\frac{\omega}{2}\right) \frac{d}{d\omega}(g_2) \right)} \\ - \dfrac{1}{2M} \end{array} \right)$$

$$(15.22a)$$

$$\lambda_{E'_2} = \frac{I_2}{2kL^3}$$

$$(15.22b)$$

$$\lambda_{E''_2} = -\frac{I_2 \left((g_1/\omega) + \left(\frac{\omega}{2}\right) \frac{d}{d\omega}(g_1/\omega) \right)}{2kL^3 \left(m/L + g_2 + \left(\frac{\omega}{2}\right) \frac{d}{d\omega}(g_2) \right)}$$

$$(15.22c)$$

For some viscoelastic (e.g., polymer) coatings, the analyte-induced changes in the flexural rigidity could have a significant effect on the resonant frequency [22], and thus is accounted for in Eq. (15.21). However, for trace chemical detection, changes in the storage and loss moduli, $\Delta E'_2$ and $\Delta E''_2$ are generally negligible and changes in the resonant frequency are only due to changes in beam mass from the sorbed analytes. Again, it is should be pointed out that the expression for the frequency sensitivity of a laterally vibrating coated beam is the same as the expression for the transversely vibrating beam [14, 20], but with different values for the hydrodynamic function, Γ_{lat}, and the moment of inertia I_{lat}.

15.3.2
Quality Factor

Coated microcantilevers operating in viscous fluids can suffer from a drastic decrease in their quality factors. This decrease in the quality factor increases the frequency noise (which is proportional to f_{res}/Q when operating in an oscillator configuration [10, 23, 24]), thus increasing the limit of detection (LOD) in bio-chemical sensing applications. The quality factor is defined as 2π times the ratio of the maximum energy stored in a resonating system to the amount of energy dissipated in one cycle [25]. However, when a resonating device is used in an oscillator configuration (or when considering the device transfer function), the quality factor is commonly obtained using the 3-dB bandwidth Δf_{3dB} of the device and is given by [17, 23]

$$Q = \frac{f_{res}}{\Delta f_{3dB}} \tag{15.23}$$

If both the Young's modulus of the polymer layer and the Reynolds number are again considered constant in the range of operational frequencies, the quality factor can be obtained from Eq. (15.1) as

$$Q = \left(2 \left(1 - \sqrt{1 - \left| \frac{(g_1/\omega)\,(EI)' + (m/L + g_2)\,(EI)''}{(m/L + g_2)\,(EI)' - (g_1/\omega)\,(EI)''} \right|} \right) \right)^{-1} \tag{15.24}$$

Assuming operation in a vacuum, g_1 and g_2 are both zero and Eq. (15.24) reduces to the one presented in [9], which incorporates only the viscoelastic losses in the polymer layer:

$$Q_0 = \left(2 \left(1 - \sqrt{1 - \frac{(EI)''}{(EI)'}} \right) \right)^{-1} \tag{15.25}$$

In low loss media, Eq. (15.24) can further be approximated as

$$Q_{approx} = \frac{(m/L + g_2)\,(EI)' - (g_1/\omega)\,(EI)''}{(g_1/\omega)\,(EI)' + (m/L + g_2)\,(EI)''} \tag{15.26}$$

By solving Eq. (15.24) or Eq. (15.26), it is possible to determine the quality factor (of different flexural, transverse, and lateral modes) of the coated cantilever, which, in turn, determines the figures of merit of the sensor.

15.4
Figures of Merit

An ideal chemical sensor should feature a high sensitivity, a low LOD, a high selectivity to the target analyte (and, thus, low cross-sensitivities to possible interferents), short response and recovery times, and a large dynamic range (i.e., being able to measure over an as wide as possible concentration range). At the same time, the chemical sensor should be easy to use, reversible (i.e., the analyte can be removed and the sensor regenerated), not exhibit any significant signal drift, and be largely insensitive to environmental changes, such as temperature or pressure. Ideally, all of these requirements should be met while ensuring low cost and low maintenance. Clearly, this is a daunting set of requirements for all chemical sensors, including the resonant chemical sensors reviewed in this chapter. In the following, the key characteristics of chemical sensors are discussed and defined in the context of mass-sensitive (resonant) chemical sensors.

The chemical sensitivity S is defined as the change of the sensor output signal, in the case of most resonant chemical sensors the change of a particular resonant frequency f_{res}, due to a change in analyte concentration c_A in the surrounding medium:

$$S = \frac{\partial f_{res}}{\partial c_A}, \quad S_R = \frac{1}{f_{res}} \frac{\partial f_{res}}{\partial c_A} \tag{15.27}$$

Often, the relative chemical sensitivity S_R is used instead of S to be able to better compare the performance of sensors with different resonant frequency. It should be noted, that the analyte concentration can be presented in various units, including mass-per-volume (e.g., $kg\,m^{-3}$), volume-per-volume (e.g., ppm-v/v often used in gas-phase chemical sensing), mass-per-mass (e.g., ppm-w/w often used in liquid-phase chemical sensing), or molar concentrations, that is, number of analyte moles per liter of sample volume.

In the following discussion, trace chemical detection is assumed and only changes in the polymer mass upon analyte sorption are accounted for. (In the more general case, where viscoelastic properties of the sensing film and/or changes in the properties of the surrounding fluid need to be considered, the chemical sensitivities (Eq. (15.27)) need to be calculated from Eq. (15.21).) As described in [26], the chemical sensitivity S (or relative chemical sensitivity S_R) may be written as the product of the gravimetric sensitivity G (or the relative gravimetric sensitivity G_R) of the coated resonant sensor, that is, the absolute (or relative) change in frequency f due to a change in coating density ρ_m, and the analyte sensitivity S_A, that is, the change in coating density ρ_m due to a change in analyte concentration c_A in the surrounding medium:

$$S = G \cdot S_A = \frac{\partial f_{res}}{\partial \rho_m} \frac{\partial \rho_m}{\partial C_A}, \quad S_R = G_R \cdot S_A = \left(\frac{1}{f_{res}} \frac{\partial f_{res}}{\partial \rho_m} \right) \frac{\partial \rho_m}{\partial C_A} \tag{15.28}$$

This way the sorption characteristics of the sensing film for the particular analyte of interest are solely described by the analyte sensitivity S_A, while the sensor response to the "physical" density change of the sensing film is described by the gravimetric sensitivities G and G_R. Instead of using the relative gravimetric sensitivity G_R, especially the literature on mass-sensitive acoustic wave sensors considers the device's sensitivity to mass loading per surface area G_M, which equals the relative gravimetric sensitivity divided by the sensing film thickness

$$G_M = \frac{1}{f_{res}} \frac{\partial f_{res}}{\partial (m_{add}/A)} = \frac{G_R}{h_2} \tag{15.29}$$

with the mass loading per area m_{add}/A and the sensing film thickness h_2. It is important to note that the sensitivity G_M is independent of the sensing film thickness, which is especially of advantage when considering biosensors with monolayer surface functionalizations or when comparing different device designs. Since the sensing film, however, affects the characteristics of the resonator, especially in case of thicker sensing films, the relative gravimetric sensitivity might be more meaningful.

If the analyte concentration is given in parts per million-volume/volume, the analyte sensitivity S_A may be calculated as [27]

$$S_A = \frac{\partial \rho_m}{\partial c_A} = \begin{cases} M_A \cdot K \cdot (p/RT) \cdot 10^{-6} & \text{in gas} \\ \rho_A \cdot K \cdot 10^{-6} & \text{in liquid} \end{cases} \tag{15.30}$$

where ρ_A is the density of the analyte (in liquid), M_A its molar mass (for the in-gas case), p, R, and T are gas pressure, gas constant, and temperature, respectively, and K is the partition coefficient (gas or liquid phase) of the particular analyte/coating combination, that is, the ratio of the steady-state analyte concentration in the sensitive film to the analyte concentration in the surrounding medium. The factor 10^{-6} accounts for the fact that c_A is given in parts per million.

A common scenario is to use mass-sensitive chemical microsensors, for example, cantilever beams, with a uniform layer sandwich structure vibrating in a lateral (in-plane) or transverse (out-of-plane) flexural mode. If such a resonant sensor is *uniformly coated* with a sensing film, the resulting relative gravimetric sensitivity G_R (and thus, S_R) is, to first order, independent of the microstructure's in-plane dimensions and the flexural eigenmode used, and only depends on the layer thicknesses and densities of the MEMS device and the sensing film as well as the sorption properties of the sensing film. Assuming that the analyte sorption only affects the density of the sensing film with thickness h_m (and not its Young's modulus), one can easily show that the relative gravimetric sensitivity of a cantilever with a uniform layer sandwich becomes [28]

$$G_R = \frac{1}{f_{res}} \frac{\Delta f_{res}}{\Delta \rho_m} = -\frac{1}{2} \frac{h_m}{\sum\limits_{i=1}^{m} h_i \, \rho_i} \tag{15.31}$$

where h_i and ρ_i are thickness and density of the ith layer. In Eq. (15.31), we assume that the base layer shown in Figure 15.2 is actually a uniform layer sandwich of layers 1 to $m-1$, coated by layer m, the sensing film. (Figure 15.2, thus, shows the case of $m = 2$, that is, the base layer is a single material, and the sensing film thickness $h_m = h_2$).

It should be emphasized that G and G_R describe the sensitivity of the resonant sensor to changes in the density of the sensing film. As has been shown above (Eq. (15.31)), *G_R generally does not benefit from device scaling, that is, the relative gravimetric and chemical sensitivities, G_R and S_R, cannot be simply improved by reducing the device dimensions, in particular the width b and length L. This is in contrast to the mass sensitivity of a resonant sensor, which can be readily increased by reducing the device dimensions.* For a mass-sensitive sensor operated at a given resonant frequency f_{res} the relative frequency change due to a change of its (effective) mass Δm, is simply given by

$$\frac{\Delta f_{res}}{f_{res}} = -\frac{1}{2}\frac{\Delta m}{m} \tag{15.32}$$

which indicates that the relative mass sensitivity can be increased by simply decreasing the resonator (effective) mass, that is, by decreasing the resonator dimensions. While such device scaling has resulted in mass resolutions in the attogram range [29, 30], *the chemical sensitivity of such nanoresonators is generally not improved compared to microresonators.*

The ultimate performance metric for any chemical sensor is its LOD, that is, the smallest detectable analyte concentration. In the case of a resonant (mass-sensitive) biochemical sensor, the limit of detection can be defined as three times the noise-equivalent analyte concentration, which itself can be approximated by the ratio of the minimal detectable frequency change Δf_{min} and the chemical sensor sensitivity S [31]:

$$\text{LOD} = 3\frac{\Delta f_{min}}{S} \tag{15.33}$$

Alternatively, the LOD might be expressed as three times the ratio of the smallest detectable relative frequency change σ_{min} to the relative chemical sensor sensitivity S_R:

$$\text{LOD} = 3\frac{\sigma_{min}}{S_R} = 3\frac{\Delta f_{min}/f}{S/f} \tag{15.34}$$

The sensor's LOD is particularly affected by the geometrical dependence of the minimal detectable frequency change Δf_{min}. Δf_{min} is generally improved (i.e., reduced) by increasing the quality factor Q of the resonance (see Eq. (15.26) and Chapter 3 on Damping). Thus, even though the relative gravimetric and chemical sensitivities of uniformly coated resonant sensors are independent of the microstructure's in-plane dimensions and the flexural eigenmode used (see Eq. (15.31)), the sensor's LOD can be improved by choosing an eigenmode or geometry that minimizes σ_{min}, that is, generally has a large Q factor. As a result, significant efforts have gone into improving the Q-factor of mass-sensitive chemical sensors in air and water, exploring in-plane [32] and torsional [33, 34] modes instead of out-of-plane flexural modes, higher-order resonance modes [17] and special device geometries such as disks [35, 36] or hammerheads [37].

The minimal detectable frequency change of a resonant sensor operated in an amplifying feedback loop is generally limited by (i) the short-term frequency stability of the sensing system and (ii) the long-term frequency drift. The frequency stability of the resonator in a feedback loop can be quantified by the Allan Variance method [38]. The Allan Variance $\sigma(\tau)$, which is a function of the measurement/integration time τ, is defined by the mean square of the differences between neighboring relative frequency measurements

$$\sigma(\tau) = \sqrt{\frac{1}{2M} \sum_{n=1}^{M-1} \left[\frac{f_{n+1} - f_n}{f_1} \right]^2} \tag{15.35}$$

where M is the total number of frequency measurements and f_n is the nth frequency measurement averaged over the time interval τ. For small integration times τ, the Allan Variance is limited by frequency noise in the resonator and feedback circuit, whereas for long integration times, the Allan Variance is limited by long-term frequency drift-caused aging and changes in the environment (e.g., temperature). Recent work on a chemical microsystem comprising silicon-based resonant microsensors and feedback circuitry implemented on a CMOS (complementary metal-oxide-semiconductor) ASIC (application specific integrated circuit) yields Allan Variances in the range of $2-4 \times 10^{-8}$ for integration times between 1 and 10 s, corresponding to frequency stabilities ≈ 10 mHz for a resonator with a resonant frequency of 380 kHz [39]. The long-term sensor drift can be reduced by, for example, using an uncoated reference resonator that exhibits similar responses to environmental changes as the coated resonator in a differential setup. Many more sophisticated schemes to reduce the long-term drift have been investigated [40, 41].

The selectivity of a chemical sensor describes its ability to detect the analyte of interest in the presence of interferents (i.e., molecules that can elicit a sensor response, but are not the compound of interest). Sensor selectivity can be generally improved by proper selection of the sensing film, for example, by opting for stronger and, thus, typically more selective binding mechanisms, but this approach often comes at the expense of reduced sensor reversibility. Examples include highly specific key–lock interactions, such as antibody–antigen binding used in biochemical sensing. In the case of chemical sensors, reversibility is often a key requirement and, thus, the selected sensing films often have a limited selectivity (see also Section 15.5). One approach to improve the selectivity of the sensing system is to perform an analyte separation, for example, using a gas chromatography (GC) column in the case of a gas sensor, prior to the analysis with the chemical sensor. This approach is widely used in laboratory-based analytical instrumentation (e.g., gas chromatography/mass spectroscopy – GC/MS – systems) and has been successfully applied in micro-GC systems [42, 43]. If analyte separation is not an option, the system selectivity can generally be improved by using arrays of sensors coated with different sensing films. Such sensor arrays can either feature only one sensing mode (e.g., an array of mass-sensitive sensors) [26], or multiple sensing modalities [44]. Hierlemann and Gutierrez-Osuna [45] have published an excellent overview of such higher-order sensing approaches. However, recent research on capacitive, calorimetric, and mass-sensitive sensor arrays also highlights that quantitative analysis of even ternary or quaternary gas mixtures remains a formidable challenge [46]. Most research in this area focuses on analyzing quasi steady-state sensor responses, but investigation of signal transients might improve sensor selectivity as well [47, 48]. In the case of (mass-sensitive) sensors coated with polymeric sensing films, the characteristic absorption and desorption time constants depend on the analyte-polymer combination and have been used to discriminate between different alcohols [47]. Because of their high mass-sensitivities and potentially high sampling rates, mass-sensitive sensors are very well suited to study sorption processes into thin sensing films in real time. When investigating transient sensor responses, the sensing system generally needs to be carefully designed, so that surface reaction and/or sorption into the sensing film are dominating the transient response (often referred to as reaction limited) but not the analyte transport to the sensing surface (often referred to as mass-transport limited). A thorough discussion of analyte transport for the case of biosensors is presented in [49].

15.5
Chemically Sensitive Layers

As seen in Figure 15.1, the sensitive layer of a chemical sensor is crucial for its chemical sensitivity, but also impacts, for example, the sensor selectivity, reversibility, and stability. Thus, it comes as no surprise that many research efforts have targeted the development of "better" sensing films for particular applications. Table 15.1 summarizes some of the inorganic and organic materials that have been explored as sensitive layers for different transducers [50].

The different types of chemical reactions explored in a sensing film range from weak physisorption to strong chemisorption, charge transfer, and chemical reactions [50]. To ensure sensor reversibility, many chemical microsensors use weak physisorptive interactions, generally at the expense of sensor selectivity (see Section 15.4). In the following section, some of the sensitive layer materials explored in connection with resonant microsensors are briefly discussed. It should be noted that sensor selectivity solely comes from the proper choice of the sensitive layer (or group of sensitive layers in the case of sensor arrays); the underlying resonant sensor measures a mass uptake and, since every analyte of interest has a mass, is non-specific.

Arguably, the most widely used class of sensing films in resonant chemical sensors is non-conducting polymers, such as polysiloxanes or polyurethanes. Polymers have been widely used for the detection of volatile organic compounds and environmental monitoring in both the gas and liquid phase. One of the key advantages of polymers is the straightforward processing: the polymer sensing film can be deposited onto the (resonant) microstructure by, for example, drop-coating,

Table 15.1 Typical sensitive layer materials for chemical sensors and target applications.

Materials	Examples	Applications
Metals	Pt, Pd, Ni, Ag, Sb, Rh	Inorganic gases (CH_4, H_2, etc.)
Ionic compounds	SnO_2, In_2O_3, $AlVO_4$, $SrTiO_3$, Ga_2O_3, ZrO_2, LaF_3, Nasicon	Inorganic gases (CO, NO_x, CH_4), exhaust gases, oxygen, ions in water
Molecular crystals	Phthalocyanines (PCs): PbPC, $LuPC_2$	Nitrogen dioxide, volatile organics
Langmuir-Blodgett films	Lipid bilayers, polydiacetylene	Organic molecules in medical applications, biosensing
Cage compounds	Zeolites, calixarenes, cyclodextrins, crown ethers	Water analysis (ions), volatile organics
Polymers	Polyurethanes, polysiloxanes, polypyrroles, polythiophenes, Nafion	Detection of volatile organics, food industry (odor and aroma), environmental monitoring
Biological entities	Phospholipids, lipids, enzymes, proteins, cells, membranes	Medical applications, biosensing, water and blood analysis, pharma screening

Source: Adapted from Ref. [50].

spin-coating, spray-coating using shadow masks, or ink-jet printing. Moreover, analytes are absorbed into the polymer volume and, thus, relatively large mass changes and sensor sensitivities can be achieved. The polymer-analyte interaction is generally by weak physisorption and, as a result, polymer-coated (resonant) sensors exhibit only limited selectivity. Nevertheless, the selectivity of polymers can be tuned by, for example, adding polymer side-chain functional groups to promote different fundamental physisorption mechanisms, such as dispersion, polar interactions, polarizability, and Lewis acid–base interactions [51]. This has been demonstrated by investigating polysiloxanes with different side-chains as sensitive films on thickness-shear mode resonators [51]. Linear solvation energy relationships allow modeling the response of polymer-coated mass-sensitive sensors [52] and thus choosing the proper film for a particular analyte class and application.

Investigators have tried different strategies to further improve the selectivity of polymer sensing films. One such strategy is to examine transient responses and the characteristic time constants of analyte absorption into the polymer film rather than steady-state signal responses to generate selectivity between analytes [47]. A second strategy that has been investigated for the generation of more selective polymer films is molecular imprinting [53–55]. In this method, a polymer layer is fabricated in such a way that spaces exist within the polymer that are tuned to interact with a specific analyte [53]. Molecularly imprinted polymers (MIPs) are an interesting strategy for improving selectivity, but have proven difficult to optimize. Two key issues are the challenges associated with clearing the imprinting agent from the fabricated film and also generating adequate interactions with analytes that may be charged or have other properties that are difficult to induce in polymer films [53].

Another material class that has been investigated as sensitive layers on resonant sensors for VOC (volatile organic compound) detection is metal organic frameworks (MOFs). MOFs consist of crystalline metal ion clusters linked by organic linkers [56]. These materials have found use in a variety of applications including the detection of chemical vapors and in humidity sensing [56–59]. MOFs have the added advantage of a large surface area and can be coated relatively thickly onto the surface of a sensor allowing for high sensitivity. Depending on the crystallization method, the materials used and the size of the pores in the film, MOFs can be made selective for different analytes [60].

Sensitive materials in biosensing applications include, for example, lipids, enzymes, and proteins (see also Chapter 16 on Resonant Biosensors). Many antibodies to proteins or other large molecules can be created in mice or other mammals for use in biotechnology applications. Essentially, the mouse is exposed to the ligand of interest and then the mouse's immune response leads to production of endogenous antibodies within the mouse that can be procured and purified. Antibodies have the advantage that they are inherently selective, and will often only bind the specific epitope to which they are sensitive. A disadvantage of using antibodies, or any protein-based sensitive layer is that the layer often requires constant wetting and can be temperature- and/or UV-sensitive. This can create issues with antibody integration into sensor platforms [53].

Aptamers are a second class of biomolecules that are frequently used as a sensitive film [61–63]. Aptamers and, more generally, ligand-binding nucleic acids can be made to interact with a wide range of molecules including biomolecules and organics. Aptamers selective to a particular analyte are found through screening the affinity of randomly generated nucleotide sequences for the analyte of interest. The specifics of the aptamer selection process can be found in [64]. Nucleotide chains can be readily manufactured to have a wide range of affinities allowing for applications in sensing of proteins and also small molecules with a variety of properties [63, 65]. Further, aptamers can be manufactured artificially allowing tailoring of their affinity for the analyte of interest. Finally, techniques for producing nucleic acid chains are readily available to produce large quantities of aptamers. Aptamers have been investigated for a variety of sensing applications including the sensing of small molecules [64, 65].

Self-assembled monolayers (SAMs) are often used to attach biomolecules to a sensor surface [66]. These chemical structures are comprised of molecules having one end that covalently bonds to a surface and another serving as a functional group that can be used to link to the molecule of interest. In particular, two popular choices are silane chemistry [67] and gold-thiol chemistry [68–70]. Silane chemistry can be used to attach molecules to the surface of silicon [71], glass/dielectric [67, 72] materials, and several other types of surfaces. Silane chemistry is versatile, but is not particularly selective with regard to the region to which the molecules will attach. On the other hand, the use of thiol chemistry to attach proteins to gold is a highly selective strategy. If patterned gold surfaces are used, then one can form a patterned protein or antibody surface. Extensive work has been done on the creation of SAMs on metals using organic compounds [66]. Since SAMs from thiolated organic compounds will form only on a gold surface, gold can be deposited on the area of the cantilever that should be functionalized and the surrounding areas of the cantilever and the die itself ideally should not see any specific protein attachment (non-specific absorption can still be an issue in this scenario). This helps to ensure maximum sensor signal because analyte will only attach to parts of system that are sensitive to it.

15.6
Packaging

The packaging requirements for chemical and biochemical sensors can vary greatly depending on the application. As with all sensor packages, a suitable biochemical sensor package must protect the sensing system from the environment while being able to sense the quantity of interest [73–75]. As a result, a chemical sensor package typically requires not only electrical interconnects, but also feedthroughs for non-electrical signals. The need for nonelectrical feedthroughs (or windows to the outside world) is fundamentally different from microelectronics packaging requirements and requires, depending on the applications, radically new packaging concepts. In addition, resonant MEMS require some sort of cavity around the vibrating microstructure. Chapter 12 focuses on ways to hermetically encapsulate resonant microstructures; the technologies highlighted in that chapter are applicable to, for example, MEMS-based high-frequency resonators and filters, as well as sensors that can be hermetically sealed, such as inertial sensors, magnetic sensors, or even optical sensors. This section will focus on how the analyte of interest can be routed to the sensor surface, while still protecting the fragile elements of the sensing system from the environment. The solutions are in many cases highly customized for a particular application and, as a result, the packaging often adds substantial cost to the biochemical sensor.

A few select example of packages for commercial chemical sensors are given; even though these sensors do not utilize resonant MEMS, the involved packaging schemes can also be applied to resonant chemical microsensors. Figaro Engineering Inc. packages most of its chemical sensors in metallic TO (transistor outline) packages, generally using a lid with a metal mesh to protect the microsystem (Figure 15.3a) [76]. Some Figaro sensors, such as the carbon monoxide sensor TGS 2442, feature chemical filters integrated into the lid. In the chemical domain, filters can reduce the influence of interference molecules on a chemical measurement. In the case of the TGS 2442, an active carbon filter reduces the influence of interference gases. While being quite modular, such metal packages are too large and costly for many applications. Today's standard for integrated circuits are plastic packages and it is thus not surprising that (chemical) sensor packages follow this trend too. As an example, Figure 15.3b shows Sensirion

(a)   (b)

Figure 15.3 (a) Figaro chemical sensors in transistor outline (TO) packages [76]. (b) Sensirion humidity/temperature sensors SHT21 and SHTC1 in dual-flat no-lead (DFN) plastic packages [77].

humidity sensors packaged in reflow-solderable dual-flat no-lead (DFN) packages [77]. The plastic encapsulation of these packages has an opening at the location of the sensors, but protects the integrated circuitry located on the same silicon chip. The smallest Sensirion humidity/temperature sensor, the SHTC1, measures only $2 \times 2 \times 0.8 \, mm^3$ and can be found in Samsung Galaxy S4 smartphones. For further protection against water, dust, and other contaminants, filter caps can be mounted on top of some of Sensirion's humidity sensors.

The packaging approach for Sensirion's humidity sensors can likely be applied to resonant chemical sensors as well. It suits applications where the analyte can diffuse to the sensing surface, such as in the case of gas sensing. In the case of resonant sensors, special care must be taken that the rather fragile resonant microstructures are not damaged during the packaging process and are protected during device operation by, for example, a filter cap. Another approach on how to protect the integrated circuitry while providing access to the sensors is shown in [78]: here, the sensor chip is flip-chip mounted onto a ceramic substrate, which has laser-cut openings at the location of the sensor surfaces. Applying a soldering ring around these openings or just applying an underfill protects the circuitry from the environment.

An alternative method is to fabricate and/or attach a manifold on top of the sensor chip that allows for isolation of the analyte from the rest of the sensor system. This technique is especially suited for liquid-phase sensing applications [10, 27]. Manifolds have the advantage that they can readily be fabricated from, for example, inexpensive plastics and are straightforward to implement. For prototyping, stereolithography is a technique that is well suited to the fabrication of packaging elements and manifolds for sensors [79]. Stereolithography tools write a pattern layer by layer into a curable resin. Feature sizes down to the hundreds-of-microns range with aspect ratios >10 are achievable [79]. The manifold can directly seal against the chip surface (Figure 15.4b), thus protecting integrated circuitry, bonding pads, and bonding wires from the liquid. Alternatively, the manifold can seal against a more conventional microelectronics package (Figure 15.4a), such as a dual-in-line package; in this case, any on-chip

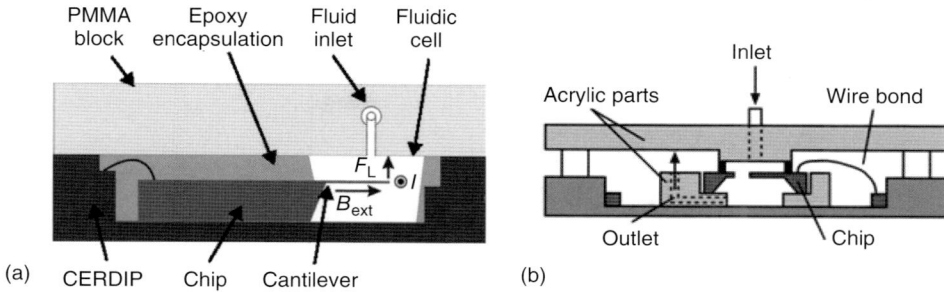

Figure 15.4 Two sensor packaging designs for liquid-phase applications using the commonly employed manifold approach: (a) manifold pressed against microelectronics package [10] and (b) manifold pressed directly against chip surface [27].

circuitry, bond pads, and bond wires can be protected by a glob top material dispensed locally on the chip surface [10]. This approach is relatively simple and builds on currently used packaging methods in the integrated circuits industry.

In all the above cases, sensor chips are packaged individually, which provides more flexibility but also increases cost. To reduce packaging costs, wafer-level packaging technologies are of interest. One interesting approach, which can be used for resonator packaging, is the use of sacrificial polymer processes, where a thermally degradable polymer is deposited under a second, more stable polymer layer. Thermal degradation of the first polymer leaves the resonator in an empty cavity. Capping the structure with a nitride or oxide can create a vacuum encapsulated structure [80].

Finally, techniques in which the resonator can be completely protected from the environment are of interest. As an example, a vacuum-encapsulated resonator was used as a humidity sensor [81]. In this case, the build-up of charge on an electrode was used to shift the resonant frequency of the electrostatic resonator. Similarly, coupled resonator structures have been investigated, where the primary resonator with driving and sensing electrodes is well protected from the environment, while the coupled (passive) resonator is in contact with the environment [82].

The discussion in this section has focused almost exclusively on the packaging of sensors using rigid materials and envisioning the sensors plugged into an electrical device such as a cellphone or hand-held reader. A future direction for sensor technologies, which will depend heavily on adequate packing, is the development of flexible and implantable sensors [73]. Both of these will be key for the deployment of biomedical sensors and in-field monitoring devices for infrastructure and security applications. The development of flexible technologies in particular poses challenges in terms of both electrical and fluid interconnections that are exciting new challenges for microsystems engineers.

15.7
Gas-Phase Chemical Sensors

The use of mass-sensitive sensors for gas sensing dates back almost 50 years, when King investigated quartz crystals coated with materials used as stationary phases in GC [83] for the detection of hydrocarbons. Acoustic wave sensors, such as quartz crystal microbalances (QCMs) and surface acoustic wave (SAW) devices, have been extensively studied for gas- and liquid-phase chemical sensing applications since the 1980s [84, 85] and, for example, the HAZMATCAD™ detectors by MSA (*http://us.msasafety.com/*) employ an array of coated SAW sensors to detect chemical warfare agents. Some of the first MEMS-based mass-sensitive chemical sensors include a resonant microbridge sensor for the detection of xylene [86], CMOS-integrated resonant cantilevers for humidity sensing [87], and a resonant silicon nitride sensor for detecting mercury vapor [7]. Over the past two decades, MEMS-based resonant microstructures and in particular resonant cantilevers have been widely researched for chemical and biochemical sensing applications and a number of extensive reviews have been published [1−3, 88−90]. In the following section, trends in gas-phase chemical sensing using resonant microstructures will be briefly highlighted.

Gas-phase chemical sensing has a broad range of applications including biochemical sample analysis [91, 92], detection of chemical warfare and bioterrorism agents [93], and environmental monitoring [94]. While in many cases, gas analysis is still done by field sampling and subsequent analysis using laboratory-based analytical instrumentation, most if not all of the above applications demand for real-time detection in the field. Combined with the miniaturization potential, microfabricated chemical sensors have received considerable attention in recent years to meet the goals of portable, low-cost gas analysis. Among the large variety of possible transduction mechanisms, mass-sensitive sensors feature a well-understood transduction mechanism, are straightforward to fabricate even in arrays, and can be easily interfaced with signal conditioning circuitry.

One of the most widely investigated MEMS-based microstructures for mass-sensitive chemical sensing are (silicon-based) rectangular cantilever beams (see Figure 15.5a) vibrating in their fundamental out-of-plane flexural mode [26, 97]. By optimizing the silicon cantilever dimensions, Q-factors around 1200 in air have been measured for this mode [98]. To further improve the quality factor and, thus, the frequency stability of the resonant sensors, more complex device geometries (see Figure 15.5b, c), such as disks [36], hammerheads [37], tuning forks [99], and I^2-BARs [96], and alternative vibration modes, such as in-plane flexural [100, 101] and torsional [33, 34] modes, have been investigated. In most cases, the resulting microstructure vibration primarily shears the surrounding fluid rather than compressing it, thus reducing the damping by the surrounding fluid and increasing the Q-factor. This way, Q-factors in the range of 4000−6000 in air can be readily achieved for uncoated resonant microstructures vibrating at in-plane resonant frequencies in the range of 500 kHz to 2 MHz [100, 102], especially in the case of thicker silicon structures. While the Q-factor for

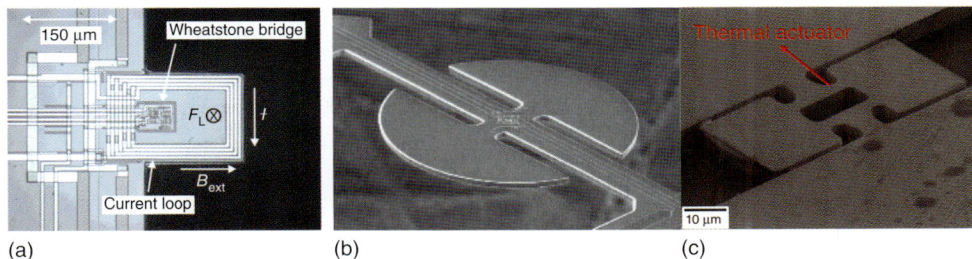

Figure 15.5 Examples of resonant microstructures used for gas-phase chemical sensing: (a) rectangular cantilever (reprinted with permission from [95]. Copyright 2005 American Chemical Society); (b) disk-type microresonator [36]; and (c) I^2-BAR resonator (© 2012 IEEE, Reprinted, with permission, from [96]).

microstructures vibrating in an in-plane flexural mode generally increases with the device thickness due to the increased volume/surface ratio, thicker resonators generally exhibit reduced gravimetric sensitivities (see Eq. (15.31)). Thus, film thicknesses including the sensing film thickness can be optimized to obtain the right balance between gravimetric (and chemical) sensitivity and short-term frequency stability in order to maximize the LOD [102].

While mass-sensitive chemical microsensors can be easily fabricated and exhibit high mass sensitivities, the underlying mechanical sensor platform does not offer any intrinsic chemical selectivity. As discussed earlier, chemical selectivity is introduced by the sensing film, which is deposited on top of the resonant microstructure. Since signal reversibility is a key feature for chemical sensors, the interaction of the sensing film with the target analyte is often based on weak chemical interactions (e.g., van der Waals forces) that are generally not specific enough to detect a target molecule in the presence of interferents. Examples include nonconducting polymer films, including polysiloxanes and polyurethanes, that are known as stationary phases in gas chromatography and are widely used as sensing films for detecting VOCs [26, 102]. To improve the system selectivity, sensor arrays can be coated with different polymeric sensing films, enabling the quantitative analysis of binary and ternary mixtures [26]. However, the limited orthogonality of the sensing films relying on weak chemical interaction makes analysis of more complex samples challenging [46]. To improve the sensor selectivity, recent work has studied the use of alternative, more selective sensing layers, such MIPs [53, 54], MOFs [56], and peptides [103]. Thereby, the idea is to mimic selective key-lock principles found in nature, such as the well-known antibody–antigen interaction. An example of the use of MIPs with resonant microsensors is the detection of the herbicide 2,4-dichlorophenoxyacetic acid using MIP-coated resonant piezoelectric micromembranes [104]. By depositing Cu_3(benzenetricarboxylate)$_2$ (Cu-BTC) MOF films onto SAW sensors, ultrasensitive humidity sensors were demonstrated [105]. Finally, peptide receptors can be functionalized on microsensor surfaces to mimic odor binding proteins. Specific peptide sequences that selectively bind 2,4-dinitrotoluene (DNT) have been immobilized on resonant cantilever surfaces and the selective DNT detection with LOD < 1 ppb could be demonstrated [106]. In the case of surface functionalization protocols that form surface monolayers, the sensitivity of the mass-sensitive sensor can be significantly improved by integrating high-surface area structures, such as carbon nanostructures or mesoporous materials, on the resonator surfaces. As an example, carbon nanostructures have been grown on silicon cantilevers using a low-temperature, plasma-assisted growth process, functionalized with peptides, and used for VOC sensing [37, 107]. Similarly, Figure 15.6 shows a resonant silicon microstructure with a mesoporous silica sensing layer deposited at its tip [58]; by surface functionalizing the mesoporous silica with hexafluoro-2-propanol, trinitrotoluene (TNT) vapor could be selectively detected with ppt-level detection limits.

System-level investigations help improve the performance of (resonant) gas sensing systems. As mentioned earlier, sensor arrays have been coated with different sorption films to improve the analyte discrimination [26, 108]. Similarly, resonant sensors and sensor arrays have been combined with other sensing modalities, such as capacitive and calorimetric sensor arrays [44, 46] and thermal conductivity detectors [109], to improve analyte identification. In the latter case, the same uncoated piezoresistive cantilever is used to investigate the viscous damping by the surrounding gas and the thermal conductivity of the surrounding gas [109]. In addition, signal transients rather than steady-state signals can be investigated to improve the analyte discrimination [48, 110]; in the case of polymer-coated microsensors, the analyte sorption into thin sensing films can be investigated in real time and has been used to distinguish different alcohols [47].

Dedicated signal processing circuitry simplifies resonant sensor operation by, for example, incorporating the sensor in an amplifying feedback loop for self-oscillation [26, 111]. The required circuitry can be co-integrated with the sensor if a CMOS process is used for system integration [26, 112], which yields

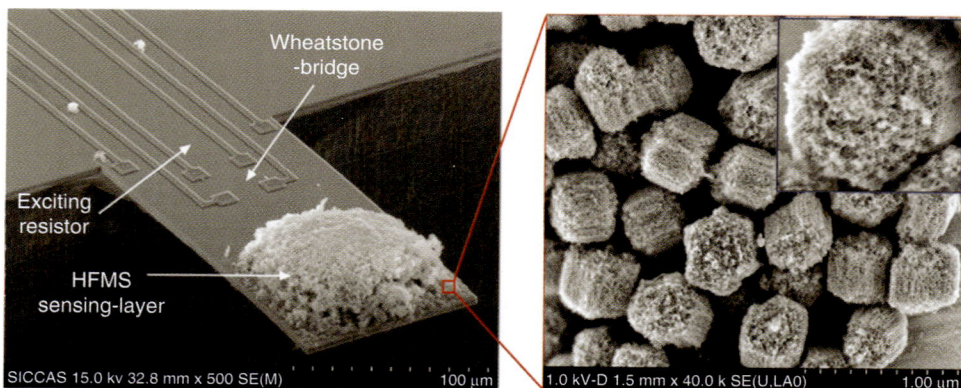

Figure 15.6 SEM (scanning electron microscope) image of silicon-based resonant cantilever with hexafluoro-2-propanol functionalized mesoporous silica sensing film (reprinted with permission from [58]. Copyright 2011 American Chemical Society).

very compact integrated gas-sensing platforms. Additional circuit elements can be implemented as needed: as an example, the closed-loop system based on the self-oscillation technique investigated in [111] features an automatic phase adjustment control loop that adjusts the phase of the main feedback loop to ensure an operating point as close as possible to the desired resonant frequency and a frequency drift compensation that compensates drift effects stemming from, for example, temperature changes using a controlled stiffness modulation scheme described in [36]. Other ways to compensate for unwanted long-term frequency drift are the use of differential setups with (uncoated) reference resonators or the investigation of two resonance modes of the same resonator [98].

Finally, mass-sensitive sensors and sensor arrays have been embedded into larger gas-phase sensing systems: nanomechanical resonators have been used for fast, gas-phase chromatographic chemical analysis, demonstrating a chromatographic analysis of 13 chemicals within a 5-s time period [29]; they form the basis for the Cantisens Research platform that enables cantilever sensing in both static and dynamic mode (Concentris, Basel, Switzerland); and they have been paired with infrared radiation sources to perform photothermal infrared absorption spectroscopy [113].

Table 15.2 summarizes the performance of select MEMS-based resonant sensors for gas-phase chemical sensing. Rather than providing a comprehensive overview, the examples are chosen to represent a wide range of MEMS-based resonator designs with QCM data as a benchmark.

15.8
Liquid-Phase Chemical Sensors

Liquid-phase chemical sensors have a variety of applications, including environmental monitoring [116] and biomedical analysis [1, 2]. Many of the issues of analyte discrimination that are important in the gas-phase applications are also important for liquid-phase applications. Liquid-phase chemical sensing has three important added complications compared to the gas-phase sensing. First, liquids can be corrosive to sensor surfaces, and can participate in electrochemical reactions with metal lines carrying electric current. This requires that liquids be isolated from the sensing regions of the sensor system (see Section 15.6). Second, liquid-phase operation of resonators can cause significant damping of the resonator signal. This can cause significant reduction in Q-factor and increase in frequency noise, which will reduce the sensor's signal-to-noise ratio. Lastly, fluid inertia increases the effective mass of the resonator, thus lowering its resonance frequency and mass sensitivity. Significant efforts have been dedicated to overcoming these problems. Figure 15.7 shows select examples of resonant microstructures that have been used for liquid-phase bio-chemical sensing.

Table 15.2 Comparison of MEMS-based and QCM-based resonant chemical gas sensor performances.

Device	Dimensions	Vibration mode	Frequency	Q in Air	f_{min} (Hz)	Mechanical sensitivity G_M (cm² g⁻¹)	Sensing film and thickness	Analyte and LOD	References
QCM	R = 6–7 mm; h = 167 µm	TSM	10 MHz	—	0.1 (0.01)	23	PEUT 0.3–0.6 µm	n-Octane 10 ppm	[114]
Cantilever	L = 150 µm; h = 9.5 µm	OOP	400 kHz	950	0.03	230	PEUT 3–4 µm	n-Octane 1 ppm	[26]
Cantilever	Beam L = 120.6 µm; b = 1.8 µm; h = 4.9 µm; End plate L = 116 µm; b = 103 µm	IP	5475 Hz	—		2073[a),b)]	Polystyrene 967 pg	Acetone 606.5 pg	[101]
Cantilever	L = 200 µm; b = 100 µm; h = 3 µm	OOP	101 kHz	—	0.4	626[a)]	Mesoporous silica 10 ng	TNT ppt range	[58]
Hammerhead	R = 200 µm; h = 10 µm	IP	370 kHz	2850	0.01	220	PIB 1.5 µm	m-Xylene 0.35 ppm	[111]
Cantilever	L = 2.5 µm; b = 0.8 µm; h = 1.3 µm	OOP	8–10 MHz	200	1.5	1645[c)]	DKAP 10 nm	DIMP 0.6 ppb	[29]
Disk resonator	R = 170 µm; h = 7 µm	IP	408 kHz	2600	0.01	310	PIB 2.6 µm	m-Xylene 0.6 ppm	[102]
CMUT	R = 9 µm; h = 0.5 µm	OOP	18 MHz	140 (parallel)		2840	PIB 50 nm	DMMP 56 ppb	[108]
Contour mode	b = 20 µm; h = 250 nm	Contour	180 MHz	1000	2.39	6135	DNA	DNT 0.7 ppb	[115]
I-Shape resonator	L = 80 µm; h = 5 µm	IP	15.5 MHz	4400	20	435	PGA 1.5 µm	Toluene 5 ppm	[96]

Polymers: PEUT, polyetherurethane; PIB, polyisobutylene; PGA, polyglycolic acid. TSM, thickness-shear mode; OOP, out-of-plane mode; IP, in-plane mode.
a) The sensing film thickness was calculated from the provided mass.
b) The sensing film was estimated to extend half the length of the cantilever.
c) An accurate density for (DKAP) modified fluorinated polyol could not be found. It was assumed to be 1 g cm⁻³ for calculation purposes.

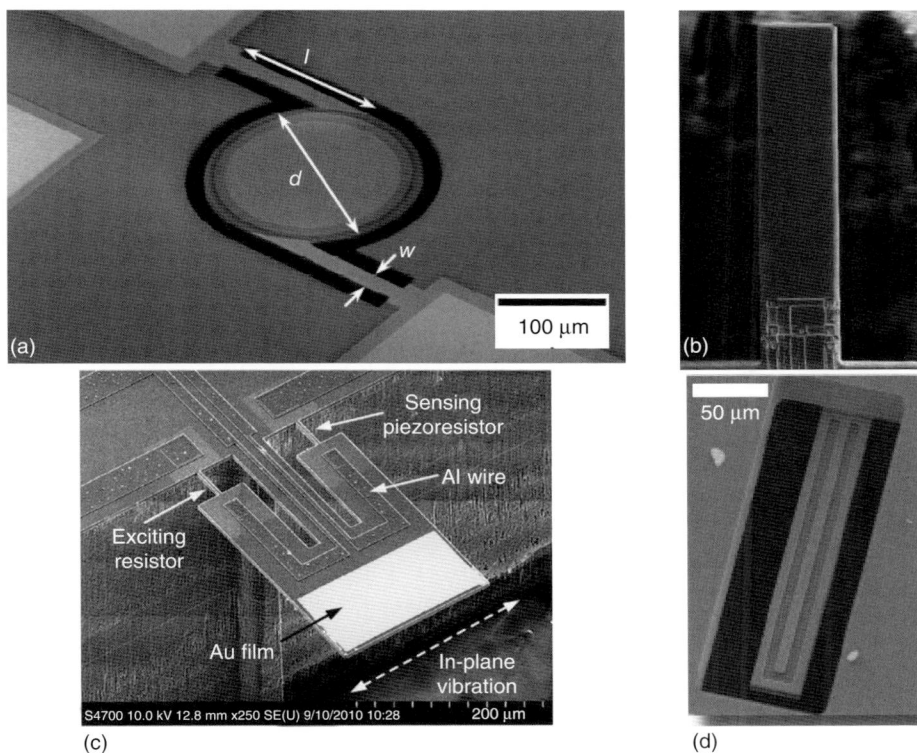

Figure 15.7 Examples of resonant microstructures used for liquid-phase chemical sensing: (a) thermally actuated microdisk (© 2012 IEEE, Reprinted, with permission, from [117]), (b) thermally actuated cantilever vibrating in in-plane flexural mode [100], (c) thermally actuated cantilever (Reprinted from [118], Copyright 2011, with permission from Elsevier), and (d) cantilever with embedded fluid channel (Reprinted by permission from Macmillan Publishers Ltd: Nature [119], copyright 2007).

15.8.1
Cantilevers

A variety of different cantilever designs and sensing systems have been employed in order to achieve low detection limits in the analysis of liquid-phase analytes. Some systems use entirely on-chip excitation and detection schemes such as piezoelectric, capacitive, or piezoresistive transducers. Other investigators have employed systems that utilize on-chip cantilevers with off-chip lasers or other components in order to measure the deflection of the cantilever. For embedded or other applications outside the laboratory the former is preferred.

As with many resonator technologies, the dynamic response of resonant cantilevers in liquid is severely damped leading to lower sensor resolution. Consequently, a sensing strategy that is sometimes chosen is to measure a baseline sensor output (i.e., resonant frequency, piezoresistor resistance, etc.);

subsequently, expose the sensor surface to liquid-phase analytes, which are absorbed by a sensitive layer or coating applied to the beam; and, finally, dry and re-measure the device's response [3, 120]. This strategy is problematic in practical applications because a washing step is required, which necessitates additional reagents, and also during the drying process particles can contaminate the device surface, leading to measurement errors. Drying the device is also not appropriate for real-time or *in-situ* measurements. Despite these issues, this technique has been used to achieve nanogram per milliliter range limits-of-detection for cancer biomarkers [121].

Silicon cantilevers have the advantage of ease of integration with CMOS technologies, and can be readily fabricated using the same tools and processes as integrated circuits. Several investigators have created silicon cantilever platforms for operation in liquid. Vancura *et al.* [10] investigated cantilevers operated in their first out-of-plane flexural mode as liquid-phase resonant chemical sensors. These devices are CMOS-compatible and use a magnetic excitation mechanism and piezoresistive detection scheme to achieve parts per million range LOD for volatile organic compounds in water [10]. A second example uses cantilevers with integrated resistors for thermal actuation and piezoresistive detection to selectively excite and sense the beam's fundamental in-plane mode resonance. These cantilevers achieved a detection limit in the 50–100 ppb range for volatile organics in water [27]. Tao *et al.* [118] investigated a variation of the in-plane sensing approach in which the structure holding the piezoresistors for deflection sensing is made significantly thinner in an attempt to improve sensor performance. These devices were able to detect *E. Coli* and mercury ions with a sensitivity of $5.5 \, \text{pg Hz}^{-1}$.

A sensing scheme that uses direct cantilever operation in a fluid, but utilizes an optical instead of integrated read-out has been demonstrated to detect nanogram per milliliter quantities of protein in fluid [67]. In this technique, a cantilever beam is piezoelectrically actuated into resonance and the resonant frequency is measured using a laser. This method gives good detection limits, but requires additional equipment that might limit its applicability for point-of-care settings. While optical detection schemes require more components than an integrated (i.e., piezoresistive or capacitive) read-out, they have been implemented with discrete components to give single-digit nanogram per milliliter detection limits for detection of biomarkers for hepatitis [122].

15.8.2
Microdisk Resonators

Microdisk resonators supported by a clamped–clamped beam and utilizing thermal actuation and piezoresistive detection have been used for liquid-phase chemical sensing [36]. Similar microdisk resonators with thermal excitation have achieved quality factors on the order of a few hundred for liquid-phase operation [35]. More recently, these resonators have been demonstrated using piezoelectric materials, yielding quality factors in water on the order of 50–70, while reducing the power consumption compared to thermally excited devices [117].

15.8.3
Acoustic Wave Sensors

Extensive research exists on creating miniaturized sensors utilizing acoustic wave devices for operation in both air and liquid [89]. The devices themselves can be fabricated using different methods: (i) from piezoelectric crystals and combined with oscillator circuitry [123], (ii) by deposition of piezoelectric materials onto CMOS oscillator circuitry [124], or (iii) by fabrication of extremely thin membranes in a post-CMOS process in the case of CMUTs (capacitive micromachined ultrasonic transducers) [125]. In a typical excitation and detection configuration, interdigitated electrodes excite and detect an acoustic wave in the piezoelectric material. The transit time of the generated wave is altered by the binding of analytes onto the device, which alters the resonant frequency of the sensor (the resonant frequency can be tracked with an oscillator circuit) [126].

One of the most common acoustic wave devices are SAW sensors. Surface acoustic waves propagate along the surface of the device, and thus, are highly sensitive to changes that occur when an analyte attaches to the device surface. SAW devices have been extensively investigated as biochemical sensors [127]. For liquid operation, the typically used Raleigh waves are too severely damped. The damping is a result of the dielectric constants of the acoustic materials in the device being significantly lower than that of water, causing the electric fields needed for wave excitation to be drawn into solution, and also a result of mode conversion due to the shear vertical component of the wave. Two possible remedies to this problem are to use piezoelectric materials with a higher dielectric constant, or to use Love-wave devices (with only a shear horizontal component of the acoustic wave) where the acoustic energy can be confined to a coating on the device's surface [127]. Love-wave chemical sensors have been used to detect volatile organics in water down to the low parts per billion range [123]. In addition, arrays of SAW devices have been integrated with CMOS circuitry to create sensor arrays for the detection of cancer biomarkers in the nanogram per milliliter range [128]. Flexural plate wave devices are a third example of acoustic wave devices that have been investigated for liquid-phase applications [129]. These devices use a piezoelectric material deposited on a thin membrane in order to excite the membrane into resonance. An example application is the detection of IgE antibodies for applications in allergy diagnosis and management [130].

Acoustic wave devices offer low detection limits for compounds in water. On the other hand, they often require the integration of nonstandard materials, such as piezoelectric materials, increasing fabrication costs; and they can be highly sensitive to properties of the surrounding medium (i.e., viscosity of the solution in which the devices are operated) [97, 129]. In addition, the high frequencies of operation can lead to interferences requiring shielding or other techniques, making system design more complex [26]. Despite these challenges, acoustic wave devices have been used to create robust sensors, and work has even been done in integrating them into handheld platforms [131].

Table 15.3 Comparison of MEMS-based and QCM-based resonant chemical gas sensor performances.

Device	Dimensions	Vibration mode	Frequency	Q in water f_{min} (Hz)	Mechanical sensitivity G_M (cm² g⁻¹)	Sensing film and thickness	Analyte and LOD	References
Cantilever	$L = 120\,\mu m$ $b = 40\,\mu m$ $h = 5\,\mu m$	IP	406 kHz	14 103	68	1 µg Mesoporous silica	Hg[a] 100 ppb	[118]
Cantilever	$L = 400\,\mu m$ $b = 75\,\mu m$ $h = 7.5\,\mu m$ [b]	IP	426 kHz	40 1.1	173	EPCO 2 µm (both sides)	VOCs 40–100 ppb	[27]
Cantilever	$L = 150\,\mu m$ $b = 140\,\mu m$ $h = 5\,\mu m$ [b]	OOP	200 kHz	10 3	222	PECH 0.8 µm	VOCs 1–10 ppm	[10]
Disk	$D = 200\,\mu m$ $h = 15\,\mu m$	IP	2.36 MHz	27 —	143	Thiols/ssDNA Monolayer	DNA	[117]
Cantilever w/Channel	$L = 200\,\mu m$ $b = 33\,\mu m$ $h = 7\,\mu m$	OOP	209.6 kHz	15 000 0.002	307[c]	IgG	Single bio-molecule	[119]
SH-SAW	—	Shear	103 MHz	30[a]	—	PECH 0.8 µm	VOCs 10 ppb	[123]
QCM	$D = 8\,mm$	TSM	10 MHz	10[d]	—	Cavitands (hydrophic cavity)	Toluene 9 ppm	[136]
Cantilever	$L = 150\,\mu m$ $b = 50\,\mu m$	OOP	≈30 kHz	—	324	*Anti*-PSA	PSA 1 ng ml⁻¹	[137]

EPCO, ethylene propylene copolymer; PECH, polyepichlorohydrin; ssDNA, single-stranded DNA; VOCs, volatile organic compounds such as toluene and xylene; PSA, prostate specific antigen; TSM, thickness-shear mode; OOP, out-of-plane mode; IP, in-plane mode.

a) Signal noise level, not Q.
b) Thickness of silicon layer only. The resonators are coated with additional passivation layers.
c) Calculation takes into account only the silicon layer thickness.
d) Frequency stability.

15.8.4
Resonators with Encapsulated Channels

A promising strategy to reduce damping for liquid-phase sensing using cantilevers has been presented in [119]. In this work, a fluid channel is routed through the interior of the cantilever itself. This allows the beam to operate in air or even vacuum giving quality factors as high as 15 000, mitigating the damping problem in fluids. This strategy makes extremely low LOD possible. Quantities as small as a single cell/particle can be measured using this approach. Recently, an integrated piezoresistive read-out has been implemented for the embedded fluid channel cantilevers [132]. Although the embedded fluid channel resonator offers single-molecule resolution, it requires a relatively complicated fabrication process and can only tolerate very low fluid flow-rates. More details on sensors with embedded channels can be found in Chapter 11.

Along the same lines, electrostatically actuated flexural plate resonators have been developed that allow for fluid to be routed through the resonator while it is operated in vacuum. The flexural plate resonators are fabricated using a wafer-bonding process and can be packaged on-chip [133]. Alternatively, a system has been developed that allows a cantilever resonator to be placed in a flow cell such that only one side of the cantilever is immersed, yielding Q-factors in the hundreds [134]. It is expected that the increased Q-factor from this approach will improve sensor resolution. In addition, rotational disk resonators similar to those discussed above have been demonstrated with encapsulated channels for liquid-phase sensing [135].

Table 15.3 summarizes the performance of select MEMS-based resonant sensors for liquid-phase chemical sensing. Again, rather than providing a comprehensive overview, the examples are chosen to represent a wide range of MEMS-based resonator designs with QCM data as a benchmark.

References

1. Waggoner, P.S. and Craighead, H.G. (2007) Micro- and nanomechanical sensors for environmental, chemical, and biological detection. *Lab Chip*, 7, 1238–1255.
2. Goeders, K.M., Colton, J.S., and Bottomley, L.A. (2008) Microcantilevers: sensing chemical interactions via mechanical motion. *Chem. Rev.*, 108, 522–542.
3. Arlett, J.L., Myers, E.B., and Roukes, M.L. (2011) Comparative advantages of mechanical biosensors. *Nat. Nanotechnol.*, 6, 203–215.
4. Janata, J. and Bezegh, A. (1988) Chemical sensors. *Anal. Chem.*, 60, 62R–74R.
5. Boisen, A., Thaysen, J., Jensenius, H., and Hansen, O. (2000) Environmental sensors based on micromachined cantilevers with integrated read-out. *Ultramicroscopy*, 82, 11–16.
6. Rogers, B., Manning, L., Jones, M., Sulchek, T., Murray, K., Beneschott, B., Adams, J.D., Hu, Z., Thundat, T., Cavazos, H., and Minne, S.C. (2003) Mercury vapor detection with a self-sensing, resonating piezoelectric

cantilever. *Rev. Sci. Instrum.*, **74**, 4899–4901.

7. Thundat, T., Wachter, E.A., Sharp, S.L., and Warmack, R.J. (1995) Detection of mercury vapor using resonating microcantilevers. *Appl. Phys. Lett.*, **66**, 1695–1697.

8. Baselt, D.R., Fruhberger, B., Klaassen, E., Cemalovic, S., Britton, C.L. Jr., Patel, S.V., Mlsna, T.E., McCorkle, D., and Warmack, B. (2003) Design and performance of a microcantilever-based hydrogen sensor. *Sens. Actuators B*, **88**, 120–131.

9. Sampath, U., Heinrich, S.M., Josse, F., Lochori, F., Dufour, I., and Rebiere, D. (2006) Study of viscoelastic effect on the frequency shift of microcantilever chemical sensors. *IEEE Trans. Ultrason. Ferroelectr. Freq. Control*, **53**, 2166–2173.

10. Vancura, C., Li, Y., Lichtenberg, J., Kirstein, K.-U., Hierlemann, A., and Josse, F. (2007) Liquid-phase chemical and biochemical detection using fully integrated magnetically actuated complementary metal oxide semiconductor resonant cantilever sensor systems. *Anal. Chem.*, **79**, 1646–1654.

11. Kanwal, R.P. (1955) Rotatory and longitudinal oscillations of axi-symmetric bodies in a viscous fluid. *Q. J. Mech. Appl. Math.*, **8**, 146–163.

12. Basak, S., Raman, A., and Garimella, S.V. (2006) Hydrodynamic loading of microcantilevers vibrating in viscous fluids. *J. Appl. Phys.*, **99**, 114906–114910.

13. Lindholm, U., Kana, D., Chu, W.-H., and Abramson, H. (1965) Elastic vibration characteristics of cantilever plates in water. *J. Ship Res.*, **9**, 11–22.

14. Cox, R., Josse, F., Wenzel, M.J., Heinrich, S.M., and Dufour, I. (2008) Generalized model of resonant polymer-coated microcantilevers in viscous liquid media. *Anal. Chem.*, **80**, 5760–5767.

15. Sader, J.E. (1998) Frequency response of cantilever beams immersed in viscous fluids with applications to the atomic force microscope. *J. Appl. Phys.*, **84**, 64–76.

16. Dufour, I., Heinrich, S.M., and Josse, F. (2007) Theoretical analysis of strong-axis bending mode vibrations of resonant microcantilever (bio)chemical sensors in gas or liquid phase. *J. Micro-electromech. Syst.*, **16**, 44–49.

17. Bargatin, I., Kozinsky, I., and Roukes, M.L. (2007) Efficient electrothermal actuation of multiple modes of high-frequency nanoelectromechanical resonators. *Appl. Phys. Lett.*, **90**, 093116-3.

18. Brumley, D.R., Willcox, M., and Sader, J.E. (2010) Oscillation of cylinders of rectangular cross section immersed in fluid. *Phys. Fluids*, **22**, 052001–052015.

19. Tuck, E.O. (1969) Calculation of unsteady flows due to small motions of cylinders in a viscous fluid. *J. Eng. Math.*, **3**, 29–44.

20. Cox, R., Josse, F., Heinrich, S.M., Brand, O., and Dufour, I. (2012) Characteristics of laterally vibrating resonant microcantilevers in viscous liquid media. *J. Appl. Phys.*, **111**, 014907–014914.

21. Sampath, U. (2005) Analytical modeling of microcantilever-based dynamic microsensors. Master of Science. Marquette University.

22. Sampath, U., Heinrich, S.M., Josse, F., Lochon, F., Dufour, I., and Rebiere, D. (2005) Study of viscoelastic effect on the frequency shift of microcantilever chemical sensors. Frequency Control Symposium and Exposition, 2005. Proceedings of the 2005 IEEE International, 2005, p. 8 pp.

23. Dufour, I., Lochon, F., Heinrich, S.M., Josse, F., and Rebiere, D. (2007) Effect of coating viscoelasticity on quality factor and limit of detection of microcantilever chemical sensors. *IEEE Sens. J.*, **7**, 230–236.

24. Fadel, L., Dufour, I., Lochon, F., and Francais, O. (2004) Signal-to-noise ratio of resonant microcantilever type chemical sensors as a function of resonant frequency and quality factor. *Sens. Actuators B*, **102**, 73–77.

25. Razavi, B. (1996) A study of phase noise in CMOS oscillators. *IEEE J. Solid-State Circuits*, **31**, 331–343.

26. Lange, D., Hagleitner, C., Hierlemann, A., Brand, O., and Baltes, H. (2002) Complementary metal oxide semiconductor cantilever arrays on a single chip: mass-sensitive detection of volatile organic compounds. *Anal. Chem.*, **74**, 3084–3095.

27. Beardslee, L.A., Demirci, K.S., Luzinova, Y., Mizaikoff, B., Heinrich, S.M., Josse, F., and Brand, O. (2010) Liquid-phase chemical sensing using lateral mode resonant cantilevers. *Anal. Chem.*, **82**, 7542–7549.

28. Beardslee, L.A., Truax, S., Su, J.J., Heinrich, S.M., Josse, F., and Brand, O. (2011) On the relative sensitivity of mass-sensitive chemical microsensors. 2011 16th International Solid-State Sensors, Actuators and Microsystems Conference, TRANSDUCERS'11, Beijing, China, June 5–9, 2011, pp. 1112-1115.

29. Bargatin, I., Myers, E.B., Aldridge, J.S., Marcoux, C., Brianceau, P., Duraffourg, L., Colinet, E., Hentz, S., Andreucci, P., and Roukes, M.L. (2012) Large-scale integration of nanoelectromechanical systems for gas sensing applications. *Nano Lett.*, **12**, 1269–1274.

30. Li, M., Myers, E.B., Tang, H.X., Aldridge, S.J., McCaig, H.C., Whiting, J.J., Simonson, R.J., Lewis, N.S., and Roukes, M.L. (2010) Nanoelectromechanical resonator arrays for ultrafast, gas-phase chromatographic chemical analysis. *Nano Lett.*, **10**, 3899–3903.

31. Ballantine, D.S., White, R.M., Martin, S.J., Ricco, A.J., Zellers, E.T., Frye, G.C., and Wohltjen, H. (1997) *Acoustic Wave Sensors, Theory, Design, and Physico-Chemical Applications*, Academic Press, San Diego, CA.

32. Beardslee, L.A., Josse, F., Heinrich, S.M., Dufour, I., and Brand, O. (2012) Geometrical considerations for the design of liquid-phase biochemical sensors using a cantilever's fundamental in-plane mode. *Sens. Actuators B*, **164**, 7–14.

33. Xiaoyuan, X., Zhixiang, Z., and Xinxin, L. (2008) A Latin-cross-shaped integrated resonant cantilever with second torsion-mode resonance for ultra-resoluble bio-mass sensing. *J. Micromech. Microeng.*, **18**, 035028.

34. Jin, D., Xinxin, L., Bao, H., Zhixiang, Z., Wang, Y., Haitao, Y., and Guomin, Z. (2007) Integrated cantilever sensors with a torsional resonance mode for ultraresoluble on-the-spot bio/chemical detection. *Appl. Phys. Lett.*, **90**, 041901–041901-3.

35. Rahafrooz, A. and Pourkamali, S. (2010) Rotational mode disk resonators for high-Q operation in liquid. Presented at the IEEE Sensors Conference, Waikoloa, HI.

36. Seo, J.H. and Brand, O. (2008) High Q-factor in-plane mode resonant mircosensor platform for gaseous/liquid environment. *J. Microelectromech. Syst.*, **17**, 483–493.

37. Beardslee, L.A., Truax, S., Lee, J.H., Pavlidis, S., Hesketh, P., Hansen, K.M., Kramer, R., and Brand, O. (2011) Selectivity enhancement strategy for cantilever-based gas-phase VOC sensors through use of peptide-functionalized carbon nanotubes. Presented at the 24th IEEE Conference on Microelectromechanical Systems, Cancun, Mexico, 2011.

38. Allan, D.W. and Barnes, J.A. (1981) A modified Allan variance with increased oscillator characterization ability. Presented at the 35th Annual Frequency Control Symposium, Philadelphia, PA, 1981.

39. Demirci, K.S., Beardslee, L.A., Truax, S., Su, J.J., and Brand, O. (2012) Integrated silicon-based chemical microsystem for portable sensing applications. *Sens. Actuators B*, **180**, 50–59

40. Seo, J.H., Demirci, K.S., Truax, S., Beardslee, L.A., and Brand, O. (2008) Tracking microresonator Q-factor in closed-loop operation. Presented at the 2008 Hilton Head Solid-State Sensors, Actuators, and Microsystems Workshop, Hilton Head, 2008.

41. Melamud, R., Chandorkar, S.A., Bongsang, K., Hyung Kyu, L., Salvia, J.C., Bahl, G., Hopcroft, M.A., and Kenny, T.W. (2009) Temperature-insensitive composite micromechanical

resonators. *J. Microelectromech. Syst.*, **18**, 1409–1419.

42. Serrano, G., Chang, H., and Zellers, E.T. (2009) A micro gas chromatograph for high-speed determinations of explosive vapors. Presented at the Transducers 15th International Conference Solid State Actuators and Microsystems, Denver, CO, 2009.

43. Kim, S.K., Chang, H., and Zellers, E.T. (2011) Microfabricated gas chromatograph for the selective determination of trichloroethylene vapor at sub-parts-per-billion concentrations in complex mixtures. *Anal. Chem.*, **83**, 7198–7206.

44. Hagleitner, C., Hierlemann, A., Lange, D., Kummer, A., Kerness, N., Brand, O., and Baltes, H. (2001) Smart single-chip gas sensor microsystem. *Nature*, **414**, 293–296.

45. Hierlemann, A. and Gutierrez-Osuna, R. (2008) Higher-order chemical sensing. *Chem. Rev.*, **108**, 563–613.

46. Jin, C., Kurzawski, P., Hierlemann, A., and Zellers, E.T. (2008) Evaluation of multitransducer arrays for the determination of organic vapor mixtures. *Anal. Chem.*, **80**, 227–236.

47. Su, J., Carron, C., Truax, S., Demirci, K.S., Beardslee, L.A., and Brand, O. (2011) Assessing polymer sorption kinetics using micromachined resonators. Solid-State Sensors, Actuators and Microsystems Conference (TRANSDUCERS), 2011 16th International, 2011, pp. 1420–1423.

48. Semancik, S. and Cavicchi, R. (1998) Kinetically controlled chemical sensing using micromachined structures. *Acc. Chem. Res.*, **31**, 279–287.

49. Squires, T.M., Messinger, R.J., and Manalis, S.R. (2008) Making it stick: convection, reaction and diffusion in surface-based biosensors. *Nat. Biotechnol.*, **26**, 417–426.

50. Hierlemann, A. (2008) *CMOS—MEMS*, Wiley-VCH Verlag GmbH, pp. 335–390.

51. Hierlemann, A., Ricco, A.J., Bodenhöfer, K., Dominik, A., and Göpel, W. (2000) Conferring selectivity to chemical sensors via polymer side-chain selection: thermodynamics of vapor sorption by a set of polysiloxanes

on thickness-shear mode resonators. *Anal. Chem.*, **72**, 3696–3708.

52. Hierlemann, A., Zellers, E.T., and Ricco, A.J. (2001) Use of linear solvation energy relationships for modeling responses from polymer-coated acoustic-wave vapor sensors. *Anal. Chem.*, **73**, 3458–3466.

53. Kryscio, D.R. and Peppas, N.A. (2012) Critical review and perspective of macromolecularly imprinted polymers. *Acta Biomater.*, **8**, 461–473.

54. Dickert, F.L., Lieberzeit, P., Miarecka, S.G., Mann, K.J., Hayden, O., and Palfinger, C. (2004) Synthetic receptors for chemical sensors—subnano- and micrometre patterning by imprinting techniques. *Biosens. Bioelectron.*, **20**, 1040–1044.

55. Vasapollo, G., Sole, R.D., Mergola, L., Lazzoi, M.R., Scardino, A., Scorrano, S., and Mele, G. (2011) Molecularly imprinted polymers: present and future prospective. *Int. J. Mol. Sci.*, **12**, 5908–5945.

56. Kreno, L.E., Leong, K., Farha, O.K., Allendorf, M., Van Duyne, R.P., and Hupp, J.T. (2012) Metal–organic framework materials as chemical sensors. *Chem. Rev.*, **112**, 1105–1125.

57. Achmann, S., Hagen, G., Kita, J., Malkowsky, I., Kiener, C., and Moos, R. (2009) Metal-organic frameworks for sensing applications in the gas phase. *Sensors*, **9**, 1574–1589.

58. Xu, P., Yu, H., and Li, X. (2011) Functionalized mesoporous silica for microgravimetric sensing of trace chemical vapors. *Anal. Chem.*, **83**, 3448–3454.

59. Venkatasubramanian, A., Lee, J.-H., Stavila, V., Robinson, A., Allendorf, M.D., and Hesketh, P.J. (2012) MOF @ MEMS: design optimization for high sensitivity chemical detection. *Sens. Actuators B*, **168**, 256–262.

60. Liu, B. (2012) Metal-organic framework-based devices: separation and sensors. *J. Mater. Chem.*, **22**, 10094–10101.

61. Cho, E.J., Lee, J.-W., and Ellington, A.D. (2009) Applications of aptamers as sensors. *Annu. Rev. Anal. Chem.*, **2**, 241–264.

62. Liu, S., Zhang, X., Luo, W., Wang, Z., Guo, X., Steigerwald, M.L., and Fang, X. (2011) Single-molecule detection of proteins using aptamer-functionalized molecular electronic devices. *Angew. Chem. Int. Ed.*, **50**, 2496–2502.

63. Famulok, M. and Mayer, G. (2011) Aptamer modules as sensors and detectors. *Acc. Chem. Res.*, **44**, 1349–1358.

64. McKeague, M. and DeRosa, M.C. (2012) Challenges and opportunities for small molecule aptamer development. *J. Nucleic Acids*, **2012**, 20.

65. Ni, X., Castanares, M., Mukherjee, A., and Lupold, S.E. (2011) Nucleic acid aptamers: clinical applications and promising new horizons. *Curr. Med. Chem.*, **18**, 4206–4214.

66. Love, J.C., Estroff, L.A., Kriebel, J.K., Nuzzo, R.G., and Whitesides, G.M. (2005) Self-assembled monolayers of thiolates on metals as a form of nanotechnology. *Chem. Rev.*, **105**, 1103–1169.

67. Ricciardi, C., Canavese, G., Castagna, R., Ferrante, I., Ricci, A., Luigi Marasso, S., Napione, L., and Bussolino, F. (2010) Integration of microfluidic and cantilever technology for biosensing applications in liquid environment. *Biosens. Bioelectron.*, **26**, 1565–1570.

68. Briand, E., Salmain, M., Compere, C., and Pradier, C.-M. (2006) Immobilization of Protein A on SAMs for the elaboration of immunosensors. *Colloids Surf. B Biointerfaces*, **53**, 215–224.

69. Jang, L.-S. and Keng, H.-K. (2008) Modified fabrication process of protein chips using a short-chain self-assembled monolayer. *Biomed. Microdevices*, **10**, 203–211.

70. Yu-Chia, T., Yi-Wen, Y., Woo-Hu, T., and Tsong-Rong, Y. (2008) Side-polished fiber immunosensor based on surface plasmon resonance for detection of Legionella pneumophila. *Proc. SPIE*, **6852**, 685206-1.

71. Towarfe, G.K., Composto, R.J., Shapiro, I.M., and Ducheyne, P. (2006) Nucleation and growth of calcium phosphate on amine-, carboxyl- and hyrdroxyl-silane self-assembled monolayers. *Biomaterials*, **27**, 631–642.

72. Tlili, A., Ali Jarboui, M., Abdelghani, A., Fathallah, D.M., and Maaref, M.A. (2005) A novel silicon nitride biosensor for specific antibody-antigen interaction. *Mater. Sci. Eng. C*, **25**, 490–495.

73. Velten, T., Ruf, H.H., Barrow, D., Aspragathos, N., Lazarou, P., Jung, E., Malek, C.K., Richter, M., Kruckow, J., and Wackerle, M. (2005) Packaging of bio-MEMS: strategies, technologies, and applications. *IEEE Trans. Adv. Packag.*, **28**, 533–546.

74. Smith, R.L. and Collins, S.D. (1988) Micromachined packaging for chemical microsensors. *IEEE Trans. Electron Devices*, **35**, 787–792.

75. Brand, O. (2008) in *Comprehensive Microsystems*, vol. 1 (eds Y. Gianchandani, O. Tabata, and H. Zappe), Elsevier, pp. 431–463.

76. Figaro *http://www.figarosensor.com/*

77. Sensirion *http://www.sensirion.com/en/technology/humidity/*

78. Song, W.H. (2005) Packaging techniques for CMOS-based chemical and biochemical microsensors. Doctor of Technical Sciences. Swiss Federal Institute of Techonology, Zurich.

79. Tse, L.A., Noh, H.-S., Seals, L., Gole, J., Rosen, D.W., and Hesketh, P.J. (2001) Fabrication of chemical sensor packaging with stereolithography. Presented at the Chemical Sensing, Olfaction, and Electronic Noses, 2001.

80. Joseph, P.J., Monajemi, P., Ayazi, F., and Kohl, P.A. (2007) Wafer-level packaging of micromechanical resonators. *IEEE Trans. Adv. Packag.*, **30**, 19–26.

81. Hennessy, R.G., Shulaker, M.M., Melamud, R., Klejwa, N., Chandorkar, S., Kim, B.S., Provine, J., Kenny, T.W., and Howe, R.T. (2010) Vacuum encapsulated resonators for humidity measurement. Hilton Head Solid-State Sensors, Actuators and Microsystems Workshop, Hilton Head, SC, 2010.

82. Lin, A.T.H., Jize, Y., and Seshia, A.A. (2012) Electrically addressed dual resonator sensing platform for biochemical detection. *J. Microelectromech. Syst.*, **21**, 34–43.

83. King, W.H. Jr., (1964) Piezoelectric sorption detector. *Anal. Chem.*, **36**, 1735–1738.

84. Grate, J.W., Martin, S.J., and White, R.M. (1993) Acoustic wave microsensors. *Anal. Chem.*, **65**, 940A–948A.

85. Grate, J.W., Martin, S.J., and White, R.M. (1993) Acoustic wave microsensors. Part II. *Anal. Chem.*, **65**, 987A–996A.

86. Howe, R.T. and Muller, R.S. (1986) Resonant-microbridge vapor sensor. *IEEE Trans. Electron Devices*, **33**, 499–506.

87. Baltes, H., Boltshauser, T., Brand, O., Lenggenhager, R., and Jaeggi, D. (1992) Silicon microsensors and microstructures. 1992 IEEE International Symposium on Circuits and Systems Proceedings, ISCAS '92, Vol. 4, pp. 1820–1823.

88. Bellan, L.M., Wu, D., and Langer, R.S. (2011) Current trends in nanobiosensor technology. *Wiley Interdiscip. Rev. Nanomed. Nanobiotechnol.*, **3**, 229–246.

89. Lucklum, R. and Hauptmann, P. (2006) Acoustic sensors the challenge behind microgravimetry. *Anal.Bioanal. Chem.*, **384**, 667–682.

90. Datar, R., Kim, S., Jeon, S., Hesketh, P., Manalis, S., Boisen, A., and Thundat, T. (2009) Cantilevers: nanomechanical tools for diagnostics. *MRS Bull.*, **34**, 449–454.

91. Chambers, S.T., Bhandari, S., Scott-Thomas, A., and Syhre, M. (2011) Novel diagnostics: progress toward a breath test for invasive Aspergillus fumigatus. *Med. Mycol.*, **49**, S54–S61.

92. van de Kant, K.D., van der Sande, L.J., Jöbsis, Q., van Schayck, O.C., and Dompeling, E. (2012) Clinical use of exhaled volatile organic compounds in pulmonary diseases: a systematic review. *Respir. Res.*, **13**, 117.

93. Gooding, J.J. (2006) Biosensor technology for detecting biological warfare agents: recent progress and future trends. *Anal. Chim. Acta*, **559**, 137–151.

94. Lieberzeit, P.A. and Dickert, F.L. (2007) Sensor technology and its application in environmental analysis. *Anal. Bioanal. Chem.*, **387**, 237–247.

95. Vancura, C., Ruegg, M., Li, Y., Hagleitner, C., and Hierlemann, A.

(2005) Magnetically actuated complementary metal oxide semiconductor resonant cantilever gas sensor systems. *Anal. Chem.*, **77**, 2690–2699.

96. Hajjam, A. and Pourkamali, S. (2012) Fabrication and characterization of MEMS-based resonant organic gas sensors. *IEEE Sens. J.*, **12**, 1958–1964.

97. Haitao, Y., Pengcheng, X., Xiaoyuan, X., Dong-Weon, L., and Xinxin, L. (2012) Micro-/nanocombined gas sensors with functionalized mesoporous thin film self-assembled in batches onto resonant cantilevers. *IEEE Trans. Ind. Electron.*, **59**, 4881–4887.

98. Naeli, K. and Brand, O. (2009) Dimensional considerations in achieving large quality factors for resonant silicon cantilevers in air. *J. Appl. Phys.*, **105**, 014908.

99. Beardslee, L.A., Lehmann, J., Carron, C., Su, J.J., Josse, F., Dufour, I., and Brand, O. (2012) Thermally actuated silicon tuning fork resonators for sensing applications in air. 2012 IEEE 25th International Conference on Micro Electro Mechanical Systems (MEMS), Piscataway, NJ, January 29–February 2, 2012, pp. 607-610.

100. Beardslee, L.A., Addous, A.M., Heinrich, S.M., Josse, F., Dufour, I., and Brand, O. (2010) Thermal excitation and piezoresistive detection of cantilever in-plane resonance modes for sensing applications. *J. Microelectromech. Syst.*, **19**, 1015–1017.

101. Bedair, S.S. and Fedder, G.K. (2004) CMOS MEMS oscillator for gas chemical detection. IEEE Sensors 2004, Vienna, Austria, October 24–27, 2004, pp. 955–958.

102. Truax, S., Demirci, K.S., Beardslee, L.A., Luzinova, Y., Hierlemann, A., Mizaikoff, B., and Brand, O. (2011) Mass-sensitive detection of gas-phase volatile organics using disk microresonators. *Anal. Chem.*, **83**, 3305–3311.

103. Kuang, Z., Kim, S.N., Crookes-Goodson, W.J., Farmer, B.L., and Naik, R.R. (2010) Biomimetic chemosensor: designing peptide recognition elements for surface functionalization of carbon nanotube

field effect transistors. *ACS Nano*, **4**, 452–458.

104. Liviu, N. and Ayela, C. (2007) Micro-machined piezoelectric membranes with high nominal quality factors in newtonian liquid media: a Lamb's model validation at the microscale. *Sens. Actuators B*, **123**, 860–868.

105. Robinson, A.L., Stavila, V., Zeitler, T.R., White, M.I., Thornberg, S.M., Greathouse, J.A., and Allendorf, M.D. (2012) Ultrasensitive humidity detection using metal–organic framework-coated microsensors. *Anal. Chem.*, **84**, 7043–7051.

106. Hwang, K.S., Lee, M.H., Lee, J., Yeo, W.-S., Lee, J.H., Kim, K.-M., Kang, J.Y., and Kim, T.S. (2011) Peptide receptor-based selective dinitrotoluene detection using a microcantilever sensor. *Biosens. Bioelectron.*, **30**, 249–254.

107. Zuniga, C., Rinaldi, M., Khamis, S.M., Johnson, A.T., and Piazza, G. (2009) Nanoenabled microelectromechanical sensor for volatile organic chemical detection. *Appl. Phys. Lett.*, **94**, 223122.

108. Park, K.K., Lee, H., Kupnik, M., Oralkan, O., Ramseyer, J.-P., Lang, H.P., Hegner, M., Gerber, C., and Khuri-Yakub, B.T. (2011) Capacitive micromachined ultrasonic transducer (CMUT) as a chemical sensor for DMMP detection. *Sens. Actuators B*, **160**, 1120–1127.

109. Loui, A., Sirbuly, D.J., Elhadj, S., McCall, S.K., Hart, B.R., and Ratto, T.V. (2010) Detection and discrimination of pure gases and binary mixtures using a dual-modality microcantilever sensor. *Sens. Actuators, A: Phys.*, **159**, 58–63.

110. Kummer, A.M., Burg, T.P., and Hierlemann, A. (2006) Transient signal analysis using complementary metal oxide semiconductor capacitive chemical microsensors. *Anal. Chem.*, **78**, 279–290.

111. Demirci, K.S., Beardslee, L.A., Truax, S., Su, J.-J., and Brand, O. (2013) Integrated silicon-based chemical microsystem for portable sensing applications. *Sens. Actuators B*, **180**, 50–59.

112. Yue, L., Vancura, C., Kirstein, K.U., Lichtenberg, J., and Hierlemann, A. (2008) Monolithic resonant-cantilever-based CMOS microsystem for biochemical sensing. *IEEE Trans. Circuits Syst. Regul. Pap.*, **55**, 2551–2560.

113. Finot, E., Rouger, V., Markey, L., Seigneuric, R., Nadal, M.H., and Thundat, T. (2012) Visible photother-mal deflection spectroscopy using microcantilevers. *Sens. Actuators B*, **169**, 222–228.

114. Bodenhöfer, K., Hierlemann, A., Noetzel, G., Weimar, U., and Göpel, W. (1996) Performances of mass-sensitive devices for gas sensing: thickness shear mode and surface acoustic wave transducers. *Anal. Chem.*, **68**, 2210–2218.

115. Rinaldi, M., Zuniga, C., and Piazza, G. (2011) SS-DNA functionalized array of ALN Contour-Mode NEMS Resonant Sensors with single CMOS multiplexed oscillator for sub-ppb detection of volatile organic chemicals. 24th IEEE International Conference on Micro Electro Mechanical Systems, MEMS 2011, Cancun, Mexico, January 23–27, 2011, pp. 976–979.

116. Ho, C.K., Robinson, A., Miller, D.R., and Davis, M.J. (2005) Overview of sensors and needs for environmental monitoring. *Sensors*, **5**, 4–37.

117. Mehdizadeh, E., Chapin, J., Gonzales, J., Rahafrooz, A., Abdolvand, R., Purse, B., and Pourkamali, S. (2012) Direct detection of biomolecules in liquid media using piezoelectric rotational mode disk resonators. Sensors, 2012 IEEE, pp. 1–4.

118. Tao, Y., Li, X., Xu, T., Yu, H., Xu, P., Xin, X., and Wei, C. (2011) Resonant cantilever sensors operated in a high-Q in-plane mode for real-time bio/chemical detection in liquids. *Sens. Actuators B*, **157**, 606–614.

119. Burg, T.P., Godin, M., Knudsen, S.M., Shen, W., Carlson, G., Foster, J.S., Babcock, K., and Manalis, S.R. (2007) Weighing of biomolecules, single cells and nanoparticles in fluid. *Nature*, **446**, 11066–11069.

120. Li, X., Yu, H., Gan, X., Xia, X., Xu, P., Li, J., Liu, M., and Li, Y. (2009) Integrated MEMS/NEMS resonant

cantilevers for ultrasensitive biological detection. *J. Sens.*, **2009**, 1–10

121. Liu, Y., Li, X., Zhang, Z., Zuo, G., Cheng, Z., and Yu, H. (2009) Nanogram per milliliter-level immunologic detection of alpha-fetoprotein with integrated rotating-resonance micro-cantilevers for early-stage diagnosis of hetocellular carcinoma. *Biomed. Microdevices*, **11**, 183–191.

122. Urey, H., Timurdogan, E., Ermek, E., Kavakli, I.H., and Alaca, B.E. (2011) MEMS biosensor for parallel and highly sensitive and specific detection of hepatitis. Presented at the 24th IEEE MEMS Conference, Cancun, Mexico, 2011.

123. Li, Z., Jones, Y., Hossenlopp, J., Cernosek, R., and Josse, F. (2005) Analysis of liquid-phase chemical detection using guided shear horizontal-surface acoustic wave sensors. *Anal. Chem.*, **77**, 4595–4603.

124. Tadigadapa, S. and Mateti, K. (2009) Piezoelectric MEMS sensors: state-of-the-art and perspectives. *Meas. Sci. Technol.*, **20**, 92001.

125. Lee, H.J., Park, K.K., Cristman, P., Oralkan, O., Kupnik, M., and Khuri-Yakub, B.T. (2009) A low-noise oscillator based on a multi-membrane CMUT for high sensitivity resonant chemical sensors. Presented at the IEEE MEMS, Serrento, Italy, 2009.

126. Kovacs, G.T.A. (1998) *Micromachined Transducers Sourcebook*, McGraw-Hill, Boston, MA.

127. Lange, K., Rapp, B.E., and Rapp, M. (2008) Surface acoustic wave biosensors: a review. *Anal. Bioanal. Chem.*, **391**, 1509–1519.

128. Tigli, O., Bivona, L., Berg, P., and Zaghloul, M.E. (2010) Fabrication and characterization of a surface-acoustic-wave biosensor in cmos technology for cancer biomarker detection. *IEEE Trans. Biomed. Circuits Syst.*, **4**, 62–73.

129. Weinberg, M., Dube, C.E., Petrovich, A., and Zapta, A.M. (2003) Fluid dampling in resonant flexural plate wave devices. *J. Microelectromech. Syst.*, **12**, 567–576.

130. Huang, I.Y. and Lee, M.C. (2008) Development of a FPW allergy biosensor for human IgE detection by MEMS and cystamine-based SAM technologies. *Sens. Actuators B*, **132**, 340–348.

131. Jeutter, D., Josse, F., Johnson, M., Wenzel, M., Hossenlopp, J., and Cernosek, R. (2005) Design of a portable guided SH-SAW chemical sensor system for liquid environments. Proceedings of the IEEE International Frequency Control Symposium and Exposition, Vancouver, BC, Canada, 2005, pp. 59-68.

132. Lee, J., Chunara, R., Shen, W., Prayer, K., Babcock, K., Burg, T.P., and Manalis, S.R. (2011) Suspended Microchannel resonators with piezoresistive sensors. *Lab Chip*, **11**, 645–651.

133. Agache, V., Blanco-Gomez, G., Cochet, M., and Caillat, P. (2011) Suspended nanochannel in MEMS plate resonator for mass sensing in liquid. Presented at the 24th IEEE MEMS Conference, Cancun, Mexico, 2011.

134. Park, J., Nishida, S., Kawaktsu, H., and Fujita, H. (2011) Novel type of microcantilever biosensor resonating at the interface between liquid and air. Presented at the 24th IEEE MEMS Conference, Cancun, Mexico, 2011.

135. Iqbal, A., Chapin, J.C., Mehdizadeh, E., Rahafrooz, A., Purse, B.W., and Pourkamali, S. (2012) Real-time bio-sensing using micro-channel encapsulated thermal-piezoresistive rotational mode disk resonators. Sensors, 2012 IEEE, pp. 1–4.

136. Ferrari, M., Ferrari, V., Marioli, D., Taroni, A., Suman, M., and Dalcanale, E. (2006) In-liquid sensing of chemical compounds by QCM sensors coupled with high-accuracy ACC oscillator. *IEEE Trans. Instrum. Meas.*, **55**, 828–834.

137. Hwang, K.S., Lee, J.H., Park, J., Yoon, D.S., Park, J.H., and Kim, T.S. (2004) In-situ quantitative analysis of a prostate-specific antigen (PSA) using a nanomechanical PZT cantilever. *Lab Chip*, **4**, 547–552.

16
Biosensors

Blake N. Johnson and Raj Mutharasan

16.1
Introduction

Biosensors based on resonant micro-electromechanical systems (MEMS), some-times called bioMEMS, are highly sensitive devices capable of label-free quanti-tative measurement of biologics such as cells, viruses, proteins, and nucleic acids. They are a platform technology and are adaptable for various diverse biosensing applications. For example, bioMEMS have facilitated new measurements includ-ing weighing of single cells and biomolecules [1, 2], growth rate monitoring of individual bacterial and fungal colonies [3–5], and single-cell growth studies [6]. They have also contributed to an improved understanding of protein immobi-lization [7, 8]. From a practical application standpoint, bioMEMS have provided means of detecting pathogens and parasites found in food matrices and source water at levels of $1-100$ cells ml^{-1} [9–15], as well as both protein [16–20] and nucleic acid biomarkers [21] in complex backgrounds.

Resonant bioMEMS design and biosensing characteristics vary widely. A large number of combinations of mechanical structures and interfacing electronics for actuation and measurement have been created. Cantilever and cantilever-like designs have been among the most commonly investigated. Cantilever sensors are traditionally used in two different modes – the static or the dynamic mode. In the former, sensor response is based on change in cantilever deflection caused by analyte binding. In the latter, sensor response is based on direct change in the resonant frequency which depends on sensor mass. We note that although cantilever-based biosensing has been achieved in the static mode, the dynamic mode approach is the focus of this chapter.

In this chapter, a detailed review of resonant MEMS-based biosensing is provided. The primary discussion is limited to cantilever and cantilever-like resonant bioMEMS as they have shown the highest potential for sensitive measurement. The chapter is organized broadly as: design variation in resonant

Resonant MEMS – Fundamentals, Implementation and Application, First Edition.
Edited by Oliver Brand, Isabelle Dufour, Stephen M. Heinrich and Fabien Josse.
© 2015 Wiley-VCH Verlag GmbH & Co. KGaA. Published 2015 by Wiley-VCH Verlag GmbH & Co. KGaA.

bioMEMS (Section 16.2), techniques for sensor functionalization and target bio-recognition (Section 16.3), sensing measurement format (Section 16.4), and application case studies (Section 16.5). In the design section, considerations of sensor length scale, geometry, and fabrication materials are provided, each of which influence biosensor preparation and response characteristics. In the techniques section, common and effective surface functionalization techniques for various bio-recognition agents, such as antibody and ssDNA, are provided along with strategies for enhancing and verifying sensor response to target binding. In the measurement format section, batch versus continuous flow assay designs are discussed. In the applications section, several illustrative case studies of practical detection of biologics are highlighted.

16.2
Design Considerations: Length Scale, Geometry, and Materials

Over the last two decades, cantilever sensors have been fabricated from a wide range of materials, in various different geometries, tested for their resonant properties, and in some cases, evaluated for biosensing [22]. The discussion in this chapter will be limited to those publications and reports that investigated biosensing.

16.2.1
Fabrication Materials

Silicon-based materials have been widely used as they exhibit relatively low dissipation and high Q-values [23–25]. In addition, they are convenient to fabricate at micro- and nano-scales as semi-conductor fabrication processes have matured over the years, and are quite well established. Besides silicon-based materials, piezoelectric and magnetoelastic materials (often metal oxides) have also been investigated [26–28], as they offer advantageous properties for both exciting and measuring cantilever resonance. Polymers with high modulus have also been examined [29, 30]. The differences in material properties among such cantilevers alter their effective mass and spring constant, and therefore, the resonant frequencies and sensitivities can differ widely even though geometry may be similar. In Tables 16.1–16.4, biosensor applications are summarized with a description of materials used. We note that although choice of material influences many design aspects including the means of sensing and actuating as well as surface functionalization, biosensing approaches have been demonstrated from devices arising from a wide range of materials. For example, silicon-based cantilevers have been used to measure the mass of a single cell [1, 24], piezoelectric-based cantilevers were shown to measure bacterial pathogen concentration continuously in liquid samples [31], metal oxide-based cantilevers were shown to detect prostate cancer biomarkers [17], and polymer-based cantilevers were shown to detect the adsorption of ssDNA [32].

Table 16.1 Applications of cantilever-like bioMEMS for whole cell detection.

Material and size	Geometry/design	Mode type, mode order, and resonant frequency	Measurement medium and Q-value	Target or application, recognition agent (RA), and sensitivity (S)	Resonant frequency measurement format	References
Silicon-based, $L = N/A$	Suspended microchannel resonator, hollow internal channels	N/A	N/A	Single cell *B. subtilis*, *E. coli*, *S. cerevisiae*, and lymphoblast growth rate, RA: buoyant mass	Continuous	[6]
Silicon-based, $L = N/A$	Suspended microchannel resonator, hollow internal channels	N/A	N/A	Growth cycle variations on mass, density, and volume in various yeast strains, RA: buoyant mass, and $S: \sim 3\,fg$	Continuous	[87]
Piezoelectric- and silicon-based, $L \sim 2\,mm$	Composite, rectangular cross-section, step-discontinuity	Mode N/A, $n = N/A$, and $f = 865\,kHz$	Liquid and $Q = 30$	*C. parvum* oocyst detection, RA: IgG and IgM, and $S: 5$ oocysts ml^{-1}	Continuous	[83]
Piezoelectric- and silicon-based, $L \sim 2\,mm$	Composite, rectangular cross-section, step-discontinuity	Mode N/A, $n = N/A$, and $f \sim 856\,kHz$	Liquid and $Q = 30$	*A. laidlawii* mycoplasma detection RA: Ig, and $S: <10^3\,CFU\,ml^{-1}$	Continuous	[88]
Piezoelectric- and silicon-based, $L = 2.7\,mm$	Composite, rectangular cross-section, step-discontinuity	Mode N/A, $n = N/A$, and $f \sim 870\,kHz$	Liquid and $Q = 26$	*G. lamblia* cyst detection, RA: Ig, and $S: <10$ cysts ml^{-1}	Continuous	[86]
Magnetoelastic- and metal-based, $L \sim 3\,mm$	Rectangular-cross-section, bimorph	Transverse, $n = 1$, and $f \sim 2.6\,kHz$	Liquid and $Q \sim 20$	*B. anthracis* detection, RA: phage, and $S: \sim 10^5\,CFU\,ml^{-1}$	Continuous	[26]
Piezoelectric- and silicon-based $L \sim 3.0\,mm$	Composite, rectangular cross-section, step-discontinuity	Mode N/A, $n = N/A$, and $f = 913\,kHz$	Liquid and $Q = 15$	*C. parvum* oocyst detection, RA: IgM, and $S: 100$ oocysts ml^{-1}	Continuous	[89]

(continued overleaf)

Table 16.1 (Continued)

Material and size	Geometry/design	Mode type, mode order, and resonant frequency	Measurement medium and Q-value	Target or application, recognition agent (RA), and sensitivity (S)	Resonant frequency measurement format	References
Silicon-based, $L = 25$–$40\,\mu m$	Single-layer, rectangular cross-section, array	Transverse, $n = 1$, and $f \sim 35\,kHz$	Liquid and $Q = N/A$	Single HeLa cell mass and RA: poly-l-lysine	Batch	[33]
Silicon-based, $L = 596\,\mu m$	Single-layer, triangular, V-shaped tip	Transverse and torsional, $n = 1$, 2, and $f \sim 30$–$285\,kHz$	N/A	Ragweed pollen detection, RA: physisorption, and S: single pollen	Batch	[90]
Silicon-based, $L = N/A$	Suspended microchannel resonator, hollow internal channels	N/A	Vacuum (liquid-filled channel) and $Q \sim 8000$–15000	E. coli and human red blood cell mass densities and RA: buoyant mass	Continuous	[91]
Silicon-based, $L = 20$–$50\,\mu m$	Single-layer, uniform rectangular cross-section	Transverse, $n = 1$, and $f = N/A$	Liquid and $Q < 5$	B. anthracis detection, RA: Ig, and S: 50 spores	Batch	[34]
Piezoelectric- and silicon-based, $L \sim 3.0\,mm$	Composite, rectangular cross-section, step-discontinuity	Mode N/A, $n = N/A$, and $f = 925\,kHz$	Liquid and $Q \sim 12$	B. anthracis detection, RA: Ig, and S: 38 spores l^{-1} (air concentration)	Continuous	[92]
Piezoelectric- and silicon-based, $L \sim 2.3\,mm$	Composite, rectangular cross-section, step-discontinuity	Mode N/A, $n = N/A$, and $f = 970\,kHz$	Liquid and $Q \sim 15$	B. anthracis detection in presence of B. cereus and B. thuringiensis, RA: Ig, and S: 333 spores ml^{-1} at 1:1000 B. anthracis: (B. cereus + B. thuringiensis)	Continuous	[93]
Piezoelectric- and silicon-based, $L \sim 2\,mm$	Composite, rectangular cross-section, step-discontinuity	Mode N/A, $n = N/A$, and $f \sim 900\,kHz$	Liquid and $Q \sim 20$–30	E. coli O157:H7 detection in food matrices, RA: Ig, and S: 1–100 cells ml^{-1}	Continuous	[9–11]

Material, length	Geometry	Mode	Medium and Q	Application and results	Operation	Ref.
Silicon-based, $L = 200\,\mu m$	Suspended microchannel resonator, hollow internal channels	Transverse, $n = N/A$, and $f \sim 200\,kHz$	Liquid-filled and $Q = 15000$	Single *E. coli* and *B. subtilis* cell mass and RA: buoyant mass	Continuous	[2]
Piezoelectric- and silicon-based, $L = 3\,mm$	Composite, rectangular cross-section, step-discontinuity	Transverse, $n = 1$, and $f = 31\,kHz$	Air and $Q = 39$	*E. coli* JM101 growth rate and RA: agar layer	Continuous	[12]
Silicon-based, $L = 500\,\mu m$	Single-layer, rectangular cross-section, array	Transverse, $n = 5$, and $f = 280\,kHz$	Air and $Q = N/A$	*B. subtilis* detection, RA: polypeptide, and S: 10^5 spores ml^{-1}	Batch	[74]
Silicon-based, $L = 250, 500\,\mu m$	Composite, rectangular cross-section, array	Transverse, $n = N/A$, and $f \sim 30 - 135\,kHz$	Air and $Q = 116$	*A. niger* and *S. cerevisiae* growth and detection, RA: concanavalin A, fibronectin, and IgG, and S: 10^3 CFU ml^{-1}	Batch	[4]
Silicon-based, $L = 500\,\mu m$	Single-layer rectangular cross-section, array	Transverse, $n = N/A$, and $f \sim 33\,kHz$	Air and $Q = N/A$	*E. coli* XL1-blue growth, RA: agar layer with and without antibiotic, and S: $\sim$100–200 cells, 50–140 pg Hz^{-1}	Continuous	[3, 94]
Piezoelectric- and silicon-based, $L = 3 - 5\,mm$	Composite, rectangular cross-section, step-discontinuity	Transverse, $n = 2$, and $f \sim 55\,kHz$	Liquid and $Q = 30 - 100$	*E. coli* O157:H7 detection in presence of *E. coli* JM101, RA: Ig, and S: 700 cells ml^{-1} at 1:1 O157:H7:JM101	Continuous	[14]
Silicon-based, $L = 78\,\mu m$	Single layer, rectangular cross-section	Transverse, $n = 1$, and $f \sim 85\,kHz$	Air and $Q \sim 55$	*L. innocua* detection, RA: Ig, and σ: 65–90 Hz pg^{-1}	Batch	[23]
Silicon-based, $L = 15 - 25\,\mu m$	Single layer, rectangular cross-section	Transverse, $n = 1$, and $f \sim 1\,MHz$	Air and $Q \sim 50$	Single cell *E. coli* O157:H7 detection, RA: Ig, and S: 1–7 Hz fg^{-1}	Batch	[1]
Silicon-based, $L = 100 - 200\,\mu m$	Single layer, rectangular cross-section	Transverse, $n = 1$, and $f \sim 33\,kHz$	Air and $Q = 5 - 8$	*E. coli* O157:H7 detection, RA: Ig, and S: $\sim$6 Hz pg^{-1}, 14.7 fg	Batch	[24]

Ig, immunoglobulin; CFU, colony-forming unit; N/A, not available.

Table 16.2 Applications of cantilever-like bioMEMS for protein detection.

Material and size	Geometry/ design	Measurement medium and Q-value	Mode type, mode order, and resonant frequency	Target or application, recognition agent (RA), and sensitivity (S)	Resonant frequency measurement format	References
Silicon-based, $L \sim 200\,\mu m$	Suspended microchannel resonator, hollow internal channels	Vacuum (liquid filled channel) and Q=N/A	Transverse, n=N/A, and $f = 200\,kHz$	Human recombinant activated leukocyte cell adhesion molecule (ALCAM) detection in undiluted serum, RA: IgG, and S: $10\,ng\,ml^{-1}$	Continuous	[97]
Silicon-based, $L = 300\,\mu m$	Single-layer, T-shaped cross-section	Air and $Q = 3500-4000$	Torsional, n=N/A, and $f = 95-99\,kHz$	Alpha-fetoprotein (AFP) detection, RA: Ig, and S: $2\,ng\,ml^{-1}$	Continuous	[44]
Silicon-based, $L = 450\,\mu m$	Single-layer, uniform-rectangular	Vacuum and Q=N/A	Transverse, $n = 1, 2$, and $f = 12-74\,kHz$	*Anti*-IgG detection, RA: IgG, and S: $0.4-2.5\,Hz\,pg^{-1}$	Batch	[68]
Piezoelectric- and silicon-based, $L = 240-30\,\mu m$	Composite, rectangular cross-section, array	N/A and $Q = 213-677$	Transverse, n=N/A, and $f \sim 0.03-3.3\,MHz$	Human IgG detection, RA: reactive thiol-based SAM, and S: $\sim 10\,pg\,Hz^{-1}$ to $3\,fg\,Hz^{-1}$	Batch	[98]
Piezoelectric- and silicon-based, $L = 100\,\mu m$	Composite, rectangular cross-section, step-discontinuity	Air and N/A	Transverse, $n = 1$, and $f \sim 1.2\,MHz$	Carcinoembryonic antigen (CEA) detection, RA: Ig, and S: 30 pM ($5\,ng\,ml^{-1}$)	Batch	[61]
Silicon-based, $L = 300\,\mu m$	Single-layer, cross-shaped	Air and $Q = 11145$	Torsional, $n = 2$, and $f = 525\,kHz$	Alpha-fetoprotein (AFP) detection, RA: Ig, and S: $\sim 9\,fg$	Batch	[99]
Silicon-based, $L = 300\,\mu m$	Single-layer, cross-shaped, T-shaped, and rectangular	Air and Q $\sim$ 1000–10 000	Transverse and torsional, $n = 1, 2$, and $f = 47-518\,kHz$	Streptavidin detection, RA: biotinylated thiol-based SAM, and S: $0.33-5.1\,Hz\,pg^{-1}$	Batch	[46]

Material, length	Geometry	Mode, frequency	Medium, Q	Application	Operation	Ref.
Metal oxide-based, $L = 150\,\mu m$	Single-layer, uniform-cross-section	Transverse, $n = 1, 2, 3$, and $f \sim 200$–$700\,kHz$	Liquid and $Q \sim 10$–20	Prostate specific antigen (PSA) detection, RA: Ig, and S: $10\,ng\,ml^{-1}$	Continuous	[17, 100]
Piezoelectric- and silicon-based, $L \sim 2.5\,mm$	Composite, rectangular cross-section, step-discontinuity	mode N/A, $n = N/A$, and $f \sim 876\,kHz$	Liquid and $Q = 28$–35	Staphylococcal enterotoxin B (SEB) detection in food matrices, RA: Ig, and S: $100\,fg$	Continuous	[96]
Piezoelectric- and silicon-based, $L \sim 2.5\,mm$	Composite, rectangular cross-section, step-discontinuity	mode N/A, $n = N/A$, and $f \sim 890\,kHz$	Liquid and $Q = 28$–31	Alpha-methylacyl-CoA racemase (AMACR) detection in urine, RA: Ig, and S: $3\,fg\,ml^{-1}$	Continuous	[84]
Silicon-based, $L = 300\,\mu m$	Single-layer, T-shaped	Torsional, $n = 1$, and $f = 97\,kHz$	Air and $Q = 3725$	Streptavidin detection, RA: biotinylated thiol-based SAM, and S: $\sim 23\,fg$	Batch	[45]
Piezoelectric- and silicon-based, $L = 150\,\mu m$	Single layer, rectangular cross-section, array	Transverse, $n = 1$, and $f = 66\,kHz$	Air and $Q = N/A$	Hepatitis C virus helicase detection, RA: RNA aptamer, and S: $100\,pg\,ml^{-1}$	Batch	[76]
Piezoresistive-based, $L = 335\,\mu m$	Single layer, triangular cross-section	Transverse, $n = N/A$, and $f = 265\,kHz$	Liquid and $Q = 294$	Egg albumin detection, RA: IgG, and S: $200\,fg\,Hz^{-1}$	Continuous	[101]
Piezoelectric-based, $L = 150\,\mu m$	Single layer, rectangular cross-section, array	Transverse, $n = 1, 2$, and $f = 77, 487\,kHz$	Air and $Q = N/A$	Myoglobin detection and RA: Ig	Batch	[102]
Piezoelectric- and metal-based, $L = 3.5\,mm$	Composite, rectangular cross-section, step-discontinuity	Transverse, $n = 3$, and $f \sim 40\,kHz$	Liquid and $Q = 43$	Human serum albumin detection and RA: various functional-headgroup SAMs	Continuous	[103]
Piezoelectric- and silicon-based, $t \sim 100\,\mu m$	Composite triangular cross-section, step-discontinuity	Transverse, $n = 1$, and $f \sim 1.3\,MHz$	Air and $Q = N/A$	Human insulin detection and RA: Ig	Batch	[39]

(continued overleaf)

Table 16.2 (*Continued*)

Material and size	Geometry/ design	Mode type, mode order, and resonant frequency	Measurement medium and Q-value	Target or application, recognition agent (RA), and sensitivity (S)	Resonant frequency measurement format	References
Piezoelectric- and silicon-based, $L = 300\,\mu m$	Composite, rectangular cross-section, step-discontinuity, array	N/A	Air and $Q = $ N/A	Myoglobin detection, RA: biotinylated Ig, and S: $1\,ng\,ml^{-1}$	Batch	[104]
Silicon-based, $L = 300\,\mu m$	Suspended microchannel resonator, hollow internal channels	Transverse, $n = $ N/A, and $f \sim 34\,kHz$	Vacuum (liquid filled channel) and $Q \sim 400\text{--}1700$	Biotinylated bovine serum albumin detection, RA: avidin, and S: $0.8\,ng\,cm^{-2}$	Continuous	[105]
Piezoelectric- and silicon-based, $L = 150\text{--}300\,\mu m$	Composite, rectangular cross-section, step-discontinuity, array	Transverse, $n = $ N/A, and $f = 16\text{--}61\,kHz$	Air and liquid and $Q = $ N/A	Prostate specific antigen (PSA) detection, RA: Ig, and S: $\sim 10\,pg\,ml^{-1}$	Batch and Continuous	[18, 19]
Piezoelectric- and silicon-based, $L = 100\,\mu m$	Composite, rectangular cross-section, array	Transverse, $n = 1$, and $f = 25\,kHz$	Air and $Q = $ N/A	C-reactive protein (CRP) detection, RA: Ig, and S: $<$ nM	Continuous	[28, 106]
Silicon-based, $L = 300\,\mu m$	Suspended microchannel resonator, hollow internal channels	Transverse, $n = $ N/A, and $f \sim 40\,kHz$	Vacuum (liquid filled channels) and $Q \sim 90$	Avidin detection, biotinylated bovine serum albumin, and S: $10^{-17}\,g\,\mu m^{-2}$	Continuous	[107]
Silicon-based, $L = 100\,\mu m$	Single layer, V-shaped cross-section	Transverse, $n = $ N/A, and $f \sim 16\,MHz$	Liquid and $Q = 625$	BRAC30 (STAR 71), *anti*-BRAC30	Continuous	[108]

Ig, immunoglobulin; CFU, colony-forming unit; N/A, not available; SAM, self-assembled monolayer.

Table 16.3 Applications of cantilever-like bioMEMS for virus detection.

Material and size	Geometry/ design	Mode type, mode order, and resonant frequency	Measurement medium and Q-value	Target or application, recognition agent (RA), and sensitivity (S)	Resonant frequency measurement format	References
Piezoelectric- and metal-based, $L = 715\,\mu m$	Composite, rectangular cross-section	Longitudinal, $n = 1$, and $f \sim 1\,MHz$	Liquid and $Q \sim 25$	White spot syndrome virus detection, RA: IgG, and S: ~ 100 virions ml^{-1}	Continuous	[111]
Silicon-based, $L = 500\,\mu m$	Single-layer, rectangular cross-section, array	Transverse, $n = 10-15$, and $f = N/A$	Liquid and $Q = N/A$	Bacterial T5 phage detection, RA: *E. coli* transmembrane protein, and S: $\sim 300\,fM$	Continuous	[112]
Silicon-based, $L = 6, 21\,\mu m$	Single-layer, rectangular cross-section	Transverse, $n = N/A$, and $f = 0.5-5.5\,MHz$	Medium N/A and $Q = 707$	Vaccinia virus detection – single virus mass determination, RA: physisorption, and S: $\sim 1-20\,fg$	Batch	[113]
Silicon-based, $L = 3-5\,\mu m$	Single-layer, rectangular cross-section, rounded-tip	Transverse, $n = N/A$, and $f \sim 1-3\,MHz$	N/A	Vaccinia virus detection – recognition layer effect on frequency change and RA: Ig	Batch	[77]
Silicon-based, $L = 6-10\,\mu m$	Single-layer, rectangular cross-section with tip paddle	Transverse, $n = N/A$, and $f \sim 5.5-10\,MHz$	Vacuum and $Q \sim 10^4$	Insect baculovirus detection, RA: Ig, and S: $\sim 0.4-1\,ag\,Hz^{-1}$, $10^5-10^7\,PFU\,ml^{-1}$	Batch	[109]
Silicon-based, $L = 3-5\,\mu m$	Single-layer, rectangular cross-section, array	Transverse, $n = 1$, and $f \sim 1.3\,MHz$	Air and $Q \sim 5-7$	Vaccinia virus detection – single virus mass determination, RA: physisorption, and S: $\sim 6.3\,Hz\,ag^{-1}$	Batch	[110]

Ig, immunoglobulin; CFU, colony-forming unit; N/A, not available; PFU, plaque-forming unit.

Table 16.4 Applications of cantilever-like bioMEMS for nucleic acid detection.

Material and size	Geometry/ design	Mode type, mode order, and resonant frequency	Measurement medium and Q-value	Target or application, recognition agent (RA), and sensitivity (S)	Resonant frequency measurement format	References
Piezoelectric- and silicon-based, $L = 3\,mm$	Composite, rectangular cross-section, step-discontinuity	mode N/A, $n = N/A$, and $f \sim 900\,kHz$	Liquid and $Q \sim 19$	*stx2* gene of pathogenic *E. coli* detection and RA: thiolated DNA probes	Continuous	[85]
Piezoelectric, $L = 3\,mm$	Single-layer, Asymmetric anchor	Transverse, $n = 2$, and $f \sim 75\,kHz$	Liquid and $Q \sim 30$	*let-7a* microRNA detection and RA: thiolated DNA probes	Continuous	[21]
Piezoelectric, $L = 3\,mm$	Single-layer, asymmetric anchor	Transverse, $n = 2$, and $f \sim 75\,kHz$	Liquid and $Q \sim 30$	*hlyA* gene of pathogenic *L. monocytogenes* detection and RA: thiolated DNA probes	Continuous	[42]
Piezoelectric- and silicon-based, $L = 90\,\mu m$	Composite, cross-section N/A, array	N/A	Liquid and $Q = N/A$	Hepatitis B Virus DNA (27-bp) detection and RA: thiolated DNA probes	Continuous	[69]
Silicon-based, $L = 83 - 124\,\mu m$	Single layer, rectangular cross-section	Transverse, $n = 1$, and $f = 167 - 365\,kHz$	Air and Liquid: $Q = N/A$	20-bp thiolated DNA detection and RA: N/A	Continuous	[122]

Piezoelectric- and silicon-based, $L = 30–50\,\mu m$	Composite, rectangular cross-section	N/A	Air and $Q=N/A$	Hepatitis B Virus DNA (243-bp) detection, RA: thiolated DNA probes, and S: fM with nanoparticle enhancement	Batch	[82]
Piezoelectric- and silicon-based, $L = 3\,mm$	Composite, rectangular cross-section, step-discontinuity	Mode N/A, $n = N/A$, and $f \sim 900\,kHz$	Liquid and $Q \sim 19$	10-bp DNA detection in human serum, RA: 15-bp thiolated DNA probe, and S: $\sim$10 aM	Continuous	[63]
Silicon-based $L = 3.5–5\,\mu m$	Single-layer, rectangular cross-section	Transverse, $n = N/A$, and $f \sim 11\,MHz$	Vacuum and $Q \sim 3000–5000$	1587-bp thiolated DNA detection and S: $\sim$0.23 ag	Batch	[70]
Silicon-based, $L = 150\,\mu m$	Single-layer, triangular cross-section	Transverse, $n = N/A$, and $f \sim 116\,kHz$	Air and $Q=N/A$	30-bp DNA detection with and without single-nucleotide polymorphism, RA: thiolated DNA and nanoparticle probe, and S: 23 pM	Batch	[25]

Ig, immunoglobulin; CFU, colony-forming unit; N/A, not available.

16.2.2
Single-Layer Geometry

Besides the variation of materials used in fabricating cantilever sensors, a large number of variations in geometric design have also been investigated. They can be categorized broadly as designs consisting of either single- or multi-layers and either uniform or non-uniform cross-section (see variety highlighted in Figure 16.1). The single-layer cantilevers often have uniform rectangular cross-section due to fabrication convenience [24, 33, 34]. Uniform geometry is sometimes desirable as first principles mathematical models can be used to facilitate design and interpret sensor response. A few examples of such uniform rectangular designs are shown in Figures 16.1b, d. Figure 16.1b shows a cantilever array, while in Figure 16.1d a single-layer piezoelectric cantilever is illustrated. The rectangular cross-section designs also vary in length-to-width aspect ratio, a property that affects device performance [35–37]. For example, a decrease in the aspect ratio was reported to improve sensitivity [38]. Such a result is

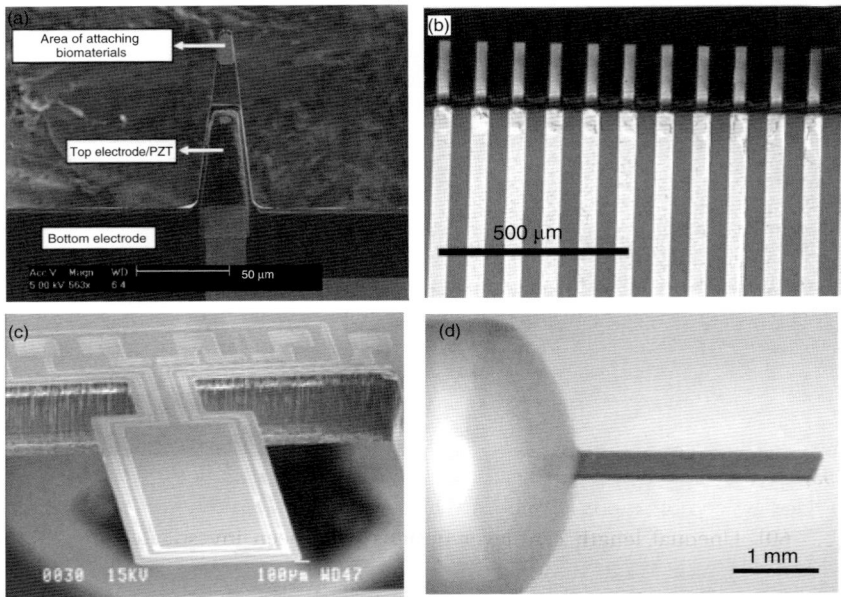

Figure 16.1 (a) Self-sensing and -exciting PZT-based triangular micro cantilever $(L-w-t = 100-30-5\,\mu m^3)$, which exhibits a fundamental resonant mode at 1.2–1.3 MHz [39]. (b) Polymer (SU-8) microcantilever arrays, which exhibit low Young's modulus, ~40 times lower than silicon [32]. (c) Paddle-shaped microcantilevers fabricated via photolithography, deep reactive ion etching, and metallization techniques [40]. (d) Single-layer uniform cross-sectioned piezoelectric macro-cantilevers fabricated with an asymmetric electrode configuration [41] (similar designs using asymmetric anchor design are described in Refs [8, 21, 42, 43]).

not surprising as the aspect ratio affects both the resonant frequency and the associated fluid-structure interaction.

Single-layer cantilever sensors with non-rectangular cross-section have been investigated by several groups to facilitate selective resonant-mode actuation, enhance Q-value, minimize fluid damping, and improve sensitivity [39, 44, 45]. For example, the triangular geometry (see Figure 16.1a), also described as V-shaped [25] and the paddle geometry (see Figure 16.1c), also described as T-shaped [40, 46, 47], have shown highly sensitive responses. They also produce resonant modes with alternative nodal distributions that stem from the non-uniform geometry. For example, T-shaped piezoelectric resonators were shown to exhibit lateral-extensional modes that had femtogram level sensitivity [47]. We also note that cantilevers with small tip extensions were investigated as a means of reducing sensor mass while still maintaining the active layer integrity for actuation [35, 48].

16.2.3
Multi-Layer Geometry

Several cantilever sensors using multi-layer designs have been investigated. The device layers often provide valuable self-actuation or self-sensing functionality, which reduces need for supporting external hardware [49, 50]. Thus, such devices are highly suitable for remote or field measurements. Device layers devoted to actuating or sensing are called active layers, while those devoted to either bonding or modulation of cantilever mass, stiffness, or geometric properties are called passive or inactive layers [15, 51]. The active layer is typically a piezoelectric [18, 19] or magnetoelastic material [26]. In some cases, the same layer can be used for both actuating and sensing purposes, as commonly found in piezoelectric cantilever sensors. Figure 16.1a shows a composite piezoelectric microcantilever with a tip-sensing region isolated from the piezoelectric layer [39]. Two-layer designs with equal length layers, commonly called unimorphs or bimorphs depending on the number of active layers involved, were among the first of such designs to be investigated [27, 49, 51–54]. They evolved to include designs in which one of the layers was shortened resulting in step-discontinuity in cross-section [55, 56]. Such modification to uniform geometry caused resonance characterized by combinations of modes that were highly sensitive to mass-changes [57–60]. Unequal length cantilever designs have been investigated by several research groups [14, 19, 28, 61]. Some multi-layer piezoelectric sensors with step-discontinuities have been shown to express high-order longitudinal-flexural modes with sub-femtogram sensitivity [60, 62, 63].

16.2.4
Length Scales

Cantilever sensors of length ranging from nanometers to millimeters have been investigated (see Tables 16.1 – 16.4). The approach of reducing size is motivated by lumped parameter design equations, which indicate that high sensitivity can be achieved if sensor mass is reduced. On the other hand, millimeter-scale sensors operating in high-order modes have also been pursued, as effective mass at high-order modes also becomes small.

Such high-order modes have provided an alternative method for achieving femtogram sensitivity without device miniaturization. Size is an important design parameter of a viable biosensor that is to detect analyte continuously in liquid environments. For example, it is well established that as the cantilever dimensions decrease, a greater relative amount of energy is dissipated by viscous damping [64], even to the point that the sensor does not resonate in liquids. Secondly, cantilever size also affects the potential for sensing the target on a practical time scale as the effective analyte collection rate decreases with decreasing device size. Increase in both energy dissipation and sensing time scale represent aspects of miniaturization that are undesirable in biosensing application. In the summary given in Tables 16.1–16.4, we see that there are both advantages and disadvantages associated with small and large cantilevers, and the choice depends on the given application pursued. Regardless of the heuristics guiding cantilever biosensor design, the goal is to develop biosensors that show: (i) sensitive resonant modes in response to molecular binding, (ii) multi-log dynamic range, (iii) reasonable detection time, (iv) ability to measure in liquids, and (v) zero false response.

16.3
Surface Functionalization: Preparation, Passivation, and Bio-recognition

Detection of cells, parasites, microbes, viruses, macromolecules, and small molecules requires first functionalizing the sensor with recognition molecules, called bio-recognition agents, which have high affinity and selectivity for the intended biological target. The recognition agent must be chemically or physically attached to the cantilever surface for transduction of the binding mass change to occur. A number of techniques for immobilizing the recognition agents to surfaces have been reported [65]. Functionalizing a biosensor surface is broadly defined as a three-step process involving: (i) preparation, (ii) immobilization, and (iii) passivation. First, the cantilever surface must be prepared via chemical modification to create a facilitating reactive layer if the cantilever lacks appropriate reactive functional groups for immobilizing the bio-recognition agent. Second, the recognition agent must be immobilized in an orientation that allows for specific binding with the intended biological target. Third, if bare reactive areas remain on the cantilever surface, they must be passivated for preventing nonspecific adsorption and potential false sensor response. The three functionalization steps can be done in either batch or flow settings. For miniaturized cantilevers, capillary incubation, inkjet printing, and contact printing techniques are common batch-mode approaches [66]. Alternatively, the functionalization steps can also be carried out directly in a flow cell immediately prior to target detection in a continuous-mode approach. We note that the continuous-mode approach is desirable as measurement of cantilever resonance properties prior to, during, and after functionalization provides for a quality control on the extent of immobilization.

16.3.1
Antibody-Based Bio-recognition

The bulk of biological targets investigated have been pathogens, parasites, biomarkers, and toxins. Therefore, many investigators have used antibodies as the selective bio-recognition agents. Antibodies have copious peripheral carboxyl and amine groups which conveniently facilitate their immobilization onto the cantilever surface. For surfaces displaying hydroxyl groups, such as silica, silanization modification is used to introduce surface amine or carboxyl groups for binding the antibody [4, 44, 67]. For gold (Au)-coated surfaces, antibody binding proteins, such as Protein A or G, have been used since they readily physisorb via nitrogenous and sulfur-containing regions [4, 67]. Immobilization of antibodies via avidin-biotin interaction has also been examined. Mercaptans with end amine or carboxyl groups have also been used on Au-coated sensors for immobilizing antibodies [44]. Less selective methods such as reactive polymer coatings have also been investigated, but they suffer the disadvantage of poor orientation of recognition agents [68]. Use of half-antibody fragments have shown very promising results as the internal sulfur groups become exposed through a reduction reaction, which then readily chemisorb onto Au <111> sites [8]. Examples of such approaches are shown schematically in Figures 16.2a, b. After immobilization, the cantilever surface is passivated using albumins, such as bovine serum albumin (BSA), or backfilling molecules, such as short mercaptans. Such a passivation step reduces potential for false sensor response by nonspecific binding, which is a concern when detecting analytes in complex biological background. Rinsing the sensor with buffer containing dilute surfactants has also been reported to yield good results for removing nonspecifically bound entities [44].

16.3.2
Nucleic Acid-Based Bio-recognition

Biosensing of nucleic acids has been primarily carried out on Au-coated sensors. The targets have been sensed in two ways. One common method uses complementary thiolated DNA probes as bio-recognition agents [25, 63, 69]. Given the sequence of the target gene or transcript is known, a linear or hairpin nucleic acid probe, typically DNA, can be readily synthesized with a modified end group that facilitates immobilization on the cantilever. After immobilizing the DNA probe, the remaining unfilled Au sites are filled with short mercaptans, such as mercaptohexanol (see Figure 16.2c) resulting in the DNA probes oriented in an upright manner. The sample containing the complementary strand is then exposed to the sensor at a suitable temperature and salt concentration for

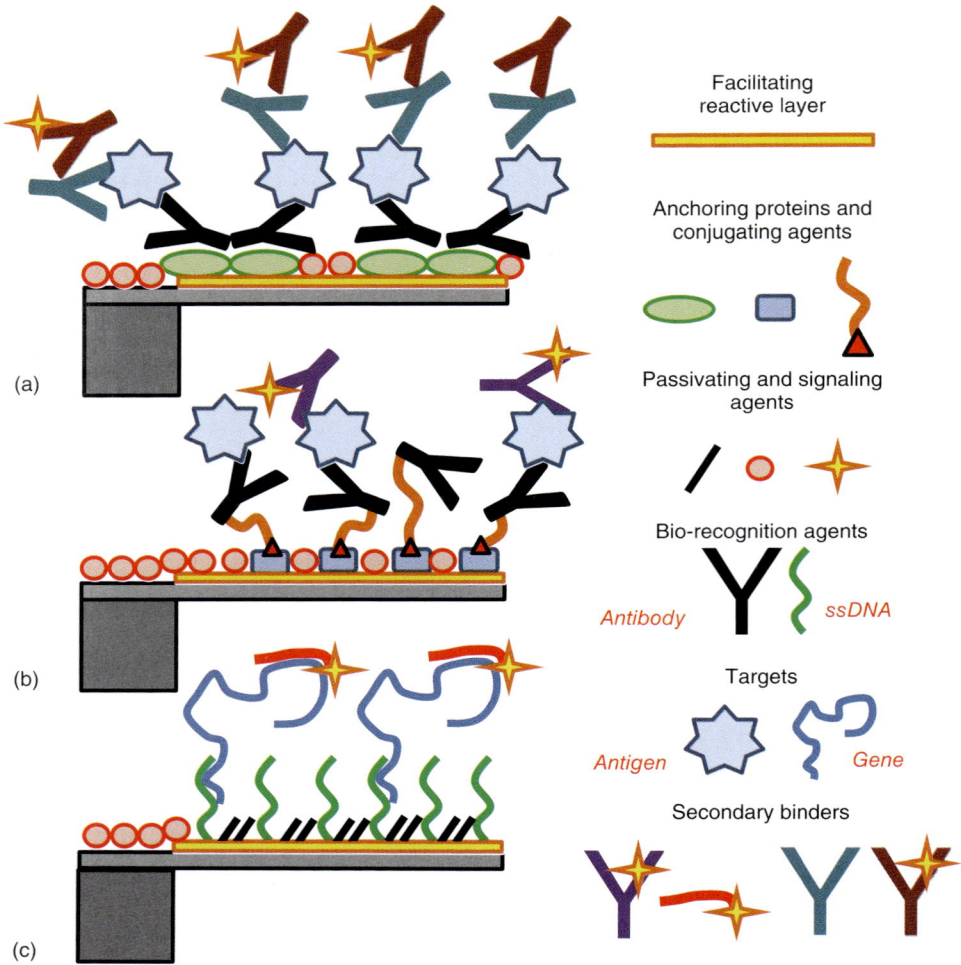

Figure 16.2 Schematic of exemplary surface functionalization strategies used in biosensing: (a) Protein G-based technique for antibody immobilization, (b) avidin-biotin-based technique for antibody immobilization, and (c) end group-based method for DNA probe immobilization. Various different strategies for secondary confirmation or verification are given. Such techniques include use of secondary and tertiary antibody (see (a) and (b)) or ssDNA binding (see (c)). Further, in addition to confirming target binding, thus decreasing the occurrences of false results, such verification steps also amplify the sensor response generated per target molecule which improves both limit of detection and dynamic range.

facilitating hybridization, which results in sensor mass increase and resonant frequency decrease. A second approach uses a polymerase-based pre-treatment step to amplify the target strand into amplicons containing a thiol end group via thiolated primers, which enables sensor response to thiol binding to serve as the detection response [70]. Both methods take advantage of the strong bond created by the sulfur-Au <111> bond [71].

It is critical to note that each of the aforementioned approaches facilitate incorporation of secondary and tertiary binding steps for confirmation or verification of target binding (see Figure 16.2). Such steps are required for assays to be highly robust and have low potential for false results, which are constraints of most practical applications (e.g., medical diagnostics, environmental monitoring, and food safety) [8, 21, 42]. Further, secondary binding species can also be labeled; they allow sensor response to be compared with a complementary, yet independent, signal (e.g., fluorescence). We note that in addition to verification by post-target selective binding reactions using antibodies or DNA probes (see Figures 16.2c), one should also examine target unbinding, or dissociation, as a useful control that should produce a recovery of sensor response. Such can often be facilitated by pH change [21]. This also has practical advantages and diagnostic significance as ability to regenerate the sensor surface *in situ* would facilitate sample re-analysis and high-throughput analysis capabilities. Ability to regenerate sensing surfaces also has potential to improve significance of biosensing results as variation in the bio-recognition layer across all samples is significantly reduced. For example, see the following references that demonstrate secondary and tertiary binding [8, 42, 72], target unbinding [21], and sensor surface regeneration [73].

16.3.3
Alternative Bio-recognition Agents

Alternative bio-recognition agents such as peptide ligands [74], phages [75], aptamers [76], and various other recognition proteins [4, 77] have been investigated *in lieu* of antibodies. Avidin was investigated as a bio-recognition agent, but the method has the severe disadvantage of requiring the target to be labeled with biotin (Tables 16.1–16.4). Electrostatic interactions between a charged surface, such as poly-lysine, and the target analyte were investigated [33], but the approach exhibited poor selectivity.

Another important practical issue is achieving reproducible surface functionalization of the bio-recognition agent, so that one sensor response can be compared with another. However, as discussed above, this requires repeatability of preparation, immobilization, and passivation steps. It is also important to note that variability exists in preparation of micro-scale surfaces themselves [78]. In one study, the measured mass and stiffness contributions varied depending on the recognition chemistry used [77]. Thus, direct comparison of sensor responses for the same target requires careful characterization of the recognition chemistry used. In Tables 16.1–16.4, we have included a description of bio-recognition agent used in the various applications reported.

16.4
Biosensing Application Formats

Resonant frequency change of a cantilever sensor in a sensing context is measured in two different ways: (i) measuring the resonant frequency at end points; namely, prior to and after contacting the sensor with the sample or (ii) by measuring the resonant frequency continuously as the sample is contacted with the sensor in a flow format. The latter is advantageous since the sensor is exposed to minimal disturbance or external contamination, and the environment of the sensor remains constant throughout the measurement. Data on continuous resonant frequency change as analyte binding occurs can also provide kinetic data unlike the first method.

16.4.1
Dip-Dry-Measure Method

The dip-dry-measure (DDM) technique is among one of the earliest and most simple biosensing measurement format used [1, 18, 40]. As the name implies, it consists of three steps: (i) immerse in the functionalized sample for an appropriate time period, (ii) remove, rinse, and dry the sensor, and (iii) measure the resonant frequency. This method relies on the state of the sensor, with and without the bound target, to be identical in order to relate the measured frequency shift to the mass change caused by target binding. In the DDM technique, control experiments for characterizing the effects of sensor drift, transfer, rinsing, and drying should be incorporated, as changes do occur naturally, especially in very highly sensitive devices. Additionally, rinsing and drying the sensor can change the state of the biochemical interactions on the surface relative to the native state. Therefore, the potential applications of the measurement may be reduced due to lack of biologically relevant binding environment. This may not be a major concern if the goal is detection of a high concentration analyte, but is important if the application is intended for characterizing interaction between biological targets, such as proteins, microbes, or DNA.

16.4.2
Continuous Flow Method

In contrast to the DDM technique, measurement under continuous sample flow conditions adds considerable validity to the results; example arrangements are given in Figure 16.3. Continuous flow measurement can be done either with or without recirculation of the binding analyte and is referred to as a recirculation mode [79] or a once-through mode [80, 81], respectively. The former is preferred for very dilute samples, and when binding rate is slow. The sample to be measured is injected into the running buffer for contacting it with the sensor. This technique has the advantages of avoiding unnecessary disturbances. It gives the user the facility to rinse the sensor *in situ* and add a confirming *in situ* binding step, analogous to a "sandwich assay," for enhancing both the sensitivity and the reliability of measurement (for example, see Figures 16.2 and 16.4a). The second binding step or target amplification can be done using either antibodies, nanoparticles, labeled-antibodies, or other proteins that have affinity for the target [25, 82]. Such assay strategies require sensor integration with a flow cell. Various strategies for introducing flow are shown schematically in Figure 16.3.

Assays that measure the resonant frequency continuously also should include controls that are on the same time scale as the sensing step. Such a measurement method adds confidence in the results by showing the sensor response is indeed due to binding and not due to sensor drift or other extraneous factors. Robustness in the measurement can be gained by repeating the experiments at various analyte concentrations in a step-wise manner. Since the bulk analyte concentration is the driving force for sensor response, it is to be expected that one should be able to observe a concentration-dependent change in sensor response (examples are shown in Figures 16.4c – e and 16.5b, c).

16.5
Application Case Studies

Resonant bioMEMS have been demonstrated for the detection of cells, pathogens, parasites, toxins, biomarkers, and nucleic acids. Practical applications in food safety, water monitoring, and medical diagnostics require the ability to sensitively detect such targets rapidly in a complex background. Common pathogens and parasites include *Escherichia coli (E. coli)* O157:H7, *Listeria monocytogenes (L. monocytogenes), Giardia lamblia (G. lamblia)*, and *Cryptosporidium parvum (C. parvum)*. Common protein targets include biomarkers and toxins, such as prostate specific antigen (PSA) and Staphylococcal enterotoxin B (SEB), respectively. Common nucleic acid targets are identifying gene sequences of pathogenic bacteria, microRNA, and genes associated with toxin production. In this section, we highlight the results of several practical biosensing applications. A detailed summary of biosensing applications can be found in Tables 16.1 – 16.4.

16.5.1
Whole Cells: Pathogens and Parasites

16.5.1.1   Foodborne Pathogen: *Escherichia coli* O157:H7
In a recent study [9, 10], macro-scale cantilever sensors called piezoelectric-excited millimeter-sized cantilever (PEMC) sensors consisting of a piezoelectric and a borosilicate glass layer with a sensing area of $4\,mm^2$ were shown to detect *E. coli* O157:H7 reliably at $50 – 100$ cells ml^{-1} in complex background. The sensor's second order resonant transverse mode was used in detection experiments. The sensor was immobilized with a polyclonal antibody specific to *E. coli*, and then was exposed to samples inoculated with live *E. coli* in various matrices

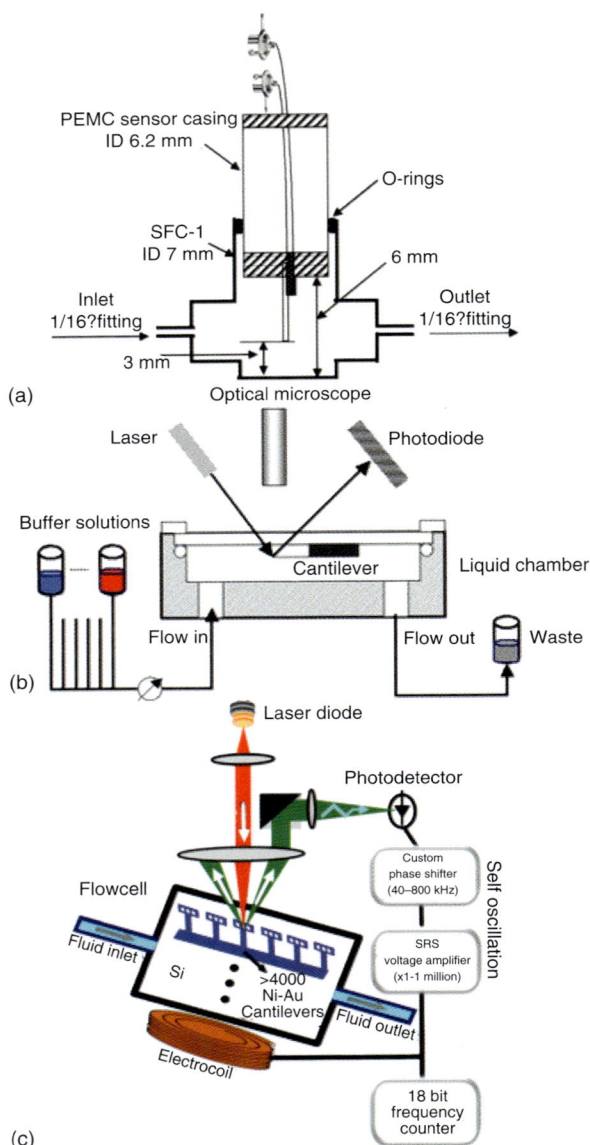

(a)

(b)

(c)

Figure 16.3 (a) Flow cell used in continuous liquid-phase detection of biologics using PEMC sensors, which facilitated detection of *Bacillus anthracis* at a sensitivity of less than 300 spores mL^{-1} in less than 1 h [79]. (b) Flow cell used in the detection of streptavidin between 1 and 10 nM using an optical measurement arrangement [80]. (c) Flow cell used for microcantilever array-based detection of Hepatitis A and C antigens in undiluted bovine serum [81].

including broth, broth plus raw ground beef, and broth plus sterile ground beef. Broth plus sterile ground beef and broth plus ground beef served as controls. The various samples were incubated at 37 °C. Periodically 1–3 ml was withdrawn from each, analyzed by plating, and tested using the PEMC sensor. The resonant frequency shift obtained for the broth plus *E. coli* samples progressively increased corresponding to 2, 4, and 6 h growth, respectively. The response to the broth plus 25 g of sterile ground beef plus 25 *E. coli* cells were 21 ± 2 Hz ($n = 2$), 37 Hz ($n = 1$), and 70 ± 2 Hz ($n = 2$) to 2, 4, and 6 h, respectively. In all cases, the three different control samples yielded no frequency change (0 ± 2 Hz; $n = 6$). The *E. coli* concentration in each of the samples was determined by counting colonies on the streaked plates as well as by pathogen modeling program.

In a later study [10], *E. coli* was detected at 10 cells ml^{-1} in spiked ground beef samples in 10 min using a higher order mode near 1 MHz in PEMC sensors. The composite PEMC sensors had a sensing area of 2 mm^2 and were prepared by immobilizing a polyclonal antibody specific to *E. coli* on the sensing surface. Ground beef (2.5 g) in 10 ml phosphate-buffered saline (PBS) was spiked with *E. coli* at an end concentration of 10^1–10^4 cells ml^{-1}. One milliliter of the supernatant was removed from the blended samples and used to perform the detection experiments. The resonant frequency changes obtained for the spiked samples were 138 ± 9, 735 ± 23, 2603 ± 51, and 7184 ± 606 Hz, corresponding to *E. coli* concentrations of 10, 10^2, 10^3, and 10^4 cells ml^{-1}, respectively. The sensor response was obtained within the first 10 min. The response to positive (*E. coli* present but antibody absent), negative (*E. coli* absent but antibody present), and buffer controls (both *E. coli* and antibody absent) were significantly lower, and were 36 ± 6, 27 ± 2, and 2 ± 7 Hz, respectively. Verification of *E. coli* attachment was confirmed by low-pH buffer release (PBS-HCl, pH 2.2), microscopy, and second antibody binding post detection.

16.5.1.2 Foodborne Pathogen: *Listeria monocytogenes*

L. monocytogenes, an important foodborne pathogen that causes a high mortality rate (~30%), was successfully detected within an hour at the infection dose limit of 10^3 cells ml^{-1} both in buffer and milk [8]. *L. monocytogenes* detection was demonstrated using an asymmetrically anchored cantilever sensor and a commercially available antibody. Sensor responses were confirmed using a secondary antibody binding step, similar to the sandwich-ELISA (enzyme-linked immunosorbent assay) as a means of signal amplification that also reduced false negatives. Detection of *L. monocytogenes* at a concentrations of 10^2 cells ml^{-1} was achieved, by incorporating a third antibody binding step (see Figure 16.4a), which is an order of magnitude smaller than the infection dose (~10^3 cells) for *L. monocytogenes*. The commercially available antibody for *L. monocytogenes* used in this work was shown to have low avidity, which partially explained the relatively low sensitivity reported in the literature for *L. monocytogenes* as compared to other pathogens in the literature. In this application, low concentration was detected in spite of a low avidity antibody because of the high sensitivity exhibited by the cantilever sensor.

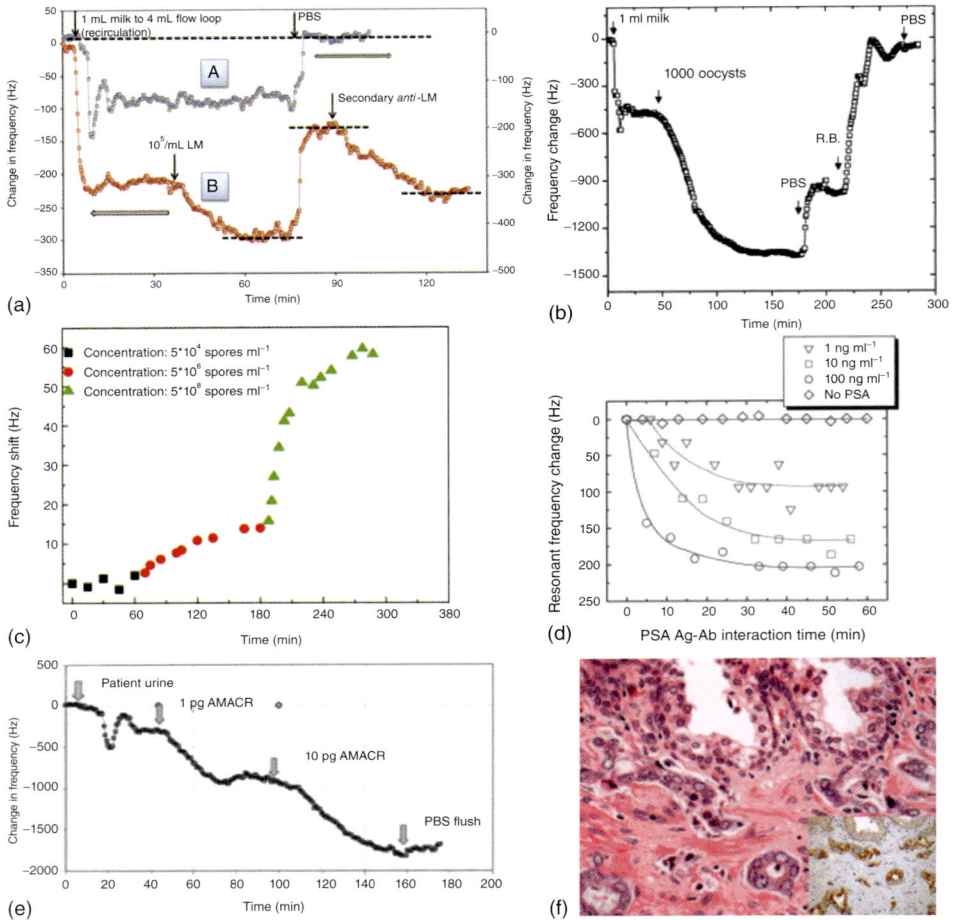

(a)

(b)

(c)

(d)

(e)

(f)

Figure 16.4 (a) Curve A shows sensor response to milk background in the absence of *L. monocytogenes*. Curve B shows a typical detection of *L. monocytogenes* in milk. One milliliters of 10^5 *L. monocytogenes* cells mL^{-1} caused the sensor resonance frequency to decrease by 104 Hz and secondary IgG binding (1 mL, 10 µg mL^{-1}) caused further 106 Hz decrease confirming *L. monocytogenes* detection [8]. (b) Piezoelectric-excited millimeter-sized cantilever (PEMC)-based detection of *C. parvum* oocysts in milk background and flow format (1 mL min^{-1}) [83]. (c) Detection of *B. anthracis* spores using magnetostrictive micro/millimeter cantilever sensor. Measurement was made in continuous fashion to demonstrate *in situ* detection capability [26]. (d) Label-free detection of a prostate-specific antigen in liquid environment using a PZT thin film cantilever [19]. (e) A macro-cantilever method for detecting a prostate cancer biomarker (alpha-methylacyl-CoA racemase; AMACR) in patient urine post biopsy (f) [84]. (f) Prostate carcinoma biopsy sample, H&E stain at 180x original magnification. The top of the image shows the outline of a benign gland. Inset shows the same tissue stained with fluorescent murine monoclonal antibody against AMACR. (180x original magnification) [84].

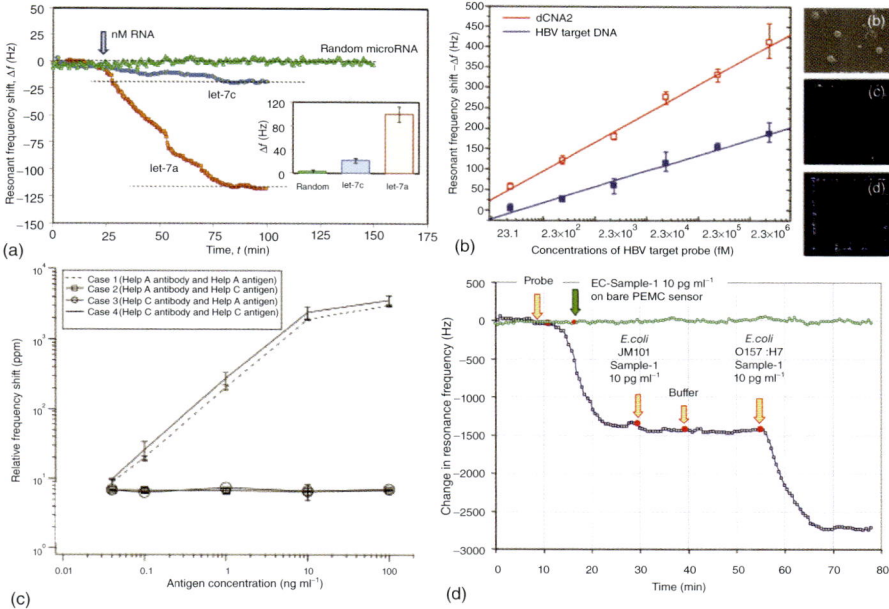

(a)

(b)

(c)

(d)

Figure 16.5 (a) Detection of microRNA in buffer and human serum using Au-coated piezoelectric cantilever sensors [21]. (b) Hepatitis B Virus DNA detection using resonant microcantilever biosensors at 23.1 fM to 2.31 nM range with silica nanoparticle enhancement [82]. (c) Label-free and real-time detection of Hepatitis A and C antigens at different concentrations in serum on cantilevers functionalized with Hepatitis antibodies [81]. (d) Response to 4450 copies of *stx2* of *E. coli* O157:H7 in both buffer and beef wash using the virulent gene *stx2* as an identifying and sensing molecule [85].

16.5.1.3 Waterborne Parasite: *Cryptosporidium parvum*

A cantilever sensor was shown to detect five *C. parvum* oocysts in 25% PBS-buffered milk sample at $1\,\mathrm{ml\,min^{-1}}$ (Figure 16.4b) [83]. To improve sensitivity, a secondary antibody (murine IgM) was used to confirm the presence of bound oocysts on the sensor. The gold-coated sensor was functionalized with Protein G, and then immobilized with goat polyclonal IgG *anti-C. parvum* for oocysts detection. In the concentration range of $50-10^4$ oocysts $\mathrm{ml^{-1}}$, the sensor response was characterized by a semi-log relationship between sensor response and *C. parvum* oocyst concentration. In 25% milk background, the binding rate was slower and total sensor response was approximately ~45% lower than in water-like buffer. Verification of detection by measurement of antigen release was also used to enhance measurement robustness, and is shown in Figure 16.4b.

16.5.1.4 Waterborne Parasite: *Giardia lamblia*

A PEMC biosensor that exhibited high-order resonant modes near $1\,\mathrm{MHz}$ immobilized with a monoclonal antibody against *G. lamblia* exhibited sensitive detection of *G. lamblia* cysts prepared in several water matrices (buffer, tap, and river water), and the detection limit was shown to be $1-10$ cysts ml^{-1}, without a pre-concentration step [86]. The antibody-immobilized sensor was exposed to $1-10^4$ *G. lamblia* cysts ml^{-1} samples in a flow arrangement. When the cysts bound to the antibody on the sensor, the resonant frequency of the cantilever sensor decreased and was recorded continuously. Positive confirmation of sensor detection responses was obtained by environmental scanning electron microscope of sensor surface after the detection experiments. A higher sample flow rate ($0.5-5.0$ ml min^{-1}) gave higher overall sensor detection response, and was interpreted by the authors as a potential convective mass transfer enhancement. Detecting as few as 10 cysts ml^{-1} was achieved in all three water matrixes tested, and significant sensor response was obtained in 15 min. The authors also showed the feasibility of analyzing at a low concentration of 1 cyst ml^{-1} in a 1 l sample at high sample flow rate of 5 ml min^{-1}. See Table 16.1 for a comprehensive summary of bioMEMS studies on whole cell detection.

16.5.2
Proteins: Biomarkers and Toxins

16.5.2.1 Prostate Cancer Biomarker: Prostate Specific Antigen

Nanomechanical Pb(Zr$_{0.52}$Ti$_{0.48}$)O$_3$ (PZT) cantilevers under controlled ambient temperature and humidity were used for label-free detection of PSA at a sensitivity as low as 10 pg ml^{-1} [19]. Concentration-dependent response in the range of $1-100$ ng ml^{-1} was reported by the authors, and is shown in Figure 16.4d. Composite layers of Ta/Pt/PZT/Pt/SiO$_2$ on a silicon-based supporting layer served for self-sensing. The resonant frequency decreased due to the specific binding of PSA to the antibody-immobilized cantilever. Theoretical and experimental analyses suggested that the minimum detection sensitivity depended on the dimensions of the PZT cantilever. The experimentally measured resonant frequency shift was larger than the theoretical value, and the reason suggested was the effect of compressive stress due to PSA–antibody interaction [18]. For the bioassay, a 20 µl liquid test cell was used. The resonant frequency response increased with PSA concentration in the reaction chamber, showing agreement with fluorescence measurements.

16.5.2.2 Prostate Cancer Biomarker: Alpha-methylacyl-CoA Racemase (AMACR)

Assay of clinical samples directly with a cantilever sensor was successfully shown in a recent study for detecting prostate cancer biomarker alpha-methylacyl-CoA racemase (AMACR) in patient urine without a sample preparation step or the use of labeled reagents [84]. Clean catch voided urine specimens were prospectively collected from five confirmed prostate cancer patients 3 weeks post-biopsy. The presence of AMACR in body fluids resulting from prostate carcinoma (biopsy shown in Figure 16.4f) was measured in a blinded manner by exposing 3 ml of urine to the *anti*-AMACR-immobilized PEMC sensor. The sensor exhibited resonance frequency decrease for each of the five patients tested: 4314 ± 35 ($n = 2$),

269 ± 17 $(n = 2)$, 977 ± 64 $(n = 3)$, 600 ± 31 $(n = 2)$, and 801 ± 81 $(n = 2)$ Hz. Positive detection was observed within ~15 min as shown in Figure 16.4e. The responses to positive (AMACR present but antibody absent), negative (AMACR absent but antibody present), and buffer controls (both AMACR and antibody absent) were negligible in comparison with patient samples, -9 ± 13, -34 ± 18, and -6 ± 18 Hz, respectively. Positive verification of AMACR attachment was confirmed by low-pH buffer release. The sensor response was quantitatively correlated to AMACR concentration in urine, and the relationship was used in developing an *in situ* calibration method for quantifying AMACR in patient urine. Estimated concentrations of 42, 2, and 3 fg ml^{-1} AMACR were calculated for the three patients' urine, while the absence of AMACR was confirmed in control urine $(n = 13)$. Because of the simplicity of measurement combined with high sensitivity and specificity, the biosensor method may be suitable in a point-of-care setting for identifying men at risk for prostate cancer.

16.5.2.3 Toxin in Source Water: Microcystin

Microcystin-leucine-arginine (MCLR) is one of the toxic microcystin congeners produced by common cyanobacteria, also known as blue-green algae. A PEMC sensor was immobilized with an antibody against MCLR via amine coupling, and then was exposed to spiked water samples in a flow format. As the toxin bound to the antibody, the resonant frequency decreased proportional to toxin concentration. Three water matrices, namely buffer, tap water, and river water, were spiked with MCLR standards and were successfully detected in the dynamic range of 1 pg ml^{-1} to 100 ng ml^{-1}. The sensor response was characterized by a log–linear relationship between resonant frequency decrease and MCLR concentration. Verification of MCLR detection was done by a sandwich binding on the sensor with a second antibody binding step to MCLR (attached in first detection step), which caused a further resonant frequency decrease. The results clearly showed that MCLR in various water samples can be detected at 1 pg ml^{-1} [95].

16.5.2.4 Toxin in Food Matrices: *Staphylococcal enterotoxin* B

Food toxin detection is another important application area from a public health perspective. Detection of the common food toxin Staphylococcal enterotoxin B (SEB) at picogram levels was investigated using PEMC sensors [62], and with a newer design, the sensor sensitivity was significantly improved to femtogram level in milk matrix [96]. The detection limit in spiked milk and apple juice samples was reported as 10 and 100 fg, respectively. Antibody-immobilized PEMC sensors were exposed to 1 ml milk or apple juice containing 10 fg to 10 ng SEB. The resonant frequency response to 100 fg, 1 pg, and 10 pg of SEB in apple juice resulted in decreases of 113 ± 18 $(n = 4)$, 308 ± 24 $(n = 4)$, and 521 ± 20 $(n = 2)$ Hz, respectively. Similar experiments in milk yielded results that were comparable to apple juice data. Sensor response was observed within the first 20 min of exposure. The responses to control samples: positive (SEB in sample, but no antibody on sensor), negative (SEB absent in sample, antibody on sensor), and buffer (SEB and sensor antibody absent) were statistically insignificant: 17 ± 10 $(n = 3)$, -9 ± 5 $(n = 3)$, and -6 ± 12 ($n = 18$) Hz, respectively. Positive verification was achieved by two methods: by observation of resonant frequency recovery when the antigen was released with a low-pH buffer, and by a second antibody binding step to SEB bound to the sensor, which caused a further resonant frequency decrease. The significance of these results is that cantilever sensors can reliably detect SEB at 10–100 fg, which under the experimental conditions correspond to concentration of 2.5 and 25 fg ml^{-1} in complex background. Another significant advantage demonstrated by the SEB case is the time to results can be less than 30 min, which should prove to be useful in practical applications. See Table 16.2 for a comprehensive summary of bioMEMS studies on protein detection.

16.5.3
Virus

Arrays of surface micromachined, antibody-coated silicon nanomechanical cantilever beams were shown to detect baculovirus in a buffer [109]. Because the sensors were small ($0.5 \mu m \times 6 \mu m$), they exhibited mass sensitivities on the order of 10^{-19} g Hz^{-1}, enabling the detection of a single virus of mass $\sim 3 \times 10^{-15}$ g. Control samples (buffer without baculovirus) gave a small response, which was attributed to nonspecific adsorption to the antibody-immobilized sensor [109].

In another virus study [110], arrays of silicon cantilevers (4–5 µm in length, 1–2 µm in width, and 20–30 nm in thickness) were shown to detect the Vaccinia virus by measuring resonant frequency shift as a function of virus concentration. The frequency spectra of the cantilever beams were measured using a laser Doppler vibrometer under ambient conditions. The results obtained by the authors clearly show that detection of a single Vaccinia virus (9.5 fg) is feasible. See Table 16.3 for a comprehensive summary of bioMEMS studies on virus detection.

16.5.4
Nucleic Acids: Biomarkers and Genes Associated with Toxin Production

16.5.4.1 RNA-Based Biomarkers: MicroRNA

MicroRNAs (miRNAs) are emerging indicators for diagnosing diseases, and thus measuring them at low concentration in a rapid and inexpensive manner is significant. The main difficulty is their short length, which makes the application of traditional molecular techniques more challenging. A recent report [21] showed a sensitive, preparation-free method for near real-time detection of microRNA in buffer and human serum using Au-coated dynamic cantilever sensors [21]. The study characterized responses to *hsa-let-7a* hybridization, an evolving cancer biomarker. High confidence of measurement was established via secondary labeled-DNA hybridization and Au nanoparticle binding in flow format using custom flow cells. The experimental investigation showed successful detection of target *let-7a* over a 6-log range (10 fM to 1 nM) with a LOD (limit of detection) of ~ 10 attomoles (~ 4 fM). The serum background was found to have negligible effect on sensitivity when compared with the results obtained in the buffer. The authors suggested that reduction or elimination of nonspecific binding was due to continuous sensor vibration. The sensor-based method demonstrated excellent selectivity for the microRNA target in comparison with similar microRNA differing by only a single nucleotide (*hsa-let-7c*) and random microRNA sequences (see Figure 16.5a).

16.5.4.2 Gene Signature of a Virus

A DNA probe for Hepatitis B Virus (HBV) targeting the viral precore/core region was used as a bio-recognition agent, and enabled detection of HBV in samples containing genomic DNA. A thiolated 35-mer capture probe was immobilized on an Au-coated microcantilever surface for hybridization with genomic extract in a batch mode. A nanoparticle-labeled 35-mer DNA strand that was complementary to the target also hybridized downstream of the probe region section. The resonance frequency difference between the bare sensor and post two-step hybridization was used as a measure of detection response. The HBV 243-mer target was detected at picomolar level without nanoparticle enhancement and at femtomolar level using the two-step hybridization approach. The authors found resonant frequency shifts to linearly correlate with HBV concentration (Figure 16.5b). Detection of Hepatitis antigen is shown in Figure 16.5c.

16.5.4.3 Toxin-Associated Genes for Pathogen Detection without DNA Amplification

PEMC sensors that exhibit femtogram and sub-femtogram level of sensitivity were successfully shown to measure the presence of identifying genes unique to two pathogens. In the first study [42], genomic extract of $\sim 10^2 - 10^3$ cells of foodborne pathogen, *L. monocytogenes*, were exposed in a flow format to sensors prepared with probes that hybridize with the virulent hemolysin gene, *hlyA*. The measurement was confirmed by two methods: the use of a fluorescent indicator for the presence of double-strand DNA (dsDNA), and the hybridization response of a secondary single-strand DNA (ssDNA) to the hanging end of the target, much like in sandwich-ELISA. Hybridization of the second ssDNA tagged to gold nanoparticles amplified sensor response, and confirmed the target hybridization to the *hlyA* probe on the sensor. The authors also showed that the detection response remained faithful, albeit slightly reduced, when the sample contained 10^4 times excess non-target genomic DNA extracted from *E. coli* JM101. Detection limit was reported as 700 copies (corresponds to 700 *L. monocytogenes* cells), and in proteinaceous and *E. coli* JM101 background the detection limit deteriorated to 7000 *L. monocytogenes* cells. Hybridization response times were ~ 90 min, which is a significant improvement over current methods if rapid results are required [42].

In the second study [85], detection of *E. coli* O157:H7 (*E. coli*) was reported in both buffer and beef wash using the virulent gene *stx2* as an identifying molecule. *E. coli* suspended in buffer, beef or beef wash was separated and genomic DNA extracted using a 30-min procedure, which provided a sample that was then exposed to the sensor immobilized with a 19-mer probe for *stx2* gene. *E. coli* wild strain JM101 subjected to the same preparation and procedure did not induce a hybridization response; nor did the genomic extract of *E. coli* with a bare sensor.

The detection limit of the sensor was estimated to be a few hundred cells, without any culture enrichment or amplification. The genomic DNA extraction time and sensor measurement times were estimated as 1–2 h. The highly sensitive polymerase chain reaction (PCR) methods require an incubation time longer than 8 h followed by a PCR assay time of 3–4 h. Thus, the sensor-based method has significant time advantage because the method does not require purification of target samples and does not require amplification or labeling steps. The PEMC sensor response to beef wash spiked with 4450 *E. coli* cells showed a response of greater than 1 kHz with a noise level of 25 Hz. Since the measurement noise was much lower than the sensor signal, the authors suggested that a much lower pathogen concentration might be detectable in beef wash. Exposure of the sensor to diluted genomic extract prepared in buffer gave reliable response to samples equivalent to 720 cells. One can improve the LOD by using multiple DNA targets. For example, if the non-coding strand for *stx2* was immobilized along with the probe used in their study, one expects to obtain a response twice as large. Further, if other *E. coli* characteristic genes are also included on the sensor, such as those that have been reported in the literature [114–119], the authors suggested that the LOD may be reduced to a few tens of cells, a sensitivity that may be competitive with PCR-based methods. We note the PEMC detection response is shown in Figure 16.5d.

It is important to point out that even though PCR method is highly sensitive, practical food matrix reduces sensitivity due to recovery and contaminating proteins [120]. PCR-based commercial instrumentation that is optimized for *E. coli* in beef samples typically give LOD as 10^4 CFU ml^{-1}, which is obtained after an incubation period of 8–24 h [120, 121]. Typical PCR step requires 3–4 h, and when combined with a conservative incubation time of 12 h, the combined assay requires a minimum of ∼15 h for detecting low concentration samples. Sensors that have high mass change sensitivity have the potential to detect these low concentration samples in a 1–2 h assay time. See Table 16.4 for a comprehensive summary of bioMEMS studies on nucleic acid detection.

16.6
Conclusions and Future Trends

A critical review of the dynamic-mode cantilever sensors for biosensing was presented with a focus on bioassay design and applications investigated. We presented comprehensive tables that summarize the reported applications, which include parameters such as resonant frequency (f), mode order (n), mode type, length scale, geometry, material, sensing medium, Q-value (Q), recognition chemistry (recognition agent (RA)), target detected, sensitivity (S), and measurement format. Biosensing applications were organized by the analyte detected.

We conclude by noting that highly sensitive and reliable cantilever sensors can contribute significantly to discovery driven research as well as practical applications. In spite of the extensive published literature, several important questions and areas remain relatively unexplored. To state a few: (i) What are the resonant properties of very high-order modes of cantilevers with modest or low aspect ratio? (ii) Are sensor non-linearities present in biosensing applications, especially when the resonant mode is a combination of two or more modes? (iii) Does vibration affect binding kinetics, binding equilibrium, and nonspecific adsorption? (iv) Can reliable methods for *in situ* renewal of biosensors be developed, especially for high-throughput applications? (v) Can very rapid resonant frequency measurement provide measures of single molecular binding events?

Acknowledgment

The authors are grateful for the generous support of NSF Grant CBET-1159841 which provided funds for the reported work.

References

1. Ilic, B., Czaplewski, D., Zalalutdinov, M., Craighead, H.G., Neuzil, P., Campagnolo, C., and Batt, C. (2001) Single cell detection with micromechanical oscillators. *J. Vac. Sci. Technol. B*, **19**, 2825–2828.

2. Burg, T.P., Godin, M., Knudsen, S.M., Shen, W., Carlson, G., Foster, J.S., Babcock, K., and Manalis, S.R. (2007) Weighing of biomolecules, single cells and single nanoparticles in fluid. *Nature*, **446**, 1066–1069.

3. Gfeller, K.Y., Nugaeva, N., and Hegner, M. (2005) Rapid biosensor for detection of antibiotic-selective growth of Escherichia coli. *Appl. Environ. Microbiol.*, **71**, 2626–2631.

4. Nugaeva, N., Gfeller, K.Y., Backmann, N., Lang, H.P., Duggelin, M., and Hegner, M. (2005) Micromechanical cantilever array sensors for selective fungal immobilization and fast growth detection. *Biosens. Bioelectron.*, **21**, 849–856.

5. Park, K., Millet, L.J., Kim, N., Li, H.A., Jin, X.Z., Popescu, G., Aluru, N.R., Hsia, K.J., and Bashir, R. (2010) Measurement of adherent cell mass and growth. *Proc. Natl. Acad. Sci. U.S.A.*, **107**, 20691–20696.

6. Godin, M., Delgado, F.F., Son, S.M., Grover, W.H., Bryan, A.K., Tzur, A., Jorgensen, P., Payer, K., Grossman, A.D., Kirschner, M.W., and Manalis, S.R. (2010) Using buoyant mass to measure the growth of single cells. *Nat. Methods*, **7**, 387–390.

7. Johnson, B.N. and Mutharasan, R. (2012) pH effect on Protein G orientation on gold surfaces and characterization of adsorption thermodynamics. *Langmuir*, **28**, 6928–6934.

8. Sharma, H. and Mutharasan, R. (2013) Rapid and sensitive immunodetection of Listeria monocytogenes in milk using a novel piezoelectric cantilever sensor. *Biosens. Bioelectron.*, **45**, 158–162.

9. Campbell, G.A., Uknalis, J., Tu, S.I., and Mutharasan, R. (2007) Detect of Escherichia coli O157:H7 in ground beef samples using piezoelectric excited millimeter-sized cantilever (PEMC) sensors. *Biosens. Bioelectron.*, **22**, 1296–1302.

10. Maraldo, D. and Mutharasan, R. (2007) 10-minute assay for detecting Escherichia coli O157: H7 in ground beef samples using piezoelectric-excited millimeter-size cantilever sensors. *J. Food Prot.*, **70**, 1670–1677.

11. Maraldo, D. and Mutharasan, R. (2007) Preparation-free method for detecting

Escherichia coli O157: H7 in the presence of spinach, spring lettuce mix, and ground beef particulates. *J. Food Prot.*, **70**, 2651–2655.

12. Detzel, A.J., Campbell, G.A., and Mutharasan, R. (2006) Rapid assessment of Escherichia coli by growth rate on piezoelectric-excited millimeter-sized cantilever (PEMC) sensors. *Sens. Actuators, B*, **117**, 58–64.

13. Mao, X., Yang, L., Su, X.-L., and Li, Y. (2006) A nanoparticle amplification based quartz crystal microbalance DNA sensor for detection of Escherichia coli O157:H7. *Biosens. Bioelectron.*, **21**, 1178–1185.

14. Campbell, G.A. and Mutharasan, R. (2005) Detection of pathogen Escherichia coli O157: H7 using self-excited PZT-glass microcantilevers. *Biosens. Bioelectron.*, **21**, 462–473.

15. Campbell, G.A. and Mutharasan, R. (2005) Escherichia coli O157: H7 detection limit of millimeter-sized PZT cantilever sensors is 700 Cells/mL. *Anal. Sci.*, **21**, 355–357.

16. Lee, S.M., Hwang, K.S., Yoon, H.J., Yoon, D.S., Kim, S.K., Lee, Y.S., and Kim, T.S. (2009) Sensitivity enhancement of a dynamic mode microcantilever by stress inducer and mass inducer to detect PSA at low picogram levels. *Lab Chip*, **9**, 2683–2690.

17. Vancura, C., Li, Y., Lichtenberg, J., Kirstein, K.U., Hierlemann, A., and Josse, F. (2007) Liquid-phase chemical and biochemical detection using fully integrated magnetically actuated complementary metal oxide semiconductor resonant cantilever sensor systems. *Anal. Chem.*, **79**, 1646–1654.

18. Lee, J.H., Hwang, K.S., Park, J., Yoon, K.H., Yoon, D.S., and Kim, T.S. (2005) Immunoassay of prostate-specific antigen (PSA) using resonant frequency shift of piezoelectric nanomechanical microcantilever. *Biosens. Bioelectron.*, **20**, 2157–2162.

19. Hwang, K.S., Lee, J.H., Park, J., Yoon, D.S., Park, J.H., and Kim, T.S. (2004) In-situ quantitative analysis of a prostate-specific antigen (PSA) using a nanomechanical PZT cantilever. *Lab Chip*, **4**, 547–552.

20. Maraldo, D., Garcia, F.U., and Mutharasan, R. (2007) 15-minute detection of a prostate cancer biomarker (AMACR) in voided urine samples using immuno-cantilever sensors. *Mod. Pathol.*, **20**, 352A.

21. Johnson, B.N. and Mutharasan, R. (2012) Sample preparation-free, real-time detection of microRNA in human serum using piezoelectric cantilever biosensors at attomole level. *Anal. Chem.*, **84**, 10426–10436.

22. Binnig, G., Quate, C.F., and Gerber, C. (1986) Atomic force microscope. *Phys. Rev. Lett.*, **56**, 930–933.

23. Gupta, A., Akin, D., and Bashir, R. (2004) Detection of bacterial cells and antibodies using surface micromachined thin silicon cantilever resonators. *J. Vac. Sci. Technol. B*, **22**, 2785–2791.

24. Ilic, B., Czaplewski, D., Craighead, H.G., Neuzil, P., Campagnolo, C., and Batt, C. (2000) Mechanical resonant immunospecific biological detector. *Appl. Phys. Lett.*, **77**, 450–452.

25. Su, M., Li, S.U., and Dravid, V.P. (2003) Microcantilever resonance-based DNA detection with nanoparticle probes. *Appl. Phys. Lett.*, **82**, 3562–3564.

26. Li, S.Q., Fu, L.L., Barbaree, J.M., and Cheng, Z.Y. (2009) Resonance behavior of magnetostrictive micro/milli-cantilever and its application as a biosensor. *Sens. Actuators, B*, **137**, 692–699.

27. Yi, J.W., Shih, W.Y., and Shih, W.H. (2002) Effect of length, width, and mode on the mass detection sensitivity of piezoelectric unimorph cantilevers. *J. Appl. Phys.*, **91**, 1680–1686.

28. Lee, J.H., Kim, T.S., and Yoon, K.H. (2004) Effect of mass and stress on resonant frequency shift of functionalized $Pb(Zr_{0.52}Ti_{0.48})O_3$ thin film microcantilever for the detection of C-reactive protein. *Appl. Phys. Lett.*, **84**, 3187–3189.

29. McFarland, A.W., Poggi, M.A., Bottomley, L.A., and Colton, J.S. (2004) Production and characterization of polymer microcantilevers. *Rev. Sci. Instrum.*, **75**, 2756–2758.

30. Bottomley, L.A., McFarland, A.W., Poggi, M.A., and Colton, J.S. (2004) Polymeric microcantilevers for sensing applications. *Abstr. Pap. Am. Chem. Soc.*, **227**, 96-PMSE.

31. Campbell, G.A. and Mutharasan, R. (2007) A method of measuring Escherichia coli O157: H7 at 1 cell/mL in 1 liter sample using antibody functionalized piezoelectric-excited millimeter-sized cantilever sensor. *Environ. Sci. Technol.*, **41**, 1668–1674.

32. Calleja, M., Nordstrom, M., Alvarez, M., Tamayo, J., Lechuga, L.M., and Boisen, A. (2005) Highly sensitive polymer-based cantilever-sensors for DNA detection. *Ultramicroscopy*, **105**, 215–222.

33. Park, K., Jang, J., Irimia, D., Sturgis, J., Lee, J., Robinson, J.P., Toner, M., and Bashir, R. (2008) "Living cantilever arrays" for characterization of mass of single live cells in fluids. *Lab Chip*, **8**, 1034–1041.

34. Davila, A.P., Jang, J., Gupta, A.K., Walter, T., Aronson, A., and Bashir, R. (2007) Microresonator mass sensors for detection of Bacillus anthracis Sterne spores in air and water. *Biosens. Bioelectron.*, **22**, 3028–3035.

35. Vazquez, J., Rivera, M.A., Hernando, J., and Sanchez-Rojas, J.L. (2009) Dynamic response of low aspect ratio piezoelectric microcantilevers actuated in different liquid environments. *J. Micromech. Microeng.*, **19**, 015020.

36. Lee, J.H., Lee, S.T., Yao, C.M., and Fang, W.L. (2007) Comments on the size effect on the microcantilever quality factor in free air space. *J. Micromech. Microeng.*, **17**, 139–146.

37. Waggoner, P.S. and Craighead, H.G. (2009) The relationship between material properties, device design, and the sensitivity of resonant mechanical sensors. *J. Appl. Phys.*, **105**, 054306.

38. Ricciardi, C., Canavese, G., Castagna, R., Ferrante, I., Ricci, A., Marasso, S.L., Napione, L., and Bussolino, F. (2010) Integration of microfluidic and cantilever technology for biosensing application in liquid environment. *Biosens. Bioelectron.*, **26**, 1565–1570.

39. Lee, Y., Lim, G., and Moon, W. (2006) A self-excited micro cantilever biosensor actuated by PZT using the mass micro balancing technique. *Sens. Actuators, A*, **130**, 105–110.

40. Fadel, L., Dufour, I., Lochon, F., and Francais, O. (2004) Signal to noise ratio of resonant microcantilever type chemical sensors as a function of resonant frequency and quality factor. *Sens. Actuators, B*, **102**, 73–77.

41. Johnson, B.N. and Mutharasan, R. (2010) Expression of picogram sensitive bending modes in piezoelectric cantilever sensors with nonuniform electric fields generated by asymmetric electrodes. *Rev. Sci. Instrum.*, **81**, 125108.

42. Sharma, H. and Mutharasan, R. (2013) hlyA gene-based sensitive detection of Listeria monocytogenes using a novel cantilever sensor. *Anal. Chem.*, **85**, 3222–3228.

43. Sharma, H., Lakshmanan, R.S., Johnson, B.N., and Mutharasan, R. (2011) Piezoelectric cantilever sensors with asymmetric anchor exhibit picogram sensitivity in liquids. *Sens. Actuators, B*, **153**, 64–70.

44. Liu, Y.J., Li, X.X., Zhang, Z.X., Zuo, G.M., Cheng, Z.X., and Yu, H.T. (2009) Nanogram per milliliter-level immunologic detection of alpha-fetoprotein with integrated rotating-resonance microcantilevers for early-stage diagnosis of heptocellular carcinoma. *Biomed. Microdevices*, **11**, 183–191.

45. Jin, D.Z., Li, X.X., Bao, H.H., Zhang, Z.X., Wang, Y.L., Yu, H.T., and Zuo, G.M. (2007) Integrated cantilever sensors with a torsional resonance mode for ultraresoluble on-the-spot bio/chemical detection. *Appl. Phys. Lett.*, **90**, 041901.

46. Xia, X.Y. and Li, X.X. (2008) Resonance-mode effect on microcantilever mass-sensing performance in air. *Rev. Sci. Instrum.*, **79**, 074301.

47. Pang, W., Yan, L., Zhang, H., Yu, H.Y., Kim, E.S., and Tang, W.C. (2006) Femtogram mass sensing platform based on lateral extensional mode piezoelectric resonator. *Appl. Phys. Lett.*, **88**, 243503.

48. Salehi-Khojin, A., Bashash, S., Jalili, N., Muller, M., and Berger, R. (2009) Nanomechanical cantilever active probes for ultrasmall mass detection. *J. Appl. Phys.*, **105**, 013506.

49. Itoh, T., Lee, C., and Suga, T. (1996) Deflection detection and feedback actuation using a self-excited piezo-electric Pb(Zr,Ti)O$_3$ microcantilever for dynamic scanning force microscopy. *Appl. Phys. Lett.*, **69**, 2036–2038.

50. Itoh, T. and Suga, T. (1996) Self-excited force-sensing microcantilevers with piezoelectric thin films for dynamic scanning force microscopy. *Sens. Actuators, A*, **54**, 477–481.

51. Shih, W.Y., Campbell, G., Yi, J.W., Mutharasan, R., and Shih, W.H. (2003) Miniaturized highly piezoelectric uni-morph cantilevers for rapid in-situ pathogen quantification. *Abstr. Pap. Am. Chem. Soc.*, **225**, 257-IEC.

52. Gaucher, P., Eichner, D., Hector, J., and Von Munch, W. (1998) Piezoelectric bimorph cantilever for actuating and sensing applications. *J. Phys. IV*, **8**, 235–238.

53. Coughlin, M.F., Stamenovic, D., and Smits, J.G. (1997) Determining spring stiffness by the resonance frequency of cantilevered piezoelectric bimorphs. *IEEE Trans. Ultrason. Ferroelectr. Freq. Control*, **44**, 730–732.

54. Smits, J.G., Dalke, S.I., and Cooney, T.K. (1991) The constituent equations of piezoelectric bimorphs. *Sens. Actuators, A*, **28**, 41–61.

55. Campbell, G.A. and Mutharasan, R. (2005) Detection and quantification of proteins using self-excited PZT-glass millimeter-sized cantilever. *Biosens. Bioelectron.*, **21**, 597–607.

56. Yi, J.W., Shih, W.Y., Mutharasan, R., and Shih, W.H. (2003) In situ cell detection using piezoelectric lead zirconate titanate-stainless steel cantilevers. *J. Appl. Phys.*, **93**, 619–625.

57. Mahmoodi, S.N. and Jalili, N. (2007) Non-linear vibrations and frequency response analysis of piezoelectri-cally driven microcantilevers. *Int. J. Nonlinear Mech.*, **42**, 577–587.

58. Mahmoodi, S.N. and Jalili, N. (2008) Coupled flexural-torsional nonlinear vibrations of piezoelectrically actu-ated microcantilevers with application to friction force microscopy. *J. Vib. Acoust*, **130**, 061003.

59. Mahmoodi, S.N. and Jalili, N. (2009) Piezoelectrically actuated microcan-tilevers: an experimental nonlinear vibration analysis. *Sens. Actuators, A*, **150**, 131–136.

60. Johnson, B.N. and Mutharasan, R. (2011) The origin of low-order and high-order impedance-coupled reso-nant modes in piezoelectric-excited millimeter-sized cantilever (PEMC) sensors: experiments and finite ele-ment models. *Sens. Actuators, B*, **155**, 868–877.

61. Lee, Y., Lee, S., Seo, H., Jeon, S., and Moon, W. (2008) Label-free detec-tion of a biomarker with piezoelectric micro cantilever based on mass micro balancing. *J. Assoc. Lab. Autom.*, **13**, 259–264.

62. Campbell, G.A., Medina, M.B., and Mutharasan, R. (2007) Detection of Staphylococcus enterotoxin B at picogram levels using piezoelectric-excited millimeter-sized cantilever sensors. *Sens. Actuators, B*, **126**, 354–360.

63. Rijal, K. and Mutharasan, R. (2007) PEMC-based method of measuring DNA hybridization at femtomolar con-centration directly in human serum and in the presence of copious noncom-plementary strands. *Anal. Chem.*, **79**, 7392–7400.

64. Sader, J.E. (1998) Frequency response of cantilever beams immersed in vis-cous fluids with applications to the atomic force microscope. *J. Appl. Phys.*, **84**, 64–76.

65. Hermanson, G.T. (1996) *Bioconjugate Techniques*, Elsevier, San Diego, CA.

66. Nugaeva, N., Gfeller, K.Y., Backmann, N., Duggelin, M., Lang, H.P., Guntherodt, H.J., and Hegner, M. (2007) An antibody-sensitized micro-fabricated cantilever for the growth detection of Aspergillus niger spores. *Microsc. Microanal.*, **13**, 13–17.

67. Maraldo, D., Rijal, K., Campbell, G., and Mutharasan, R. (2007) Method for label-free detection of femtogram

quantities of biologics in flowing liquid samples. *Anal. Chem.*, **79**, 2762–2770.

68. Oliviero, G., Bergese, P., Canavese, G., Chiari, M., Colombi, P., Cretich, M., Damin, F., Fiorilli, S., Marasso, S.L., Ricciardi, C., Rivolo, P., and Depero, L.E. (2009) A biofunctional polymeric coating for microcantilever molecular recognition (vol 630, pg 161, 2008). *Anal. Chim. Acta*, **655**, 92.

69. Zheng, S., Choi, J.H., Lee, S.M., Hwang, K.S., Kim, S.K., and Kim, T.S. (2011) Analysis of DNA hybridization regarding the conformation of molecular layer with piezoelectric microcantilevers. *Lab Chip*, **11**, 63–69.

70. Ilic, B., Yang, Y., Aubin, K., Reichenbach, R., Krylov, S., and Craighead, H.G. (2005) Enumeration of DNA molecules bound to a nanomechanical oscillator. *Nano Lett.*, **5**, 925–929.

71. Love, J.C., Estroff, L.A., Kriebel, J.K., Nuzzo, R.G., and Whitesides, G.M. (2005) Self-assembled monolayers of thiolates on metals as a form of nanotechnology. *Chem. Rev.*, **105**, 1103–1169.

72. Johnson, B.N. and Mutharasan, R. (2013) A cantilever biosensor-based assay for toxin-producing cyanobacteria microcystis aeruginosa using 16S rRNA. *Environ. Sci. Technol.*, **47**, 12333–12341.

73. Johnson, B.N. and Mutharasan, R. (2013) Regeneration of gold surfaces covered by adsorbed thiols and proteins using liquid-phase hydrogen peroxide-mediated UV-Photooxidation. *J. Phys. Chem. C*, **117**, 1335–1341.

74. Dhayal, B., Henne, W.A., Doorneweerd, D.D., Reifenberger, R.G., and Low, P.S. (2006) Detection of Bacillus subtilis spores using peptide-functionalized cantilever arrays. *J. Am. Chem. Soc.*, **128**, 3716–3721.

75. Fu, L.L., Li, S.Q., Zhang, K.W., Chen, I.H., Barbaree, J.M., Zhang, A.X., and Cheng, Z.Y. (2011) Detection of Bacillus anthracis spores using phage-immobilized magnetostrictive milli/micro cantilevers. *IEEE Sens. J.*, **11**, 1684–1691.

76. Hwang, K.S., Lee, S.M., Eom, K., Lee, J.H., Lee, Y.S., Park, J.H., Yoon, D.S., and Kim, T.S. (2007) Nanomechanical microcantilever operated in vibration modes with use of RNA aptamer as receptor molecules for label-free detection of HCV helicase. *Biosens. Bioelectron.*, **23**, 459–465.

77. Gupta, A.K., Nair, P.R., Akin, D., Ladisch, M.R., Broyles, S., Alam, M.A., and Bashir, R. (2006) Anomalous resonance in a nanomechanical biosensor. *Proc. Natl. Acad. Sci. U.S.A.*, **103**, 13362–13367.

78. Mertens, J., Calleja, M., Ramos, D., Taryn, A., and Tamayo, J. (2007) Role of the gold film nanostructure on the nanomechanical response of microcantilever sensors. *J. Appl. Phys.*, **101**, 034904.

79. Campbell, G.A. and Mutharasan, R. (2006) Detection of Bacillus anthracis spores and a model protein using PEMC sensors in a flow cell at 1 mL/min. *Biosens. Bioelectron.*, **22**, 78–85.

80. Shu, W., Laue, E.D., and Seshia, A.A. (2007) Investigation of biotin–streptavidin binding interactions using microcantilever sensors. *Biosens. Bioelectron.*, **22**, 2003–2009.

81. Timurdogan, E., Alaca, B.E., Kavakli, I.H., and Urey, H. (2011) MEMS biosensor for detection of Hepatitis A and C viruses in serum. *Biosens. Bioelectron.*, **28**, 189–194.

82. Cha, B.H., Lee, S.M., Park, J.C., Hwang, K.S., Kim, S.K., Lee, Y.S., Ju, B.K., and Kim, T.S. (2009) Detection of Hepatitis B Virus (HBV) DNA at femtomolar concentrations using a silica nanoparticle-enhanced microcantilever sensor. *Biosens. Bioelectron.*, **25**, 130–135.

83. Xu, S. and Mutharasan, R. (2010) Detection of Cryptosporidium parvum in buffer and in complex matrix using PEMC sensors at 5 oocysts/mL. *Anal. Chim. Acta*, **669**, 81–86.

84. Maraldo, D., Garcia, F.U., and Mutharasan, R. (2007) Method for quantification of a prostate cancer

biomarker in urine without sample preparation. *Anal. Chem.*, **79**, 7683–7690.

85. Rijal, K. and Mutharasan, R. (2013) A method for DNA-based detection of E. coli O157:H7 in proteinous background using piezoelectric-excited cantilever sensors. *Analyst*, **138**, 2943–2950.

86. Xu, S. and Mutharasan, R. (2010) Rapid and sensitive detection of Giardia lamblia using a piezoelectric cantilever biosensor in finished and source waters. *Environ. Sci. Technol.*, **44**, 1736–1741.

87. Bryan, A.K., Goranov, A., Amon, A., and Manalis, S.R. (2010) Measurement of mass, density, and volume during the cell cycle of yeast. *Proc. Natl. Acad. Sci. U.S.A.*, **107**, 999–1004.

88. Xu, S., Sharma, H., and Mutharasan, R. (2010) Sensitive and selective detection of mycoplasma in cell culture samples using cantilever sensors. *Biotechnol. Bioeng.*, **105**, 1069–1077.

89. Campbell, G.A. and Mutharasan, R. (2008) Near real-time detection of Cryptosporidium parvum oocyst by IgM-functionalized piezoelectric-excited millimeter-sized cantilever biosensor. *Biosens. Bioelectron.*, **23**, 1039–1045.

90. Xie, H., Vitard, J., Haliyo, S., and Regnier, S. (2008) Enhanced sensitivity of mass detection using the first torsional mode of microcantilevers. *Meas. Sci. Technol.*, **19**, 055207.

91. Godin, M., Bryan, A.K., Burg, T.P., Babcock, K., and Manalis, S.R. (2007) Measuring the mass, density, and size of particles and cells using a suspended microchannel resonator. *Appl. Phys. Lett.*, **91**, 123121.

92. Campbell, G.A., Delesdernier, D., and Mutharasan, R. (2007) Detection of airborne Bacillus anthracis spores by an integrated system of an air sampler and a cantilever immunosensor. *Sens. Actuators, B*, **127**, 376–382.

93. Campbell, G.A. and Mutharasan, R. (2007) Method of measuring Bacillus anthracis spores in the presence of copious amounts of Bacillus thuringiensis and Bacillus cereus. *Anal. Chem.*, **79**, 1145–1152.

94. Gfeller, K.Y., Nugaeva, N., and Hegner, M. (2005) Micromechanical oscillators as rapid biosensor for the detection of active growth of Escherichia coli. *Biosens. Bioelectron.*, **21**, 528–533.

95. Ding, Y.J. and Mutharasan, R. (2011) Highly sensitive and rapid detection of microcystin-LR in source and finished water samples using cantilever sensors. *Environ. Sci. Technol.*, **45**, 1490–1496.

96. Maraldo, D. and Mutharasan, R. (2007) Detection and confirmation of staphylococcal enterotoxin B in apple juice and milk using piezoelectric-excited millimeter-sized cantilever sensors at 2.5 fg/mL. *Anal. Chem.*, **79**, 7636–7643.

97. von Muhlen, M.G., Brault, N.D., Knudsen, S.M., Jiang, S.Y., and Manalis, S.R. (2010) Label-free biomarker sensing in undiluted serum with suspended microchannel resonators. *Anal. Chem.*, **82**, 1905–1910.

98. Shin, S., Kim, J.P., Sim, S.J., and Lee, J. (2008) A multisized piezoelectric microcantilever biosensor array for the quantitative analysis of mass and surface stress. *Appl. Phys. Lett.*, **93**, 102902.

99. Xia, X.Y., Zhang, Z.X., and Li, X.X. (2008) A Latin-cross-shaped integrated resonant cantilever with second torsion-mode resonance for ultra-resoluble bio-mass sensing. *J. Micromech. Microeng.*, **18**, 035028.

100. Li, Y., Vancura, C., Kirstein, K.U., Lichtenberg, J., and Hierlemann, A. (2008) Monolithic resonant-cantilever-based CMOS microsystem for biochemical sensing. *IEEE Trans. Circuits Syst.*, **55**, 2551–2560.

101. Hosaka, S., Chiyoma, T., Ikeuchi, A., Okano, H., Sone, H., and Izumi, T. (2006) Possibility of a femtogram mass biosensor using a self-sensing cantilever. *Curr. Appl. Phys.*, **6**, 384–388.

102. Hwang, K.S., Eom, K., Lee, J.H., Chun, D.W., Cha, B.H., Yoon, D.S., and Kim, T.S. (2006) Dominant surface stress driven by biomolecular interactions in the dynamical response of nanomechanical microcantilevers. *Appl. Phys. Lett.*, **89**, 173905.

103. Campbell, G.A. and Mutharasan, R. (2006) Use of piezoelectric-excited

millimeter-sized cantilever sensors to measure albumin interaction with self-assembled monolayers of alkanethiols having different functional headgroups. *Anal. Chem.*, **78**, 2328–2334.

104. Kang, G.Y., Han, G.Y., Kang, J.Y., Cho, I.H., Park, H.H., Paek, S.H., and Kim, T.S. (2006) Label-free protein assay with site-directly immobilized antibody using self-actuating PZT cantilever. *Sens. Actuators, B*, **117**, 332–338.

105. Burg, T.P., Mirza, A.R., Milovic, N., Tsau, C.H., Popescu, G.A., Foster, J.S., and Manalis, S.R. (2006) Vacuum-packaged suspended microchannel resonant mass sensor for biomolecular detection. *J. Microelectromech. Syst.*, **15**, 1466–1476.

106. Lee, J.H., Yoon, K.H., Hwang, K.S., Park, J., Ahn, S., and Kim, T.S. (2004) Label free novel electrical detection using micromachined PZT monolithic thin film cantilever for the detection of C-reactive protein. *Biosens. Bioelectron.*, **20**, 269–275.

107. Burg, T.P. and Manalis, S.R. (2003) Suspended microchannel resonators for biomolecular detection. *Appl. Phys. Lett.*, **83**, 2698–2700.

108. Tamayo, J., Humphris, A.D.L., Malloy, A.M., and Miles, M.J. (2001) Chemical sensors and biosensors in liquid environment based on microcantilevers with amplified quality factor. *Ultramicroscopy*, **86**, 167–173.

109. Ilic, B., Yang, Y., and Craighead, H.G. (2004) Virus detection using nanoelectromechanical devices. *Appl. Phys. Lett.*, **85**, 2604–2606.

110. Gupta, A., Akin, D., and Bashir, R. (2004) Single virus particle mass detection using microresonators with nanoscale thickness. *Appl. Phys. Lett.*, **84**, 1976–1978.

111. Capobianco, J.A., Shih, W.H., Leu, J.H., Lo, G.C.F., and Shih, W.Y. (2010) Label free detection of white spot syndrome virus using lead magnesium niobate-lead titanate piezoelectric microcantilever sensors. *Biosens. Bioelectron.*, **26**, 964–969.

112. Braun, T., Ghatkesar, M.K., Backmann, N., Grange, W., Boulanger, P., Letellier, L., Lang, H.P., Bietsch, A., Gerber, C., and Hegner, M. (2009) Quantitative time-resolved measurement of membrane protein-ligand interactions using microcantilever array sensors. *Nat. Nanotechnol.*, **4**, 179–185.

113. Johnson, L., Gupta, A.T.K., Ghafoor, A., Akin, D., and Bashir, R. (2006) Characterization of vaccinia virus particles using microscale silicon cantilever resonators and atomic force microscopy. *Sens. Actuators, B*, **115**, 189–197.

114. Trabulsi, L.R., Keller, R., and Gomes, T.A.T. (2002) Typical and atypical enteropathogenic Escherichia coli. *Emerging Infect. Dis.*, **8**, 508–513.

115. Yaron, S. and Matthews, K.R. (2002) A reverse transcriptase-polymerase chain reaction assay for detection of viable Escherichia coli O157:H7: investigation of specific target genes. *J. Appl. Microbiol.*, **92**, 633–640.

116. Fortin, N.Y., Mulchandani, A., and Chen, W. (2001) Use of real-time polymerase chain reaction and molecular beacons for the detection of Escherichia coli O157:H7. *Anal. Biochem.*, **289**, 281–288.

117. Wang, G.H., Clark, C.G., and Rodgers, F.G. (2002) Detection in Escherichia coli of the genes encoding the major virulence factors, the genes defining the O157:H7 serotype, and components of the type 2 Shinga toxin family by multiplex PCR. *J. Clin. Microbiol.*, **40**, 3613–3619.

118. Franz, E., Klerks, M.A., Vos, O.J.D., Termorshuizen, A.J., and Bruggen, A.H.C.v. (2007) Prevalence of Shiga toxin-producing Escherichia coli stx1, stx2, eaeA, and rfbE genes and survival of E. coli O157:H7 in manure from organic and low-input conventional dairy farms. *Appl. Environ. Microbiol.*, **73**, 2180–2190.

119. Mora, A., Leon, S.L., Blanco, M., Blanco, J.E., Lopez, C., Dahbi, G., Echeita, A., Gonzalez, E.A., and Blanco, J. (2007) Phage types, virulence genes and PFGE profiles of Shiga toxin-producing Escherichia coli O157:H7 isolated from raw beef, soft cheese and vegetables in Lima (Peru). *Int. J. Food Microbiol.*, **114**, 204–210.

120. Roda, A., Mirasoli, M., Roda,
B., Bonvicini, F., Colliva, C., and
Reschiglian, P. (2012) Recent developments in rapid multiplexed bioana-
lytical methods for foodborne pathogen
bacteria detection. *Microchim. Acta*,
178, 7–28.

121. Lu, J.Z., Tang, M.J., Liu, H.L., Huang,
L.H., Wan, Z.G., Zhang, H., and Zhao,
F. (2012) Comparative evaluation of a
phage protein ligand assay with VIDAS
and BAX methodology for detection
of Escherichia coli O157:H7 using a
standard nonproprietary enrichment
broth. *J. AOAC Int.*, **95**, 1669–1671.

122. Kim, S., Yi, D.C., Passian, A., and
Thundat, T. (2010) Observation
of an anomalous mass effect in
microcantilever-based biosensing
caused by adsorbed DNA. *Appl. Phys.
Lett.*, **96**, 153703.

17
Fluid Property Sensors

Erwin K. Reichel, Martin Heinisch, and Bernhard Jakoby

17.1
Introduction

This chapter treats the measurement of physical fluid properties using miniaturized devices. The most practical approach is to measure the frequency response of immersed or partly exposed mechanical resonators. Depending on the technology used, the mechanical response can be measured optically or in the electrical domain utilizing piezoelectric, piezoresistive, or electrodynamic coupling mechanisms. Spurious signal components have to be handled accordingly to enable a proper interpretation of the data. The fluid parameters that are accessible by this method are the mass-density, the complex viscosity at the frequency of oscillation (or equivalently, the linear viscoelastic storage and loss moduli), and the speed of sound if compressibility of the fluid is of relevance in the considered frequency range. The specific form of the fluid-structure interaction as a consequence of the resonance mode shape determines how these parameters affect the sensor response function and whether a separation of the individual parameters is possible or not.

As the sensor principles rely on dynamic methods, the fluid density always appears in the reaction force. Often, an apparently added mass term is introduced, which has to be handled with care. Only in cases where the vibrating structure has holes or gaps where trapped liquid moves completely in-phase with the rest of the structure, the amount of added mass can be identified independently of viscous effects. This effect is known as *liquid trapping*. In all other cases, the complete velocity field around the structure has to be known to calculate the apparently added mass contributing to the fluid-reaction force. A highly accurate method of density measurement is to fill a tube and measure the shift in resonance frequency, see [1] for a laboratory instrument, and [2] for a micro-electromechanical systems (MEMS) version of this principle. Measurements of contamination and concentration, for example, salinity, alcohol and sugar in water, can be realized by measuring the fluid density.

Viscosity is an important property for the processing and application of liquids, or soft matter in general. In the laboratory, a variety of viscometric and rheometric

Resonant MEMS – Fundamentals, Implementation and Application, First Edition.
Edited by Oliver Brand, Isabelle Dufour, Stephen M. Heinrich and Fabien Josse.
© 2015 Wiley-VCH Verlag GmbH & Co. KGaA. Published 2015 by Wiley-VCH Verlag GmbH & Co. KGaA.

methods are available, often requiring delicate instrumentation and skilled operators. Some process in-line sensors are available but restricted to simple (Newtonian) liquids and limited in viscosity range. Solvents and pure, low viscous, oils are examples of Newtonian liquids, where a single viscosity parameter is sufficient for characterizing the fluid's behavior in the whole range of applicable shear rates and flow conditions. Suspensions like paint, emulsions, surfactant and lipid solutions are examples for fluids exhibiting various kinds of non-Newtonian effects [3]. For these materials, a wide spectrum of rheological properties can be measured. A first extension is considering a shear-rate dependent viscosity [4]. Increasing viscosity with shear-rate is known as the *shear-thickening* effect, the opposite as shear-thinning—both phenomena are observed, for example, in concentrated suspensions [5]. Some fluids need to overcome a certain stress threshold, known as the *yield stress*, to make a transition from elastic to viscous behavior. Common examples of yield-stress fluids are mayonnaise, toothpaste, and various gels. These materials do not behave like common liquids as they can easily maintain a deformed surface and it requires an extra effort to make them flow.

For larger deformations, the methods of finite strain theory have to be adopted, where the description of the deformation in terms of the common (infinitesimal) strain tensor is replaced by the Cauchy-Green deformation tensor (or, similarly, by the Finger tensor [4, 6]). For the example of a simple shear deformation, these tensors contain quadratic terms in shear strain. In terms of these finite strain tensors, comparatively simple constitutive models (representing so-called neo-Hookean materials) covering non-linear viscoelastic effects, can be established. An example of such effects would be the appearance of normal stress components in the direction of shear gradient and orthogonal to it. A well-known effect related to normal stresses is the Weissenberg effect where a liquid polymer solution is drawn into the direction of an immersed spinning rod rather than being thrown outward [4]. Often these materials exhibit such a complex behavior only at sufficiently large deformation or rate of deformation, leading to non-linear rheological effects, which include the dependence of the constitutive parameters on the amplitude of the strain or stress, memory effects (such as hysteresis), and explicit shear-time dependence like thixotropy [7]. Nevertheless, measuring at very low deformation amplitudes but over a sufficiently large frequency range provides valuable information about the structure or condition. Using resonators as sensor devices for liquid property measurements has proven to be a successful approach and constitutes an active field of research with some issues yet to be resolved [8].

In miniaturized devices, viscous damping usually dominates the energy loss by radiation of compressional acoustic waves [9]. The reason is that the wavelength of pressure waves at resonance is usually larger than the length scale of typical resonators, such that radiation phenomena can often be neglected. Exceptions are specifically designed MEMS transducers based on relatively large membranes like capacitive micromachined ultrasonic (CMUT) devices [10]. Using compressional waves, in addition to the previously mentioned parameters, the speed of sound and bulk viscosity can be measured [11–13]. General methods applicable for density, speed of sound, and acoustic impedance sensing based on ultrasonic pressure waves are described in [14, 15].

17.2
Definition of Fluid Properties

17.2.1
Rheological Properties

The basic theory of rheology can be found in several textbooks, for example, [3, 4, 6]. A brief overview of the concepts relevant to the following examination is outlined here.

Stress and strain are spatially localized measures described by second rank tensors. The strain tensor is symmetric by definition, while symmetry of the stress tensor can be proven for most cases [16]. This reduces the number of independent components of each tensor to six, allowing to write them as six component vectors in reduced (Voigt) notation. A constitutive material model relates stress to strain. Without long-range interactions, this is a fourth rank tensor in the general cases, which reduces to a six-by-six matrix under the mentioned symmetry conditions [17].

Rheology is the field studying material models of deformable matter. In rheological measurements, the applied deformation is chosen to selectively measure certain components in the material model. Best known is the shear rheometer, where the sample is deformed by an almost constant deformation (shear) gradient. In the case of a simple liquid, the relation between applied shear stress and resulting velocity gradient yields a constant viscosity parameter η. This model was established by Newton, hence the name Newtonian liquid:

$$T_{xz} = \eta \frac{\partial v_x}{\partial z} = \eta \dot{\gamma} \tag{17.1}$$

where T_{xz} is the shear stress, v_x the velocity in x (flow) direction, z is called *the gradient direction*, and y the direction of vorticity. The shear rate $\dot{\gamma}$ is the gradient of the deformation velocity, the dot expresses derivation with respect to time. In a microscopic theory, this model requires the rate of momentum exchange between fluid particles (stress relaxation) to be instantaneous. A measure for this assumption to be valid is the Debora number $De = t_c/t_p$, relating the stress relaxation time t_c to timescale of observation t_p, which has to be small for fluid-like behavior. This requirement holds best for low viscous fluids, where t_c is small.

Even in steady flow, the viscosity is not necessarily a constant parameter but generally depends on the shear rate $\dot{\gamma}$. Such fluids are described by a shear-rate dependent viscosity $\eta(\dot{\gamma})$, represented by the flow curve, which is measured pointwise by applying a steady flow at different well-defined shear-rates like in the Couette (concentric rotating cylinders) rheometer, see Section 17.2.3.

For small deviations around an equilibrium state, the material model can be linearized. As a consequence, there is a well-defined relation between the time-dependent response, measured, for example, by step-strain or step-stress methods, and the response to fixed-frequency sinusoidal excitation [4, 6]. The time-domain response is related to the frequency response by the Fourier transform. Additionally, the Kramer-Kronig relations are fulfilled since the material response has to be causal.

Considering shear deformation only, the material can be described either by storage and loss moduli, G' and G'', respectively, or complex viscosity $\eta(\omega) = \eta' + j\eta''$, where all parameters are frequency dependent in general. With oscillatory measurements, the values obtained at single frequencies can be combined to obtain a rheological spectrum.

Depending on the interaction between fluid particles, several extensions to the Newtonian fluid model are possible. Considering a finite rate of momentum exchange, leads to the Maxwell model. With a single relaxation time τ the model reads,[1]

$$\eta(\omega) = \frac{\eta_0}{1 + j\omega\tau} \tag{17.2}$$

$$G(\omega) = j\omega\eta = \frac{1}{1/G_0 - j/(\omega\eta_0)} \tag{17.3}$$

where η_0 ist the steady-shear viscosity, and $G_0 = \eta_0/\tau$ the high-frequency modulus.

The Maxwell model is suitable for materials that behave like liquids in steady shear deformation, and elastic at high-frequency oscillations. A dual model to the Maxwell fluid is the Kelvin-Voigt model, which behaves elastic at slow deformation but dissipates more energy at high frequencies. The latter is especially applicable for solids with a certain absorption of acoustic (ultrasonic) waves. These two linear models can be combined and extended in multi-component models corresponding to different relaxation mechanisms [18]. The representation as a continuous relaxation spectrum is possible as well.

An interesting, surprisingly simple, empirical relation between the flow curve and the frequency-dependent viscosity was found by Cox and Merz [19]. This rule states that $\eta(\dot{\gamma}) \approx |\eta(\omega)|_{\omega=\dot{\gamma}}$, that is, the viscosity value at a certain shear rate $\dot{\gamma}$ is approximately equal to the magnitude of the complex spectral component $\eta(\omega)$ at $\omega = \dot{\gamma}$. It is valid for polymer melts and some other materials, but has its limitations for complex structured matter, such as emulsions and suspensions.

The frequency dependence of the complex-valued viscosity (or equivalently, the shear moduli) means that the material, for example, a liquid, partly exhibits elastic response. For elastic materials in shear deformation, normal stresses in the direction of the velocity gradient are observed. The corresponding material parameters are the normal stress coefficients, relating the normal stress differences to the

---

1) In rheology literature, complex valued material parameters are denoted by an asterisk, for example, $G^*(\omega)$ and $\eta^*(\omega)$. This notation is not adopted here to avoid confusion with the complex conjugate. Instead, the quantities are distinguished by explicitly stating the dependent variable, either $\dot{\gamma}$ for shear rate dependence, or ω for frequency dependence.

square of the shear strain. These normal stress differences can be measured with additional force transducers in the rheometer.

The temperature dependence of viscosity for simple liquids follows an exponential rule in the form of the Arrhenius relation. Thermo-rheological complex (TRC) fluids deviate from this rule. To derive the parameters of the exponential equation, representing the Arrhenius law, requires measurements at two temperatures at least, [20],

$$\eta(T) = \eta(T_0)e^{-b(T-T_0)}$$

where $\eta(T_0)$ is the viscosity at reference temperature T_0, b the exponential coefficient in $1/K$, and T the temperature. Then the viscosity at other temperatures can be calculated. For polymer melts or other fluids with a glass transition point, the Williams-Landel-Ferry (WLF) model has to be used [20].

The variety of nonlinear and non-Newtonian effects in materials is manifold and usually specific measurement geometries are set up to measure specific parameters. Quantification of such effects is non-trivial using laboratory methods. The development of miniaturized sensor devices for this purpose is a problem yet to be solved.

17.2.2
Time-Harmonic Deformation

Neglecting all non-linear rheological effects and the convective term in the Navier-Stokes equations leads to a linear theory of fluid-structure interaction, which is applicable for microstructures oscillating at low amplitudes. It also serves as a first perturbation to study the effects of acoustic streaming induced by vibrating structures in liquids [21, 22]. Linearization means that a sinusoidal stress at one frequency gives a sinusoidal strain at this excitation frequency only. Then, complex notation can be introduced to describe the time-dependence of stress and strain. The material parameters, that is, the components of the stiffness matrix, generally are frequency dependent, complex values, describing the linear viscoelastic behavior (see also Section 17.2.1).

A non-linear effect often observed in micromachined resonators is the Duffing behavior arising from an elastic restoration force proportional to the third power of the deflection (in addition to the linear relation). Experimentally this yields an apparent frequency response that depends on the direction of sweep (from high to low frequencies or in the other direction) and a strong dependence on excitation amplitude [23].

17.2.3
Classical Methods for Measuring Fluid Properties

In Figure 17.1 an overview of rheometric laboratory methods is given. The flow in certain drag flow rheometers, that is, sliding plate or rotating concentric cylinders, cone and plate, generates a uniform shear rate throughout the volume (except for

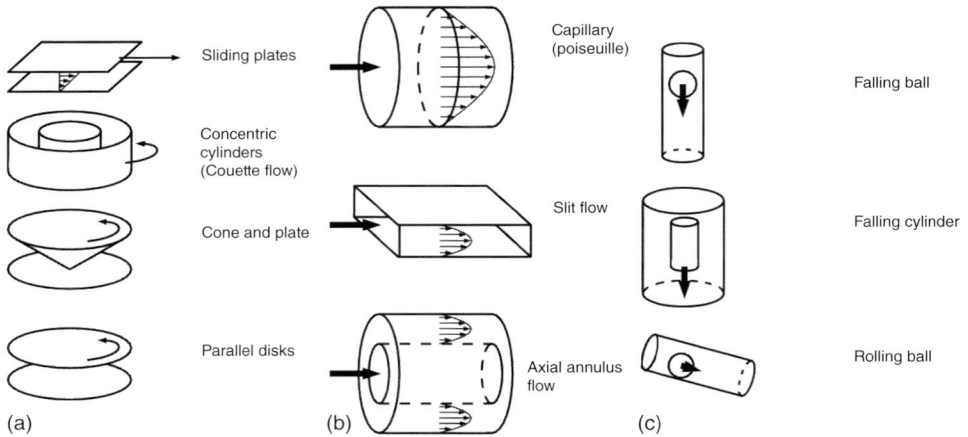

Figure 17.1 Overview of rheometric methods: (a) drag-flow closely approximating pure shear deformation, (b) pressure-driven flows, and (c) using falling or rolling objects. Figure from Ref. [24]

boundary effects), and is therefore suitable for measuring the flow curve $\eta(\dot{\gamma})$. Parallel disks are used mainly for practical reasons, since an error in the adjustment of the gap affects the measurement less than in the case of the cone and plate, but the shear rate is not homogeneous anymore. For approximately linear materials, this is acceptable in oscillatory shear experiments where $G'(\omega)$ and $G''(\omega)$ are measured. Pressure-driven flows are used in capillary rheometers, where it is necessary to know the flow profile for calculating the viscosity from the pressure drop over a certain length of flow. The advantage is that these rheometers can be used in a continuous flow, which closely resembles the actual process, for example, polymer extrusion. Falling or rolling-body viscometers are suitable for low viscous liquids. Clearly, the flow around the falling body is not homogeneous and calibration with known fluids is mandatory.

17.2.4
Miniaturized Rheometers

The macroscopically moving parts found in laboratory instruments do not allow a straight-forward miniaturization of rheometers to the microscale. However, for pressure-driven flows this has been realized, for example, in [25]. For dynamic measurements, the use of resonators is advantageous since these consist of solid structures set into vibration by integrated transduction mechanisms. When considering mechanical resonances, the knowledge of resonance frequency and bandwidth is usually sufficient to determine the material properties, without requiring absolute force or displacement measurements. Resonators are widely used as very sensitive devices for mass deposition, for example, in the form of the quartz crystal microbalance (QCM). Operation in liquid requires a deeper analysis of the fluid-structure interaction. The damping can easily get excessive so that resonance peaks vanish in the measurement noise or other spurious signal components. Nevertheless they are sensitive for small changes at low viscosity values, making them ideal for condition monitoring of fuels, lubricants, solvents and the like.

17.3
Resonator Sensors

17.3.1
Excitation and Readout

Various physical effects can be used to excite a resonant vibration of a mechanical system. In piezoelectric materials, stresses are generated from an applied electrical field. This is very effective for thin-film structures, where low voltage signals are sufficient to obtain high field strengths. The reciprocal effect allows the measurement of the deformation via piezoelectrically generated charges. Usually, the electrical impedance spectrum is measured and linked to the properties of the material in contact with the mechanical resonator.

Alternatively, electromagnetic (Lorentz) forces can be used to excite the vibration. In contrast to piezoelectric transduction, it is the current instead of the electric field, that generates a force when an external magnetic field is applied. The vibration induces an electric field, which can be measured as a voltage drop over dedicated readout lines or indirectly, as an alternation of the electrical impedance of the excitation path.

Optical readout is possible when the vibrating structure is exposed or visible through a transparent liquid [26, 27]. Laser Doppler anemometry is regularly used for analyzing the modes of vibration in laboratory setups.

17.3.2
Eigenmode Decomposition

Mechanical resonators are distributed systems with infinite degrees of freedom. Some parts are connected to a support or substrate, which can be viewed as rigid, while others are intentionally designed to deform under the action of the excitation force. Mathematically, the dynamic deformation is described by partial differential equations with respective boundary conditions for the clamping. The equations are linear for small deformation amplitudes. In that case, the decomposition in eigenmodes is possible, where each mode is associated with a single eigenvalue, or equivalently, an eigenfrequency. Also, a lumped mass m_i and lumped stiffness coefficient k_i can be assigned from the solution of the mode shape (determined up to an arbitrary constant for the scaling of the dynamic variable $q(t)$, carrying the dimension of a length). For simple geometries like strings, membranes, singly or doubly suspended beams, analytical solutions for the mode shapes are available. In other cases, numerical methods like the finite element analysis (FEA) are used to calculate mode shapes and eigenfrequencies. The solution of the mode shape can be represented by a normalized (dimensionless) displacement vector $\hat{\mathbf{u}}$ at each point in the geometry. The normalization is preferably done with respect to the maximum of the mode shape displacement, like a vibrational antinode. The equation of motion is decomposed in a series of harmonic (second order lumped element) oscillators driven by an excitation force F_i:

$$m_i \ddot{q} + d_i \dot{q} + k_i q = F_i \tag{17.4}$$

where $q(t)$ is the dynamic variable and the dot denotes derivation with respect to time t. F_i is the projection of the total excitation force to the mode shape [23].

From the solution of the equation of motion, mechanical energy terms can be derived. The eigenmodes, given as the displacement field $\hat{\mathbf{u}}_i(\vec{r})$, form an orthogonal system, enabling the representation as given in the following [23, 28]. The meaning of m_i and k_i as effective mass and stiffness becomes clear by considering the elastic (potential) and kinetic energies, $(E_{\text{pot}})_i$ and $(E_{\text{kin}})_i$, respectively, for each eigenmode i [29, 30],

$$(E_{\text{pot}})_i = \frac{1}{2}[q_i(t)]^2 \underbrace{\int \nabla_s \hat{\mathbf{u}}_i(\vec{r}) : \Re\left\{\underline{\underline{c}}\right\} : \nabla_s \hat{\mathbf{u}}_i(\vec{r}) \mathrm{d}V}_{k_i} \tag{17.5}$$

$$(E_{\text{kin}})_i = \frac{1}{2}[\dot{q}_i(t)]^2 \underbrace{\int \rho_r |\hat{\mathbf{u}}_i(\vec{r})|^2 \mathrm{d}V}_{m_i} \tag{17.6}$$

where ∇_s denotes the symmetric divergence, $\Re\left\{\underline{\underline{c}}\right\}$ is the real part of the complex stiffness matrix $\underline{\underline{c}}$, and : represents the double dot or double scalar product, according to the notation used in [28].

Intrinsic losses (in the bulk of the material) can be obtained from the imaginary part of the stiffness matrix, $\Im\left\{\underline{\underline{c}}\right\}$, by Auld [17]

$$(P_{\text{diss}})_i = [\dot{q}_i(t)]^2 \frac{1}{2} \underbrace{\int \nabla_s \hat{\mathbf{u}}_i(\vec{r}) : \Im\left\{\underline{\underline{c}}\right\} : \nabla_s \hat{\mathbf{u}}_i(\vec{r}) \mathrm{d}V}_{d_i} \tag{17.7}$$

where d_i is the (intrinsic) damping coefficient.

The above energy considerations are conducive if the deformation field is bounded by a surface with zero power flow, that is, either rigid or stress-free boundaries. Damping due to the immersion in liquid is introduced afterwards in the form of fluid reaction forces.

17.3.3
Electrical Equivalent Circuit

In the following section, the mode number index i used in the previous section will be omitted. The analysis can be performed for each mode number separately.

Especially for devices using piezoelectric or electromagnetic transduction, the derivation of an electrical equivalent circuit model is advantageous. Using complex notation (time dependence of $e^{j\omega t}$), and introducing the velocity $v(t) = \dot{q}(t) = ve^{j\omega t}$ for time-harmonic motion, Eq. (17.4) reads

$$v\left(j\omega m + d + \frac{k}{j\omega}\right) = F \tag{17.8}$$

For piezoelectric transduction, the applied voltage and the associated electric current are linearly related to the transduction force and the displacement by means of the piezoelectric coupling parameters [17]. Inserting this in Eq. (17.8), a transfer function in terms of voltage and current can be obtained from which an electrical equivalent circuit can be derived. A typical example is the Butterworth-Van Dyke circuit for quartz resonators [28].

In case of electromagnetic transduction, the force F is proportional to the applied current, the external magnetic field B, and an effective length l,

$$F = BlI$$

The induced voltage over the same conductive line of effective length l is proportional to the displacement velocity v,

$$U = Blv$$

In terms of U and I, this yields the equivalent parallel RLC circuit:

$$U\left[j\omega \underbrace{\frac{m}{(Bl)^2}}_{C_{\mathrm{p}}} + \underbrace{\frac{d}{(Bl)^2}}_{1/R_{\mathrm{p}}} + \frac{1}{j\omega}\underbrace{\frac{k}{(Bl)^2}}_{1/L_{\mathrm{p}}}\right] = I \tag{17.9}$$

See Refs. [31, 32] for examples of the equivalent circuit modeling of devices featuring electromagnetic transduction.

When separate excitation and readout lines or electrodes are used, the same principle can be applied, yielding a transconductance as transfer function. Additional purely electric components can appear in the complete equivalent circuit, like the capacitance of the electrodes in piezoelectric transducers, or series inductance and resistance of conductive lines in the electromagnetic case.

17.3.4
Damping

The damping coefficient d requires more attention since it can have multiple causes, which in most cases can be assumed additive. The intrinsic damping originates from dissipative processes in the resonator material. These can become especially important in devices fabricated in polymer technology or which have special coatings. In crystalline materials, they usually can be neglected. Another mechanism of energy loss is the imperfect clamping of the structure. In that case acoustic power is radiated into the support structure. For fluid property sensors, the most important contribution to the damping comes from the viscous dissipation in the surrounding liquid. The exact dependence of the frequency response on material parameters $\eta(\omega)$ and ρ has to be derived from the analysis of the fluid-structure interaction [30, 33–35].

17.3.5
Fluid-Structure Interaction

A force is acting back on the resonator surface if immersed in, or partly exposed to, a liquid, which is proportional to the surface velocity, but can have in-phase and out-of-phase components. The integral over the exposed surface can be expressed by a fluid impedance[2] Z_{fl}, added to the bracketed expression in the resonator equation (17.8):

$$v\left(j\omega m + d + \frac{k}{j\omega} + Z_{\mathrm{fl}} \right) = F \tag{17.10}$$

The real part of Z_{fl} increases the damping d, and the imaginary part can be interpreted as a change either in m or in k. Usually the former is used since the notion of an added mass seems more intuitive.

Often, an idealization of the geometry has to be performed to derive the dependence of the fluid reaction force on the material parameters. The exact solution of Z_{fl} can be derived numerically by boundary element methods (BEM), or FEA. While the former is not very accessible for sensor designers, the latter can easily become computationally too expensive due to the different length scales involved. The viscous penetration depth (or shear-wave decay length) of a planar shear wave [36],

$$\delta = \sqrt{\frac{2\eta}{\omega\rho}} \tag{17.11}$$

dictates the discretization of the fluid volume in the boundary layer of the solid-fluid contact (as a rule of thumb, at least ten discretization points should be used to fit within δ).

The dependence of Z_{fl} on the fluid parameters viscosity η and density ρ depends on the mode shape and frequency of the oscillating structure. For simplified geometries like slender cantilevers, doubly clamped beams [35, 37], or spheres, closed form solutions are available. Numerical modeling, like the FEA carried out in [34, 38], can be used but is usually computationally expensive due to the mentioned requirement on discretization. In order to simplify the expression while maintaining reasonable generality, we assume a relation of the form

$$Z_{\mathrm{fl}}(\omega) = j\omega\rho V_{\mathrm{fl}} + \omega\rho\delta\frac{(1+j)}{2}A_{\mathrm{fl}} + \omega\rho\delta^2 L_{\mathrm{fl}} \tag{17.12}$$

---

2) The term *fluid impedance* Z_{fl} is derived from the acoustic impedance where it is the pressure divided by velocity times area. In the electrical equivalent models, it corresponds to an electrical impedance in the case of piezoelectric coupling, but to an electrical admittance in the case of electromagnetic coupling, see Eq. (17.9).

This form is deduced from analytical solutions for the shapes mention above, and general considerations. V_{fl} is an effective fluid volume moving in-phase with the structure, A_{fl} is an effective interaction surface of shear-waves, and L_{fl} is an effective length scale of other viscous dissipation mechanisms.

The angular resonance frequency and Q-factor are then calculated by

$$\frac{1}{\omega_r^2} = \frac{1}{\omega_0^2} \left(1 + \frac{1}{\omega_r m} \Im \left\{ Z_{fl} \right\} \right) \tag{17.13}$$

$$\frac{1}{Q} = \frac{\omega_r}{\omega_0} \left[\frac{1}{Q_0} + \frac{1}{\omega_0 m} \Re \left\{ Z_{fl} \right\} \right] \tag{17.14}$$

where $1/Q_0$ is the intrinsic damping accounting for losses of the unloaded resonator. Note that the equation for ω_r is implicit. Here, the resonance frequency is the frequency where the imaginary part of the force-velocity transfer function, Eq. (17.10), becomes zero. Considering only a weak frequency-dependence of Z_{fl}, the velocity amplitude has a maximum at approximately ω_r. An alternative definition would be the frequency where the displacement amplitude has its maximum. The difference is small for high Q-factors, that is, especially in the case when measuring low viscous liquids.

With the linearization of the Navier-Stokes equations, the fluid impedance can be calculated by frequency domain analysis as implemented in common FEA solvers. The equations are equivalent to the elastic equations using complex material parameters. In a compressible viscous fluid with viscosity η, second (voluminal) viscosity λ, and compressibility ζ, the fluid stiffness parameters are:

$$c_{11} = c_{22} = c_{33} = j\omega(\lambda + 2\eta) + \frac{1}{\zeta} \tag{17.15}$$

$$c_{12} = c_{13} = c_{31} = j\omega\lambda + \frac{1}{\zeta} \tag{17.16}$$

$$c_{44} = c_{55} = c_{66} = j\omega\eta \tag{17.17}$$

All other components are zero or given by symmetry [24, 39]. In most MEMS devices, the wavelength of compressional acoustic waves is much larger than the geometry so that the liquid can be considered incompressible. Then the displaced fluid can be derived from an irrotational potential flow solution coupled to a solution of the rotational velocity field [40]. The rotational part is necessary to fulfill the no-slip boundary condition at solid-fluid interfaces.

Most models for the fluid-structure interaction rely on the assumption of small amplitudes and neglect convection completely. In some cases, the oscillation drives a visible steady flow which is known as *acoustic streaming* [21, 22]. The modeling route for this kind of flow starts with the analysis of the linear flow as sketched before. From this velocity field, the second order force terms can be derived which enter the convective Navier-Stokes equations averaged over a full period of oscillation of the linear flow.

17.4
Examples of Resonant Sensors for Fluid Properties

The variety of MEMS devices reported for fluid property sensing is exceptionally broad and thus only some characteristic examples of their types are discussed at this point. Thus, we refer to selected works, which are considered to be representative in their field (without claiming completeness).

From a mechanical point of view, three major resonator principles are considered: Singly clamped, doubly (or multiply) clamped, and membrane-based devices. These different types may be operated in in-plane, out-of-plane or torsional modes. Independent of design, mode of operation and basic measuring principle, the resonance characterizing parameters resonance frequency f_r and quality factor Q are both dependent on the liquid's viscosity η and mass density ρ. Lately, generalized expressions for f_r and Q were established for resonant devices (assuming a constant ambient temperature) and read

$$f_r \approx \frac{1}{\sqrt{C_{f_0} + C_{f_\rho}\, \rho + C_{f_{\eta\rho}}\, \sqrt{\eta\, \rho}}} \qquad (17.18)$$

$$\frac{1}{Q} \approx D_0 + D_\eta\, \eta + D_{\eta\rho}\, \sqrt{\eta\, \rho} \qquad (17.19)$$

Here, C_{f_0}, C_{f_ρ}, $C_{f_{\eta\rho}}$, D_0, D_η and $D_{\eta\rho}$ are coefficients that can be determined by evaluating f_r and Q in (at least) three liquids. These simple expressions are mathematically valid for in-plane oscillating plates, oscillating spheres and laterally oscillating cylinders, but it was experimentally shown that these equations can also be applied for other geometries. The associated work and the derivation of the above equations will be published elsewhere, see [41].

In the following section, examples of typical resonant MEMS devices are briefly discussed. The review of literature showed that in general, the resonator principles, their fabrication techniques, as well as their excitation and read-out mechanisms are explained. In an experimental part, mostly, frequency responses and (or) associated resonance frequencies and quality factors in air and the liquids are given to show the sensors' characteristics. However, despite the precise description and documentation, it is difficult to compare the performances and characteristics of sensors of the same or different types. Obvious quantities for this purpose of course would be accuracy and sensitivity of the devices.

In principle, it is certainly possible to determine and indicate the sensors' accuracies and sensitivities (for comparison purpose). However, these quantities might not be the best choice allowing an intuitive and meaningful characterization of viscosity sensors. This is substantiated by the following arguments: The term *accuracy* (e.g., given with a percentage value) can be misleading. It can be the deviation of a single measurement result from the real value; it can be the repeatability of multiple measurements under the same measuring conditions or it can also address the long-term stability of the device, just to name some examples. In order to determine an overall accuracy value, extensive measurement series in preferably enclosed, temperature and pressure-controlled setups are necessary. Specifying an absolute or relative sensitivity X_y and $S_{X,y}$, respectively with

$$X_y(\eta, \rho) == \frac{\partial X}{\partial y} \qquad \text{and} \qquad S_{X,y}(\eta, \rho) = \left| \frac{\partial X}{\partial y} \cdot \frac{y}{X} \right| \qquad (17.20)$$

where X stands either for f_r or Q and y for η or ρ is only possible for given, discrete values for η and ρ (e.g., for water) or otherwise, the functions for $X(\eta, \rho)$ and $S_{X,y}(\eta, \rho)$ have to be given. Both possibilities are not very descriptive as the first approach describes the sensitivity for a specific liquid only and the second approach involves non-intuitive functions, which are dependent on two variables. Furthermore, for both types of sensitivities (absolute and relative), four sensitivities have to be given ($f_{r_\eta}, f_{r_\rho}, Q_\eta, Q_\rho$ and $S_{f_r,\eta}, S_{f_r,\rho}, S_{Q,\eta}, S_{Q,\rho}$) to describe the sensor's characteristics.

A possible sensor characterization process, which does not involve too much effort, but which would give useful insights about the particular sensor characteristics and performance could be based on three types of experimental investigations:

1) Determination of f_r and Q in (at least) three liquids and subsequent determination of the six parameters C_{f_0}, C_{f_ρ}, $C_{f_{\eta\rho}}$, D_0, D_η and $D_{\eta\rho}$ in Eqs. (17.18), and (17.19). The values for f_r and Q in air are mostly also of interest. The approach or technique on how f_r and Q were evaluated is also important to be defined. The ambient temperature and associated viscosity and mass density values have to be given in any case.

2) Determination of the resonator's f_{r-} and Q-cross-sensitivity to temperature. This characterization is very important as the temperature measurement (which is necessary in any case for reliably measuring η and ρ) is subjected to a certain accuracy and thus, directly limits the sensor's accuracy (see discussion below).

3) Continuous determination of f_r and Q under as stable as possible conditions over a long period of time (e.g., for 24 h or longer). These measurements show the spread of obtained values for f_r and Q and furthermore reveal the stability of the device. Especially doubly clamped devices might be subjected to internal stresses, which might relax over time, which detunes the sensor and thus limits its accuracy.

Such a characterization process based on these three (or similar) experimental investigations and characterizations would be an approach enabling comparison of sensors of the same or different types. Furthermore, assuming the availability of such data, such a specification facilitates comparing and distinguishing the own work from others.

The literature review of resonant MEMS devices for viscosity and mass density showed that, in most cases, the devices' cross-sensitivities of f_r and Q are not investigated. However, this investigation is extremely important, particularly if the sensor's resolution or accuracy is of interest. This statement is explained by the following example: In [42] a suspended silicon based microchannel resonator is reported. The device features a mass density resolution of 0.01 kg cm^{-3} with a sensitivity of 16 Hz (kg/ cm^{-3})$^{-1}$, and an operation (resonance) frequencies of about 130 kHz. Taking into account the relative cross sensitivity of the resonance frequencies to temperature of a bare silicon cantilever and a gold-coated cantilever, which are $\Delta f_r/(f_r\ \Delta T) \approx -30 \cdot 10^{-6}$ K^{-1} [43] and $-100 \cdot 10^{-6}$ K^{-1} [44] respectively, allows estimating the devices cross-sensitivity to temperature. In this particular case, this would be $\Delta f_r/\Delta T = -(3.9 \ldots 13)$ Hz K^{-1}. To achieve the aforementioned resolution for mass density, the accuracy of the temperature would have to be in the order of (0.01 ... 0.04) K (assuming similar temperature dependencies as found for cantilever devices in literature). This example is given to show that the required accuracy of the temperature measurement can be directly estimated when aiming at a certain accuracy (e.g., in mass density) and knowing the device's resonance frequency dependence to temperature. The example furthermore shows and substantiates the addressed need for precise temperature measurements and description of the sensor's temperature dependence.

In the following, the discussed sensor concepts for liquid property sensing include conventional cantilevers (which seems to be the mostly investigated device), U-shaped cantilevers, tuning forks, doubly clamped beams, shear vibrating devices, membranes, discs, and other principles. For the actuation of the vibration, piezoelectric, electromagnetic, thermal, and external vibrational actuation are used. Concerning read-out of the sensor's mechanical oscillation, piezoresistive, capacitive, inductive, and optical methods are mainly reported.

17.4.1
Microacoustic Devices

Even though they are most often not considered MEMS devices, due to the related technology and interaction mechanisms, we briefly discuss fluid property sensing using so-called microacoustic sensors. These devices emerged from the utilization of piezoelectric media in electronic components where, by means of the piezoelectric effect, vibrations in the utilized crystal are excited. Resonances associated with these mechanical vibrations can feature high Q-factors, which makes these devices attractive for the realization of filters or resonators. If the vibrating structure gets in contact with a fluid the vibration characteristics are affected by the fluid properties as discussed above. The associated sensor principles can similarly be realized using alternative actuation mechanisms (i.e., other than the piezoelectric effect). Many of the issues and aspects that are encountered with microacoustic devices hold in a similar manner for MEMS resonators.

A very common technique for sensing physical fluid properties with microacoustic devices (and in particular viscosity) is that of bringing a shear-vibrating surface in contact with the liquid, for example,, by using a thickness shear mode (TSM) quartz resonator and related devices [45–52] acoustic plate mode devices [53] or Love wave sensors [54–56]. The basic interaction mechanism relies on the contact of a surface vibrating in the in-plane direction with the fluid to be sensed.

The resonant or wave propagation characteristics of the device (e.g., resonance frequency, Q-factor and wave velocity) will be affected by the complex characteristic acoustic shear wave impedance of the liquid (see also Section 17.3.5), which, in turn, allows the determination of the viscosity and/or density of the liquid. For resonant sensors in highly viscous liquids, readout can be an issue since due to the higher damping for higher viscosities the sensor cannot be used as the frequency-determining element in an oscillator circuit. To overcome this, dedicated readout circuits and algorithms for damped resonators have been investigated recently [57−66].

The shear wave penetrating the liquid is attenuated featuring a penetration depth δ as given in Eq. (17.11), which, at relatively high frequencies (at least compared to the frequencies of laboratory viscometers which do not exceed ∼100 Hz), can be very low, for example, in the submicron range. For instance, at 5 MHz the penetration depth for water would be in the order of 0.25 microns, which means that only a thin layer of liquid is being sensed. For liquids featuring a microstructure with characteristic dimensions above δ, the measured viscosity will be different from that obtained by laboratory instruments, see, for example, the cases of emulsions, microemulsions and zeolite synthesis solutions discussed, for example, in [67−70].

The devices discussed above use in-plane motion of the surface being in contact with the liquid. Out-of-plane vibrations generally lead to radiation of pressure waves, which leads to spurious damping. This is the reason why classical surface acoustic wave (SAW) devices [71] are scarcely used as fluid sensors. An exception is when the slowness of the guided wave prevents pressure wave radiation into the adjacent fluid, such as, for example, in the Lamb wave devices discussed in [72].

Pressure waves can be utilized, however, to sense the so-called longitudinal viscosity, (which is associated with the second coefficient of viscosity or the bulk viscosity) as the latter determines the attenuation of pressure waves propagating in the fluid. In [11−13] setups utilizing piezoceramic transducers are discussed for this purpose considering also the sensing of the acoustic pressure wave impedance as well as the sound velocity of the liquid.

17.4.2
MEMS Devices

In the following, selected MEMS devices for mass density and viscosity sensing are discussed. Photographs of associated devices are shown in Figure 17.2 and a comparative overview is given in Tables 17.1 and 17.2.

17.4.2.1 Cantilever Devices

The micro-cantilever is the most reported resonant device for mass density and viscosity sensing applications, see, for example, [76−79] and references cited there. The read-out mechanisms involve piezoresistive, impedance spectroscopic and optical principles. In general, optical read-out mechanisms are very commonly used as they allow very precise amplitude measurements which

Figure 17.2 Examples for resonant MEMS mass density and viscosity sensors

are moreover convenient and straight-forward to implement. However, when it comes to measurements in liquids, it is not only to the obvious arguments that optical mechanisms might involve serious drawbacks. The obvious points are that optical read-out allows measurements in transparent liquids only and that the complete setup has to be mounted on a vibration compensated setup, which makes the application as a sensor difficult. The more important point (which is not evident at first sight) however is, that in general the optical investigation is performed under non-enclosed conditions, which in general involves evaporation of the liquid. This in turn, implies non-stable conditions and a change of the liquid's physical properties during the measurement. For example, evaporating acetone (without blowing onto the liquid), changes its temperature for more than $-5°C$ in respect to the ambient temperature.

Figure 17.2a shows a pair of two identical silicon micro-machined cantilevers. The structure on the right hand side, completely released from the bulk material acts as cantilever and is operated in an out-of-plane mode. The second device on the left hand side serves for reference measurements accounting for the influence of cross-sensitivities on the measured signal. The device was designed to be actuated electromagnetically and read-out piezoresistively. In [73] three different geometries of cantilevers are investigated, where the "second" cantilever yields resonance frequencies of about 5 kHz in liquids, see Table 17.1. There, L, W, T denote the cantilevers' lengths, widths, and thicknesses, respectively. In [80] and [34] the operation of cantilevers at higher vibration modes is discussed.

Table 17.1 Resonant MEMS devices for viscosity and mass density sensing. o-o-p, out-of-plane; L-F, Lorentz force; L, length; W, width; T, thickness; CL, cantilever; i-p, in-plane.

Device	Material(s)	Operational mode	f_r (air) (kHz)	f_r (liquid) (kHz)	Q (air)	Q (liquid)	Actuation	Readout	η (mPas)	Dimensions (µm)	Year	References
Cantilever	Si/ Metall	o-o-p	206.5	93.4 (water)	900	17 (water)	L-F	Stress sensitive PMOS Transistors	0.9 (water)	L 200 W 50 … 186 T 8.2	2007	[95]
	Si with conductive paths	o-o-p	—	5 (CL 2.) (water)	—	—	L-F	Piezores./ Optical	0.9 … 20	1.: 2810×100×20 2.: 1440×285×20 3.: 500×100×20	2012	[73]
	Silicon	o-o-p (fundamental)	17.31	5.39 (water)	55.5	2 (water)	Thermal	Thermal noise spectrum	0.3 … 2.5	397×29×2	2002	[96]
	Si with Au paths	6 Modes	—	2 … 55	—	—	Photo-thermal	Optical	1 … 20	500×100×4	2013	[80]
	—	8 Modes	6.23 … 988.83	1.08 … 283.83 (water)	16 … 455.2 (water)	1.5 … 21.4 (water)	None (noise)	Optical	0.9 (water)	519.6×47.1×1.17	2005	[34]
	Si/Au	i-p	50 … 2200	—	4200	67	Thermal	Piezores.	0.9 (water)	L 200 … 1000 W 45 … 90 T 12	2010	[97]
U-Shaped CL	Si with Au paths	o-o-p	8	3	—	—	L-F	Optical	05 … 400	1500×1100 W 100 T 15	2005	[27]
	Si with Au paths	o-o-p	20	12.5 (water)	—	25 … 30	L-F	Inductive (motion ind. voltage)	1 … 1.3	1600×1600 W 200 T 70	2013	[81]
Tuning Fork	Quartz crystal	Fundamental (antiphase, in-plane)	32.7	29.8 (Hep.) 28.25 (N35)	—	163.76 (Hep.) 14.75 (N35)	Piezoel.	Impedance Spectroscopy	0.38 … 55.2	1000	2014	[84]

Table 17.2 Resonant MEMS devices for viscosity and mass density sensing. o-o-p, out-of-plane; L-F, Lorentz force; L, length; W, width; T, thickness; D, diameter; i-p, in-plane.

Device	Material(s)	Operational mode	f_r (air) (kHz)	f_r (liquid) (kHz)	Q (air)	Q (liquid)	Actuation	Readout	η (mPas)	Dimensions (µm)	Year	References
Bridge	Si with Al paths	o-o-p	5.8 … 68.7	2.5 … 57	210 … 881	0.4 … 10	L-F	Inductive, optical	0.2 … 103.9	L 1500 … 5000 / W 30, 50 / T 20	2007	[86]
	Si with Au paths	o-o-p	96.7	31 (toluene) 24 (octanol)	—	3.5 (toluene) 1 (octanol)	L-F	Optical	0.57 … 7.37	350×50×1.3	2007	[26]
Suspended plate	Si with Al paths	i-p	19.47	13.29 (water)	330	3.47 (water)	L-F	Piezores	0.89 … 81.5	plate: 100×100 / overall: 1000 / T 20	2012	[91]
Mid-point Supported plate	AlN (extensional)	i-p	3700	3700	3000	100 (water) 18 (51.15 mPas)	Piezoel.	Impedance spectroscopy	0.9 … 51.15	1000×125x1	2012	[92]
Rotational disk	Si (rotational)	i-p	1770 … 8410	1770 … 8410	600 … 11700 (heptane)	50 … 304	Thermal	Impedance spectroscopy	0.38	D 100 … 200 / T 5 … 20	2011	[75]
Semicircular disks	Si (parylene coated)	i-p (rotational)	300 … 1000	300 … 1000	1200, 5000	100 (water)	Thermal	Piezores	0.9 (water)	D 240 … 300 / T 8	2008	[93]
Suspended microchannel	SiN	o-o-p	143	130.4	—	10400 (ethanol)	External (piez. act.)	Optical	0.8 … 10	20×200 / cross-sec: 4x3	2013	[42]
Vibrating diaphragm	Parylene (on Si)	o-o-p	—	1200	—	—	Electromag.	Capacitive	(glucose solutions)	400×400	2009	[94]

17.4.2.2   U-Shaped Cantilevers

A special class of cantilevers are so called U-shaped cantilevers, which have been reported, for example, in [27] and [81]. Figure 17.2b shows the tip of a U-shaped cantilever carrying a gold path for electromagnetic actuation. For read-out, optical, piezoresistive and inductive methods by means of a motion-induced voltage can be implemented. In Table 17.1 the overall sizes are given. W and T denotes the width and the thickness of the cross-sectional dimensions.

17.4.2.3   Tuning Forks

Commercially available quartz tuning forks have successfully been applied for viscosity and mass density sensing in liquids, see, for example, [82–84]. As for microacoustic TSM resonators, viscosity and mass density are measured by evaluating the immersed tuning fork's impedance spectrum, see also [45].

17.4.2.4   Doubly-Clamped Beam Devices

Similar to vibrating cantilevers, straight doubly clamped vibrating beams (also termed *clamped-clamped beams*, *bridges*, or *wires*) can be used for viscosity sensing [85–88]. An example for the latter is depicted in Figure 17.2c. The bridges can be excited using Lorentz-forces and the readout can be performed inductively. Similar to a guitar string, such bridge or wire-based devices can also be detuned, which has been shown in the macroscopical scale using tungsten wires [89]. The advantage of these devices in comparison with conventional and U-shaped cantilevers is the alignment of the straight conductor path, which allows for a straightforward magnet-assembly design necessary for Lorentz-force actuation. However, doubly clamped devices are subjected to high cross-sensitivities of their resonance frequencies to temperature due to thermal stresses within the structure, which complicate precise and reliable measurements. The modeling of such devices is presented, for example, in [31, 35, 37, 90].

17.4.2.5   In-Plane Resonators

Examples for in-plane oscillating resonators, designed to preferentially excite shear waves in the liquid are suspended platelets [74, 91], see Figure 17.2d, mid-point suspended plates [92], rotational disks [75] see Figure 17.2e and semicircular disks [93]. Figure 17.2d shows a scanning electron microscopy image of the aforementioned suspended silicon-based plate, which is actuated by means of Lorentz-forces on AC-currents in an external magnetic field and read-out with two piezoresistors implemented in the supporting beams. In the overview in Table 17.2, D denotes the diameter of the disk-based devices.

17.4.2.6   Other Principles

Other devices that are not assigned to a general principle are a suspended microchannel [42], which is a miniaturized version of a commercially available concept for mass density sensing, see [1] and a vibrating diaphragm with capacitive read-out [94].

17.4.3
Comparison

17.5
Conclusions

Vibrating resonant MEMS devices are very well suited for determining mechanical liquid properties as their resonance properties, for example, the frequency response, are affected upon immersion into the liquid to be investigated. A number of different technologies, actuation and readout mechanisms have successfully been employed. By designing the device appropriately and utilizing selected vibration modes, the characteristics of the influence of a desired physical parameter can be tuned. However, most often it is not possible to determine a targeted physical quantity by evaluating a single particular resonance parameter (e.g., resonance frequency or Q-factor). Thus, establishing a valid model describing the underlying relations is essential for obtaining reliable measurement results.

References

1. Kratky, O., Leopold, H., and Stabinger, H. (1979) Apparatus for density determination. US Patent 4,170,128.
2. Enoksson, P., Stemme, G., and Stemme, E. (1995) Fluid density sensor based on resonance vibration. *Sens. Actuators, A*, **47** (1), 327–331.
3. Larson, R.G. (1998) *The Structure and Rheology of Complex Fluids*, Oxford University Press.
4. Macosko, C.W. (1994) *Rheology: Principles, Measurements, and Applications*, John Wiley & Sons, Inc.
5. Mewis, J. and Wagner, N.J. (2013) *Colloidal Suspension Rheology*, Cambridge University Press.
6. Giesekus, H. (1994) *Phänomenologische Rheologie: Eine Einführung*, 1st edn, Springer.
7. Barnes, H.A. (1997) Thixotropy - a review. *J. Non-Newtonian Fluid Mech.*, **70** (1), 1–33.
8. Jakoby, B., Beigelbeck, R., Keplinger, F., Lucklum, F., Niedermayer, A., Reichel, E.K., Riesch, C., Voglhuber-Brunnmaier, T., and Weiss, B. (2010) Miniaturized sensors for the viscosity and density of liquids – performance and issues. *IEEE Trans. Ultrason. Ferroelectr. Freq. Control*, **57** (1), 111–120.
9. Beigelbeck, R. and Jakoby, B. (2004) A two-dimensional analysis of spurious compressional wave excitation by thickness-shear-mode resonators. *J. Appl. Phys.*, **95** (9), 4989–4995.
10. Oralkan, O., Ergun, A.S., Johnson, J.A., Karaman, M., Demirci, U., Kaviani, K., Lee, T.H., and Khuri-Yakub, B.T. (2002) Capacitive micromachined ultrasonic transducers: next-generation arrays for acoustic imaging? *IEEE Trans. Ultrason. Ferroelectr. Freq. Control*, **49** (11), 1596–1610.
11. Antlinger, H., Clara, S., Beigelbeck, R., Cerimovic, S., Keplinger, F., and Jakoby, B. (2012) Sensing the characteristic acoustic impedance of a fluid utilizing acoustic pressure waves. *Sens. Actuators, A*, **186**, 94–99.
12. Antlinger, H., Clara, S., Beigelbeck, R., Cerimovic, S., Keplinger, F., and Jakoby, B. (2012) An acoustic transmission sensor for the characterization of fluids in terms of their longitudinal viscosity. *Procedia Eng.*, **47**, 248–252.
13. Beigelbeck, R., Antlinger, H., Cerimovic, S., Clara, S., Keplinger, F., and Jakoby, B. (2013) Resonant pressure wave setup for simultaneous sensing of longitudinal viscosity and sound velocity of liquids. *Meas. Sci. Technol.*, **24** (12), 125101.

14. Puttmer, A., Hauptmann, P., and Henning, B. (2000) Ultrasonic density sensor for liquids. *IEEE Trans. Ultrason. Ferroelectr. Freq. Control*, **47** (1), 85–92.

15. Greenwood, M.S., Skorpik, J.R., Bamberger, J.A., and Harris, R. (1999) On-line ultrasonic density sensor for process control of liquids and slurries. *Ultrasonics*, **37** (2), 159–171.

16. Landau, L.D., Pitaevskii, L.P., Kosevich, A.M., and Lifshitz, E.M. (1986) *Theory of Elasticity*, Course of Theoretical Physics, vol. 7, 3rd edn, Butterworth-Heinemann.

17. Auld, B.A. (1973) *Acoustic Fields and Waves in Solids*, vol. I, John Wiley & Sons, Inc.

18. Kirschenmann, L. and Pechhold, W. (2002) Piezoelectric Rotary Vibrator (PRV)—a new oscillating rheometer for linear viscoelasticity. *Rheol. Acta*, **41** (4), 362.

19. Cox, W.P. and Merz, E.H. (1958) Correlation of dynamic and steady flow viscosities. *J. Polym. Sci.*, **28** (4), 619–622.

20. Arrhenius equation. (2014, October 15). In *Wikipedia, The Free Encyclopedia*. Retrieved 12:19, November 11, 2014, from *http://en.wikipedia.org/w/index.php?title=Arrhenius_equation&oldid=629740452*.

21. Friend, J. and Yeo, L.Y. (2011) Microscale acoustofluidics: microfluidics driven via acoustics and ultrasonics. *Rev. Mod. Phys.*, **83**, 647–704.

22. Nyborg, W.L. (1958) Acoustic streaming near a boundary. *J. Acoust. Soc. Am.*, **30** (4), 329–339.

23. Weaver, W. Jr., Timoshenko, S.P., and Young, D.H. (1990) *Vibration Problems in Engineering*, John Wiley & Sons, Inc.

24. Reichel, E.K. (2012) Dynamic methods for viscosity and mass-density sensing. PhD thesis, Johannes Kepler University Linz, Austria.

25. Pipe, C.J. and McKinley, G.H. (2009) Microfluidic rheometry. *Mech. Res. Commun.*, **36** (1), 110–120.

26. Riesch, C., Jachimowicz, A., Keplinger, F., Reichel, E.K., and Jakoby, B. (2007) A novel sensor system for liquid properties based on a micromachined beam and a low-cost optical readout. IEEE Sensors 2007, IEEE, pp. 872–875.

27. Agoston, A., Keplinger, F., and Jakoby, B. (2005) Evaluation of a vibrating micromachined cantilever sensor for measuring the viscosity of complex organic liquids. *Sens. Actuators, A*, **123**, 82–86.

28. Auld, B.A. (1973) *Acoustic Fields and Waves in Solids*, vol. II, John Wiley & Sons, Inc.

29. Reichel, E.K. and Jakoby, B. (2012) *Microsensors Based on Mechanically Vibrating Structures*, Springer.

30. Morand, H.J.-P. and Ohayon, R. (1995) *Fluid Structure Interaction*, John Wiley & Sons, Inc.

31. Reichel, E., Riesch, C., Keplinger, F., and Jakoby, B. (2008) Resonant measurement of liquid properties in a fluidic sensor cell. Eurosensor XXII Proceedings, pp. 540–543.

32. Reichel, E.K., Riesch, C., Keplinger, F., Kirschhock, C.E., and Jakoby, B. (2010) Analysis and experimental verification of a metallic suspended plate resonator for viscosity sensing. *Sens. Actuators, A*, **162** (2), 418–424.

33. Lighthill, J. (2001) *Waves in Fluids*, Cambridge University Press.

34. Maali, A., Hurth, C., Boisgard, R., Jai, C., Cohen-Bouhacina, T., and Aimé, J.-P. (2005) Hydrodynamics of oscillating atomic force microscopy cantilevers in viscous fluids. *J. Appl. Phys.*, **97** (7), 074907.

35. Sader, J.E. (1998) Frequency response of cantilever beams immersed in viscous fluids with applications to the atomic force microscope. *J. Appl. Phys.*, **84** (1), 64–76.

36. Landau, L.D. and Lifshitz, E.M. (2013) *Fluid Mechanics: Landau and Lifshitz: Course of Theoretical Physics*, vol. 6, Elsevier.

37. Weiss, B., Reichel, E.K., and Jakoby, B. (2008) Modeling of a clamped–clamped beam vibrating in a fluid for viscosity and density sensing regarding compressibility. *Sens. Actuators, A*, **143** (2), 293–301.

38. Basak, S., Raman, A., and Garimella, S.V. (2006) Hydrodynamic loading of microcantilevers vibrating in viscous fluids. *J. Appl. Phys.*, **99** (11), 114906.

39. Voglhuber-Brunnmaier, T. (2014) The modeling of acoustic fluidic sensors using spectral methods. PhD thesis, Johannes Kepler University Linz, Austria.

40. Reichel, E.K., Riesch, C., Keplinger, F., and Jakoby, B. (2009) Modeling of the fluid-structure interaction in a fluidic sensor cell. *Sens. Actuators, A*, **156** (1), 222–228.

41. Heinisch, M., Voglhuber-Brunnmaier, T., Reichel, E.K., Dufour, I., and Jakoby, B. (2014) Reduced order models for resonant viscosity and mass density sensors. Sensors and Actuators A: Physical, Volume 220, December 1, 76–84, ISSN 0924-4247, *http://dx.doi.org/10.1016/j.sna.2014.09.006*. (*http://www.sciencedirect.com/science/article/pii/S0924424714004014*)

42. Khan, M., Schmid, S., Larsen, P.E., Davis, Z.J., Yan, W., Stenby, E.H., and Boisen, A. (2013) Online measurement of mass density and viscosity of PL fluid samples with suspended microchannel resonator. *Sens. Actuators, B*, **185**, 456–461.

43. Wasisto, H.S., Merzsch, S., Waag, A., Uhde, E., Salthammer, T., and Peiner, E. (2013) Airborne engineered nanoparticle mass sensor based on a silicon resonant cantilever. *Sens. Actuators, B*, **180**, 77–89.

44. Sandberg, R., Svendsen, W., Mølhave, K., and Boisen, A. (2005) Temperature and pressure dependence of resonance in multi-layer microcantilevers. *J. Micromech. Microeng.*, **15** (8), 1454.

45. Martin, S., Frye, G., and Wessendorf, K. (1994) Sensing liquid properties with thickness-shear mode resonators. *Sens. Actuators, A*, **44** (3), 209–218.

46. Reed, C., Kanazawa, K.K., and Kaufman, J. (1990) Physical description of a viscoelastically loaded at-cut quartz resonator. *J. Appl. Phys.*, **68** (5), 1993–2001.

47. Josse, F., Shana, Z.A., Radtke, D.E., and Haworth, D.T. (1990) Analysis of piezoelectric bulk-acoustic-wave resonators as detectors in viscous conductive liquids. *IEEE Trans. Ultrason. Ferroelectr. Freq. Control*, **37** (5), 359–368.

48. Lucklum, R., Behling, C., Cernosek, R.W., and Martin, S.J. (1997) Determination of complex shear modulus with thickness shear mode resonators. *J. Phys. D: Appl. Phys.*, **30** (3), 346.

49. Johannsmann, D., Mathauer, K., Wegner, G., and Knoll, W. (1992) Viscoelastic properties of thin films probed with a quartz-crystal resonator. *Phys. Rev. B*, **46** (12), 7808.

50. Hu, Y., French, LA Jr., Radecsky, K., da Cunha, M.P., Millard, P., and Vetelino, J.F. (2004) A lateral field excited liquid acoustic wave sensor. *IEEE Trans. Ultrason. Ferroelectr. Freq. Control*, **51** (11), 1373–1380.

51. Thalhammer, R., Braun, S., Devcic-Kuhar, B., Groschl, M., Trampler, F., Benes, E., Nowotny, H., and Kostal, P. (1998) Viscosity sensor utilizing a piezoelectric thickness shear sandwich resonator. *IEEE Trans. Ultrason. Ferroelectr. Freq. Control*, **45** (5), 1331–1340.

52. Jakoby, B., Klinger, F.P., and Svasek, P. (2005) A novel microacoustic viscosity sensor providing integrated sample temperature control. *Sens. Actuators, A*, **123**, 274–280.

53. Ricco, A. and Martin, S. (1987) Acoustic wave viscosity sensor. *Appl. Phys. Lett.*, **50** (21), 1474–1476.

54. Jakoby, B. and Vellekoop, M.J. (1998) Viscosity sensing using a love-wave device. *Sens. Actuators, A*, **68** (1), 275–281.

55. Herrmann, F., Hahn, D., and Büttgenbach, S. (1999) Separate determination of liquid density and viscosity with sagittally corrugated love-mode sensors. *Sens. Actuators, A*, **78** (2), 99–107.

56. Turton, A., Bhattacharyya, D., and Wood, D. (2005) High sensitivity love-mode liquid density sensors. *Sens. Actuators, A*, **123**, 267–273.

57. Eichelbaum, F., Borngräber, R., Schröder, J., Lucklum, R., and Hauptmann, P. (1999) Interface circuits for quartz-crystal-microbalance sensors. *Rev. Sci. Instrum.*, **70** (5), 2537–2545.

58. Schröder, J., Borngräber, R., Lucklum, R., and Hauptmann, P. (2001) Network analysis based interface electronics for quartz crystal microbalance. *Rev. Sci. Instrum.*, **72** (6), 2750–2755.

59. Riesch, C. and Jakoby, B. (2007) Novel readout electronics for thickness shear-mode liquid sensors compensating for spurious conductivity and capacitances. *IEEE Sens. J.*, **7** (3), 464–469.

60. Ferrari, M., Ferrari, V., and Kanazawa, K. (2008) Dual-harmonic oscillator for quartz crystal resonator sensors. *Sens. Actuators, A*, **145**, 131–138.

61. Doerner, S., Schneider, T., Schroder, J., and Hauptmann, P. (2003) Universal impedance spectrum analyzer for sensor applications. Proceedings of IEEE Sensors, 2003, vol. 1, pp. 596–599.

62. Schaefer, R., Doerner, S., Lucklum, R., and Hauptmann, P. (2004) Single board impedance analyzer and transient analysis of QCR sensor response. Proceedings of the 2004 IEEE International Frequency Control Symposium and Exposition, 2004, IEEE, pp. 795–799.

63. Schnitzer, R., Reiter, C., Harms, K.-C., Benes, E., and Groschl, M. (2006) A general-purpose online measurement system for resonant baw sensors. *IEEE Sens. J.*, **6** (5), 1314–1322.

64. Niedermayer, A.O., Reichel, E.K., and Jakoby, B. (2009) Yet another precision impedance analyzer (YAPIA)-readout electronics for resonating sensors. *Sens. Actuators, A*, **156** (1), 245–250.

65. Jakoby, B., Art, G., and Bastemeijer, J. (2005) Novel analog readout electronics for microacoustic thickness shear-mode sensors. *IEEE Sens. J.*, **5** (5), 1106–1111.

66. Niedermayer, A., Voglhuber-Brunnmaier, T., Sell, J., and Jakoby, B. (2012) Methods for the robust measurement of the resonant frequency and quality factor of significantly damped resonating devices. *Meas. Sci. Technol.*, **23** (8), 085107.

67. Jakoby, B. and Vellekoop, M.J. (2004) Physical sensors for water-in-oil emulsions. *Sens. Actuators, A*, **110** (1), 28–32.

68. Jakoby, B., Ecker, A., and Vellekoop, M.J. (2004) Monitoring macro- and microemulsions using physical chemosensors. *Sens. Actuators, A*, **115** (2), 209–214.

69. Jakoby, B. and Dörr, N. (2004) Monitoring phase transitions in microemulsions using impedance and viscosity sensors. IEEE Sensors conference.

70. Follens, L.R.A., Reichel, E.K., Riesch, C., Vermant, J., Martens, J.A., Kirschhock, C.E.A., and Jakoby, B. (2009) Viscosity sensing in heated alkaline zeolite synthesis media. *Phys. Chem. Chem. Phys.*, **11** (16), 2854–2857.

71. D'amico, A. and Verona, E. (1989) Saw sensors. *Sens. Actuators*, **17** (1), 55–66.

72. Vellekoop, M., Lubking, G., Sarro, P., and Venema, A. (1994) Evaluation of liquid properties using a silicon lamb wave sensor. *Sens. Actuators, A*, **43** (1), 175–180.

73. Dufour, I., Maali, A., Amarouchene, Y., Ayela, C., Caillard, B., Darwiche, A., Guirardel, M., Kellay, H., Lemaire, E., Mathieu, F., et al. (2011) The microcantilever: a versatile tool for measuring the rheological properties of complex fluids. *J. Sens.*, **2012**, 1–9.

74. Riesch, C., Reichel, E., Jachimowicz, A., Schalko, J., Hudek, P., Jakoby, B., and Keplinger, F. (2009) A suspended plate viscosity sensor featuring in-plane vibration and piezoresistive readout. *J. Micromech. Microeng.*, **19** (7), 075010.

75. Rahafrooz, A. and Pourkamali, S. (2011) Characterization of rotational mode disk resonator quality factors in liquid. 2011 Joint Conference of the IEEE International Frequency Control and the European Frequency and Time Forum (FCS), IEEE, pp. 1–5.

76. Youssry, M., Belmiloud, N., Caillard, B., Ayela, C., Pellet, C., and Dufour, I. (2011) A straightforward determination of fluid viscosity and density using microcantilevers: from experimental data to analytical expressions. *Sens. Actuators, A*, **172** (1), 40–46.

77. Youssry, M., Lemaire, E., Caillard, B., Colin, A., and Dufour, I. (2012) On-chip characterization of the viscoelasticity of complex fluids using microcantilevers. *Meas. Sci. Technol.*, **23** (12), 125306.

78. Lemaire, E., Caillard, B., Youssry, M., and Dufour, I. (2013) High-frequency viscoelastic measurements of fluids based on microcantilever sensing: new modeling and experimental issues. *Sens. Actuators, A*, **201**, 230–240.

79. El Bsat, M.N., Yaz, E.E., Schneider, S.C., Dufour, I., and Josse, F. (2013) Nonlinear estimation of fluid properties using the

time domain response of a vibrating microcantilever. *Int. J. Model. Simul.*, **33** (4), *10.2316/Journal.205.2013.4.205-5789.*

80. Bircher, B.A., Duempelmann, L., Renggli, K., Lang, H.P., Gerber, C., Bruns, N., and Braun, T. (2013) Real-time viscosity and mass density sensors requiring microliter sample volume based on nanomechanical resonators. *Anal. Chem.*, **85** (18), 8676–8683.

81. Rust, P., Cereghetti, D., and Dual, J. (2013) A micro-liter viscosity and density sensor for the rheological characterization of dna solutions in the kilo-hertz range. *Lab Chip*, **13** (24), 4794–4799.

82. Matsiev, L., Bennett, J., and Kolosov, O. (2005) High precision tuning fork sensor for liquid property measurements. IEEE Ultrasonics Proceedings, vol. 3, pp. 1492–1495.

83. Zhang, J., Dai, C., Su, X., and O'Shea, S.J. (2002) Determination of liquid density with a low frequency mechanical sensor based on quartz tuning fork. *Sens. Actuators, B*, **84** (2), 123–128.

84. Toledo, J., Manzaneque, T., Hernando-García, J., Vázquez, J., Ababneh, A., Seidel, H., Lapuerta, M., and Sánchez-Rojas, J. (2014) Application of quartz tuning forks and extensional microresonators for viscosity and density measurements in oil/fuel mixtures. *Microsyst. Technol.*, **20** (4-5), 945–953.

85. Sullivan, M., Harrison, C., Goodwin, A.R., Hsu, K., and Godefroy, S. (2009) On the nonlinear interpretation of a vibrating wire viscometer operated at large amplitude. *Fluid Phase Equilib.*, **276** (2), 99–107.

86. Etchart, I., Chen, H., Dryden, P., Jundt, J., Harrison, C., Hsu, K., Marty, F., and Mercier, B. (2008) Mems sensors for density–viscosity sensing in a low-flow microfluidic environment. *Sens. Actuators, A*, **141** (2), 266–275.

87. Riesch, C., Reichel, E.K., Keplinger, F., and Jakoby, B. (2008) Characterizing vibrating cantilevers for liquid viscosity and density sensing. *J. Sens.*, **2008**.

88. Riesch, C., Jachimowicz, A., Keplinger, F., Reichel, E.K., and Jakoby, B. (2008) A micromachined doubly-clamped beam rheometer for the measurement of

89. Heinisch, M., Reichel, E., Dufour, I., and Jakoby, B. (2012) Tunable resonators in the low khz range for viscosity sensing. *Sens. Actuators, A*, **186**, 111–117.

90. Reichel, E.K., Riesch, C., Weiss, B., and Jakoby, B. (2008) A vibrating membrane rheometer utilizing electromagnetic excitation. *Sens. Actuators, A*, **145**, 349–353.

91. Cerimovic, S., Beigelbeck, R., Antlinger, H., Schalko, J., Jakoby, B., and Keplinger, F. (2012) Sensing viscosity and density of glycerol–water mixtures utilizing a suspended plate mems resonator. *Microsyst. Technol.*, **18** (7-8), 1045–1056.

92. Manzaneque, T., Ruiz, V., Hernando-García, J., Ababneh, A., Seidel, H., and Sánchez-Rojas, J. (2012) Characterization and simulation of the first extensional mode of rectangular micro-plates in liquid media. *Appl. Phys. Lett.*, **101** (15), 151904.

93. Seo, J.H. and Brand, O. (2008) High-factor in-plane-mode resonant microsensor platform for gaseous/liquid environment. *J. Microelectromech. Syst.*, **17** (2), 483–493.

94. Huang, X., Li, S., Schultz, J., Wang, Q., and Lin, Q. (2009) A capacitive MEMS viscometric sensor for affinity detection of glucose. *J. Microelectromech. Syst.*, **18** (6), 1246–1254.

95. Vančura, C., Dufour, I., Heinrich, S.M., Josse, F., and Hierlemann, A. (2008) Analysis of resonating microcantilevers operating in a viscous liquid environment. *Sens. Actuators, A*, **141** (1), 43–51.

96. Boskovic, S., Chon, J., Mulvaney, P., and Sader, J. (2002) Rheological measurements using microcantilevers. *J. Rheol. (1978-present)*, **46** (4), 891–899.

97. Beardslee, L.A., Addous, A.M., Heinrich, S., Josse, F., Dufour, I., and Brand, O. (2010) Thermal excitation and piezoresistive detection of cantilever in-plane resonance modes for sensing applications. *J. Microelectromech. Syst.*, **19** (4), 1015–1017.

18
Energy Harvesting Devices

Stephen P. Beeby

18.1
Introduction

This chapter is concerned with mechanical energy harvesting devices, which are designed to convert ambient kinetic energy into electrical energy. Such harvesters can provide a localized power source for remote or autonomous electronic systems and are used to augment or replace battery power supplies. Energy harvesting is attractive because it can avoid the need to replace batteries and, if designed correctly, can provide a power supply for the lifetime of the system [1]. A block diagram of a typical autonomous system powered by energy harvesting is shown in Figure 18.1. In this, the electrical output of the energy harvester goes to the power-conditioning electronics located in the energy and systems management block. This delivers the conditioned power to the energy storage and also controls the switching on of the various load electronic modules. Systems such as these are typically duty cycled, that is, the load electronics are switched on only when sufficient energy is available in the energy storage [2].

Mechanical energy harvesting can be used in many applications where sufficient kinetic energy exists in a form that is suitable for harvesting. Vibrations are an obvious form of kinetic energy that can be found in many applications including machinery and equipment, vehicles, rail rolling stock, and power tools. One potential application of mechanical energy harvesters is portable electronic devices carried by people. In this case, the kinetic energy comes from the movement of the person and this takes a very different form in comparison with, for example, machinery vibrations. Human motion is characterized by large amplitude, low frequency displacements and this is one of the most challenging applications of mechanical energy harvesting.

MEMS technology is clearly an attractive technology for the type of system shown in Figure 18.1. MEMS offer the potential for integrating the mechanical harvester alongside integrated electronics and energy storage. Numerous types

Resonant MEMS – Fundamentals, Implementation and Application, First Edition.
Edited by Oliver Brand, Isabelle Dufour, Stephen M. Heinrich and Fabien Josse.
© 2015 Wiley-VCH Verlag GmbH & Co. KGaA. Published 2015 by Wiley-VCH Verlag GmbH & Co. KGaA.

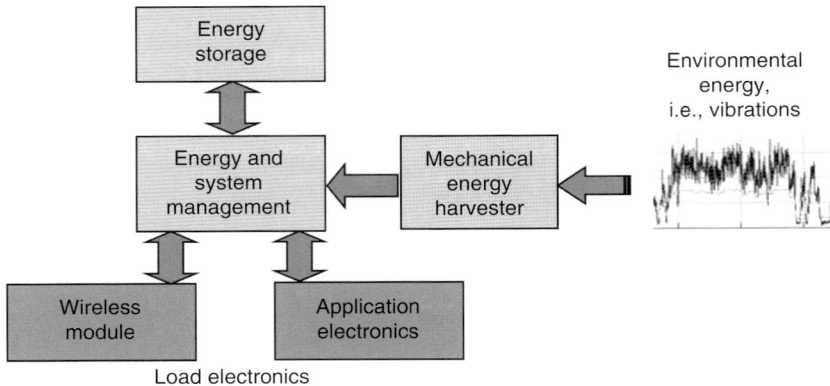

Figure 18.1 Autonomous electronics system powered by mechanical energy harvesting.

of transduction mechanisms can be realized with MEMS technology and the fabrication processes provide low-cost mass manufacture of the devices and systems. The mechanical properties of MEMS materials such as single crystal silicon or polysilicon are also ideally suited to this type of device. On the downside, however, miniaturized MEMS mechanical structures are associated with high-resonant frequencies. Also, given the typically small device dimensions, displacement amplitudes, and inertial masses are going to be small, which limits the mechanical energy that can be captured by the harvester. These factors will be highlighted by the analysis in the following section.

This chapter will discuss the application of resonant mechanical systems in energy harvesting and will further explain the implications of MEMS technology. The chapter will provide basic formulas that characterize the devices and will discuss strategies for maximizing the kinetic energy that can be captured by the harvester from different forms of ambient energy. The transduction mechanisms available to MEMS devices will be discussed in Section 18.3. These factors will be illustrated by example MEMS mechanical energy harvesters from the literature and the chapter will conclude with a comparison of MEMS devices realized to date.

18.2
Generic Harvester Structures

In order to convert kinetic energy, a mechanical energy harvester requires a mechanical structure that captures the ambient energy and couples it effectively to some form of transducer that does the energy conversion. The design of the mechanical system should maximize the coupling between the kinetic energy source and the transduction mechanism and will depend entirely upon the characteristics of the environmental motion.

Inertial frame

Spring element

k

m   z(t)

Inertial mass

C_T

Damper (transducer
+ parasitic)

y(t)

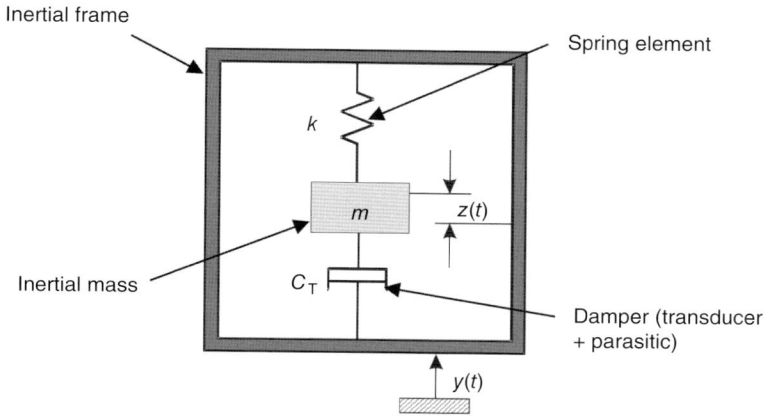

Figure 18.2   Model of a linear, inertial generator.

18.2.1
Inertial Energy Harvesters

Vibration energy is best suited to inertial generators with the mechanical component attached by a spring element to an inertial frame, which acts as the fixed reference frame (see Figure 18.2). This approach enables the generator to be simply placed in a suitable location on the host structure. The inertial frame transmits the external vibrations of amplitude $y(t)$ to a suspended inertial mass producing a relative displacement between them, $z(t)$. Energy is extracted from the inertial mass by the transduction mechanism, which converts the mechanical motion into electrical energy. The transducer produces a damping effect on the inertial mass which along with other parasitic damping effects limits the maximum displacement of the inertial mass. This is a straightforward second-order spring-mass-damper system that possesses a resonant frequency, which must typically be designed to match a characteristic frequency of the application environment. The resonant approach magnifies the environmental vibration amplitude by the quality factor (Q-factor, Q_T where the subscript T denotes total) of the resonant system. This inertial approach is required in the majority of vibration applications where absolute displacements $y(t)$ are very low. Take an example of sinusoidal vibrations of peak acceleration, A, equal to $1\,g$ (where $g = 9.81\,\mathrm{m\,s^{-2}}$) at $100\,\mathrm{Hz}$. Peak amplitude (Y) is given by Eq. (18.1) where ω is the radial frequency and this gives an amplitude of just $25\,\mu\mathrm{m}$. A Q_T of 100 would give an inertial mass displacement of $2.5\,\mathrm{mm}$, greatly improving the coupling efficiency of the transducer.

$$Y = A/\omega^2 \tag{18.1}$$

The theory of inertial-based generators is well documented [3] and assumes the harvester is driven by a harmonic base excitation $y(t) = Y\sin(\omega t)$. This will cause the inertial mass, m, to move out of phase with the inertial frame, resulting in a net displacement, $z(t)$, between the mass and the frame. The average power dissipated within the damper (i.e., the power extracted by the transduction mechanism and the power lost through parasitic damping mechanisms) is given by

$$P_{av} = \frac{m\xi_T Y^2 \left(\frac{\omega}{\omega_n}\right)^3 \omega^3}{\left[1 - \left(\frac{\omega}{\omega_n}\right)^2\right]^2 + \left[2\xi_T \left(\frac{\omega}{\omega_n}\right)\right]^2} \tag{18.2}$$

where ξ_T is the total damping ratio given by $x_T = c_T/(2\,mw_n)$ where c_T is the damping coefficient and ω_n the natural frequency. Since this equation is valid for steady-state conditions, P_{av} is equal to the average kinetic energy supplied per second by the application vibrations. The maximum power dissipation within the generator occurs when the device is operated at ω_n and Eq. (18.2) can be represented by Eq. (18.3). This equation highlights the benefits of larger inertial mass displacements, which are discussed further below.

$$P_{av} = \frac{m\omega_n^3 Y z_{max}}{2} \tag{18.3}$$

Since ξ_T comprises mechanical (parasitic) damping (ξ_m) and electrical damping (ξ_e) and maximum power dissipation within the generator occurs when the device is operated at ω_n, Eq. (18.2) can also be represented by Eq. (18.4):

$$P_{av} = \frac{mY^2\omega_n^3}{4(\xi_e + \xi_m)} \tag{18.4}$$

Since energy harvesting is concerned with delivering electrical energy, the power delivered to the electrical domain is given by Eq. (18.5):

$$P_e = \frac{\xi_e}{(\xi_e + \xi_m)} \frac{mY^2\omega_n^3}{4(\xi_e + \xi_m)} \tag{18.5}$$

Maximum power is delivered into the electrical domain when $\xi_e = \xi_m$, which gives Eq. (18.6):

$$P_{emax} = \frac{mY^2\omega_n^3}{16\xi_m} \tag{18.6}$$

From Eq. (18.1) we know peak acceleration $A = Y\omega^2$ and therefore P_{emax} can be given by

$$P_{emax} = \frac{mA^2}{16\omega_n\xi_m} \tag{18.7}$$

Since the open circuit Q factor, $Q_{OC} = 1/(2\xi_m)$, Eq. (18.7) then becomes

$$P_{emax} = \frac{mA^2}{8\omega_n}Q_{OC} \tag{18.8}$$

Equation (18.8) is a practical form of the equation to use since all the variables can be easily measured. What do these equations tell us? Firstly and most clearly, the power delivered into the electrical domain is proportional to the inertial mass and Q_{OC}. The implications for MEMS generators are obvious in the case of the inertial mass. As the harvester shrinks in size, the inertial mass will scale with the linear dimension cubed, so halving a device in size will reduce the electrical power by a factor of 8. Equation (18.8) also suggests that Q_{OC} should be maximized, that is, parasitic damping minimized. However, as the parasitic damping decreases the inertial mass amplitude will increase, but in practice, this will be limited to a maximum value (z_{max}), which is dependent upon the harvester size, the design of the spring element and the fatigue strength of the spring material. High inertial mass amplitudes may also introduce nonlinear effects such as spring stiffening, which would result in a resonant frequency change with amplitude. Therefore, in practice, especially in the case of small devices, it may be best to maximize the electrical damping and minimize parasitic damping due to the restricted z_{max} for a particular device. This decision also depends upon the characteristics of the mechanical excitation (Y and ω), but if this approach is taken then clearly $\xi_m \neq \xi_e$ and the maximum theoretical power given by Eqs. (18.6)–(18.8) cannot be obtained. So the small size of MEMS structures may mean optimum power output cannot be achieved.

Another consequence of the Q-factor is its effect on the bandwidth of the harvester which is shown in Figure 18.3. A high Q_T (e.g., $Q_T = 500$ in Figure 18.3) gives a sharp, narrow resonant peak. This gives a high peak power but variations in

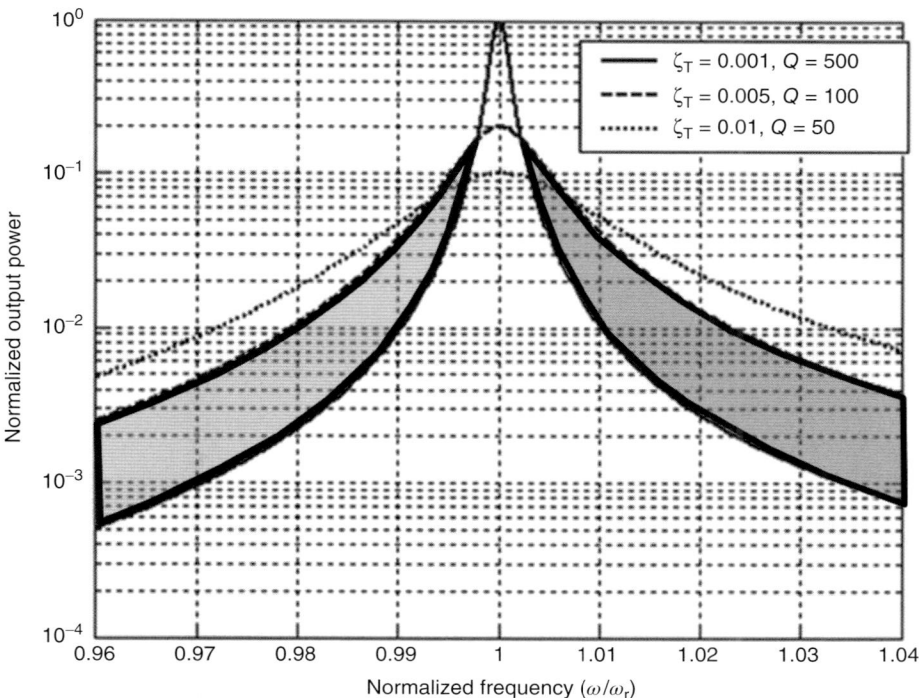

Figure 18.3 The influence of total Q-factor on power output and bandwidth.

frequency result in a sharp drop-off in the power harvested. Reducing Q_T reduces the peak output power but broadens the width of the peak reducing the effect of changes in frequency on the power harvested. A lower-Q harvester can produce higher levels of output power at some off-resonant frequencies. This is shown in Figure 18.3 where a Q_T of 100 means a peak power five times lower than for $Q_T = 500$ but the power in the shaded areas is greater for the lower Q device. In addition to z_{max}, the design of the energy harvester must clearly take into consideration the nature of the environmental vibrations and the expected variations in frequency and amplitude that may be expected together with the electrical power required by the load electronics. This highlights the fact that for practical applications of energy harvesting, surveys of the application environment are essential in order to achieve a useful harvester that delivers the required power in all cases.

While the power delivered to the electrical domain is interesting, what we really need to know is the power delivered to the load. This takes into consideration losses in the electrical domain from, for example, the resistance of a coil in an electromagnetic harvester. For an electromagnetic generator with a coil resistance R_C connected to a resistive load R_L, the maximum power in the load is given by Eq. (18.9):

$$P_{\mathrm{Lmax}} = \frac{mA^2}{8\omega_n} Q_{OC} \frac{R_L}{R_L + R_C} \tag{18.9}$$

Equation (18.9) indicates that having R_L much greater than R_C ensures that most of the energy is delivered to the load. Increasing R_L decreases electrically induced damping and increases the output voltage, but there will be an optimum resistive load to maximize power. This is discussed further in Section 18.3.

18.2.2
Direct Force Energy Harvesters

Direct force type harvesters do not exploit the principle of resonance and rely on exploiting the relative force between two structures to generate mechanical energy [4]. This approach requires the harvester to be either permanently attached to both structures or attached to one and periodically come into contact with the second. Energy can be extracted from the applied force between the two structures, or energy can be obtained from a displacement within the harvester caused by the force. The simplest examples of this type of harvester are shoe-based generators that harvest energy from the footfall of individuals as they walk. The weight of the individual is concentrated at the contact between the shoe and ground, causing large forces in the footwear. Piezoelectric materials can harvest the mechanical motion from the direct application of these direct forces [5]. This type of application is not ideally suited to MEMS since directly applying large forces to microme-chanical structures is likely to lead to fracture.

18.2.3
Broadband Energy Harvesters

The limitations of the linear resonant approach are the need to operate at a particular frequency and the effect that variations in the excitation frequency have on
the power output. These can result in particularly challenging requirements for
MEMS energy harvesters where the resonant frequency of MEMS structures will
tend to be higher than the target application frequency. Also, the restrictions on
z_{max} for MEMS devices discussed above mean linear resonant operation may not
be the best approach for MEMS. The requirement to operate at resonance and limitations on bandwidth can be overcome by using different structures or operating
strategies [6]. The simplest example of this would be to increase the level of damping to broaden the bandwidth and offset the reduction in peak power by increasing
the size of the inertial mass. Practical commercially proven industrial generators such as the PMG FSH [7] free-standing vibration energy harvester offered by
Perpetuum Ltd are based on this principle. This is an electromagnetic energy harvester that produces up to 20 mW and has a variable bandwidth depending upon
the driving amplitudes. For example, at 0.5 g_{RMS} the full-width half maximum current bandwidth is 15 Hz on a 50-Hz center frequency. This bandwidth falls to 2 Hz
at 0.02 g_{RMS}. The device is 68 mm in diameter and 63.3 mm high and has an overall
mass of 1.075 kg. In applications where there are no space constraints and a large
harvester can be employed, it is clear that a very wide operational bandwidth can
be achieved with a high power output. This approach is, however, not applicable
to MEMS implementations, which are limited in size.

An alternative approach is to use an array of resonant structures in the harvester, each tuned to a different frequency. By selecting the frequencies carefully
with respect to the center frequency and bandwidth of each structure, the output
of each element can overlap, producing a peak power output over a broader
frequency range (see Figure 18.4). This approach is certainly suitable for MEMS
implementations although it clearly increases the overall size of the harvester
and power output at any frequency is typically provided by one or two resonant
elements with the remainder being largely inactive. Device complexity is also
increased and process yield becomes an important consideration since each
element must be produced reliably and accurately for the system to function
correctly.

Tuning approaches can also be adopted, whereby the center frequency of a single
resonant structure can be adjusted by mechanical forces. Changing the mechanical stress in a spring element has the effect of altering the spring constant of the
structure and therefore the resonant frequency of the system. These forces can be
applied directly using a mechanical system [8] or indirectly using magnetic forces
[9]. The cantilever demonstrated by Eichhorn *et al.* [8] uses two arms to apply a
compressive axial stress to the end of a cantilever beam which reduces the resonant frequency. The mechanical force results from the manual adjustment of a
screw. Zhu *et al.* [9] used a tuning magnet placed on the tip of a cantilever, which
was aligned to a fixed tuning magnet attached to the reference frame with the

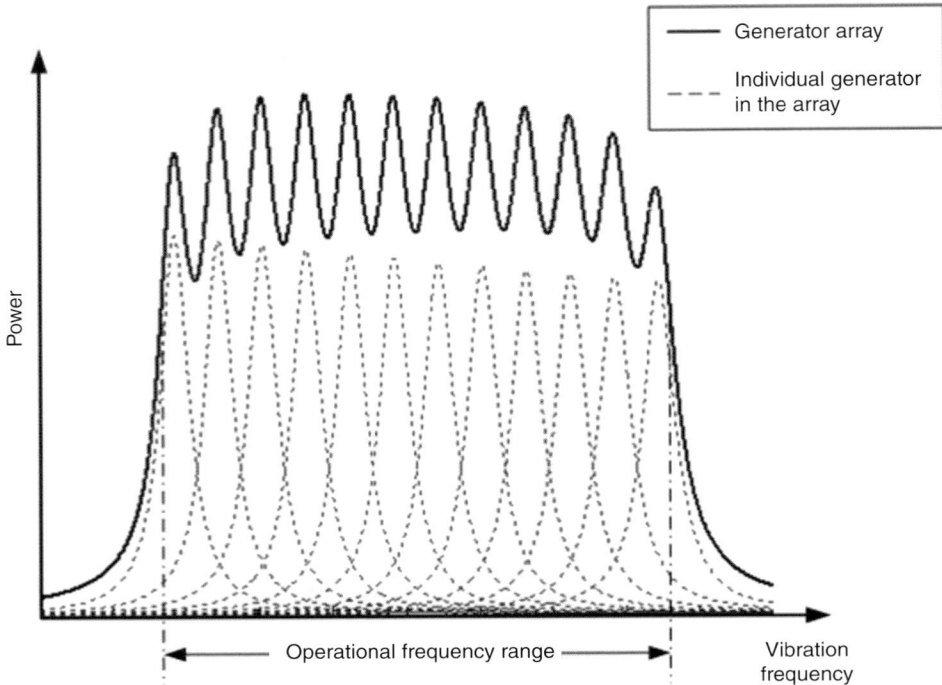

Figure 18.4   Generator array power spectrum.

opposite poles facing each other. The attractive force between the magnets produced an axial tensile force on the cantilever, increasing the effective stiffness of the beam and therefore the resonant frequency. The magnitude of the tensile force can be adjusted by varying the gap between the tuning magnets and this approach provides a non-contact method for applying such tuning forces. The tuning magnet position can be autonomously adjusted by a control circuit, which can all be powered by the energy harvester, producing a self-powered tunable system [10]. Tuning mechanisms have yet to be successfully demonstrated at the MEMS scale.

Multimodal or coupled oscillator structures can also be used to broaden the practical bandwidth of the harvester. Multimodal harvesters are designed to extract energy from more than one resonant frequency of a mechanical structure. The structure must be designed to achieve the desired frequencies and the transduction mechanisms must be coupled to each mode shape to ensure effective energy conversion at each frequency [11]. Coupled oscillator structures consist of more than one inertial mass and spring element to form an overall structure with a wider range of resonant modes. A dual-mass harvester with two spring elements has been characterized by Tang and Zuo [12]. This analysis claimed that more electrical power can be generated with the dual mass approach compared with a single-degree-of-freedom linear harvester, but results are dependent upon damping ratio, mass ratio, and tuning ratio. For both these approaches, obtaining

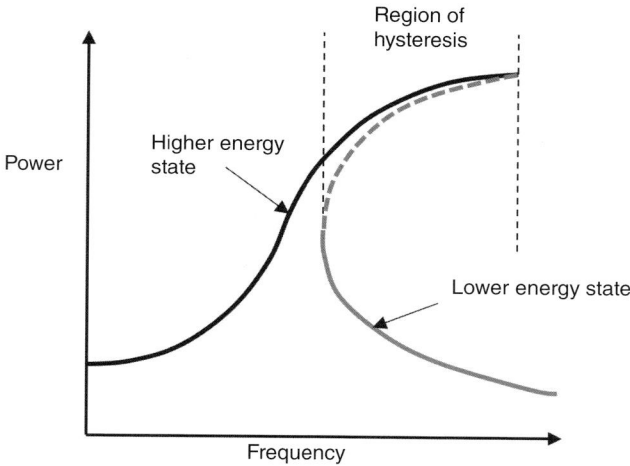

Figure 18.5 Nonlinear harvester output showing region of hysteresis.

a suitable frequency response that will actually generate more power from real application vibrations is a considerable challenge. Also, these approaches add to the size and weight of the harvester.

Nonlinear harvesters that deliberately introduce spring hardening or softening effects can be used such that the resonant frequency increases or decreases with amplitude [13]. Rather than use the conventional second-order model as used above (Eq. (18.2) onwards), non-linear generators are represented by a Duffing's type equation as presented in Ref. [6]. The harvester in Ref. [13] exhibits the hardening-spring nonlinear effect and various numerical and analytical studies have shown that a device with this behavior has a larger bandwidth over which power can be harvested. The maximum amount of power harvested by a hardening-spring nonlinear system is the same as the maximum power harvested by a linear system, although this possibly occurs at a different frequency depending on the nonlinearity. The bandwidth of the hardening system depends on the damping ratio, the nonlinearity, and the input acceleration. Systems exhibiting sufficient non-linear behavior and driven by suitably high-amplitude vibrations can enter a region of hysteresis where the harvester can operate at a high- or low-energy state (see Figure 18.5). Such nonlinear harvesters are, however, not straightforward to realize or operate. The practical output power and bandwidth of the nonlinear generators depend on whether the frequency of external vibrations are increasing or decreasing toward the resonant frequency. For the hardening-spring nonlinearity, this approach will only produce an increase in bandwidth if the frequency is increasing toward resonance and providing the driving vibrations are of sufficient amplitude. In this case, the harvester will maintain the high energy state over a wide frequency range (compared to linear generators) until dropping off to the low-energy state. An example of this has been demonstrated by a nonlinear MEMS piezoelectric harvester driven by a sinusoidal excitation [14]. Nonlinear harvesters operated in the nonlinear region will also be unstable and the frequency at which this drop-off to the lower energy state occurs will often be variable. Even when driven by a pure sinusoidal vibration source, the frequency at which this happens will show variations between experiments and devices. However, real-world stochastic vibrations are not pure sinusoids and random variations in amplitude and frequency will make such nonlinear harvester behavior even more chaotic and unrepeatable.

Another type of nonlinear approach presented in the literature is the use of bistable structures that switch between two equilibrium states. The nonlinear behavior can be introduced by placing a spring structure, for example, a fixed–fixed beam, in compression. When the compressive force exceeds the maximum buckling load, the structure deforms into one of two equilibrium states. Another method for achieving bistability is to use magnets, one on the end of a cantilever and a second placed next to the first with like poles facing each other, that is, generating a repulsive force [15]. When the gap between the magnets becomes sufficiently small, two magnetic potential wells are introduced, which become the equilibrium states of the cantilever. Whichever technique is used to realize the bistability, the harvester will function when external displacements of sufficient acceleration cause the structure to switch equilibrium states. This can happen at any frequency producing the broadband response. Bistable MEMS structures have been demonstrated (e.g., [14, 16]) and are claimed to perform better than linear harvesters when driven by random stochastic vibrations. Example MEMS devices will be discussed further in Section 18.4.

The final approach for increasing the operation bandwidth is to use a mechanical stopper that limits the displacement of an otherwise linear harvester [17]. As the vibration frequency is increased, approaching the resonant frequency of the harvester, the increasing amplitude causes the cantilever to strike the stopper, which alters the frequency response of the structure. While this approach does increase the bandwidth in the case of an increasing frequency, it does not work in the case of a decreasing frequency. It also results in a reduced power output due to the reduced maximum amplitude. Most importantly, however, since energy harvesting is intended to be a long-lifetime power supply solution, techniques that result in mechanical wear are not advisable.

18.2.4
Frequency Conversion

Solutions to overcome the tendency for MEMS structures to typically resonate at higher frequencies than experienced in real world energy harvesting applications have also been widely researched. There are three main approaches for coupling low-frequency excitations into higher frequency structures as described in what follows.

The plucking approach is analogous to a guitar string being plucked by a plectrum during playing. In the case of energy harvesting, the high-frequency resonant structure is displaced by either a mechanical or magnetic force and released to vibrate at its natural frequency. An example of the mechanical approach is the energy harvesting knee brace developed at Cranfield University, United Kingdom [18]. This harvester is mounted at the knee and consists of four piezoelectric bimorphs effectively attached to the lower leg and a series of ring-mounted plectra attached to the top part of the leg. As the knee bends, the relative angular motion causes the plectra to pluck the bimorphs and hence energy can be harvested. In order to address the concerns about component wear, magnetic forces can be used to pluck the harvester. One version of this approach uses a circular magnet suspended between two other magnets within a Teflon tube [19]. Two piezoelectric cantilevers are positioned outside the tube with magnets mounted on the tips of the beam. The suspended magnets oscillate within the tube at $10-22$ Hz and the magnetic forces pluck the external cantilevers, which vibrate at 240 Hz.

High-frequency mechanical structures can also be driven into resonance by a sharp impulse or shock arising from an impact. An example of this technique was demonstrated by Renaud *et al.* [20] who used a steel capsule-shaped missile, which was free to move along a Teflon channel. The missile stuck piezoelectric bimorphs located at each end of the channel, which resonated at around 800 Hz. The example harvester was shown to generate up to 600 μW for a device of dimensions $3.5\,\text{cm} \times 2\,\text{cm} \times 2\,\text{cm}$, weighing 60 g and excited at 10 Hz with a large 10 cm displacement. An earlier incarnation of this approach was reported by Cavallier *et al.* [21] and was developed as part of the EU-funded VIBES project. This approach relied on four small steel ball bearings which were free to bounce within a chamber striking piezoelectric cantilevers. This was tested at 6 Hz and each impact typically excited a 10 kHz decaying voltage output from a piezoelectric beam. This approach is clearly able to convert low-frequency vibrations of sufficient amplitude into higher-frequency responses, but again there is a concern about the longevity of such harvesters in practice.

A bistable structure, similar to those discussed in Section 18.2.3, has also been used to provide an impulse excitation to a piezoelectric cantilever [22]. A bistable structure consisting of a fixed–fixed beam in compression, having an inertial mass and a piezoelectric cantilever located at the center of the beam, was driven at 12 Hz. A $24\,\text{m\,s}^{-2}$ amplitude of acceleration was found to be sufficient for the structure to snap between equilibrium states, which produced a decaying oscillation in the cantilever at its resonant frequency of 230 Hz. This demonstrated that the principle works but the excitation amplitude required is much higher than that which occurs in typical application vibrations.

18.3
MEMS Energy Harvester Transduction Mechanisms

Whatever type of mechanical structure is employed to capture the external kinetic energy, some form of transduction mechanism is required in order to convert energy from the mechanical to the electrical domain. The transducer must be designed with the mechanical structure to ensure that effective coupling occurs.

The transduction mechanism itself can generate electricity by exploiting the strain in the mechanical structure or relative displacement occurring within the system. The strain effect exploits the deformation of the mechanical structure and typically employs active (e.g., piezoelectric) materials to generate electrical energy. In the case of relative displacement, either velocity, or position can be used. Velocity is typically associated with electromagnetic transduction, while relative position is associated with electrostatic transduction. Each transduction mechanism exhibits different damping characteristics and this should be taken into consideration when modeling the generators. As discussed in the following sections, some forms of transduction are better suited to MEMS implementations than others.

18.3.1
Piezoelectric Transduction

The piezoelectric effect arises when certain materials subject to mechanical strain become electrically polarized, producing a net charge across the material. This method of operation is ideally suited to kinetic energy harvesting. Conversely, these materials deform when exposed to an electric field and can be used in this manner as an actuator. Piezoelectric materials are widely available in many forms, including single crystal (e.g., quartz), piezoceramic (e.g., lead zirconate titanate or PZT), thin film (e.g., sputtered zinc oxide), screen-printable thick-films based upon piezoceramic powders and polymeric materials such as polyvinylidenefluoride (PVDF).

Piezoelectric materials exhibit anisotropic characteristics, that is, the properties of the material differ depending upon the direction of forces and orientation of the polarization and electrodes. The level of piezoelectric activity of a material is defined by a series of constants used in conjunction with the axes notation. In MEMS, piezoelectric materials are typically deposited as thin films with the piezoelectric material deposited onto a substrate and sandwiched between two electrodes as shown in Figure 18.6. This means they are polarized through the thickness of the film, which is the 3 direction. The piezoelectric strain constant, d, relates the strain developed to the applied field [23]. Piezoelectric generators that rely on a compressive strain applied perpendicular to the electrodes exploit the d_{33} coefficient of the material, that is, the force is also applied in the 3 direction. Typically, in the case of MEMS piezoelectric films, or piezoelectric elements bonded onto substrates, the elements are coupled with a longitudinal strain that is parallel to the electrodes (the 1 direction) as the substrate bends during cyclical motion. This approach utilizes the d_{31} coefficient. Such an arrangement provides mechanical amplification of the applied stresses due to the thickness of the substrate. The d_{31} coefficient is, however, typically smaller than the d_{33} coefficient. A typical arrangement for a cantilever-based piezoelectric energy harvester is shown in Figure 18.6.

In addition to the d coefficient, another important constant determining the generation of electrical power is the electro-mechanical coupling coefficient, k. This describes the efficiency with which the energy is converted by the material

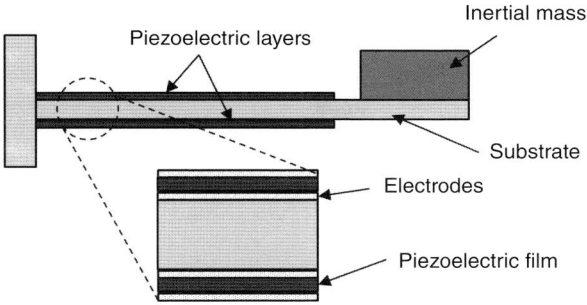

Figure 18.6 Cross section of a cantilevered piezoelectric harvester with a close-up of the piezoelectric layers and associated electrodes.

between the electrical and mechanical domains. It is given by Eq. (18.10), where W_e and W_m are electrical and mechanical energy respectively and the 33 and 31 notation applies to k as described above.

$$k^2 = \frac{W_e}{W_m} \tag{18.10}$$

The performance of piezoelectric materials in energy harvesting applications is commonly determined by the figure of merit (FOM) given in Eq. (18.11) [24], where e_{31} is the charge generated for the applied strain in the 1 direction and ε_r is the relative permittivity.

$$\text{FOM} = e_{31}^2/e_r \tag{18.11}$$

Table 18.1 summarizes the typical properties of common piezoelectric materials and gives the calculated FOM. In terms of MEMS technology, thin-film PZT and AlN are common piezoelectric materials. It is interesting to note that, despite the inferior piezoelectric properties of AlN, its reduced permittivity means that for energy harvesting, AlN films are comparable to thin-film PZT. AlN has the advantage that it is much more compatible with microelectronic processes than PZT. Table 18.1 also quotes the piezoelectric voltage constant g_{31}, which relates the electric field produced by a mechanical stress for a film poled in the 3 direction and stressed in the 1 direction.

Piezoelectric properties do vary with age, stress, and temperature, which will reduce the performance of a device. The aging rate of a piezoceramic is dependent on the construction methods and the material type. The changes in the material tend to be logarithmic with time; thus, the material properties stabilize with age. The aging process is accelerated by the stress applied to the material and this should be considered in cyclically loaded energy harvesting applications. The maximum temperature of a piezoelectric material is limited to the Curie point above which it becomes de-poled. The application of stress can lower this Curie temperature making high-stress, high-temperature applications challenging. Soft piezoceramic compositions, such as PZT-5H, are more susceptible to stress-induced changes than the harder compositions such as PZT-5A.

Table 18.1 Typical material properties of common piezoelectric materials [1, 25, 26].

Property	PZT-5H1 (bulk)	BaTiO$_3$	PVDF	Thin film PZT	Thin film AlN
d_{33} (10^{-12} C N^{-1})	620	86	−33	300	5
d_{31} (10^{-12} C N^{-1})	−250	−35	23	−120	1.5
g_{31} (10^{-3} Vm N^{-1})	−8.7	5	0.2	0.014−0.057	—
e_{31} (10^{-12} C m^{-2})	−16	−4	0.074	−8 to −12	−1.4
k_{31}	0.35	0.21	0.15	0.15	6.5
ε_r	3400	1700	12	1000	10
FOM	0.075	0.009	0.0005	0.06−0.14	0.20

With regard to fundamental scaling effects, piezoelectric transduction is well suited to MEMS implementations. There are numerous processes for the deposition of high-quality piezoelectric films and there is considerable ongoing research to improve the piezoelectric properties of thin film materials, for example, [27, 28] which will benefit MEMS energy harvesting solutions.

18.3.2
Electromagnetic Transduction

Electromagnetic induction is based on the principle of Faraday's law, which is that the voltage, or electromotive force (emf), induced in a circuit is proportional to the time rate of change of the magnetic flux linkage of that circuit. In basic terms, this means that a voltage is generated in a conductor when it is placed in a varying magnetic field. The conductor typically takes the form of a coil and the electricity is generated by either the relative movement of the magnet and coil and/or because of changes to the flux gradient. In the former case, the amount of electricity generated depends upon the strength of the magnetic field, the velocity of the relative motion, and the number of turns of the coil.

Power is extracted from an electromagnetic generator by connecting the coil to a load resistance, R_{load}, and allowing a current to flow in the coil. The coil current creates its own magnetic field, which opposes the original magnetic field that generated the emf. The interaction between these two magnetic fields gives rise to an electromagnetic force, F_{em}, which opposes the motion of the energy harvester. It is the work done against F_{em} that is transformed into electrical energy. The electromagnetic force is proportional to the current and hence the velocity. It is expressed as the product of an electromagnetic damping, c_e, and the velocity (dx/dt) as shown in Eq. (18.12).

$$F_{em} = c_e \frac{dx}{dt} \tag{18.12}$$

In order to deliver the maximum electrical power to the load, it is important to maximize the electromagnetic damping coefficient, c_e, which is given by Eq. (18.13).

$$c_{\mathrm{e}} = \frac{(NlB)^2}{R_{\mathrm{load}} + R_{\mathrm{coil}} + j\omega L_{\mathrm{coil}}} \tag{18.13}$$

where N is the number of turns in the generator coil, l is the side length of the coil (assumed square), and B is the flux density to which it is subjected and R_{coil} and L_{coil} are the coil resistance and coil inductance respectively. In practice, at typical vibration levels the imaginary component of the coil impedance can be ignored but it may become a factor at frequencies above 1 kHz. Equation (18.13) shows that R_{load} can be used to adjust c_{e} to match the parasitic damping coefficient, c_{p}, and therefore maximize power, although this must be done with the coil parameters in mind. It can be shown that the maximum average power delivered to the load can be found from Eq. (18.14) [3].

$$P_{\mathrm{eload\ max}} = \frac{mA^2}{16\,\xi_{\mathrm{m}}\omega_{\mathrm{n}}} \left(1 - \frac{R_{\mathrm{coil}}}{R_{\mathrm{load}}}\right) \tag{18.14}$$

It is clear from Eq. (18.14) that the coil resistance will limit the maximum energy delivered to the load and should therefore be minimized. This raises the issue of the scalability of electromagnetic energy harvesting. For MEMS devices, the coil will be formed by microfabrication processes (metal and insulating film deposition, photolithography, and etching). The number of turns, the thickness of the conductor, and the width of the tracks will determine both the electromagnetic damping and the coil resistance. Electromagnetic damping increases with N^2 but the limitations of the planar fabrication processes typically means that in practice, the resistance of the microcoil can increase with N^3. This is a generalization and depends upon the exact capabilities of the fabrication processes but is a reasonable approximation for typical process capabilities. The net result is that increasing coil turns actually decreases electromagnetic damping but it would increase coil voltage, which is an important consideration. Coupled with this, deposited magnetic materials are approximately a factor of 2 weaker than their bulk counterpart. It is clear that electromagnetic transduction does not scale down well and MEMS kinetic energy harvesters should avoid electromagnetic transduction.

18.3.3
Electrostatic Transduction

Electrostatic transduction exploits the relative movement between the electrodes of a capacitor, which causes a periodic change in capacitance. The electrodes can be charged by periodic connection to a voltage source or by the use of electrets. The harvested energy arises from the work done against the electrostatic force between the plates [4]. There are two types of electrostatic harvesters: charge-constrained and voltage-constrained. The charge-constrained type operates with a fixed charge, while the voltage-constrained operates with a fixed voltage. The electrostatic forces in each case are calculated as follows.

If the charge, Q, on the plates is held constant, the perpendicular force between the plates is given by

$$F = Q\frac{d}{\varepsilon A} \tag{18.15}$$

where d is the gap between the electrodes, A is the surface area, and ε is the permittivity of free space. If the voltage, V, between the plates is held constant, the perpendicular force between the plates is given by

$$F = 0.5\varepsilon A\frac{V^2}{d^2} \tag{18.16}$$

In the case of the voltage-constrained operation, the net energy delivered to the charge reservoir is given by Eq. (18.17) [29] where C_{max} and C_{min} are the maximum and minimum capacitances at the extremes of the electrode displacements and V_{max} is the maximum allowable voltage. In a charge-constrained electrostatic harvester, the net energy is given by Eq. (18.18), where V_{start} is the voltage to which the electrodes are initially charged.

$$E_{net} = 0.5(C_{max} - C_{min})V_{max}^2 \tag{18.17}$$

$$E_{net} = 0.5(C_{max} - C_{min})V_{max}V_{start} \tag{18.18}$$

There are three types of electrode arrangement for electrostatic harvesters and these configurations are well suited to MEMS structures [1]: in-plane overlap varying (Figure 18.7), in-plane gap closing (Figure 18.8), and out-of-plane gap closing (Figure 18.9).

Note that both in-plane configurations create two variable capacitors with the capacitances 180° out of phase. Maximum power generation occurs for very small dielectric gaps in which case the source impedance can be very high resulting in poor power delivery. In MEMS, the separation between the two plates is typically

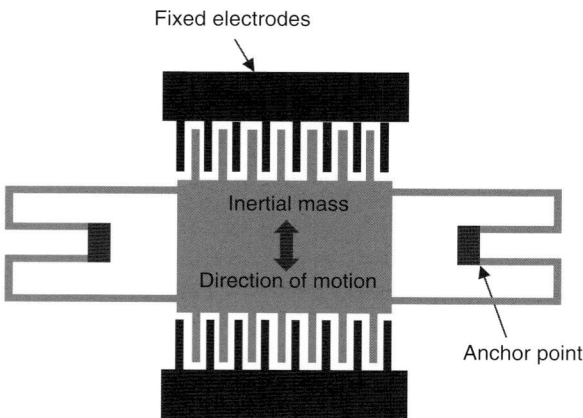

Figure 18.7 In-plane overlap varying electrode arrangement (plan view).

Fixed electrodes

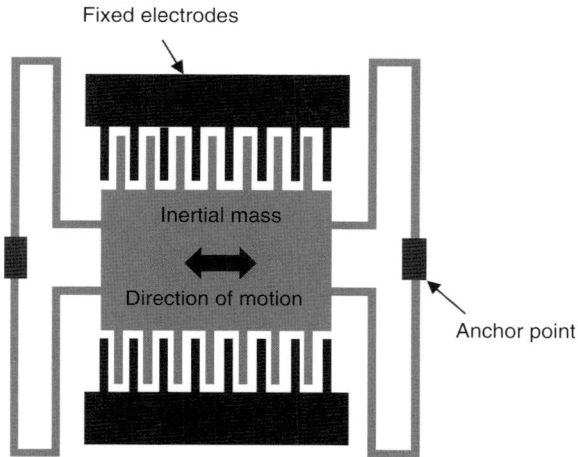

Figure 18.8 In-plane gap closing electrode arrangement (plan view).

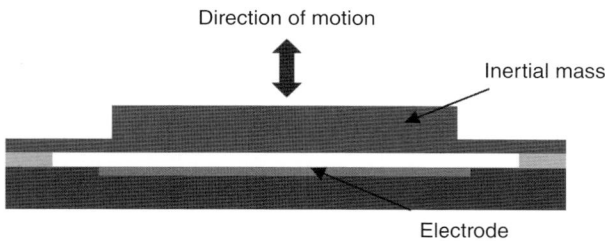

Figure 18.9 Out-of-plane gap closing electrode arrangement (side view).

very small (nm to mm range), which makes this type of transducer an attractive option for MEMS harvesters.

18.3.4
Other Transducer Materials

Magnetostrictive materials are active materials that interact with magnetic fields, that is, they expand when subjected to a magnetic field which is known as the Joule effect or magnetostriction. This occurs because magnetic domains in the material align with the magnetic field. Conversely, when the material is strained (stretched or compressed by an external force), its magnetic energy changes and this is known as the Villari effect. Magnetostrictive materials contain magnetic domains, which realign under the application of a magnetic field and this realignment alters the size of the material. Magnetostrictive materials such as Terfenol-D can generate far larger strains than piezoelectric materials. When strained, the forced realignment of the domains alters the magnetic field, which can be coupled to a coil in order to harvest the mechanical energy [30]. The coupling to a coil introduces many of the issues discussed in Section 18.3.2 and makes MEMS technology unsuitable for this approach. An alternative method is to bond a magnetostrictive material to a piezoelectric and expose the composite structure to a varying magnetic field. In this approach, external kinetic energy can be captured by a moving permanent magnet, which introduces a strain in the magnetostrictive material which in turn strains the piezoelectric material. The piezoelectric material then provides the electrical output and this approach *is* potentially suitable for MEMS implementation [31].

Another type of material recently being applied to the use of energy harvesting is ferroelectrets (also known as piezoelectrets). These materials are essentially polymer dielectrics with a foam-like structure, that is, they contain voids. The polymer dielectric material must be one that can become charged (e.g., by corona poling) and retain the charge, that is, act as an electret. The charge is distributed across the void interfaces, creating an electric field within the void. As external forces are applied to the material and the voids change shape, the charge distribution and therefore the electric field are altered and a net flow of charge can be captured from the material as a whole. This mechanism for charge generation is totally different from that of piezoelectric materials but the macroscopic effect is the same, enabling properties such as d_{33} to be measured. These materials are available as foams and can exhibit d_{33} values far in excess of polymer ferroelectric materials such as PVDF. Cellular polypropylene foam, for example, can exhibit d_{33} values of up to $600\,\mathrm{pC\,N^{-1}}$ [32]. More recently, ferroelectrets have been microengineered using common polymer MEMS materials and processes [33]. With regards to the application of these materials in energy harvesting, there remains some questions regarding longevity and how long material can maintain the charge. Nevertheless, an early MEMS device that uses ferroelectrets has been presented [34] and the high coefficients of the polymer materials compared to alternative polymers could make it an attractive approach in some applications.

Another class of polymer materials that can be used in energy harvesting applications is electroactive polymers (EAPs). There are numerous examples of these types of materials from dielectric elastomers to ionic polymer gels and these can exploit electrostriction (the coupling between strain and polarization), piezoelectricity or electrostatic effects to harvest energy [35]. Various EAP's have been compared for use in energy harvesting applications by Jean-Mistral *et al.* [36], which concludes that dielectric polymers in particular look promising. However, the implementation of these materials in MEMS is relatively immature as is their actual use in energy harvesting applications.

18.4
Review and Comparison of MEMS Energy Harvesting Devices

This section presents a review of some MEMS kinetic energy harvesters presented in the literature that demonstrate many of the principles discussed previously.

Table 18.2 Comparison of different MEMS kinetic energy harvesters from the literature.

Lead author generator[a] and year	Frequency (Hz)	Acceln (ms^{-2})	Inertial mass (g)	Volume (cm^3)	Power (µW)	NPD ($kgs\,cm^{-3}$)	Details	References
Mitcheson ES 2003	30	50	0.1	0.075	3.7	0.02	MEMS ES harvester designed for low frequency high amplitude applications	[37]
Roundy PZ 2004	120	2.5	9.15	1	375	60	A miniature PZ cantilever generator using bulk piezoelectric material	[38]
Jeon PZ 2005	13 900	106.8	2.2e-07	0.000027	1	3.3	MEMS PZ harvester with sol gel PZT and interdigital electrodes. No added inertial mass	[39]
Beeby EM 2007	52	0.589	0.66	0.15	46	884	Miniature EM generator using conventional magnets and coil	[40]
Marzencki PZ 2008	1496	19.6	8.9e-4	0.005	0.8	0.42	MEMS PZ harvester with thin film AlN and integrated power management circuit	[41]
Sari EM 2008	4.2–5e3	490	1.3e-5	1.4	0.4	1.2e-6	MEMS array of parylene cantilevers of different lengths	[42]
Elfrink PZ 2009	325	17.2	0.07	0.27	85	1.06	Fully vacuum encapsulated packaged AlN MEMS harvester	[43]
Wang EM 2009	55	14.9	0.031	0.13	0.61	0.02	MEMS coil and spring structure, bulk NeFeB magnet	[44]
Yang ES 2010	63	2.5	2.2e-3	0.04	0.35	1.4	A rotary MEMS ES harvester fabricated using an SOI wafer	[45]
Miller PZ 2011	232	0.29	3.1e-3	0.0037	0.0017	5.46	MEMS PZ cantilever harvester designed for low frequency	[46]
Guilllemet ES 2013	150	9.81	0.066	0.042	2.2	0.54	Batch fabricated MEMS electrostatic harvesters with wide bandwidth operation	[47]

a) Generators are labeled by technology: EM, electromagnetic; PZ, piezoelectric; ES, electrostatic.

A selection of MEMS devices has been presented in Table 18.2, which also contains two larger, non-MEMS devices for comparison. The first non-MEMS device is a piezoelectric cantilever generator using bulk piezoelectric materials and a tungsten inertial mass [32]. The second is a miniature electromagnetic generator that is probably as small as can be made using standard fabrication processes such as traditional wound coils [34]. The table also includes a basic metric (normalized power density, NPD) used to provide a comparison among the devices. The NPD is the power output for a given volume but, since power output varies with the square of acceleration, A, the calculated NPD is given by $P/(A^2V)$, where P is the stated power output and V is the reported volume of the generator.

A basic comparison shows that the bulk devices have the highest NPD. This reflects that the use of bulk materials and tungsten inertial masses improves transducer performance and power densities compared to their MEMS counterparts. The MEMS electromagnetic generators (Sari and Wang) both exhibit relatively a low NPD, which is due to the limitations of the fabrication processes and, in the case of Sari, the fact that it comprises an array of cantilevers increases the size of the device, while limiting the output to one or two cantilevers at a time. The cantilevers used in the array also lack added inertial masses, which reduce the mechanical energy available in each cantilever.

The piezoelectric MEMS generators show promising performance and one of the reasons the NPD values are not higher is the reduced size of the inertial mass, which is made from silicon. The density of silicon is over seven times smaller than that of tungsten, which was used in the bulk devices included in the table. Of the devices presented here, the early work by Jeon demonstrated that interdigital electrode structures are better than a standard capacitor electrode structure because it enables the 33-mode material properties to be exploited [33]. This is true only if the lithography used to define the electrode pattern has high-enough resolution to minimize the width of the tracks and maximize the number of overlapping fingers. The device presented by Elfrink *et al.* [37] is the most complete MEMS energy harvester in that it is a fully packaged device including vacuum encapsulation of the mechanical structure (see Figure 18.10). The packaging has increased the device volume and therefore the NPD is lower than other unpackaged devices. In reality, any practical device will have to be packaged and therefore this device is more indicative of the final size of any MEMS solution. This harvester is being developed for powering tire-pressure monitoring systems.

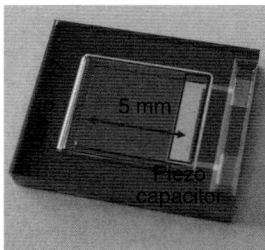

Figure 18.10 Fully packaged piezoelectric MEMS cantilever harvester from IMEC [26].

Electrostatic transduction is well suited to MEMS technologies due to the small electrode gaps that can be realized. However, the practicalities of cyclically charging the electrodes make implementing these devices challenging. Nevertheless, the two later harvesters presented in Table 18.2 ("Yang ES" and "Guillemet ES") provide useful levels of power output, which certainly demonstrate the feasibility and potential of this approach.

18.5
Conclusions

The motivation for operating an energy harvester at resonance is clear but using a basic linear resonant structure also has some limitations. Other operating strategies such as nonlinear structures, tuning resonant frequencies, or using some form of frequency conversion, are available. The relative suitability of these techniques to MEMS design and fabrication constraints have to be considered along with a detailed knowledge of the real-world vibrations for which the harvester is being designed. It is essential for any designer of a kinetic energy harvester to understand the application spectrum and to simulate devices using real vibration data. Nonlinear approaches such as the use of bistable structures may work better than linear resonant harvesters with random stochastic vibrations, while the opposite is true for harvesters tested with a sinusoidal excitation. In reality, vibrations are highly unlikely to be either of these types. If the reader requires further information on real vibration data, then an applications database of downloadable vibration data is available from the UK's Energy Harvesting Network [48]. This data can be used to replicate the vibrations on laboratory shaker rigs, enabling devices to be tested as if they are actually deployed in the application. Calibration of the shaker (acceleration levels at different frequencies) is essential to ensure accurate reproduction.

MEMS energy harvesters have been demonstrated and research continues to improve performance and address issues such as operating frequencies. It is clear that either electrostatic or piezoelectric transduction can scale down in size and be used in MEMS harvesters. Electromagnetics, on the other hand, do not scale well and are not suitable for MEMS implementations. Scaling laws also mean that the small size of MEMS harvesters fundamentally reduces the energy that can be harvested and also leads to higher resonant frequencies than typically found in applications. The low density of silicon means that it is not a good material for the inertial mass, and techniques to increase the effective density would improve the power generated and lower the frequency of operation. The printed tungsten film used in the inertial mass of a larger-size, screen-printed energy harvester demonstrates an available approach to achieve this [49]. Developing a technique to increase the inertial mass by a factor of 2 may potentially double the output power of the harvester and such a step change in performance is unlikely to be realized by optimizing a transduction method or associated material.

References

1. Beeby, S.P., Tudor, M.J., and White, N.M. (2006) Energy harvesting vibration sources for microsystems applications. *J. Meas. Sci. Technol.*, **17**, R175–R195.

2. Torah, R., Glynne-Jones, P., Tudor, M., O'Donnell, T., Roy, S., and Beeby, S. (2008) Self-powered autonomous wireless sensor node using vibration energy harvesting. *Meas. Sci. Technol.*, **19** (12), Article no. 125202.

3. Stephen, N.G. (2006) On energy harvesting from ambient vibration. *J. Sound Vib.*, **293**, 409–425.

4. Mitcheson, P.D., Yeatman, E.M., Rao, G.K., Holmes, A.S., and Green, T.C. (2008) Energy harvesting from human and machine motion for wireless electronic devices. *Proc. IEEE*, **96** (9), 1457–1486.

5. Rocha, J.G., Goncalves, L.M., Rocha, P.F., Silva, M.P., and Lanceros-Méndez, S. (2010) Energy harvesting from piezoelectric materials fully integrated in footwear. *IEEE Trans. Ind. Electron.*, **57** (3), 813–819.

6. Zhu, D., Tudor, J., and Beeby, S. (2010) Strategies for increasing the operating frequency range of vibration energy harvesters: a review. *Meas. Sci. Technol.*, **21** (2), article no. 022001.

7. Perpetuum. PMG FSH Datasheet, *http://www.perpetuum.com/fsh.asp* (accessed 26 April 2013).

8. Eichhorn, C., Goldschmidtboeing, F., and Woias P. (2008) A frequency tunable piezoelectric energy converter based on a cantilever beam. Proceedings of the PowerMEMS 2008, Sendai, Japan, November 9–12, pp. 309–312.

9. Zhu, D., Roberts, S., Tudor, J., and Beeby, S. (2010) Design and experimental characterization of a tunable vibration-based electromagnetic microgenerator. *Sens. Actuators, A: Phys.*, **158** (2), 284–293.

10. Ayala Garcia I., Zhu D., Tudor J., and Beeby S. (2009) Autonomous tunable energy harvester. Proceedings of the PowerMEMS 2009, Washington DC, December 1–4, pp. 49–52.

11. Tadesse, Y., Zhang, S., and Proya, S. (2009) Multimodal energy harvesting system: piezoelectric and electromagnetic. *J. Intell. Mater. Syst. Struct.*, **20**, 625–632.

12. Tang, X. and Zuo, L. (2011) Enhanced vibration energy harvesting using dual-mass systems. *J. Sound Vib.*, **330** (21), 5199–5209.

13. Barton, D., Burrow, S., and Clare, L. (2010) Energy harvesting from vibrations with a nonlinear oscillator. *J. Vib. Acoust.*, **132** (2), 021009.

14. Marzencki, M., Defosseux, M., and Basrour, S. (2009) MEMS vibration energy harvesting devices with passive resonance frequency adaptation capability. *J. Microelectromech. Syst.*, **18** (6), 1444–1453.

15. Andò, B., Baglio, S., Trigona, C., Dumas, N., Latorre, L., and Nouet, P. (2010) Non-linear mechanism in MEMS devices for energy harvesting applications. *J. Micromech. Microeng.*, **20** (12), 125020.

16. Andò, B., Baglio, S., L'Episcopo, G., and Trigona, C. (2012) Investigation on mechanically bistable MEMS devices for energy harvesting from vibrations. *J. Microelectromech. Syst.*, **21** (4), 779–790.

17. Soliman, M.S.M., Abdel-Rahman, E.M., El-Saadany, E.F., and Mansour, R.R. (2008) A wideband vibration-based energy harvester. *J. Micromech. Microeng.*, **18** (11), 115021.

18. Pozzi, M. and Zhu, M. (2011) Plucked piezoelectric bimorphs for knee-joint energy harvesting: modelling and experimental validation. *Smart Mater. Struct.*, **20**, 055007.

19. Tang, Q.C., Yang, Y.L., and Xinxin, L. (2011) Bi-stable frequency up-conversion piezoelectric energy harvester driven by non-contact magnetic repulsion. *Smart Mater. Struct.*, **20** (12), 125011.

20. Renaud, M., Fiorini, P., van Schaijk, R., and van Hoof, C. (2009) Harvesting energy from the motion of human limbs: the design and analysis of an impact-based piezoelectric generator. *Smart Mater. Struct.*, **18**, 035001.

21. Cavallier, B., Berthelot, P., Nouira, H., Foltete, E., Hirsinger, L., and Ballandras, S. (2005) Energy harvesting using vibrating structures excited by shock.

Proceedings of the IEEE Ultrasonics Symposium 2005, Vol. 2, pp. 943–945.

22. Jung, S. and Yun, K. (2010) A wideband energy harvesting device using snap-through buckling for mechanical frequency-up conversion. Proceedings of the IEEE 23rd International Conference on MEMS, pp. 1207–1210.

23. IEEE ANSI/IEEE Std 176-1987. (1988) *IEEE Standard on Piezoelectricity*, Institute of Electrical and Electronics Engineers, New York.

24. Elfrink, R., Kamel, T.M., Goedbloed, M., Matova, S., Hohlfeld, D., van Andel, Y., and van Schaijk, R. (2009) Vibration energy harvesting with aluminum nitride-based piezoelectric devices. *J. Micromech. Microeng.*, **19**, 094005.

25. Morgan (0000) Piezoelectric Ceramics Brochure, MTC ElectroCeramics, *http://www.morganelectroceramics.com/resources/literature/* (accessed 6 September 2013).

26. van Schaijk, R. (2011) Emerging Self Powered Systems. Keynote presentation at Energy Harvesting 2011, London, UK, *http://eh-network.org/events/dissemination2011/presentations/Rob%20van%20Schaijk.pdf* (accessed 20 May 2013).

27. Matloub, R., Artieda, A., Sandu, C., Milyutin, E., and Muralt, P. (2011) Electromechanical properties of Al0.9Sc0.1N thin films evaluated at 2.5 GHz film bulk acoustic resonators. *Appl. Phys. Lett.*, **99**, 092903.

28. Nguyen, M.D., Nguyen, C.T.Q., Trinh, T.Q., Nguyen, T., Pham, T.N., Rijnders, G., and Vu, H.N. (2013) Enhancement of ferroelectric and piezoelectric properties in PZT thin films with heterolayered structure. *Mater. Chem. Phys.*, **138** (2-3), 862–869.

29. Meninger, S., Mur-Miranda, J., Lang, J., Chandrakasan, A., Slocum, A., Schmidt, M., and Amirtharajah, R. (2001) Vibration to electric energy conversion. *IEEE Trans. Very Large Scale Integr. VLSI Syst.*, **9**, 64–76.

30. Wang, L. and Yuan, F.G. (2008) Vibration energy harvesting by magnetostrictive material. *Smart Mater. Struct.*, **17**, 045009.

31. Lafont, T., Gimeno, L., Delamare, J., Lebedev, G.A., Zakharov, D.I., Viala, B., Cugat, O., Galopin, N., Garbuio, L., and Geoffroy, O. (2012) Magnetostrictive–piezoelectric composite structures for energy harvesting. *J. Micromech. Microeng.*, **22**, 094009.

32. Bauer, S., Multhaupt, R., and Sessler, G. (2004) Ferroelectrets: soft electroactive foams for transducers. *Phys. Today*, **57** (2), 37–43.

33. Wang, J.-J., Hsu, T.-H., Yeh, C.-N., Tsai, J.-W., and Su, Y.-C. (2012) Piezoelectric polydimethylsiloxane films for MEMS transducers. *J. Micromech. Microeng.*, **22**, 015013.

34. Feng, Y., Hagiwara, K., Iguchi, Y., and Suzuki, Y. (2012) Trench-filled cellular parylene electret for piezoelectric transducer. *Appl. Phys. Lett.*, **100**, 262901.

35. Lallart, M., Cottinet, P.J., Guyomar, D., and Lebrun, L. (2012) Electrostrictive polymers for mechanical energy harvesting. *J. Polym. Sci., Part B: Polym. Phys.*, **50**, 523–535.

36. Jean-Mistral, C., Basrour, S., and Chaillout, J.-J. (2010) Comparison of electroactive polymers for energy scavenging applications. *Smart Mater. Struct.*, **19**, 085012.

37. Mitcheson, P., Stark, B., Miao, P., Yeatman, E., Holmes, A., and Green, T. (2003) Analysis and optimisation of MEMS on-chip power supply for self powering of slow moving sensors. Proceedings of the Eurosensors XVII, Guimaraes, Portugal, pp. 30–31.

38. Roundy, S. and Wright, P.K. (2004) A piezoelectric vibration based generator for wireless electronics. *Smart Mater. Struct.*, **13**, 1131–1142.

39. Jeon, Y.B., Sood, R., Jeong, J.-H., and Kim, S.G. (2005) MEMS power generator with transverse mode thin film PZT. *Sens. Actuators, A*, **122**, 16–22.

40. Beeby, S.P., Torah, R.N., Tudor, M.J., Glynne-Jones, P., O'Donnell, T., Saha, C.R., and Roy, S. (2007) A micro electromagnetic generator for vibration energy harvesting. *J. Micromech. Microeng.*, **17**, 1257.

41. Marzencki, M., Ammar, Y., and Basrour, S. (2008) Integrated power harvesting system including a MEMS generator

and a power management circuit. *Sens. Actuators, A*, **145–146**, 363–370.

42. Sari, I., Balkan, T., and Kulah, H. (2008) An electromagnetic micro power generator for wideband environmental vibrations. *Sens. Actuators, A: Phys*, **145–146**, 405–413.

43. Elfrink, R., Pop, V., Hohlfeld, D., Kamel, T.M., Matova, S., de Nooijer, C., Jambunathan, M., Goedbloed, M., Caballero, L., Renaud, M., Penders, J., and van Schaijk, R. (2009) First autonomous wireless sensor node powered by a vacuum-packaged piezoelectric MEMS energy harvester. Proceedings of IEEE International Electron Devices Meeting (IEDM), pp. 1–4.

44. Wang, P., Tanaka, K., Sugiyama, S., Dai, X., Zhao, X., and Liu, J. (2009) A micro electromagnetic low level vibration energy harvester based on MEMS technology. *Microsyst. Technol.*, **15**, 941–951.

45. Yang, B., Lee, C., Kotlanka, R.K., Xie, J., and Lim, S.P. (2010) A MEMS rotary

comb mechanism for harvesting the kinetic energy of planar vibration. *J. Micromech. Microeng.*, **20**, 065017.

46. Miller, L.M., Halvorsen, E., Dong, T., and Wright, P.K. (2011) Modeling and experimental verification of low-frequency MEMS energy harvesting from ambient vibrations. *J. Micromech. Microeng.*, **21**, 045029.

47. Guilllemet, R., Basset, P., Galayko, D., Cottone, F., Marty, F., and Bourouina T. (2013) Wideband MEMS electrostatic vibration energy harvesters based on gap-closing interdigited combs with a trapezoidal cross section. Proceedings of the 26th IEEE International Conference on MEMS, pp. 817–820.

48. Energy Harvesting. EH Network Data Repository, *http://eh-network.org/data/* (accessed 13 May 2013).

49. Zhu, D., Beeby, S.P., Tudor, M.J., and Harris, N.R. (2011) A credit card sized self-powered smart sensor node. *Sens. Actuators, A: Phys.*, **169** (2), 317–325.

Index

Resonant MEMS – Fundamentals, Implementation and Application, First Edition.
Edited by Oliver Brand, Isabelle Dufour, Stephen M. Heinrich and Fabien Josse.
© 2015 Wiley-VCH Verlag GmbH & Co. KGaA. Published 2015 by Wiley-VCH Verlag GmbH & Co. KGaA.